Metal Ions in Life Sciences

Volume 7

Organometallics in Environment and Toxicology

Metal Ions in Life Sciences

edited by

Astrid Sigel,[1] Helmut Sigel,[1] and Roland K. O. Sigel[2]

[1] *Department of Chemistry*
Inorganic Chemistry
University of Basel
Spitalstrasse 51
CH-4056 Basel, Switzerland

[2] *Institute of Inorganic Chemistry*
University of Zürich
Winterthurerstrasse 190
CH-8057 Zürich, Switzerland

Volume 7

Organometallics in Environment and Toxicology

DE GRUYTER

First published by the Royal Society of Chemistry in 2010. Publicatio

ISBN: 978-1-84755-177-1
ISSN: 1559-0836
DOI: 10.1039/9781849730822

A cataloque record for this book is available from the British Library

ISBN 978-3-11-044280-9
e-ISBN (PDF) 978-3-11-043660-0
Set-ISBN (Print + Ebook) 978-3-11-043661-7

Library of Congress Cataloging-in-Publication Data
A CIP catalog record for this book has been applied for at the L
Congress.

Bibliographic information published by the Deutsche Nati
The Deutsche Nationalbibliothek lists this publication in the D
Nationalbibliografie; detailed bibliographic data are available
at http://dnb.dnb.de.

Cover image: The figure on the cover shows Figure 1 of Chap
Holger Hintelmann.

www.degruyter.com

Historical Development and Perspectives of the Series *Metal Ions in Life Sciences**

It is an old wisdom that metals are indispensable for life. Indeed, several of them, like sodium, potassium, and calcium, are easily discovered in living matter. However, the role of metals and their impact on life remained largely hidden until inorganic chemistry and coordination chemistry experienced a pronounced revival in the 1950s. The experimental and theoretical tools created in this period and their application to biochemical problems led to the development of the field or discipline now known as *Bioinorganic Chemistry*, *Inorganic Biochemistry*, or more recently also often addressed as *Biological Inorganic Chemistry*.

By 1970 *Bioinorganic Chemistry* was established and further promoted by the book series *Metal Ions in Biological Systems* founded in 1973 (edited by H.S., who was soon joined by A.S.) and published by Marcel Dekker, Inc., New York, for more than 30 years. After this company ceased to be a family endeavor and its acquisition by another company, we decided, after having edited 44 volumes of the *MIBS* series (the last two together with R.K.O.S.) to launch a new and broader minded series to cover today's needs in the *Life Sciences*. Therefore, the Sigels new series is entitled

Metal Ions in Life Sciences.

After publication of the first four volumes (2006–2008) with John Wiley & Sons, Ltd., Chichester, UK, we are happy to join forces now in this still new endeavor with the Royal Society of Chemistry, Cambridge, UK; a most experienced Publisher in the *Sciences*.

---

*Reproduced with some alterations by permission of John Wiley & Sons, Ltd., Chichester, UK (copyright 2006) from pages v and vi of Volume 1 of the series *Metal Ions in Life Sciences* (MILS-1).

The development of *Biological Inorganic Chemistry* during the past 40 years was and still is driven by several factors; among these are (i) the attempts to reveal the interplay between metal ions and peptides, nucleotides, hormones or vitamins, etc., (ii) the efforts regarding the understanding of accumulation, transport, metabolism and toxicity of metal ions, (iii) the development and application of metal-based drugs, (iv) biomimetic syntheses with the aim to understand biological processes as well as to create efficient catalysts, (v) the determination of high-resolution structures of proteins, nucleic acids, and other biomolecules, (vi) the utilization of powerful spectroscopic tools allowing studies of structures and dynamics, and (vii), more recently, the widespread use of macromolecular engineering to create new biologically relevant structures at will. All this and more is and will be reflected in the volumes of the series *Metal Ions in Life Sciences*.

The importance of metal ions to the vital functions of living organisms, hence, to their health and well-being, is nowadays well accepted. However, in spite of all the progress made, we are still only at the brink of understanding these processes. Therefore, the series *Metal Ions in Life Sciences* will endeavor to link coordination chemistry and biochemistry in their widest sense. Despite the evident expectation that a great deal of future outstanding discoveries will be made in the interdisciplinary areas of science, there are still "language" barriers between the historically separate spheres of chemistry, biology, medicine, and physics. Thus, it is one of the aims of this series to catalyze mutual "understanding".

It is our hope that *Metal Ions in Life Sciences* proves a stimulus for new activities in the fascinating "field" of *Biological Inorganic Chemistry*. If so, it will well serve its purpose and be a rewarding result for the efforts spent by the authors.

Astrid Sigel, Helmut Sigel
Department of Chemistry
Inorganic Chemistry
University of Basel
CH-4056 Basel
Switzerland

Roland K. O. Sigel
Institute of Inorganic Chemistry
University of Zürich
CH-8057 Zürich
Switzerland

October 2005
and October 2008

Preface to Volume 7

Organometallics in Environment and Toxicology

Organometallic compounds contain per definition a metal-carbon bond. Therefore, the present Volume 7 is related to the preceding Volume 6, *Metal-Carbon Bonds in Enzymes and Cofactors*, which, as follows from its title, focused on living organisms. Now the focus is on the role that organometal(loid)s play in the environment and in toxicology; naturally, here again living systems are involved in the synthesis, transformation, remediation, detoxification, etc.

Volume 7 opens with two general chapters; first, environmental cycles of elements, which involve organometal(loid)s, thus enhancing the element mobility, are discussed, and next the analysis and quantification of the pertinent species are critically reviewed. Knowledge of the total concentration of a metal(loid) reveals little about its possible environmental mobility, toxicity or biochemical activity; hence, it is necessary to determine the actual chemical form of the compound under investigation.

The discovery that the biologically active forms of vitamin B_{12}, e.g., its coenzyme 5'-deoxyadenosylcobalamin and the corresponding methylcobalamin, are all compounds with a cobalt-carbon bond, opened up a new area in organometallic chemistry (*MILS-6*). In fact, the cobalt-containing corrin-like (B_{12}) cofactor is similar to the nickel coenzyme F_{430} involved in bacterial methane formation as is pointed out in Chapter 3. Furthermore, it is now recognized that methanogens are obligate anaerobes that are responsible for all biological methane production on earth (ca. 10^9 tons per year).

In Chapters 4 and 5 the organic derivatives of tin and lead, their synthesis, use, environmental distribution, and their toxicity are summarized. The next two chapters deal with organoarsenicals, their distribution and

Metal Ions in Life Sciences, Volume 7 Edited by Astrid Sigel, Helmut Sigel, and Roland K. O. Sigel

Published by the Royal Society of Chemistry, www.rsc.org DOI: 10.1039/9781849730822-FP007 transformation in the environment, their uptake, metabolism and toxicity, including an evaluation of their adverse effects on human health. Chapter 8 is devoted to a further metalloid: Antimony has no known biological role and has largely been overlooked as an element of environmental concern though its biomethylation most probably occurs. Yet, the concentrations of methylated antimony species in the environment are low and thus it seems unlikely that they could be of any great concern.

In contrast to arsenic and antimony, no methylated bismuth species have ever been found in surface waters and biota. However, as reported in Chapter 9, volatile monomethyl-, dimethyl-, and trimethylbismuthine have been produced by some anaerobic bacteria and methanogenic archaea in laboratory culture experiments, and indeed, trimethylbismuthine has been detected in landfill and sewage sludge fermentation gases. Bismuth is an element that is relatively non-toxic to humans but it is toxic to some prokaryotes.

Selenium, which is treated in Chapter 10, has one of the most diverse organic chemistries. It is also one of the few elements that may biomagnify in food chains. It is generaly assumed that organic selenium species exist in ambient waters, soils, and sediments, and that they play a key role in bioaccumulation. In contrast, the diversity of organotellurium compounds is small; so far it is limited in the environment to simple methylated tellurides.

Chapters 11 and 12 are devoted to mercury: The most important mercury species in the environment is clearly monomethylmercury, which is normally not released into the environment, but formed by natural processes, mainly via methylation of Hg(II) by bacteria. Its biomagnification potential is enormous; it is accumulated by more than 7 orders of magnitude, i.e., from sub ng/L concentrations to over 10^6 ng/kg in piscivorous fish. Thus, it is of main concern for human health, especially because methylmercury is a very potent neurotoxin; its mechanisms of toxicity are discussed including neurodegerative disorders like Parkinson's and Alzheimer's disease.

The two terminating Chapters 13 and 14 are again of a more general nature. First the environmental bioindication, biomonitoring, and bioremediation with all their consequences are considered; this is followed by an account of methylated metal(loid) species in humans. Interestingly, arsenic, bismuth, selenium, and probably also tellurium have been shown to be enzymatically methylated in the human body; such methylation has not yet been demonstrated for antimony, cadmium, germanium, indium, lead, mercury, thallium, and tin, although the latter elements can be biomethylated in the environment. The assumed and proven health effects caused by alkylated metal(loid) species are emphasized.

Astrid Sigel
Helmut Sigel
Roland K. O. Sigel

Contents

Metal Ions in Life Sciences, Volume 7 Edited by Astrid Sigel, Helmut Sigel, and Roland K. O. Sigel

Published by the Royal Society of Chemistry, www.rsc.org DOI: 10.1039/9781849730822-FP009

Contributors to Volume 7

Numbers in parentheses indicate the pages on which the authors' contributions begin.

Henry G. Abadin Agency for Toxic Substances and Disease Registry (ATSDR), US Dept. of Health and Human Services, Division of Toxicology, 1600 Clifton Road, F-62, Atlanta, GA 30333, USA (153)

Michael Aschner Department of Pediatrics, Pharmacology, and the Kennedy Center for Research on Human Development, Vanderbilt University School of Medicine, 2215-B Garland Avenue, 11415 MRB IV, Nashville, TN 37232-0414, USA, Fax: +1-615-936-4080 <michael.aschner@vanderbilt.edu> (403)

Sandra Ceccatelli Karolinska Institute, Department of Neuroscience, SE-17177 Stockholm, Sweden <sandra.ceccatelli@ki.se> (403)

William R. Cullen Chemistry Department, University of British Columbia, Vancouver, BC, V6T 1Z1, Canada <wrc@chem.ubc.ca> (165)

Mishtu Dey Department of Biological Chemistry, University of Michigan Medical School, 1150 W. Medical Center Dr., 5301 MSRB III, Ann Arbor, MI 48109-0606, USA; Current address: Department of Chemistry, Massachusetts Institute of Technology, 77 Massachusetts Ave., Cambridge, MA 02139, USA (71)

Roland A. Diaz-Bone Institute of Environmental Analytical Chemistry, University of Duisburg-Essen, Universitätsstrasse 3–5, D-45141 Essen, Germany <roland.diaz@uni-due.de> (231)

Elke Dopp University Hospital Essen, Institute of Hygiene and Occupational Medicine, Hufelandstrasse 55, D-45122 Essen, Germany, Fax: +49-201-723-4546 <elke.dopp@uni-due.de> (231)

Jörg Feldmann Trace Element Speciation Laboratory (TESLA), College of Physical Science, University of Aberdeen, Meston Walk, Aberdeen, AB24 3UE, Scotland, UK, Fax: +44-1224-272-921 <j.feldmann@abdn.ac.uk> (319)

Montserrat Filella Institute F.-A. Forel, University of Geneva, Route de Suisse 10, CH-1290 Versoix, Switzerland, Fax: +41-22-379-6069 <montserrat.filella@unige.ch> (267, 303)

Tamás Gajda Department of Inorganic and Analytical Chemistry, University of Szeged, P.O. Box 440, H-6701 Szeged, Hungary, Fax: +36-62-420-505 <tamas.gajda@chem.u-szeged.hu> (111)

Christopher F. Harrington Trace Element Laboratory, Centre for Clinical Sciences, Faculty of Health and Medical Sciences, University of Surrey, Guildford, GU2 7XH, UK <chris.harrington@royalsurrey.nhs.uk> (33)

Holger Hintelmann Department of Chemistry, Trent University, 1600 West Bank Drive, Peterborough, ON, K9J 7B8, Canada, Fax: +1-705-748-1625 <hhintelmann@trentu.ca> (365)

Alfred V. Hirner Institute of Analytical Chemistry, University Duisburg-Essen, Universitätsstrasse 3–5, D-45141 Essen, Germany, Fax: +49-201-183-3951 <alfred.hirner@uni-due.de> (465)

Attila Jancsó Department of Inorganic and Analytical Chemistry, University of Szeged, P.O. Box 440, H-6701 Szeged, Hungary <jancso@chem.u-szeged.hu> (111)

Richard O. Jenkins Faculty of Health and Life Sciences, De Montfort University, The Gateway, Leicester, LE1 9BH, UK <roj@dmu.ac.uk> (33)

Andrew D. Kligerman National Health and Environmental Effects Research Laboratory, Office of Research and Development, U.S. Environmental Protection Agency, Research Triangle Park, NC 27709, USA <kligerman.andrew@epamail.epa.gov> (231)

Iris Koch Environmental Sciences Group, Royal Military College of Canada, Kingston, Ontario K7K 7B4, Canada <koch-i@rmc.ca> (165)

Xianghui Li Department of Biological Chemistry, University of Michigan Medical School, 1150 W. Medical Center Dr., 5301 MSRB III, Ann Arbor, MI 48109-0606, USA (71)

Natalia Onishchenko Karolinska Institute, Department of Neuroscience, SE-17177 Stockholm, Sweden <natalia.onishchenko@ki.se> (403)

Hana R. Pohl, Agency for Toxic Substances and Disease Registry (ATSDR), US Dept. of Health and Human Services, Division of Toxicology, 1600 Clifton Road, F-62, Atlanta, GA 30333, USA, Fax: +1-770-488-4178 <hrp1@cdc.gov> (153)

Stephen W. Ragsdale Department of Biological Chemistry, University of Michigan Medical School, 1150 W. Medical Center Dr., 5301 MSRB III, Ann Arbor, MI 48109-0606, USA, Fax: +1-734-763-4581 <sragsdal@umich.edu> (71)

Kenneth J. Reimer Environmental Sciences Group, Royal Military College of Canada, Kingston, Ontario K7K 7B4, Canada, Fax: +1-613-541-6596 <reimer-k@rmc.ca> (165)

Albert W. Rettenmeier Institute of Hygiene and Occupational Medicine, University of Duisburg-Essen, D-45122 Essen, Germany, Fax: +49-201-183-3951 <albert.rettenmeier@uni-due.de> (465)

John S. Thayer Department of Chemistry, University of Cincinnati, 203 Crosley Tower, PO Box 210172, Cincinnati, OH 45221-0172, USA, Fax: +1-513-556-9239 <thayerj@ucmail.uc.edu> (1, 435)

Daniel S. Vidler Medical Toxicology Centre, University of Newcastle, Wolfson Unit, Claremont Place, Newcastle upon Tyne, NE2 4AA, UK <daniel.vidler@ncl.ac.uk> (33)

Dirk Wallschläger Environmental & Resource Sciences Program and Department of Chemistry, Trent University, 1600 West Bank Dr., Peterborough, ON K9J 7B8, Canada, Fax: +1-705-748-1569 <dwallsch@trentu.ca> (319)

Yuzhen Zhou Department of Biological Chemistry, University of Michigan Medical School, 1150 W. Medical Center Dr., 5301 MSRB III, Ann Arbor, MI 48109-0606, USA (71)

Titles of Volumes 1–44 in the *Metal Ions in Biological Systems* Series

edited by the SIGELs
and published by Dekker/Taylor & Francis (1973–2005)

Volume 1: **Simple Complexes**
Volume 2: **Mixed-Ligand Complexes**
Volume 3: **High Molecular Complexes**
Volume 4: **Metal Ions as Probes**
Volume 5: **Reactivity of Coordination Compounds**
Volume 6: **Biological Action of Metal Ions**
Volume 7: **Iron in Model and Natural Compounds**
Volume 8: **Nucleotides and Derivatives: Their Ligating Ambivalency**
Volume 9: **Amino Acids and Derivatives as Ambivalent Ligands**
Volume 10: **Carcinogenicity and Metal Ions**
Volume 11: **Metal Complexes as Anticancer Agents**
Volume 12: **Properties of Copper**
Volume 13: **Copper Proteins**
Volume 14: **Inorganic Drugs in Deficiency and Disease**
Volume 15: **Zinc and Its Role in Biology and Nutrition**
Volume 16: **Methods Involving Metal Ions and Complexes in Clinical Chemistry**
Volume 17: **Calcium and Its Role in Biology**
Volume 18: **Circulation of Metals in the Environment**
Volume 19: **Antibiotics and Their Complexes**
Volume 20: **Concepts on Metal Ion Toxicity**
Volume 21: **Applications of Nuclear Magnetic Resonance to Paramagnetic Species**
Volume 22: **ENDOR, EPR, and Electron Spin Echo for Probing Coordination Spheres**
Volume 23: **Nickel and Its Role in Biology**

Contents of Volumes in the *Metal Ions in Life Sciences* Series

edited by the SIGELs

Volumes 1–4

published by John Wiley & Sons, Ltd., Chichester, UK (2006–2008)

<www.Wiley.com/go/mils>

and from Volume 5 on

by the Royal Society of Chemistry, Cambridge, UK (since 2009)

<www.rsc.org/shop/books/series/85.asp?seriesid=85>

Volume 1: Neurodegenerative Diseases and Metal Ions

1. The Role of Metal Ions in Neurology. An Introduction
   Dorothea Strozyk and Ashley I. Bush
2. Protein Folding, Misfolding, and Disease
   Jennifer C. Lee, Judy E. Kim, Ekaterina V. Pletneva, Jasmin Faraone-Mennella, Harry B. Gray, and Jay R. Winkler
3. Metal Ion Binding Properties of Proteins Related to Neurodegeneration
   Henryk Kozlowski, Marek Luczkowski, Daniela Valensin, and Gianni Valensin
4. Metallic Prions: Mining the Core of Transmissible Spongiform Encephalopathies
   David R. Brown
5. The Role of Metal Ions in the Amyloid Precursor Protein and in Alzheimer's Disease
   Thomas A. Bayer and Gerd Multhaup
6. The Role of Iron in the Pathogenesis of Parkinson's Disease
   Manfred Gerlach, Kay L. Double, Mario E. Götz, Moussa B. H. Youdim, and Peter Riederer

Volume 2: Nickel and Its Surprising Impact in Nature

Volume 3: The Ubiquitous Roles of Cytochrome P450 Proteins

Volume 4: Biomineralization. From Nature to Application

Volume 5: Metallothioneins and Related Chelators

Volume 6: Metal-Carbon Bonds in Enzymes and Cofactors

Volume 7: Organometallics in Environment and Toxicology
(this book)

Volume 8: Metal Ions in Toxicology: Effects, Interactions, Interdependencies
(tentative contents)

Volume 9: Structural and Catalytic Roles of Metal Ions in RNA
(tentative contents)

Comments and suggestions with regard to contents, topics, and the like for future volumes of the series are welcome.

Met. Ions Life Sci. **2010**, *7*, 1–32

1

Roles of Organometal(loid) Compounds in Environmental Cycles

John S. Thayer

Department of Chemistry, University of Cincinnati, Cincinnati OH 45221-0172, USA
< thayerj@ucmail.uc.edu >

Metal Ions in Life Sciences, Volume 7 Edited by Astrid Sigel, Helmut Sigel, and Roland K. O. Sigel

Published by the Royal Society of Chemistry, www.rsc.org DOI: 10.1039/9781849730822-00001

ABSTRACT: Organo compounds form an integral part of the environmental cycles of metals and metalloids. For phosphorus, selenium, and (possibly) arsenic, they are biochemical necessities. For others, they create enhanced mobility and altered biological effects. Investigations in this area grew out of human introduction of these compounds or their precursors into the natural environment.

KEYWORDS: anthropogenic sources · bioalkylation · biomethylation · environmental movement · food chains · food webs · metal-carbon bonds · toxic gases · volatilization

1. INTRODUCTION

1.1. Concepts and Terminology

An excellent definition of the subject of this article appears in [1]:

> *The term "biogeochemical cycle" is used here to mean the study of the transport and transformation of substances in the natural environment . . .*

and the term "cycle" has been defined as [2]:

> *A single complete execution of a periodically repeated phenomenon . . .*

Biogeochemical cycles involving organometal(loid)s have been discussed elsewhere [3–8]. In principle, all elements on this planet comprise one complex gigantic supercycle, with the components moving and transforming in varying ways, rates, places [8]. Additional material arrives from outer space as meteorites, dust or other cosmic "debris", while other material vanishes by escape into space or undergoes nuclear transformation (radioisotopes). For simplicity, the cycles of individual elements are considered in isolation, with these cycles being broken down into "mini-cycles", limited to isolated ecosystems. In addition to elements, certain compounds also have individual cycles; methane and water are the two most common examples.

The term "biogeochemical" indicates a particular combination of changes. "Geo", referring to the planet Earth, refers to physical changes (volatilization, melting, dissolution, precipitation, etc.). Terrestrial cycles having exclusively physical changes are rare; the noble gases are the primary examples. They circulate through the atmosphere, dissolve in water, get trapped in the earth's crust and form clathrates [9,10]. Noble gas clathrates have been proposed for Mars [11] and Titan [12]. "Geochemical" cycles involve both physical and chemical changes without involvement of living organisms. Many examples are known on Earth, and a cycle for methane has been proposed for Titan [13].

The prefix "bio" indicates the effects of living organisms. These effects are both physical and chemical. Physical effects would involve uptake, excretion, and transport (most organisms are mobile, and their movements carry along elements and compounds within them). Chemical effects involve uptake, formation, sequestration and/or decomposition of compounds, either by metabolism of individual organisms or by ingestion of foods containing such compounds.

Actual cycles are mixtures of biotic and abiotic processes. Sorting out the relative contributions of components is never easy. Introduction of one or more organic groups onto a metal or metalloid changes physical and chemical properties, often drastically, resulting in changes to the element's cycle.

1.2. Consequences of Organo Substituents

As illustration of the effects of organo substituents, consider a quantity of tetramethylsilane, $(CH_3)_4Si$, in a glass tube. Here is an inorganic silicon compound (or more likely a mixture), with silicon-oxygen bonds and an organosilicon compound with silicon-carbon bonds. Their physical properties are so different that it is very easy to tell them apart!

Most elements form bonds to carbon. Organometal(loid)s with biological significance occur for most heavier main-group elements, and some are known for transition metal compounds. Metal(loid)-carbon bonds in these compounds show a slight polarity [M(δ+)–C(δ–)], have varying bond energies, and usually display low chemical reactivity. Metalloids in nature exist predominantly as oxides or oxyanions, frequently in highly polymerized forms. Metals occur as oxides or sulfides (occasionally as selenides), usually solids, with high melting points. Solubility in water varies from substantial to negligible.

Substitution of organic groups for inorganic groups causes marked changes in melting (m.p.) and boiling points (b.p.). Table 1 illustrates such changes for selected organotin compounds. Notice that the largest changes occur when the first and the last alkyl groups are introduced, such as when trimethyltin fluoride (m.p. 375 °C) is converted to tetramethyltin (m.p. –54 °C). A smaller, yet still substantial, change occurs for the corresponding chlorides. These changes arise from decreased intermolecular attraction. Unlike halogens, oxygen, nitrogen or sulfur, alkyl groups have no non-bonding electron pairs; their intermolecular attractive forces are quite weak, as illustrated by the fact that "peralkyl" compounds of these elements are gases or volatile liquids at ordinary temperature. This effect is greatest for the methyl group.

Solubility patterns also change with organo substitution. As the number and/or size of the organic ligand(s) increases, the solubility in water usually falls and the solubility in hydrocarbons grows.

1.3. Specific Effects of Organometal(loid)s in Biogeochemical Cycles

By definition, all these compounds comprise part of the carbon cycle. They also belong to the cycle(s) of the metal(loid)(s). The presence of metal(loid)-carbon bonds opens up additional physical or chemical pathways not otherwise available. The volatility of such compounds (cf. Sections 1.2 and 3.2) compared to the inorganic analogs facilitates their mobility.

Introduction of xenobiotic organometal(loid)s, whether accidently or deliberately, affects the elemental cycles involved. Some compounds (e.g., methylmercuric derivatives [14]), which form naturally at very low levels, may

Table 1. Melting and boiling points of selected organotin compounds.[a]

Compound	Melting Point (°C)	Boiling Point (°C)
$SnCl_4$	–33	114.15
CH_3SnCl_3	53	nr
$(CH_3)_2SnCl_2$	107–108	333
$(CH_3)_3SnCl$	42	249/13.5 torr
$(CH_3)_4Sn$	–54	78
SnF_4	442	
CH_3SnF_3	321–327 d	nr
$(CH_3)_2SnF_2$	360	nr
$(CH_3)_3SnF$	375 d	nr
$C_2H_5SnCl_3$	–10	196–198
$(C_2H_5)_2SnCl_2$	84–85	277
$(C_2H_5)_3SnCl$	15.5	210
$(C_2H_5)_4Sn$	~–130	175
$C_4H_9SnCl_3$	nr	93/10 torr
$(C_4H_9)_2SnCl_2$	43	135/10 torr
$(C_4H_9)_3SnCl$	nr	98/0.45 torr
$(C_4H_9)_4Sn$	nr	145/10 torr
$C_6H_5SnCl_3$	<25	142–143
$(C_6H_5)_2SnCl_2$	42–44	333
$(C_6H_5)_3SnCl$	106	249
$(C_6H_5)_4Sn$	225	>420

[a]All temperatures were collected from *Dictionary of Organometallic Compounds*, Vol. 2, Chapman & Hall, London, 1984.
nr: not reported, d: with decomposition

be generated in enormous quantities due to addition of massive quantities of substrates, that natural mechanisms for their control are overwhelmed.

Other organometal(loid)s may be totally foreign to the natural environment (e.g., tri-*n*-butyltin [15,16] and tetraethyllead [17]). These can ordinarily be degraded, but often remain long enough to become toxic to organisms.

2. FORM AND DISTRIBUTION OF ORGANOMETAL(LOID)S

2.1. Biogenic Sources

2.1.1. *Biological Methylation*

Biological methylation (usually contracted to *biomethylation*) designates processes in which a methyl group undergoes transfer by enzymes (*methyltransferases*) onto a metal or metalloid atom [6,7,14,18,19]. Biomethylation mostly commonly occurs in sediments from bacterial action [18,19]; however, fungi and algae are also known to cause biomethylation [19]. Symbiotic bacteria in termites [20] and in the rhizospheres of plants [21] can also perform biomethylation.

2.1.2. Biological Alkylation

Biological alkylation (usually contracted to *bioalkylation*) in the broadest sense would include biomethylation, but in common usage, this term specifically refers to transfer of alkyl groups other than methyl. Bioalkylation processes are more diverse and varied than biomethylation, and are found mostly for non-metals and metalloids [5,22]. Examples of compounds formed by bioalkylation include arsenobetaine [23–25], selenomethionine, telluromethionine, phosphinothricin (Figure 1), and adenosylcobalamin (vitamin B_{12}) (see Figure 2 in Section 4.2.2).

2.1.3. Other Biogenic Organometal(loid)s

There are no reports of biological arylation (*bioarylation*) – enzymatic introduction of an aryl group onto a metal or metalloid. Given the diversity of both organisms and biochemical reactions, it is quite likely this reaction

$(CH_3)_3As^+CH_2CO_2^-$
Arsenobetaine

$CH_3SeCH_2CH_2CH(NH_2)CO_2H$
Selenomethionine

$ClCH{=}CH_2AsCl_2$
Lewisite

$ClCH{=}CH_2PO_3H_2$
Ethephon

$CH_3P(:O)(F)OCH(CH_3)_2$
Sarin

$CH_3P(:O)(F)OCH(CH_3)C(CH_3)_3$
Soman

$HO_2CCH_2NHCH_2PO_3H_2$
Glyphosate

$HO_2CCH_2N(CH_2PO_3H_2)_2$
Glyphosine

$CH_3P(:O)(OH)CH_2CH_2CH(NH_2)CO_2H$
Glufosinate

$CH_3P(:O)(OH)CH_2CH_2CH[NHC(:O)]CO_2H$
Phosphinothricin

2-$CH_3CH_2HgSC_6H_4CO_2^-Na^+$
Thiomersal

$(HO)_2P(:O)HC$—CH_3 (with O bridging the two carbons)
Fosfomycin (phosphonomycin)

Figure 1. Formulas of compounds named in the text.

may eventually be discovered. Demethylation and dealkylation are biological processes by which organic groups bonded to metal(loids) may be removed, thereby generating new organometal(loid)s. Metal carbonyls have been reported in landfill [26] and sewage [27] gases. Whether these are biogenic or not remains to be determined.

2.2. Anthropogenic Sources

2.2.1. *Introduction*

Most problems arising from organometal(loid) compounds in the natural environment have resulted from human sources. Some biocidal organometal(loid)s have been deliberately introduced, usually for agricultural or pesticidal purposes. Others have appeared by unintentional introduction, as in discarded wastes.

An indirect anthropogenic source has been the discharge of inorganic substances which became substrates leading to biogenic organometals. Mercury is the outstanding example in this category (cf. Section 1.3). The use of plants and microorganisms to remove toxic oxides (e.g., As, Se, etc.) from soils [21] might be another example of this type, even though the methylated compounds formed are less toxic than the inorganic substrates.

Anthropogenic substrates, whether inorganic or organometal(loid), can also undergo speciation by abiotic reactions. This aspect has been less investigated than the other processes mentioned, and the degree of its importance still remains to be determined (cf. Section 2.3).

2.2.2. *Biocidal Organometal(loid)s*

2.2.2.1. Organotin Compounds. Tri-*n*-butyltin compounds were used in antifouling coatings for ocean-going vessels, intended to protect their surfaces from growth of algae, barnacles, etc. These compounds leached out into the surrounding waters to build up a small, highly concentrated layer of tri-*n*-butyltin that repelled free-swimming precursors to barnacles from settling [15,16]. Unfortunately, dissolved tri-*n*-butyltin compounds proved considerably more stable than had been expected. They settled into sediments and were absorbed by shellfish and other marine invertebrates, especially in harbors [5–7,22]. Widespread poisoning resulted, devastating shellfish populations and life-forms (including humans!) dependent on them. Tri-*n*-butyltin compounds were replaced by triphenyltin compounds; these, along with octyltin compounds (used for other purposes), have also been detected in marine sediments [15]. Triorganotins are successively converted to di- and monoorganotin derivatives [5–7,15] and eventually to "inorganic tin" (oxide, sulfide, etc.). The rates for these dealkylation processes are not at all uniform, allowing the intermediate species to accumulate and undergo subsequent biomethylation; methylbutyltin compounds have been reported [28].

2.2.2.2. Tetraethyllead. For many years, tetraethyllead and tetramethyllead were used as gasoline additives, and still are in some countries. Such usage often led to their escape into the environment, either by incomplete combustion or by gasoline leakage. Natural degradation of these compounds proceeded as with tin – successive loss of alkyl groups. Triethyl- and trimethyllead compounds occur in the environment [6,7,29]. These compounds remain a problem, especially since they have been reported in unexpected locations: alpine snow [30], Greenland snow [31], and French wines [32]!

2.2.2.3. Nerve Gases. Several organophosphorus and organoarsenic compounds have been used, or are stored for possible use, as toxic nerve gases [21,33]. Increased terrorist use of compounds such as sarin (Figure 1) [34], and problems of leakage from containers of stored gases [33] have raised concerns about these materials and their potential for widespread poisonings.

2.2.2.4. Agricultural and Biocidal Applications. Organo derivatives of phosphorus and arsenic have various agricultural uses [5]. Glyphosate [35,36], glyphosine, and glufosinate [37,38] (cf. Figure 1) are used as herbicides. Ethephon (cf. Figure 1) is used to promote uniform ripening in fruits [39]. Salts of methylarsonic and dimethylarsinic (cacodylic) acids are also used in agriculture [40]. The agricultural organoarsenical roxarsone (4-hydroxy-3-nitrophenylarsonic acid) is widely used (1100 tons annually) as an additive to poultry feed [41,42], raising health and pollution concerns because roxarsone undergoes biotransformation, initially to 4-hydroxy-3-aminophenylarsonic acid [43] and subsequently to arsenite and arsenate [43–45]. Since poultry litter/manure is widely used as fertilizer, the presence of arsenic oxyanions (generated by the poultry) provides an entry route for these toxic arsenic species into soils and subsequently into food webs. Sodium methylarsonate is used as a pesticide, and sodium dimethylarsinate is used as a defoliant [40]. Phenylmercuric acetate is still occasionally used in agriculture as an antitranspirant [46].

2.2.2.5. Other. Silicones [poly(dimethylsiloxanes)] provide the primary example for this category [21,47–49]. They primarily enter as discarded industrial wastes [47] or by leaching from certain antifouling paints (a minor source). While not toxic, silicones can affect the physical properties of systems [47]. They appear in landfill or digester gases [48,49], causing problems for the uses of such gases as fuels. Silicones undergo biodegradation [37,50], eventually forming SiO_2, CO_2, and water, but this does not occur uniformly and gives intermediates.

Another example is the pyridine complex of triphenylborane, $(C_6H_5)_3B \cdot NC_5H_5$, which in recent years has become a widely used antifouling agent [51,52]. Like tri-*n*-butyltin compounds, this borane leaches out from coatings on ships' hulls, fishing nets, and other surfaces continuously exposed to water. In an abiotic degradation study [51], decomposition occurred, but recovery of undecomposed borane ranged from 63 to 97%. Whether this compound or related species also used as antifouling agents will become an environmental health hazard remains to be seen; phenylboronic acid, $C_6H_5B(OH)_2$, shows biological effects in plants [53,54], so the possibility cannot be ruled out.

The compound thiomersal (sodium ethylmercurithiosalicylate; Figure 1) has been used as a preservative for vaccines and medicines since the 1930s [55,56]. Waste water containing this compound transports it into the environment. It can be degraded by bacteria [55] and may be the source of ethylmercury reported in human hair [57].

In recent years, pentamethylcyclopentadienylmanganese tricarbonyl has been used as a gasoline additive, and, along with decomposition products, it enters the environment [58–61] (cf. Section 4.4.8).

2.2.3. *Introduction of Organometal(loid) Precursors*

Organometal(loid) compounds can form in the natural environment, most commonly by biomethylation, less frequently by bioalkylation or other processes [3,5,14,62]. As previously mentioned, large quantities of an inorganic substrate introduced into natural systems can generate large quantities of their organo derivatives. Mercury is the prime example. Initially at Minamata Bay (Japan) [63] and subsequently at numerous other locations, mercury-containing substrates have entered natural waters, usually as wastes or tailings from mines [64–71].

Another source of precursors are landfills. In recent years, discarded materials from semiconductors, computers, and other instruments containing electronics have been buried in pits, providing new substrates for metal-carbon bond formation [72,73].

In addition to methylation, carbonylation (either biotic or abiotic) might occur. The two metal carbonyls $Mo(CO)_6$ and $W(CO)_6$ have been reported in landfill gases [26]. These two, along with $Ni(CO)_4$ and $Fe(CO)_5$, were also detected in sewage gases [27].

Table 2. Environmental abiotic alkylation of inorganic mercury.

Alkylating Agent	Reference
Acetic acid	[77]
Methyltin compounds	[76,77]
Methylcobaloxime	[76,81,82]
Triethyllead compounds	[78]
Rhine River sediments	[80]

2.3. Abiotic Transalkylation

Alkyl-metal bonds can form independently of biogenic sources. Active metal-carbon bonds (e.g., Grignard reagents) have been used to synthesize organometal(loid)s for over 150 years. Transalkylation reactions provide a widespread example, e.g.,

$$R_2Hg + HgCl_2 \rightarrow 2RHgCl$$

and are widespread in organometallic chemistry [74]. Most such studies have been studied in the gas phase or in organic solvents. However, such exchange can occur in aqueous media, and reports indicate that methyl exchange does occur in the natural environment [75–80]. Methyl and other alkyl groups bonded to lead have high reactivity [78,80] and readily transfer to other metals. Tin is less reactive in this respect, but it still transfers its alkyl groups to mercury [16,76,77], which is probably the strongest alkyl acceptor among the heavier metals (Table 2). Sn(II) will accept methyl groups from methylcobalamin in aqueous systems [81], as will Hg(II) [82].

Of course, transalkylation of any atom causes dealkylation of the donor atom, whether biotic or abiotic. Most dealkylation studies reported have focused their attention on biotic sources. However, abiotic alkyl exchange, involving formation or breaking of metal(loid)-carbon linkages, also occurs. These deserve more attention.

3. ENVIRONMENTAL TRANSPORT

3.1. Introduction

As mentioned in Section 1.2, introduction of one or more organic group(s) onto a metal(loid) alters the properties of the product, which, in turn, affects its mobility. Solubility and volatility are the properties most affected. Physical processes of elements (melting/freezing; boiling/liquefying; sublimation/

deposition) and of compounds (dissolution/precipitation/vaporization), and chemical processes (decomposition; dissociation/association; etc.) all change when organic groups are introduced. The biological effects also change.

Transport of organometal(loid)s through the environment may be divided into abiotic and biotic. The former involves simple physical transport through movement of air, water, ground, etc. Movement through the atmosphere has been studied the most and will be considered in detail in Section 3.2. The latter involves movement of organisms that have acquired organometal(loid)s, either by absorption or adsorption, from their surroundings.

3.2. Atmospheric Movement

Biomethylation and volatilization of arsenic was demonstrated by the work of Frederick Challenger [83–85], which in turn grew out of earlier work [83]. This led subsequently to investigations into the biomethylation of other elements (cf. Section 2.1.1). Microorganisms are the primary sources for this [86,87].

Numerous volatile organometal(loid)s have been detected in landfills, sewage sludges, municipal waste, etc.; certain representative examples are shown in Tables 3 and 4 [5,88–100]. Nor are the permethyl compounds the

Table 3. Selected examples of biogenic volatile organometal(loid)s detected in landfills, sewage and wastes involving elements from groups 12, 15, and 16.

	Compounds	Samples Tested[a]	References
Hg	$(CH_3)_2Hg$	GG, LG, LL, MW, SS	[62,88–93]
	CH_3Hg^b	GG, LG, LL, SS	[62,88–93]
As	$(CH_3)_3As$	GG, GW, LG, SS	[89,94–97]
	$(CH_3)_2As^b$	GG, GW, SS	[89,94–97]
	CH_3As^b	GG, GW, SS	[89,94–97]
Sb	$(CH_3)_3Sb$	FG, GG, GW, LG, SS	[89,94–98]
	$(CH_3)_2Sb^b$	GG, GW, SS	[89,94–98]
	CH_3Sb^b	GG, GW, SS	[89,94–98]
Bi	$(CH_3)_3Bi$	FG, LG, SS	[89,94–98]
Se	$(CH_3)_2Se$	GG, SS	[84,96]
	$(CH_3)_2Se_2$	GG, SS	[96]
Te	$(CH_3)_2Te$	GW, SS	[96]

[a] Sources: FG: fermentation gas; GG: geothermal gases; GW: geothermal waters; LG: landfill gases; LL: landfill leachates; SS: sewage sludge; WM: waste materials
[b] Inorganic group(s) attached to these compounds have been omitted.

Table 4. Selected examples of biogenic volatile group 14 organometal(loid)s detected in landfills, sewage and wastes.

	Compound	Source[a]	References
Ge	$(CH_3)_3Ge^b$	GW	[89]
	$(CH_3)_2Ge^b$	GW	[89]
	CH_3Ge^b	GW	[89]
Sn	$(CH_3)_4Sn$	FG, LG, MW, SS	[90,96,98–100]
	$(CH_3)_3Sn^b$	LG, LL, MW	[90,92,99,100]
	$(CH_3)_2Sn^b$	LG, LL, MW	[90,92,99,100]
	CH_3Sn^b	LG, LL	[90,92,99,100]
	$(C_2H_5)_3Sn^b$	LL	[100]
	$(C_2H_5)_2Sn^b$	LL	[100]
	$C_2H_5Sn^b$	LL	[100]
	$(C_4H_9)_3Sn^b$	LL	[93,100]
	$(C_4H_9)_2Sn^b$	LL	[93,100]
	$C_4H_9Sn^b$	LL	[90,93,99,100]
	$C_6H_5Sn^b$	LG	[99]
	$(C_8H_{17})_2Sn^b$	LL	[93]
	$C_8H_{17}Sn^b$	LL	[90,93]
	$(C_2H_5)_2(CH_3)_2Sn$	LG	[98,99]
	$C_2H_5(CH_3)_3Sn$	LG	[99]
	n-$C_3H_7(CH_3)_3Sn$	LG	[99]
	i-$C_3H_7(CH_3)_3Sn$	LG	[99]
	$C_4H_9(CH_3)_3Sn$	LG	[99]
Pb	$(CH_3)_4Pb$	LG	[89]

For footnotes [a] and [b] see Table 3.

only volatile organometal(loid)s. Mixed alkyl species of tin and lead have been reported in the atmosphere [101–103]. Organometal chlorides have been detected in the atmosphere above seawater [104].

Biogenically formed organometal(loid) hydrides have also been reported: As [96,105], Sb [97], Sn [99], among others. Interestingly, methylbismuth hydrides were not reported under conditions where the arsenic and antimony analogs formed [97]; this might be due to the low stability of the Bi-H bond. Phosphine occurs in the natural environment [106], and methylphosphine, CH_3PH_2, formed when simulated lightning struck sodium phosphate in the presence of methane [107]. So far, no reports of naturally occurring mono- or dimethylphosphines have appeared; methylphosphonates undergo phosphorus-carbon bond cleavage in the ocean to form methane [108,109].

Organometal(loid) volatilization by plants, both terrestrial and aquatic, is discussed elsewhere [21].

3.3. Biological Movement

Elemental cycling on lifeless planets occurs solely through physical and chemical processes (cf. Section 1.1). On Planet Earth, living organisms play a crucial role, as shown by the presence of dioxygen in our atmosphere [110]. Biomethylation, bioalkylation, biodemethylation, and other biological processes, by their very definition, require organisms to perform them. All organisms on this planet, even humans, belong to one or more food chains/webs. Ingestion of organisms by other organisms transports any organometal(loid)s within, however formed. Concentrations become enhanced (biomagnification) as compounds move along a chain/web, finally reaching toxic levels.

Another factor, not fully realized or explored, is the mobility of most living organisms. Some, like migrating birds, fishes, mammals or insects, can travel hundreds, even thousands, of miles. Wherever they go, the contents of their bodies go also. If they die far from their starting points, any organometal(loid)s they carry re-enter the environment at that point. How important this might be to the cycling of elements and compounds has not yet been, and may never be, fully determined. It is a factor, however, that must be kept considered.

4. SPECIFIC ELEMENTS AND CYCLES

4.1. Introduction

All elements belong to natural cycles, and all cycles comprise a "supercycle". All organometal(loid)s belong to the carbon cycle, and are also part of the cycles of metal(loid)s involved. The presence of organic groups (cf. Section 1.3) changes both physical and chemical properties of elements to which they are bonded. Only a small proportion of the atoms of any element, even carbon, are part of an organometal(loid) compound. Yet the smallness of this portion does not mean that it is insignificant!

Whether they are part of an organism's biochemistry, an inert addition, or a deadly toxin, organometal(loid)s will be a part of the cycling process, and the importance of their roles may be far larger than the magnitude of their concentration.

4.2. Three Transition Metals

4.2.1. Introduction

When biologically important organometal(loid)s are discussed, they are almost always compounds of the main group elements; even mercury is usually considered more of a main group element than a transition element. The only such metal usually considered is cobalt. Yet in recent years, evidence has been growing that at least two others may also fit into this category: iron and nickel. All three of these metals form metalloenzymes; the ones mentioned in this article have an elaborate chelating arrangement with one active site on the metal [111] and they all form and break metal-carbon linkages. The proportion of each metal present in these metalloenzymes is tiny compared to the total quantity of the metal on this planet; yet these enzymes are (literally) vitally necessary to organisms.

4.2.2. Cobalt

A cobalt atom is the active site of vitamin B_{12}, whose structure is shown in Figure 2. The chemistry of vitamin B_{12} has been extensively studied [112–116], and involves breaking and/or reforming Co-C linkages at a single

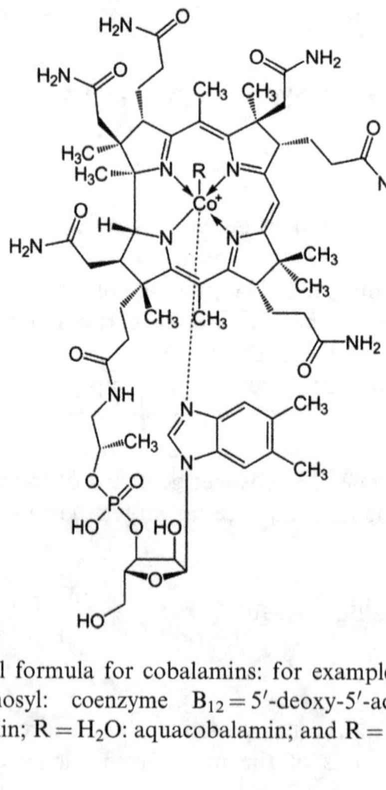

Figure 2. Structural formula for cobalamins: for example, R = CN: vitamin B_{12}; R = 5′-deoxy-5′-adenosyl: coenzyme B_{12} = 5′-deoxy-5′-adenosylcobalamin; R = CH_3: methylcobalamin; R = H_2O: aquacobalamin; and R = HO: hydroxocobalamin.

coordination site on cobalt. Cobalamins exist in various forms, depending on the group R (Figure 2): methylcobalamin, with its Co-CH_3 linkage [117], is the most relevant for the purposes of this article. This molecule acts as a methyltransferase [117] and is closely tied to the environmental formation of methylmercury [18,118]. Cobalamins are synthesized by microbes [119] but can be taken up by other organisms [120]. Vitamin B_{12} can act abiotically in the environment [81,82].

4.2.3. *Nickel*

Nickel has received growing attention in recent years and has a more substantial importance than previously realized [121]. Much of the work has been done on coenzyme F_{430} [122–124]. Formation of a Ni-CH_3 linkage on this coenzyme has been experimentally verified [125–127]. This coenzyme, also named methylcoenzyme M reductase, occurs in the semifinal step of the anaerobic genesis of methane, and is thus crucial in the cycle of that compound. A Ni-CH_3 linkage has also been used to model acetylcoenzyme A synthesis [128]. The molecules carbon monoxide dehydrogenase [129–131] and acetylcoenzyme A synthase [131,132] form Ni-CO linkages as reaction intermediates, which are used by anaerobic microorganisms both as a carbon source and as an energy source (CO is oxidized to CO_2) [132]. In a model study, methylcobalamin was found to methylate the nickel atom of (triphos)Ni(PPh_3) [133] (triphos = 1,1,4,7,7-pentaphenyl-1,4,7-triphospha-*n*-heptane).

Nickel tetracarbonyl, $Ni(CO)_4$, is a volatile and very toxic nickel derivative [134]. It has been detected in sewage gas [27] and occurs as an intermediate in the Mond process for the separation of nickel from cobalt. A review of nickel in the environment reported that, while nickel tetracarbonyl contributed to health problems, it was not found in the natural environment [135]. Considering that $Ni(CO)_4$ forms readily from nickel metal and carbon monoxide, and that nickel occurs as a component of electronic waste discards [72], this compound may play a more important role in environmental cycling than has been realized.

4.2.4. *Iron*

A toxic, and little discussed, organometallic compound is carboxyhemoglobin, containing a Fe-CO bond. This bond, and its strength, has resulted in many cases of carbon monoxide poisoning [136]. The kinetics of its buildup in human blood have been investigated [137]. Carbon monoxide also interacts with Fe atoms in hydrogenase enzymes [138–140] and in mitochondrial cytochrome *c* oxidase [141]. Like nickel, iron readily reacts with carbon monoxide to form $Fe(CO)_5$ [142], and has been reported in sewage gas [27]. This compound was less stable than nickel tetracarbonyl, especially in the presence of water [27]. What part the iron carbonyls and other iron-carbon intermediates might play in the environmental cycling of iron remains to be determined, but they are certainly important parts of the carbon cycle.

4.3. Intensively Investigated Elements

4.3.1. Mercury

Mercury is *the* element whose organo derivatives have led to the extensive growth of interest in environmental cycles. The tragic cases of mercury poisoning [14,63,143] in the second half of the 20th century and the realization that mercury was being methylated by environmental organisms [14,62,88] has generated an enormous research effort. Substantial quantities of mercury, both metal and compounds, have been introduced into the environment, usually through water (see Section 2.2.3). In addition to previously mentioned mine tailings, dental wastewater has become a significant mercury source [144,145]. Numerous biogeochemical "mini-cycles" for mercury have been proposed, of which only a few are mentioned here [146–150].

Methyl derivatives have important roles in this cycle: dimethylmercury is a volatile gas (cf. Table 3) that can escape into, and diffuse through, the atmosphere; monomethylmercury can have various inorganic groups attached. It has a lower affinity for humic substances than Hg(II) [151], which diminishes its ability to be adsorbed, and, as CH_3HgCl, has some volatility and appreciable solubility in lipids. Elemental mercury also adsorbs onto sediments, where it can be oxidized and methylated, or be solely methylated [152]. Experimental evidence indicates that there may be a linear relationship between inorganic mercury deposition and methylmercury bioaccumulation [153]. So many factors, including reservoir eutrophication [154], affect the rate and degree of mercury methylation that research will quite likely continue for many years.

4.3.2. Tin

Investigation into the environmental cycling of tin has arisen because of the use of tri-*n*-butyltin in antifouling paints (cf Section 2.2.2.1.) and their entry into the natural environment, along with other, less widespread, sources. Tri-*n*-butyltin can undergo successive debutylation [155]; however, butyltin species can also undergo biomethylation to produce mixed methylbutyltin compounds [28,156]. These have also been reported in landfill gases, along with tetramethyltin [97]. Organotin-containing sludges are often added to soils as fertilizers, which has led to research on the degradation of the tin species present. Bacteria cause biodegradation [157,158], but many organotin compounds remain unchanged over long periods of time [159–163]. Like mercury, tin and its organo derivatives will be investigated for many years to come.

4.3.3. Lead

Lead resembles tin in the sense that organo derivatives of both elements were introduced into the environment unintentionally. For many years, tetraethyllead and tetramethyllead were used as gasoline additives [17], and entered the environment in exhaust fumes. Consequently, methyl- and ethyllead derivatives have been studied for years [17,29–32]. These tend to occur in a wider variety of environments than do organotins, in snows [31,32], forest floors [164,165], urban dust [166], urban atmosphere [101,167], in landfill emissions [90], and in plant leaves [168]. A wide variety of biological/environmental reference samples have been proposed [169]. Like tin analogs, organolead compounds have been used in antifouling paints and as rodent repellants [170].

Fewer organolead compounds have been detected than organotins; trimethyllead, triethyllead, and their dialkyl counterparts are the major ones. Tetraalkylleads, including some mixed compounds [17], also occur. Triphenyllead acetate was formerly used in biocidal preparations [171,172], but has not been reported in the environment. The role of organoleads in the environmental cycling of lead appears to be more limited than for mercury or tin, due to the instability of monoalkyllead(IV) compounds and the lability of the lead-carbon bond, mentioned in Section 2.3. Biomethylation of lead has not been unequivocally established, and its possible role in environmental cycling remains uncertain. As long as alkyllead compounds are used as gasoline additives, their derivatives will continue to be detected in the environment.

4.3.4. Phosphorus

Until recently, the proposed environmental cycle for phosphorus included only inorganic phosphorus(V) compounds: mono- and polyphosphoric acids, their salts, their esters, etc. [1]. Developing realization of the existence of phosphonic acids [172,173] and other organophosphorus compounds formed by biosynthesis [174–176], including phosphonolipids (phosphono analogs of phospholipids [177]), has forced a revision of this viewpoint, although the extent of their contribution has yet to be determined.

Compounds of phosphorus in lower oxidation states have also been reported in the environment [178], especially phosphine [106], which may be formed biotically [178] or abiotically [179,180]. Except for the artificial nerve gases mentioned previously, phosphine appears to be the principal volatile phosphorus compound. There are no reports of methyl- , dimethyl- or trimethylphosphine in the environment, although a laboratory study indicated that both phosphine and methylphosphine formed when phosphate in a reducing medium received "simulated lightning" [107].

Phosphonates appear to be the predominant form of organophosphorus compounds in the environment, and play a role in phosphorus cycling in an anoxic marine basin [181]. They occur much more commonly in organisms than the organometals previously discussed in this section, and, in that sense, play a bigger role in the natural cycle.

4.3.5. *Arsenic*

Arsenic is much more similar to phosphorus in its organo derivatives than it is to the true metals. The environmental changes [182] and toxicity [183] are discussed elsewhere. Biomethylation of inorganic arsenic has already been mentioned [82–84]. Heat-resistant fungi volatilized arsenic [184], and counts of arsenic-methylating bacteria could be used to estimate the gasification potential of soil [185]. Microbes volatilized arsenic from retorted shale [186]. Bioalkylation is more extensive and important for arsenic than for most other elements. Arsenobetaine (Figure 1) is probably the best known example, and is found in many organisms, though the mechanism for its formation is not yet fully known [187]. Numerous arsenolipids of generic formula $(CH_3)_2As(:O)R$ (R = long chain fatty acid) have been reported [188].

The environmental chemistry of arsenic has been reviewed [189,190], and organoarsenic compounds play a major part. As the extensive research in this area continues, more surprises and unexpected compounds are likely to emerge.

4.3.6. *Selenium*

Selenium is similar to arsenic in the types of organo compounds found in the environment [191,192]. Methylselenium compounds (Me_2Se, Me_2Se_2, Me_2SeO, etc.) are usually found in water, soils or atmosphere, while more complex organoselenium compounds, such as selenomethionine (Figure 1), occur inside organisms [193]. Plants have been used to remove toxic selenium dioxide from soils by converting it to volatile Me_2Se [21]. The biochemistry of selenium parallels that of its lighter congenor sulfur, and mixed sulfur-selenium compounds are known [192]. Like arsenic, the organo chemistry of selenium should continue to expand.

4.4. Less Studied Elements

4.4.1. Antimony

As might be expected, biomethylation of antimony parallels that of arsenic [193,194]. Investigations received an impetus from the possibility that trimethylantimony might be connected with sudden infant death syndrome [193]. Thus far, only methylstibines have been reported in the environment [193–200], although a stibolipid was generated by the diatom *Thalassiosira nana* under laboratory conditions [196]. Like arsenic, methylantimony compounds can accumulate in terrestrial plants [199], and will form in sediments and sludges [198,200,201]. A lot more will be discovered as research continues in this area.

4.4.2. Tellurium

Tellurium, being a heavier congenor of selenium, has a very similar organo chemistry [61]. A strain of *Penicillium* methylated both selenium and tellurium [202], but biomethylation of tellurium required the presence of selenium [202]. Microbes also methylated tellurite salts [203–205]; this may contribute to the resistance of such species to tellurite toxicity [204]. A comparative study showed that rats metabolized both selenium and tellurium [206]. Both produced the cation $(CH_3)_3E^+$($E = Se, Te$), but for tellurium, this was the sole product; for selenium, it was a minor product with the major product being a selenosugar. Fungi were able to incorporate tellurium into amino acids, including telluromethionine [207]. Telluromethionine has been used in heteroatomic biochemical studies of methionine [208]. The organo derivatives of tellurium are likely to play a less significant role in the biogeochemical cycling of this element than do the corresponding compounds of selenium, but they will play some role.

4.4.3. Germanium

Germanium is an enigma with respect to its methyl derivatives in the natural environment. The limited quantity of information has been reviewed

[61,209,210]. Almost all reports on methylgermanium species have been for water samples, and they show the mono- and dimethylgermanium species only; no trimethylgermanium has been reported despite specific efforts to find it [211–213]. Concentrations of monomethylgermanium show a remarkable constancy, independent of depth, in natural waters [61,209,214]. The Ge/Si ratio shows little variation in water [61,209], and germanium may be absorbed as a "superheavy isotope" of silicon [61]. This view is consistent with the reported Ge/Si ratio in plant phytoliths [215] and C/Si/Ge bioisosterism [216]. The absence of trimethylgermanium in waters, and tetramethylgermane in gases is puzzling, being such a contrast to the tin and lead counterparts. Trimethlgermanium has been found to form in an anaerobic sewage digester [217]. Possibly the reported toxicity of trimethylgermyl complexes towards fungi and bacteria may be related to this [217]. In any event, the considerable uncertainty should encourage further research in this area.

4.4.4. Thallium

In a recent review of thallium in the natural environment [218], there is barely a mention of organothallium compounds. Thallium is a toxic metal – more toxic than its periodic table neighbors mercury and lead – and is a concern for public health [219]. Trimethylthallium is unstable under natural conditions, and the only environmental organothallium species reported to date is $(CH_3)_2Tl^+$. Several reports on this ion have been published [220–225,61]. Both Tl^+ and $(CH_3)_2Tl^+$ underwent bioaccumulation by algae, diatoms, and plankton [224,225], though the bioconcentration factor was greater for Tl^+. These observations suggest that dimethylthallium could enter a food chain/web and undergo biomagnifications. The only toxicity study reported [226] indicated that Tl^+ was considerably more toxic towards mice than $(CH_3)_2Tl^+$. There are some ominous possibilities about dimethylthallium ion in the environment that should encourage further research.

4.4.5. Bismuth

Only methylbismuth species [61,89,94–98,227] have been reported in the environment. Trimethylbismuth, the predominant product, has been detected in various gases from sewage, etc. (cf. Table 3), and volatilized from alluvial soil [228] and human feces [228]. While considerably more restricted in occurrence than the methyl analogs of arsenic and antimony, methylbismuth compounds may have a wider range of occurrence than is now known. The increasing quantity of bismuth entering landfills and waste dumps will provide additional substrate to generate volatile trimethylbismuth, providing ample reason for additional research.

4.4.6. *Polonium*

Polonium possesses only radioactive isotopes; ^{210}Po, with a half-life of 138.4 days, is the one most studied. Its organic chemistry is much less extensive (and much less studied) than that of its congenors selenium and tellurium. While this element occurs in nature, its only environmental organo compound is gaseous $(CH_3)_2Po$ [61,229]. While this compound has not been isolated, the similarity of polonium to tellurium in biovolatilization [229] and the volatile compound formed from reaction of methylB_{12} and a polonium species [230] strongly indicate the probability of its formation. Polonium undergoes bioaccumulation in marine birds [231]. Certainly the formation of dimethylpolonium will facilitate movement through the environment, and the possible risks deserve further research.

4.4.7. *Cadmium*

The literature on environmental organocadmium compounds is very sparse [232–234,61]. Thus far, the only species reported are CH_3Cd^+ and (probably) $(CH_3)_2Cd$. The former has been detected in polar ocean water, indicating a biogenic origin. Cadmium-containing waste is being added to the environment in large quantities [235]. How significant the methylation of cadmium will contribute to this elemental cycle remains to be determined.

4.4.8. *Silicon and Boron*

These elements have already been discussed in Section 2.2.2.5. Polymethylsiloxanes occur in landfill and digester gases [49,235,236] and may cause problems in the use of such gases as fuels [235,236]. Such gases can escape into the atmosphere, or, more slowly, by water or liquids. Except for phenylboranes, there do not seem to be organoboron compounds entering the environment. No evidence for biomethylation of either element has been claimed. The most likely conditions for that to occur would be for electron-rich compounds (e.g., silicides, borides) to be exposed to anaerobic bacteria under anoxic conditions. Even without biomethylation, the introduction of polydimethylsiloxanes can contribute to the silicon cycle, if only as a source of silicon dioxide.

4.4.9. *Molybdenum, Tungsten, and Manganese*

The hexacarbonyls of molybdenum and tungsten have already been mentioned. Both, molybdenum and tungsten, form metalloenzymes [237,238], of which nitrogenase is probably the best known. What roles their metal carbonyls may have in the environmental cycle of these metals, only future research will reveal.

Methylcyclopentadienylmanganese tricarbonyl, $CH_3C_5H_4Mn(CO)_3$, has been used as a gasoline additive (cf. Section 2.2.2). Most of it enters the environment as "inorganic manganese", but spillage and other sources may allow some of the original compound to escape unaltered [61]. If extensively used, this compound could add appreciably to branches of the manganese cycle [61]. Various possibilities for metal carbonyls in environmental cycling exist.

5. CONCLUSIONS

Formation and existence of organometal(loid)s comprise an important part in the environmental cycling of elements. Probably the most important part is the enhancement of mobility; volatility and altered solubility are the major changes. Permethylmetal(loid) compounds are the most notable, but mixed organometal(loid) hydrides and chlorides also volatilize. Enhanced solubility in lipids or water facilitates environmental transport, especially inside organisms. The presence of organo groups also changes adsorption on surfaces, especially in soils, sediments, and sludges.

Organometal(loid)s have different effects on many organisms, compared to their inorganic counterparts. They can be ingested more easily and move more readily along food chains/webs, undergoing biomagnifications. Many such compounds are toxic, most notably methylmercurials. The widespread poisonings that have resulted from them has resulted in extensive research. In fact, the great majority of research on organometal(loid)s and cycling has resulted from human introduction of such compounds (intentionally or inadvertently) in agriculture, pesticides, nerve gases, etc., emphasizing the most toxic.

Total research on this subject continues to expand at an impressive rate. The more that is learned, the more unanswered questions appear! Speciation studies proliferate, and new techniques are developed to investigate them. More and more "mini-cycles" are appearing. Applied research, dedicated to controlling and reversing the effects of these compounds, is also growing, as are kinetic and mechanistic studies. Roles for organometal intermediates will be found, their importance not measured by their transience. Work on organometal(loid)s in the environment and in living organisms appears likely to continue and expand for the foreseeable future.

ACKNOWLEDGMENTS

The author expresses his gratitude and appreciation to the hard-working staff of the R. E. Oesper Chemistry-Biology Library of the University of Cincinnati for their valuable assistance in searching out references.

REFERENCES

1. S. S. Butcher, R. J. Charlson, G. H. Orians and G. V. Wolfe, (Ed.), *Global Biogeochemical Cycles*, Academic Press, San Diego (CA, USA), 1992, p. 1.
2. *The American Heritage Dictionary*, 2nd College edn., Houghton Mifflin, Boston, 1982.
3. P. M. H. Kroneck, in *Biogeochemical Cycles of Elements*, Vol 43 of *Metal Ions in Biological Systems*, Ed. A. Sigel, H. Sigel and R. K. O. Sigel, Taylor & Francis, Boca Raton (FL, USA), 2005, pp. 1–8.
4. R. M. Harrison, in *Principles of Environmental Chemistry*, Ed. R. M. Harrison, RSC Publ., Cambridge, UK, 2007, pp. 314–346.
5. J. S. Thayer, *Environmental Chemistry of the Heavy Elements*, VCH, New York, 1995, (a) pp. 75–94; (b) pp. 43–48.
6. P. J. Craig, G. Eng and R. O. Jenkins, in *Organometallic Compounds in the Environment*, 2nd edn, Ed. P. J. Craig, J. Wiley & Sons, Chichester, UK, 2003, pp. 1–55.
7. P. J. Craig and R. O. Jenkins, in *Organic Metal and Metalloid Species in the Environment*, Ed. A. V. Hirner and H. Emons, Springer, Berlin, 2004, pp. 1–15.
8. *The Major Biogeochemical Cycles and Their Interactions*, (SCOPE 21), Ed. B. Bolin and R. B. Cook, J. Wiley & Sons, Chichester, UK, 1983.
9. G. R. Dickens and B. M. Kennedy, *Proc. Ocean Drilling Program: Scientific Results*, 2000, **164**, 165–170.
10. C. P. McKay, K. P. Hand, P. T. Doran, D. T. Andersen and J. C. Priscu, *Geophys. Res. Lett.*, 2003, **30**, 351–354.
11. D. Musselwhite and J. I. Lunine, *J. Geophys. Res.*, 1995, **100**, 23301–23306.
12. C. Thomas, S. Picaud, O. Mousis and V. Ballenegger, *Planetary Space Sci.*, 2008, **56**, 1607–1617.
13. J. I. Lunine and S. K. Atreya, *Nature Geoscience*, 2008, **1**, 159–164.
14. *Mercury Pollution: Integration and Synthesis*, Ed. C. J. Watras and J. W. Huckabee, Lewis, Boca Raton (FL, USA), 1994.
15. *Organotin: Environmental Fate and Effects*, Ed. M. A. Champ and P. F. Seligman, Chapman & Hall, London, 1996.
16. T. Gadja and A. Jancsó, Chapter 4 of this book.
17. J. Yoshinaga, in *Organometallic Compounds in the Environment*, 2nd edn., Ed. P. J. Craig, J. Wiley & Sons, Chichester, UK, 2003, pp. 151–194.
18. J. S. Thayer, *Appl. Organometal. Chem.*, 2002, **16**, 677–691.
19. R. Bentley and T. G. Chasteen, *Microbiol. Molec. Biol. Rev.*, 2002, **66**, 250–271.

20. U. Limper, B. Knopf and H. König, *J. Appl. Entomol.*, 2008, **132**, 168–176.
21. J. S. Thayer, Chapter 13 of this book.
22. J. S. Thayer, in *Biological Properties of Metal Alkyl Derivatives*, Vol. 29 *of Metal Ions in Biological Systems*, Ed. H. Sigel and A. Sigel, Marcel Dekker, New York, 1993, pp. 1–36.
23. M. Slekovec, W. Goessler and K. J. Irgolic, *Chem. Spec. Bioavailab.*, 1999, **11**, 115–123.
24. A. Geiszinger, W. Goessler, D. Kuehnelt, K. Francesconi and W. Kosmus, *Environ. Sci. Technol.*, 1998, **32**, 2238–2243.
25. A. W. Ritchie, J. S. Edmonds, W. Goessler and R. O. Jenkins, *FEMS Microbiol. Lett.*, 2004, **235**, 95–99.
26. J. Feldmann and W. R. Cullen, *Environ. Sci. Technol.*, 1997, **31**, 2125–2129.
27. J. Feldmann, *J. Environ. Monit.*, 1999, **1**, 33–37.
28. A. J. Vella and J. P. T. Adami, *Appl. Organometal. Chem.*, 2001, **15**, 901–906.
29. H. G. Abadin and H. R. Pohl, Chapter 5 of this book.
30. M. Heisterkamp, K. van der Velde, C. Ferrari, C. F. Boutron and F. C. Adams, *Environ. Sci. Technol.*, 1999, **33**, 4416–4421.
31. R. Łobiński, C. F. Boutron, J.-P. Candelone, S. Hong and J. Szpunar-Łobińska, *Environ. Sci. Technol.*, 1994, **28**, 1459–1466.
32. P.-L. Teissedre, R. Łobiński, M.-T. Cabanis, J. Szpunar-Łobińska, J.-C. Cabanis and F. C. Adams, *Sci. Total Environ.*, 1994, **153**, 247–253.
33. S. S. Talmadge, A. P. Watson, V. Hauschild, N. B. Munro and J. King, *Curr. Org. Chem.*, 2007, **11**, 285–298.
34. H. Morita, N. Yanagisawa and T. Nakajima, *Lancet*, 1995, **346**, 290.
35. J. E. Franz, M. K. Mao and J. A. Sikarski, *Glyphosate: A Unique Global Herbicide*, ACS Monograph 189, American Chemical Society, Washington, 1997.
36. S. O. Duke and S. B. Powles, *Pest Management Sci.*, 2008, **64**, 319–325.
37. R. G. Hall, *Phosphorus, Sulfur and Silicon*, 2008, **183**, 258–265.
38. W. J. Everman, C. R. Mayhew, J. D. Burton, A. C. York and J. W. Wilcut, *Weed Science*, 2009, **57**, 1–5.
39. R. L. Hilderbrand, in *The Role of Phosphonates in Living Systems*, Ed. R. L. Hilderbrand, CRC Press, Boca Raton (FL, USA), 1983, pp. 139–170.
40. R. Sierra-Alvarez, U. Yenal, J. A. Field, M. Kopplin, A. J. Gandolfi and J. R. Garbarino, *J. Agric. Food Chem.*, 2006, **54**, 3959–3966.
41. D. H. Chapman and D. B. Johnson, *Poult. Sci.*, 2002, **81**, 356–364.
42. B. Hileman, *Chemical & Engineering News*, April 9, 2007, 34–35.
43. I. Cortinas, J. A. Field, M. Kopplin, J. R. Garbarino, A. J. Gandolfi and R. Sierra-Alvarez, *Environ. Sci. Technol.*, 2006, **40**, 2951–2957.
44. B. P. Jackson, J. C. Seaman and P. M. Bertsch, *Chemosphere*, 2006, **65**, 2028–2034.
45. J. F. Stolz, E. Perera, B. Kilonzo, B. Kail, B. Crable, E. Fisher, M. Ranganathan, L. Wormer and P. Basu, *Environ. Sci. Technol.*, 2007, **41**, 818–823.
46. N. C. Rajapakse, J. W. Kelly and D. W. Reed, *Scientia Horticulturae*, 1990, **43**, 307–312.
47. D. Graiver, K. W. Farminer and R. Narayan, *J. Polymers and the Environment*, 2003, **11**, 129–136.

48. S. C. Popat and M. A. Deshusses, *Environ. Sci. Technol.*, 2008, **42**, 8510–8515.
49. A. Ohannessian, V. Desjardin, V. Chatain and P. Germain, *Water Sci. Technol.*, 2008, **58**, 1775–1781.
50. J. C. Carpenter, J. A. Della and S. B. Dorn, *Environ. Sci. Technol.*, 1995, **29**, 864–868.
51. X. Zhou, H. Okamura and S. Nagata, *Chemosphere*, 2007, **67**, 1904–1910.
52. K. V. Thomas, *Biofouling*, 2001, **17**, 73–86. *Chem. Abstracts*, 2001, **135**, 340403b.
53. M. P. Raghavendra and V. Prakash, *J. Agric. Food Chem.*, 2002, **50**, 6037–6041.
54. E. Bassil, H. Hu and P. H. Brown, *Plant Physiol.*, 2004, **136**, 3383–3395.
55. R. Fortunato, J. G. Crespo and M. A. M. Reis, *Water Res.*, 2005, **39**, 511–522.
56. D. A. Geier, L. K. Sykes and M. R. Geier, *J. Toxicol. Environ. Health, B*, 2007, **10**, 575–596.
57. D. Gibičar, M. Logar, N. Horvat, A. Marn-Pernat, R. Ponikvar and M. Horvat, *Anal. Bioanal. Chem.*, 2007, **388**, 329–340.
58. J. Zayed, *Amer. J. Ind. Medicine*, 2001, **39**, 426–433.
59. M. P. Walsh, *Amer. J. Ind. Medicine*, 2007, **50**, 853–860.
60. S. A. Watmough, M. C. Emers and P. J. Dillon, *Appl. Geochem.*, 2007, **22**, 241–1247.
61. J. Feldmann, in *Organometallic Compounds in the Environment*, 2nd edn., Ed. P. J. Craig, J Wiley & Sons, Chichester, UK, 2003, pp. 353–358.
62. H. Hintelmann, Chapter 11 of this book.
63. P. A. D'Itri and F. M. D'Itri, *Mercury Contamination: A Human Tragedy*, J. Wiley & Sons, New York, 1977.
64. O. Malm, *Environ. Res., Sect. A*, 1998, **77**, 73–78.
65. J. E. Gray, V. F. Labson, J. N. Weaver and D. P. Krabbenhoft, *Geophys. Res. Lett.*, 2002, **29**, 20/1–20/4.
66. J. E. Gray, I. A. Greaves, D. M. Bustos and D. P. Krabbenhoft, *Environ. Geol.*, 2003, **43**, 298–307.
67. J. E. Gray, M. E. Hines, P. L. Higueras, I. Adatto and B. K. Lasorsa, *Environ. Sci. Technol.*, 2004, **38**, 4285–4292.
68. J. E. Gray and M. E. Hines, *Appl. Geochem.*, 2006, **21**, 1819–1820.
69. S. A. Shaw, T. A. Al and K. T. B. MacQuarrie, *Appl. Geochem.*, 2006, **21**, 1986–1998.
70. L. R. P. de Andrade Lima, L. A. Bernardez and L. A. D. Barbosa, *J. Hazardous Mat.*, 2008, **150**, 747–753.
71. S. Winch, T. Praharaj, D. Fortin and D. R. S. Lean, *Sci. Total Environ.*, 2008, **392**, 242–251.
72. A. Ghaly, in *Proc. Intern. Conf. Waste Technology and Management*, 2008, 347–358.
73. J. Feldmann, in *Biogeochemistry of Environmentally Important Trace Elements*, Ed. Y. Cai and O. C. Braids, ACS Symposium Series 835, American Chemical Society, Washington, DC, 2003, pp. 128–150.

74. K. Moedritzer, in *Advances in Organometallic Chemistry*, Ed. F. G. A. Stone and R. West, Academic Press, New York, 1968, Vol. 6, pp. 171–271.
75. K. Gardfeldt, J. Munthe, D. Stromberg and O. Lindquist, *Sci. Total Environ.*, 2003, **304**, 127–136.
76. V. Celo, S. L. Scott and D. R. S. Lean, *RMZ – Materials and Geoenvironment*, 2004, **51**, 915–918. *Chem. Abstracts*, 2005, **142**, 322090n.
77. V. Celo, S. L. Scott and D. R. S. Lean, *RMZ – Materials and Geoenvironment*, 2004, **51**, 919–923. *Chem. Abstracts*, 2005, **142**, 322091p.
78. M. Hempel, J. Kuballa and E. Jantzen, *Fresenius J. Anal. Chem.*, 2000, **51**, 470–475.
79. V. Minganti, R. Capelli, G. Drava and R. De Pellegrini, *Chemosphere*, 2007, **67**, 1018–1024.
80. R. Falter and R.-D. Wilken, *Vom Wasser*, 2000, **90**, 217–231. *Chem. Abstracts*, 2002, **138**, 28711z.
81. B. Chen, Q. Zhou, J. Liu, D. Cao, T. Wang and G. Jiang, *Chemosphere*, 2007, **68**, 414–419.
82. B. Chen, T. Wang, Y. Yin, B. He and G. Jiang, *Appl. Organometal. Chem.*, 2007, **21**, 462–467.
83. F. E. Challenger, in *Organometals and Organometalloids Occurrence and Fate in the Environment*, Ed. F. E. Brinckman and J. M. Bellama, ACS Symposium Series 82, American Chemical Society, Washington, 1978, 1–22.
84. R. Bentley and T. G. Chasteen, *Microbiol. Mol. Biol. Rev.*, 2002, **66**, 250–271.
85. T. G. Chasteen and R. Bentley, *App. Organomet. Chem.*, 2003, **17**, 201–211.
86. J. Meyer, A. Schmidt, K. Michalke and R. Hensel, *System. Appl. Microbiol.*, 2007, **30**, 229–238.
87. J. Meyer, A. Schmidt, K. Michalke and R. Hensel, *System. Appl. Microbiol.*, 2008, **31**, 81–87.
88. R. P. Mason and J. M. Benoit, in *Organometallic Compounds in the Environment*, 2nd edn., Ed. P. J. Craig, J. Wiley & Sons, Chichester, UK, 2003, pp. 57–99.
89. A. V. Hirner, J. Feldmann, E. Krupp, R. Grümping, R. Goguel and W. R. Cullen, *Org. Geochem.*, 1998, **29**, 1765–1778.
90. B. Michalzik, G. Ilgen, F. Hertel, S. Hantsch and B. Bilitewski, *Waste Managem.*, 2007, **27**, 497–509.
91. J. Sommar, X. Feng and O. Lindqvist, *Appl. Organometal. Chem.*, 1999, **13**, 441–445.
92. S. E. Lindberg, G. Southworth, E. M. Prestbo, D. Wallschläger, M. A. Bogle and J. Price, *Atm. Environm.*, 2005, **39**, 249–258.
93. G. Ilgen, D. Glindemann, R. Herrmann, F. Hertel and J. -H. Huang, *Waste Managem.*, 2008, **28**, 1518–1527.
94. D. Kuehnelt and W. Goessler, in *Organometallic Compounds in the Environment*, 2nd edn., Ed. P. J. Craig, J. Wiley & Sons, Chichester, UK, 2003, pp.223–275.
95. R. A. Diaz-Bone, B. Menzel, A. Barrenstein and A. V. Hirner, in *Organic Metal and Metalloid Species in the Environment*, Ed. A. V. Hirner and H. Emons, Springer, Berlin, 2004, pp. 97–112.

96. K. Michalke, E. B. Wickenheiser, M. Mehring, A. V. Hirner and R. Hensel, *Appl. Environ. Microbiol.*, 2000, **66**, 2791–2796.
97. S. Maillefer, C. R. Lehr and W. R. Cullen, *Appl. Organometal. Chem.*, 2003, **17**, 154–160.
98. J. Feldmann, I. Koch and W. R. Cullen, *Analyst*, 1998, **123**, 815–820.
99. S. K. Mitra, K. Jiang, K. Haas and J. Feldmann, *J. Environ. Monit.*, 2005, **7**, 1066–1068.
100. P. Pinel-Raffaitin, P. Rodríguez-González, M. Ponthieu, D. Amouroux, I. Le Hecho, L. Mazeas, O. F. X. Donard and M. Potin-Gautier, *J. Anal. Atomic Spectrom.*, 2007, **26**, 258–266.
101. C. Pécheyran, B. Lalére and O. F. X. Donard, *Environ. Sci. Technol.*, 2000, **34**, 27–32.
102. A. J. Vella and R. Vassallo, *Appl. Organometal. Chem.*, 2002, **16**, 239–244.
103. E. Tessier, D. Amouroux and O. F. X. Donard, *Biochem.*, 2002, **59**, 161–181.
104. Z. Mester and R. E. Sturgeon, *Environ. Sci. Technol.*, 2002, **36**, 1198–1201.
105. C. Yuan, X. Lu, J. Quin, B. P. Rosen and X. C. Le, *Environ. Sci. Technol.*, 2008, **42**, 3201–3206.
106. Z. Feng, X. Song and Z. Yu, *Chemosphere*, 2008, **73**, 519–525.
107. D. Glindemann, M. Edwards and O. Schrems, *Atm. Environ.*, 2004, **38**, 6867–6874.
108. E. D. Ingall, *Nature Geoscience*, 2008, **1**, 419–420.
109. D. M. Karl, L. Beversdorf, K. M. Björkman, M. J. Church, A. Martinez and E. F. DeLong, *Nature Geoscience*, 2008, **1**, 473–478.
110. N. Lane, *Oxygen:The Molecule that Made the World*, Oxford University Press, Oxford, 2002.
111. S. W. Ragsdale, *Chem. Rev.*, 2006, **106**, 3317–3337.
112. B. Kräutler, in *Metal-Carbon Bonds in Enzymes and Cofactors*, Vol. 6 of *Metal Ions in Life Sciences*, Ed. A. Sigel, H. Sigel and R. K. O. Sigel, Royal Society of Chemistry, Cambridge, UK, 2009, pp. 1–51.
113. K. P. Jensen and U. Ryde, *J. Am. Chem. Soc.*, 2005, **127**, 9117–9128.
114. J. C. Escalante-Semerena, *J. Bacteriol.*, 2007, **189**, 4555–4560.
115. H. Weissbach, *J. Biol. Chem.*, 2008, **283**, 23497–23504.
116. Y.-F. Zhang and G. Ning, *Expert Opinion on Investigational Drugs*, 2008, **17**, 953–964.
117. R. G. Matthews, in *Metal-Carbon Bonds in Enzymes and Cofactors*, Vol. 6 of *Metal Ions in Life Sciences*, Ed. A. Sigel, H. Sigel and R. K. O. Sigel, Royal Society of Chemistry, Cambridge UK, 2009, pp. 53–113.
118. S. M. Chemaly, *South Afric. J. Sci.*, 2002, **98**, 568–572.
119. H.-H. Martens, H. Barg, M. J. Warren and D. Jahn, *Appl. Microbiol. Biotechnol.*, 2002, **58**, 275–285.
120. M. T. Croft, A. D. Lawrence, E. Raux-Deery, M. J. Warren and A. G. Smith, *Nature*, 2005, **438**, 90–93.
121. *Nickel and its Surprising Impact in Nature*, Vol. 2 *of Metal Ions in Life Sciences*, Ed. A. Sigel, H. Sigel and R. K. O. Sigel, J. Wiley & Sons, Chichester, UK, 2007.
122. M. Dey, X. Li, Y. Zhou and S. W. Ragsdale, Chapter 3 of this book.
123. B. Jaun and R. K. Thauer, in Ref. 121, pp. 323–356.

124. B. Jaun and R. K. Thauer, in *Metal-Carbon Bonds in Enzymes and Cofactors*, Vol. 6 of *Metal Ions in Life Sciences*, Ed. A. Sigel, H. Sigel and R. K. O. Sigel, Royal Society of Chemistry, Cambridge UK, 2009, pp. 115–131.
125. D. Hinderberger, R. P. Piskorski, M. Goenrich, R. K. Thauer, A. Schweiger, J. Harmer and B. Jaun, *Angew. Chem. Int. Ed.*, 2006, **45**, 3602–3607.
126. N. Yang, M. Reiher, M. Wang, J. Harmer and E. C. Duin, *J. Am. Chem. Soc.*, 2007, **129**, 11028–11029.
127. M. Kumar, D. Qiu, T. G. Spiro and S. W. Ragsdale, *Science*, 1995, **270**, 628–630.
128. W. G. Dougherty, K. Rangan, M. J. O'Hagan, G. P. A. Yap and C. G. Riordan, *J. Am. Chem. Soc.*, 2008, **130**, 13510–13511.
129. P. A. Lindahl, in *Metal-Carbon Bonds in Enzymes and Cofactors*, Vol. 6 of *Metal Ions in Life Sciences*, Ed. A. Sigel, H. Sigel and R. K. O. Sigel, Royal Society of Chemistry, Cambridge, UK, 2009, pp. 133–149.
130. J. Seravalli and S. W. Ragsdale, *Biochem.*, 2008, **47**, 6770–6781.
131. P. A. Lindahl and D. E. Graham, in Ref. 121, pp. 357–416.
132. E. Oelgeschläger and M. Rother, *Arch. Microbiol.*, 2008, **190**, 257–269.
133. N. A. Eckert, W. G. Dougherty, G. P. A. Yap and C. G. Riordan, *J. Am. Chem. Soc.*, 2007, **129**, 9286–9287.
134. K. S. Kasprzak and K. Salnikow, in Ref. 121, pp. 619–660.
135. *Nickel in the Environment*, Ed. J. O. Nriagu, J. Wiley & Sons, New York, 1980.
136. L. D. Prockop and R. I. Chichkova, *J. Neurol. Sci.*, 2007, **262**, 122–130.
137. W. G. Lloyd and D. R. Rowe, *Environ. Sci. Technol.*, 1999, **33**, 782–785.
138. J. C. Fontecilla-Camps, in *Metal-Carbon Bonds in Enzymes and Cofactors*, Vol. 6 of *Metal Ions in Life Sciences*, Ed. A. Sigel, H. Sigel and R. K. O. Sigel, Royal Society of Chemistry, Cambridge, UK, 2009, pp. 151–177.
139. J. W. Peters, in *Metal-Carbon Bonds in Enzymes and Cofactors*, Vol. 6 of *Metal Ions in Life Sciences*, Ed. A. Sigel, H. Sigel and R. K. O. Sigel, Royal Society of Chemistry, Cambridge, UK, 2009, pp. 179–217.
140. S. Shima, R. K. Thauer and U. Ermler, in *Metal-Carbon Bonds in Enzymes and Cofactors*, Vol. 6 of *Metal Ions in Life Sciences*, Ed. A. Sigel, H. Sigel and R. K. O. Sigel, Royal Society of Chemistry, Cambridge, UK, 2009, pp. 219–239.
141. C. E. Cooper and G. C. Brown, *J. Bioenerg. Biomembr.*, 2008, **40**, 533–539.
142. L. Q. Xu, V. L. Zholobenko, L. M. Kustov and W. M. H. Sachtler, *J. Molecul. Catalysis*, 1993, **83**, 391–395.
143. M. Aschner, Chapter 12 of this book.
144. X. Zhao, K. J. Rockne, J. L. Drummond, R. K. Hurley, C. W. Shade and R. J. M. Hudson, *Environ. Sci.Technol.*, 2008, **42**, 2780–2786.
145. L. Barregard, *Environ. Res.*, 2008, **107**, 4–5.
146. M. M. Benjamin and B. D. Honeyman, in Ref. 1, pp. 342–346.
147. N. J. O'Driscoll, R. Rencz and D. S. Lean, in *Biogeochemical Cycles of Elements*, Vol. 43 of *Metal Ions in Biological Systems*, Ed. A. Sigel, H. Sigel and R. K. O. Sigel, Taylor & Francis, Boca Raton (FL, USA), 2005, pp. 221–238.
148. C. R. Hammerschmidt, W. F. Fitzgerald, C. H. Lamborg, P. H. Balcom and C. Mao Tseng, *Environ. Sci. Technol.*, 2006, **40**, 1204–1211.

149. S. E. Rothenberg, R. F. Ambrose and J. A. Jay, *Environ. Pollut.*, 2008, **154**, 32–45.
150. J. B. Shanley, M. A. Mast, D. H. Campbell, G. R. Aiken, D. P. Krabbenhoft, R. J. Hunt, J. F. Walker, P. F. Schuster, A. Chalmers, B. T. Aulenbach, N. E. Peters, M. Marvin-DiPasquale, D. W. Clow and M. M. Shafer, *Environ. Pollut.*, 2008, **154**, 143–154.
151. E. Tipping, *Appl. Geochem.*, 2007, **22**, 1624–1635.
152. A. Bouffard and M. Amyot, *Chemosphere*, 2009, **74**, 1098–1103.
153. D. M. Orihel, M. J. Paterson, P. J. Blanchfield, R. A. Bodaly and H. Hintelmann, *Environ. Sci. Technol.*, 2007, **41**, 4952–4958.
154. J. E. Gray and M. E. Hines, *Chem. Geology*, 2009, **258**, 157–167.
155. J. E. Landmeyer, T. L. Tanner and B. E. Watt, *Environ. Sci. Technol.*, 2004, **38**, 4106–4112.
156. A. J. Vella and R. Vassallo, *Appl. Organometal. Chem.*, 2002, **16**, 239–244.
157. S. K. Dubey and U. Roy, *Appl. Organometal. Chem.*, 2003, **17**, 3–8.
158. F. Suehiro, T. Kobayashi, L. Nonaka, B. C. Tuyen and S. Suzuki, *Microbial. Ecology*, 2006, **52**, 19–25.
159. K. Saeki, A. Nabeshima, T. Kunito and Y. Oshima, *Chemosphere*, 2007, **68**, 1114–1119.
160. N. Voulvoulis and J. N. Lester, *Sci. Tot. Environ.*, 2006, **371**, 373–382.
161. C. Marcic, I. Le Hecho, L. Denaix and G. Lespes, *Chemosphere*, 2006, **65**, 2322–2332.
162. S. Dubascoux, G. Lespes, L. Denaix and M. P. Gautier, *Appl. Organometal. Chem.*, 2008, **22**, 481–487.
163. E. Tessier, D. Amouroux, A. Morin, L. Christian, E. Thybaud, E. Vindimian and O. F. X. Donard, *Sci. Total Environ.*, 2007, **388**, 214–233.
164. J. -H. Huang and E. Matzner, *Biogeochem.*, 2004, **71**, 125–139.
165. J. -H. Huang, K. Kalbitz and E. Matzner, *Soil Sci. Soc. Am. J.*, 2008, **72**, 978–984.
166. T. Yabutani, J. Motonaka, K. Inagaki, A. Takatsu, T. Yarita and K. Chiba, *Anal. Sci.*, 2008, **24**, 791–794.
167. J.-H. Huang and O. Klemm, *Atmos. Environ.*, 2004, **38**, 5013–5023.
168. R. Van Cleuvenbergen, D. Chakraborti and F. Adams, *Anal. Chim. Acta*, 1990, **228**, 77–84.
169. N. Poperechna and K. G. Heumann, *Anal. Chem.*, 2005, **77**, 511–516.
170. J. S. Thayer, *Organometallic Compounds and Living Organisms*, Academic Press, Orlando (FL, USA), 1984.
171. B. Williams, L. G. Dring and R. T. Williams, *Toxicol. Appl. Pharmacol.*, 1978, **46**, 567–578.
172. R. L. Hilderbrand and T. O. Henderson, in *The Role of Phosphonates in Living Systems*, Ed. R. L. Hilderbrand, CRC Press, Orlando (FL, USA), 1983, pp. 5–31.
173. B. Nowack, in *Phosphorus in Environmental Technology: Principles and Applications*, Ed. E. Valsami-Jones, IWA Publishing, London, 2004, pp. 147–173.
174. J. S. Thayer, *Appl. Organometal. Chem.*, 1989, **3**, 203–209.

175. H. Seto and T. Hidaka, in *Dynamic Aspects of Natural Products Chemistry*, Ed. K. Ogura and U. Sankawa, Kodansha Ltd., Tokyo, 1997, pp. 145–162.
176. H. Seto and T. Kuzuyama, *Nat. Prod. Rep.*, 1999, **16**, 589–596.
177. Kh. S. Mukhamedova and A. I. Glushenkova, *Chem. Natural Compounds*, Plenum, New York, 2001, pp. 329–341.
178. G. Hanrahan, T. M. Salmassi, C. S. Khachikan and K. L. Foster, *Talanta*, 2005, **66**, 435–444.
179. R. O. Jenkins, T. -A. Morris, P. J. Craig, A. W. Ritchie and N. Ostah, *Sci. Total Environ.*, 2000, **250**, 73–81.
180. D. Glindemann, F. Eismann, A. Bergmann, P. Kuschk and U. Stottmeister, *Environ. Sci. Pollut. Res.*, 1998, **5**, 71–74.
181. C. R. Benitez-Nelson, L. O'Neill, L. C. Kolowith, P. Pellechia and R. Thunell, *Limnol. Oceanogr.*, 2004, **49**, 1593–1604.
182. K. J. Reimer, Chapter 6 of this book.
183. E. Dopp, A. D. Kligerman and R. A. Diaz-Bone, Chapter 7 of this book.
184. S. Čerňanský, M. Ulrik, J. Ševc and M. Khun, *Environ. Sci. Pollut. Res.*, 2007, **14**(Special Issue 1), 31–35.
185. S. M. A. Islam, K. Fukushi, K. Yamamoto and G. C. Saha, *Arch. Environ. Contam. Toxicol.*, 2007, **52**, 332–338.
186. R. A. Sanford and D. A. Klein, *Appl. Organometal. Chem.*, 1987, **2**, 159–170.
187. A. W. Ritchie, J. S. Edmonds, W. Goessler and R. O. Jenkins, *FEMS Microbiol. Lett.*, 2004, **235**, 95–99.
188. A. Rumpler, J. S. Edmonds, M. Katsu, K. B. Jensen, W. Goessler, G. Raber, H. Gunnlaugsdottir and K. A. Francesconi, *Angew. Chem. Int. Ed.*, 2008, **47**, 2665–2667.
189. *Arsenic in the Environment*, Vol. 26 *in Advances in Environmental Science and Technology*, Ed. J. O. Nriagu, J. Wiley & Sons, New York, 1994.
190. P. Bhattacharya, A. H. Welch, K. G. Stollenwerk, M. J. McLaughlin, J. Bundschuh and G. Panaullah, *Sci. Total Environ.*, 2007, **379**, 109–120.
191. P. M. Fox, D. L. LeDuc, H. Hussein, Z.-q. Lin and N. Terry, in *Biogeochemistry of Environmentally Important Trace Elements*, Ed. Y. Cai and O. C. Braids, ACS Symposium Series 835, American Chemical Society, Washington DC, 2003, pp. 339–354.
192. D. A. Martens and D. L. Suarez, in *Biogeochemistry of Environmentally Important Trace Elements*, Ed. Y. Cai and O. C. Braids, ACS Symposium Series 835, Amererican Chemical Society, Washington DC, 2003, pp. 355–369.
193. P. J. Craig and W. A. Maher, in *Organometallic Compounds in the Environment*, 2nd edn., Ed. P. J. Craig, J. Wiley & Sons, Chichester, UK, 2003, pp. 391–398.
194. P. Andrewes and W. R. Cullen, in *Organometallic Compounds in the Environment*, 2nd edn., Ed. P. J. Craig, J. Wiley & Sons, Chichester, UK, 2003, pp. 277–303.
195. M. Filella, Chapter 8 of this book.
196. A. A. Benson, in *The Biological Alkylation of Heavy Elements*, Ed. P. J. Craig and F. Glockling, Royal Society of Chemistry, London, 1988, pp. 135–137.

197. S. Junyapoon, K. D. Bartle, A. B. Ross and M. Cooke, *Intern. J. Environ. Anal. Chem.*, 2002, **82**, 47–59.
198. L. Duester, J. P. M. Vink and A. V. Hirner, *Environ. Sci. Technol.*, 2008, **42**, 5866–5871.
199. R. Miravet, E. Bonilla, J. López-Sánchez and R. Rubio, *J. Environ. Monit.*, 2005, **7**, 1202–1213.
200. S. Wehmeier and J. Feldmann, *J. Environ. Monit.*, 2005, **7**, 1194–1199.
201. L. Duester, L. M. Hartmann, L. Luemers and A. V. Hirner, *Appl. Organometal. Chem.*, 2007, **21**, 441–446.
202. R. W. Fleming and M. Alexander, *Appl. Microbiol.*, 1972, **24**, 424–429.
203. J. W. Swearingen, M. A. Araya, M. F. Plishker, C. P. Saavedra, C. Vásquez and T. G. Chasteen, *Anal. Biochem.*, 2004, **331**, 106–114.
204. M. A. Araya, J. W. Swearingen, M. F. Plishker, C. P. Saavedra, T. G. Chasteen and C. C. Vásquez, *J. Biol. Inorg. Chem.*, 2004, **9**, 609–615.
205. P. R. L. Ollivier, A. S Bahrou, S. Marcus, T. Cox, T. M. Church and T. E. Hanson, *Appl. Environ. Microbiol.*, 2008, **74**, 7163–7173.
206. Y. Ogra, R. Kobayashi, K. Ishiwata and K. T. Suzuki, *J. Anal. At. Spectrom.*, 2007, **22**, 153–157.
207. S. E. Ramadan, A. A. Razak, A. M. Ragab and M. El-Meleigy, *Biol. Trace Elem. Res.*, 1989, **20**, 225–232.
208. Y. Ogra, T. Kitaguchi, N. Suzuki and K. T. Suzuki, *Anal. Bioanal. Chem.*, 2008, **390**, 45–51.
209. B. L. Lewis, M. O. Andreae, P. N. Froehlich and R. A. Mortlock, *Sci. Total Environ.*, 1988, **73**, 107–120.
210. B. L. Lewis, M. O. Andreae and P. N. Froehlich, *Marine Chem.*, 1989, **27**, 179–200.
211. G. B. Jiang and F. C. Adams, *J. Chromatog. A*, 1997, **759**, 119–125.
212. M. J. Ellwood and W. A. Maher, *J. Anal. At. Spectrom.*, 2002, **17**, 197–203.
213. K. Jin, Y. Shibata and M. Morita, *Anal. Chem.*, 1991, **63**, 986–989.
214. S. J. Santosa, S. Wada, H. Mokudai and S. Tanaka, *Appl. Organometal. Chem.*, 1997, **11**, 403–414.
215. S. W. Blecker, S. L. King, L. A. Derry, O. A. Chadwick, J. A. Ippolito and E. F. Kelly, *Biogeochem.*, 2007, **86**, 189–199.
216. R. Tacke, T. Heinrich, T. Kornek, M. Merget, S. A. Wagner, J. Gross, C. Keim, G. Lambrecht, E. Mutschler, T. Beckers, M. Bernd and T. Reissmann, *Phosphorus, Sulfur and Silicon*, 1999, **150–151**, 69–87.
217. M. Swami and R. V. Singh, *Phosphorus, Sulfur and Silicon*, 2008, **183**, 1350–1364.
218. *Thallium in the Natural Environment*, Vol. 29 *of Advances in Environmental Science & Technology*, Ed. J. O. Nriagu, J. Wiley & Sons, New York, 1998.
219. A. L. J. Peter and T. Viraraghavan, *Environ. Internat.*, 2005, **31**, 493–501.
220. O. F. Schedlbauer and K. G. Heumann, *Anal. Chem.*, 1999, **71**, 5459–5464.
221. O. F. Schedlbauer and K. G. Heumann, *Appl. Organometal. Chem.*, 2000, **14**, 330–340.
222. L. Ralph and M. R. Twiss, *Bull. Environ. Contam. Toxicol.*, 2002, **68**, 261–268.

223. B. S. Twining, M. R. Twiss and N. S. Fisher, *Environ. Sci. Technol.*, 2003, **37**, 2720–2726.
224. M. R. Twiss, B. S. Twining and N. S. Fisher, *Can. J. Fish Aquat. Sci.*, 2003, **60**, 1369–1375.
225. M. R. Twiss, B. S. Twining and N. S. Fisher, *Environ. Toxicol. Chem.*, 2004, **1**, 968–973.
226. J. M. Morgan, *Dissertation Abstracts International B*, 1981, **41**, 2578.
227. M. Filella, Chapter 9 of this book.
228. K. Michalke, A. Schmidt, B. Huber, J. Meyer, M. Sulkowski, A. V. Hirner, J. Boertz, F. Mosel, P. Dammann, G. Hilken, H. J. Hedrich, M. Dorsch, A. W. Rettenmeier and R. Hensel, *Appl. Environ. Microbiol.*, 2008, **74**, 3069–3075.
229. N. Hussain, T. G. Ferdelman, T. M. Church and G. W. Luther, *Aquatic Geochem.*, 1995, **1**, 175–188.
230. N. Momoshima, L. -X. Song, S. Osaki and Y. Maeda, *Environ. Sci. Technol.*, 2001, **35**, 2956–2960.
231. B. Skwarzec and J. Fabisiak, *J. Environ. Radioactivity*, 2007, **93**, 119–126.
232. E. Dopp, L. M. Hartmann, A.-M. Florea, A. W. Rettenmeier and A. V. Hirner, *Crit. Rev. Toxicol.*, 2004, **34**, 301–333.
233. R. Pongratz and K. G. Heumann, *Chemosphere*, 1999, **39**, 89–102.
234. R. Pongratz and K. G. Heumann, *Anal. Chem.*, 1996, **68**, 1262–1266.
235. L. Keller and P. H. Brunner, *Ecotoxicol. Environ. Safety*, 1983, **7**, 141–150.
236. S. C. Popat and M. A. Deshusses, *Environ. Sci. Technol.*, 2008, **42**, 8510–8515.
237. R. R. Mendel, *Dalton Trans.*, 2005, 3404–3409.
238. R. R. Mendel, *J. Exp. Botany*, 2007, **58**, 2289–2296.

Met. Ions Life Sci. **2010**, *7*, 33–69

2

Analysis of Organometal(loid) Compounds in Environmental and Biological Samples

Christopher F. Harrington,[a] ***Daniel S. Vidler,***[b] ***and Richard O. Jenkins***[c]

[a]Trace Element Laboratory, Centre for Clinical Science, Faculty of Health and Medical Sciences, University of Surrey, Guildford GU2 7XH, UK
<chris.harrington@royalsurrey.nhs.uk>

[b]Medical Toxicology Centre, University of Newcastle, Wolfson Unit, Claremont Place, Newcastle upon Tyne, NE2 4AA, UK
<daniel.vidler@ncl.ac.uk>

[c]Faculty of Health and Life Sciences, De Montfort University, The Gateway, Leicester LE1 9BH, UK
<roj@dmu.ac.uk>

Metal Ions in Life Sciences, Volume 7 Edited by Astrid Sigel, Helmut Sigel, and Roland K. O. Sigel

Published by the Royal Society of Chemistry, www.rsc.org DOI: 10.1039/9781849730822-00033

ABSTRACT: Measurement of the different physicochemical forms of metals and metalloids is a necessary pre-requisite for the detailed understanding of an element's interaction with environmental and biological systems. Such chemical speciation data is important in a range of areas, including toxicology, ecotoxicology, biogeochemistry, food safety and nutrition. This chapter considers developments in the speciation analysis of organometallic compounds (OMCs), focusing on those of As, Hg, Se and Sn. Typically, organometallic analysis requires a chromatographic separation prior to analyte detection and gas chromatography (GC), high performance liquid chromatography (HPLC) or capillary electrophoresis (CE) can serve this purpose. Following separation, detection is achieved using element specific detectors (ESDs) such as inductively coupled plasma mass spectrometry (ICP-MS), inductively coupled plasma optical emission spectroscopy (ICP-OES), atomic fluorescence spectrometry (AFS), atomic absorption spectrometry (AAS) or atmospheric pressure ionization mass spectrometry (API-MS). Techniques employing a vapor generation (VG) stage prior to detection are also discussed. Complementary structural and quantitative data may be acquired through the combination of elemental and molecular mass spectrometry. The advantages and disadvantages of the various analytical systems are discussed, together with issues related to quantification and quality management.

KEYWORDS: chemical speciation · ESI-MS/MS · ICP-MS · organometallics · vapor generation

1. INTRODUCTION

Measurement of the total concentration of a metal(loid) in a particular sample matrix reveals little about its possible environmental mobility, toxicity or biochemical activity. In environmental terms, the total concentration gives no indication of persistence, or biogeochemical state. Equally, in an organism or biological sample, it gives no information on essentiality, toxicity, or the risk and site of bioaccumulation [1]. To provide this information it is necessary to determine the actual chemical form of the metal(loid) under investigation. Three important categories can be defined: *organometallic compounds*, which arise when a metal(loid) forms a covalent bond with carbon; the *oxidation state* of a particular metal(loid); and *metalloproteins* incorporating a metal, which is often redox active. Chemical speciation is defined by IUPAC [2] as: the "distribution of an element amongst defined chemical species in a system" and chemical speciation analysis as the "analytical activities of identifying and/or measuring the quantities of one or more individual chemical species in a sample". This chapter deals solely with the analysis of OMCs, the first class of chemical species.

A good example in toxicology of the importance of measuring more than just the total concentration of an element is the As containing OMC arsenobetaine (AB) (trimethylarsonioacetate). This compound is widely distributed in marine organisms, such as fish and shellfish, which consequently contain a relatively high total As concentration ($mg\,kg^{-1}$) compared to seawater ($\mu g\,kg^{-1}$) [3]. Inorganic As is both an acute and chronic toxicant to humans, but in contrast AB is considered non-toxic [4]. Therefore, if only the total As content of fish or seafood is measured an incorrect impression of the human health risk would be apparent. Conversely, a significant proportion of the Hg content of edible fish is present as a methylmercury (MeHg) complex and this particular species is more toxic than inorganic mercury (Hg(II)), with the ability to cross both the blood-brain barrier and between mother and unborn child, leading to an accumulation of MeHg in fetal blood [5]. It is for this reason that women have been advised to restrict their consumption of certain fish and marine animals during pregnancy [6].

From an analytical perspective, the important characteristics of organometallic analysis include: the structural identification of the metal(loid) species; its accurate measurement in the presence of other interfering compounds; and that the sum concentration of the metal(loid) species present equals the total concentration, i.e., a mass balance for the element can be determined for each analytical step of the process. This last point is particularly significant because it sets the area apart from other analytical measurements. The analytical methodology used can be characterized as having a number of interrelated steps: *sample collection and storage*, to gather representative samples of the material under investigation and store under conditions where the species are stable; *sample extraction*, to remove the species of interest from the sample matrix; *clean-up and preconcentration*, to isolate the species from matrices with the potential to affect the measurement or when the analyte concentration is low; *analysis*, which involves calibration, replication, use of quality control (QC) measures, suitable blanks and control samples. The whole process should ideally be incorporated into a quality assurance (QA) framework.

2. SAMPLE PREPARATION

2.1. Introduction

The majority of quantitative analytical methods for biological and environmental samples require liquid samples for analysis, which necessitates extraction of the analyte from solid samples. The actual protocols used will be dependent on: the types of samples being analyzed; the chemical species of interest; and the analytical instrumentation available. The overarching aim is to quantitatively remove the analyte species from the sample matrix and determine its concentration and identity, without loss or conversion into a different species.

2.2. Sample Storage

Careful storage of the sample prior to its analysis is important because species transformations can occur at this stage. The storage conditions used will depend on the material and how long it is to be stored for. Only a few studies have looked closely at these requirements. The effect of storage conditions (temperature, time, and use of stabilizing additive) on the stability of As species in human urine is a good example [7]. All the species were stable for up to two months when stored at 4 or –20 °C, but for longer storage periods analyte transformations occurred, which were found to be dependent on the sample matrix.

2.3. Extraction Methods

The methods available for the extraction of OMCs from environmental and biological samples have employed basic, acidic or enzymatic conditions. To improve the extraction efficiency, microwave assisted extraction (MAE) in open or closed vessels or high pressure solvent extraction with heat, termed accelerated solvent extraction (ASE), have been used. Table 1 presents extraction methods used for specific OMCs.

The alkaline extraction methods generally use either 20–25% tetramethylammonium hydroxide (TMAH) in water [8–10] or methanol [11], or aqueous or methanolic 25% potassium hydroxide [12–17]. TMAH extraction methods have gained popularity for the extraction of Hg species from biological materials. This is partly because these methods were thought to retain the original mercury speciation present in the sample. However, the use of TMAH has been implicated in the artefactual formation of MeHg in fish extracts due to the methylation of Hg(II). Investigation of the transalkylation of Hg compounds in biological materials as a function of sample preparation conditions [8], using ^{198}Hg enriched MeHg and ^{201}Hg enriched Hg(II) spikes, showed that up to 11.5% of Hg(II) was methylated and up to 6.3% of MeHg was demethylated. It was concluded that methylation was taking place after the dissolution stage, probably at or after the sample

Table 1. Examples of different extraction protocols used for different OMCs.

OMCs, Sample Matrix, CRM	Extraction, Clean-up Method and Derivatization Method	Comment	Ref.
TML DORM-2, CRM 463, CRM 422, CRM 477, CRM 278, mussels, prawns, tuna fish, plaice, and pollock	1. Mix sample (0.2–1 g), spike solution ($Me_3{}^{206}PbI$) and 25% (w/v) aqueous TMAH (3–4 mL) and then shake (2–3 hours) 2. Acetate buffer and nitric acid are then added to achieve pH 5–6 3. Add aqueous 2% (w/v) $NaBEt_4$ (0.5 mL) and hexane (0.5 mL) 4. Shake reaction mixture (10 min) and recover hexane phase, following centrifugation 5. Analyze hexane phase by GC-ICP-MS	ssIDMS used for calibration Recovery: none of the biological reference materials were certified for TML. Validation was performed with CRM 605 (urban dust), recovery of 101%	[104]
MBT, DBT, and TBT CRM 477 (mussel tissue), BCR-710 (oyster tissue)	1. Mix BCR-710 (0.1 g) with 25% TMAH (4 mL) and ^{119}Sn enriched butyltin species **OR** mix CRM 477 (0.1 g) with 3:1 solution of glacial acetic acid and methanol and ^{119}Sn enriched butyltin species 2. Microwave assisted extraction (70 °C/4 min) 3. Derivatize a portion (0.5 mL) of this extract 4. To 0.5 mL of extract add sodium acetate buffer (4 mL) and adjust mixture to pH 5 with conc. HCl 5. Add aqueous 2.5% (w/v) $NaBEt_4$ (0.5 mL) and hexane (1 mL) 6. Shake reaction mixture (4 min) and recover hexane phase	ssIDMS used for calibration Recovery: MBT, 102%; DBT, 101 %, TBT, 93%. (recovery data for CRM 477) TBT, 98% (recovery data for BCR-710)	[105]

Table 1. (*Continued*)

OMCs, Sample Matrix, CRM	Extraction, Clean-up Method and Derivatization Method	Comment	Ref.
	7. Clean-up of hexane phase on Florisil 8. Pre-concentrate hexane extract using a N_2 stream prior to analysis by GC-MS		
MBT, DBT, TBT, MPT, DPT, and TPT Milk fish (*Chanos chanos*), NIES 11 (freeze-dried)	1. Dried and homogenized fish samples (0.1 g) were digested with 3% (w/v) KOH (5 mL) for 60 min at 60 °C 2. The digests were mixed with phosphate buffer (pH 6) in a volumetric flask 3. *Iso*-octane (0.5 mL) and 1% (w/v) $NaBEt_4$ (1 mL) were added and the reaction mixture shaken for 1 hour 4. Water was then added to elevate the *iso*-octane phase into the flask neck, from where it was recovered. Aliquots of the *iso*-octane phase were then analyzed by GC-FPD	TPrT served as internal standard Recovery: quantitative recovery was achieved for NIES 11 spiked with the 6 organotin species. For unspiked NIES 11 the TBT recovery was 104%	[106]
Se-Met Antarctic krill	1. Homogenized, lyophilized krill samples and Pronase E were suspended in Tris buffer (pH 7.5) 2. Digests were incubated at 37 °C for 24 hours, with shaking 3. Extracts were centrifuged to isolate supernatants 4. Supernatants were diluted with nitric acid and then filtered prior to Se-Met determination 5. Analyze by HPLC-ICP-MS	Recovery of Se-Met from krill using Pronase E with ultrasonication sonication was achieved in 15 minutes, however 24 hours were required without ultrasonication	[107]

Table 1. (*Continued*)

OMCs, Sample Matrix, CRM	Extraction, Clean-up Method and Derivatization Method	Comment	Ref.
Se-Met, Se-Me-Cys Potatoes (selenized)	1. Potato skin and flesh were worked-up separately 2. Samples were freeze-dried, ground, and stored at −80 °C in darkness 3. Extraction of water soluble Se species was achieved using either ASE, or extraction into boiling water 4. Protein-bound Se species were initially extracted with protease/lipase, followed by digestion of any residue from the first enzyme treatment with Driselase 5. Analyze by HPLC-ICP-MS and/or HPLC-ESI-MS/MS	Illustrates the complementary use of HPLC-ICP-MS and HPLC-ESI-MS/MS with the aim of identifying unknown Se species in potatoes	[108]
Sb(V), Sb(III) and unknown Sb-containing species Algae and mollusc	1. Samples were lyophilized and then the following extraction media were evaluated: (a) water at room temperature; (b) water at 90 °C; (c) methanol; (d) 0.1 M EDTA, pH 4.5; (e) 0.1 M citric acid, pH 2 2. Extractions were performed with shaking for 30 mins. Supernatants were then filtered and subjected to SPE (C_{18}). Analyze by HPLC-HG-AFS, or in the case of citric acid containing extractions HPLC-UV-HG-AFS	Sb(III) is readily oxidized to Sb(V) during sample preparation. Addition of EDTA to the extraction solvent reduced the occurrence of this artefact Recovery was quantitative for both Sb(V) and Sb(III) when EDTA extraction was used	[109]

Table 1. (*Continued*)

OMCs, Sample Matrix, CRM	Extraction, Clean-up Method and Derivatization Method	Comment	Ref.
MeHg, Hg(II), TMT, DMT, MMT, MBT, DBT, TBT, TML, DML DORM-2, CRM 710, CRM 477, BCR-605	Biological CRMs were prepared as follows: 1. CRMs (0.3 g) are mixed with 25% (m/v) aqueous TMAH (5 mL) 2. Following manual shaking (5 mins) the mixture is subjected to MAE (40W/2min) 3. Extracts are bulked to 25g and then frozen (−20 °C) 4. Extracts are buffered to achieve pH 5 in a headspace vial 5. Add aqueous 0.5% (w/v) $NaBEt_4$ (0.2 mL), seal vial and stir reaction vigorously while exposing SPME fibre to headspace at 25 °C 6. Desorb SPME fibre in GC injection port, analysis by SPME-GC-ICP-MS	TMAH may potentially degrade MeHg to Hg(II). Analysis of CRM 710 (oyster) produced a MeHg recovery of about 70% Due to the use of SPME, no organic solvent is required for extraction of derivatization reaction products	[12]
As(III), As(V), MMA and DMA Candidate RMs, Spanish white rice, Basmati rice, and NIST SRM 1568a Rice Flour	Spanish and Basmati rice samples were ground and sieved prior to extraction. In combination with sonication for 60 seconds at room temperature, the following extraction media were compared: 1. Aqueous methanol (100% water, 100% methanol and 50/50, water/methanol) 2. TMAH (1 and 2% solutions) 3. Enzymatic hydrolysis (protease XIV only, α-amylase only, both enzymes together in sequence)	1% (w/v) TMAH can extract 70% of As from rice, however, TMAH can cause the oxidation of As(III) to As(V). The best recovery (80%, total As basis) was achieved by using protease XIV in combination with α-amylase	[110]

Table 1. *(Continued)*

OMCs, Sample Matrix, CRM	Extraction, Clean-up Method and Derivatization Method	Comment	Ref.
	4. Analysis was performed by HPLC-ICP-MS		
As(III), As(V), DMA, several arsenosugars Ribbon kelp (*Algaria marginata, Sargassum muticum*)	Accelerated solvent extraction 1. Freeze-dried and homogenized seaweed samples 2. Mix sample with glass beads (dispersion media) 3. 3 sequential extraction cycles with water/methanol (30%/70%) at 500 psi and ambient temperature 4. Evaporate ASE extract to dryness under N_2 at 50 °C 5. Reconstitute in water 6. SPE on C_{18} phase 7. Analyze by HPLC-ICP-MS or HPLC-ESI-MS/MS	Solid phase extraction using a C_{18} phase was applied to the clean-up of ASE extracts Dispersion media reduces risks clogging of ASE cell by rehydrated freeze-dried seaweed. Less than optimal recovery of arsenicals, attributed to cellulose's resistance to ASE under the conditions studied	[111]

MBT, monobutyltin; DBT, dibutyltin; TBT, tributyltin; TPrT, tripropyltin; CRM, certified reference material; TML, trimethyllead; TMAH, tetramethylammonium hydroxide; FPD, flame photometric detector; Se-Met, selenomethionine; Se-Cys, selenocysteine; Se-Me-Cys, Se-methylselenocysteine; EDTA, ethylenediamine-*N,N,N′,N′*-tetraacetic acid; TMT, trimethyltin; DMT, dimethyltin; MMT, monomethyltin; DML, dimethyllead; RMs, reference materials.

extracts were pH adjusted to render them amenable to HPLC separation. Due to the low levels of Hg(II) in the fish samples studied the effect of adventitious methylation was concluded to be insignificant for the determined MeHg content. Similar observations were made regarding the potential for TMAH to degrade MeHg to Hg(II) when analysis of oyster material produced a MeHg recovery of about 70% [12]. The presence of TMAH in fish extracts has been reported to confound reversed-phase retention times of arsenicals [13] potentially affecting identification.

Acid extraction of MeHg is generally performed using HCl [9,11,14–16] with subsequent partitioning of MeHg into an organic solvent. In the presence of acid, there is a possibility that arsenosugars may degrade to dimethylarsinic acid (DMA) [17]. Phosphoric acid is known to break As-S bonds and so has great potential to alter the As speciation of a sample if used [18]. For this reason, milder enzyme-based extraction methods have been developed and successfully applied to As speciation [19]. Trifluoroacetic acid has been applied to As extraction from rice as it is readily capable of carbohydrate hydrolysis [20,21]. Whilst MAE of As species from seafood provided a high recovery, the same approach when applied to seaweed was less effective [22]. In this case ultrasonic extraction was found to be more appropriate. To aid MeHg extraction from biological samples, ultrasonication of both acid [9,11] and alkaline extracts [8,11,23,24] has been reported.

One of the advantages offered by using enzymes is that they are specific in their action, and therefore the problems encountered when using other methods such as the formation of artefacts, are unlikely to occur [25]. Due to the high protein content of fish, enzyme-based extractions using trypsin have been successfully used for As species without species interconversion [26,27]. Extraction of As species from rice has been achieved using a mixture of pepsin and pancreatin enzymes [20], but the high chloride content of the pepsin digestion solution confounded determination of total As in the extract, so a mass balance could not be estimated.

Both open and closed vessel MAE systems for the extraction of organotin compounds (OTCs) from biological samples have gained popularity due to the high speed with which samples can be processed [13]. MAE of As(III), monomethylarsonic acid (MMA), DMA, and As(V) from algal samples has been compared with ultrasonic extraction [28]. With water used as the extraction solution, MAE performed better than sonication, but three sequential extractions were employed on each sample. Recovery experiments using algal samples spiked with As(III), MMA, DMA, and As(V) were used to show that no species interconversions were occurring. Extraction of As species from fish has been achieved using MAE into TMAH [13] and mixtures of methanol and water [13,29,30]. For quantitative extraction of OMCs from fish with MAE, it is necessary for the extraction solvent to be near or at its boiling point [13,31]. Closed vessel MAE has been used to accelerate organomercury extractions [15], as have open vessel systems [10]. Mild conditions are necessary for the extraction of MeHg from biological materials, otherwise decomposition can occur. The conditions found within a closed vessel are harsher than those produced in an open one which is operating at atmospheric pressure. If the extraction is too aggressive, the Hg speciation information can be lost, either in part or completely [32]. For example, the use of concentrated HCl to extract MeHg from biological materials using MAE has been shown to rapidly decompose MeHg to Hg(II) [10].

2.4. Sample Clean-up

Solid phase extraction (SPE) using a C_{18} phase was applied to the clean-up of ASE extracts of seaweed prior to analysis by HPLC-ICP-MS [33]. For the LC-ESI-MS determination of arsenosugars in oyster extracts it was necessary to use preparative anion exchange followed by size exclusion chromatography [17]. Without this the matrix effect produced a recovery by external calibration that was half of that achieved with standard additions.

Problems associated with the use of organic solvents for the extraction of MeHg from acidic biological sample extracts include the formation of emulsions [14]. This is due in part to the high levels of fat present in certain types of fish samples. Removal of the lipid content of samples high in fat prior to extraction is recommended, to reduce the risk of emulsification. Defatting of fish samples with acetone has been reported before As speciation analysis [31]. Prior to the MAE of As species from nuts the ground samples were defatted by shaking in a chloroform/methanol solution [34].

The use of solid phase microextraction (SPME) has gained in popularity. It has been used as an alternative to extracting mercury derivatives into an organic phase for subsequent introduction into the GC [35,36]. Poor precision was a feature of early SPME work which was considered the main drawback to this mode of sample introduction. Improvements to the fibres used has encouraged more workers to use this solvent-free approach, and IDMS calibration has further reduced the repeatability problems experienced initially [37].

3. SAMPLE ANALYSIS

3.1. Introduction

State-of-the-art techniques for the analysis of OMCs in environmental and biological samples are based on coupling powerful separation technology to molecular or elemental based detection systems. The separation methods used include: GC, HPLC, CE or supercritical fluid chromatography (SFC). Element-specific detectors (ESDs) include: AAS, AFS, ICP-OES or ICP-MS. The most important molecular detectors are based on mass spectrometry, particularly atmospheric pressure ionization techniques (ESI-MS/MS, APCI-MS/MS) and conventional GC-MS/MS.

3.2. Methods Based on Elemental-Specific Detection

Investigations using the hyphenation of GC [38] or HPLC [39] to ESD were first carried out in the late 1970s and early 1980s. Refinement of the approach has taken place since then and other separation methods, such as CE and SFC have been developed. Early reviews of different separation approaches coupled to ESD or MS included the use of GC [40], HPLC [41], and SFC [42]. Element-specific detectors such as ICP-MS or techniques based on AAS or AFS are used because of their analyte specificity, provision of quantitative data using elemental standards and potential to provide suitable limits of detection (LODs) for environmental and biological samples. In practice, AAS is generally not sensitive enough without VG to be used for real samples and AFS, whilst offering suitable LODs for speciation studies [43], is limited to elements forming stable hydrides or elemental species. ICP-MS provides the most versatile detection system because it can be coupled to numerous different chromatography techniques, delivers suitable LODs, offers a long linear calibration range (although this may be limited by the separation technique), is tolerant to complex matrices, offers multi-elemental and isotopic analysis and provides quantitation based on elemental standards.

Common problems involving ESD include: identification of unknowns through a lack of standards; unrecognized coelution of different species containing the same metal(loid); and retention times affected by sample matrices. One of the first major issues that became apparent was the difficulty in identification of unknown species and the inherent possibility of misidentification. This is one of the main drivers for the development of complementary methods based on molecular MS. Identification using ESDs relies on the availability of authentic molecular standards of high purity which are used as retention time markers. However, even when these are available it is possible to make wrong assignments, particularly if the spiking procedure is not carried out with care. A good example of this relates to the misidentification of organotin compounds in the marine environment [44]. In this case a number of techniques based on sample derivatization followed by GC separation (GC-QF-AAS, GC-FPD, GC-AES, and GC-MS) were used to identify the compound responsible for a peak eluting between the derivatives of monobutyltin (MBT) and dibutyltin (DBT). It had initially been proposed that the peak was due to the presence of a mixed methylbutyltin compound, which would have indicated that an important transformation pathway was operating in the biogeochemistry of OTCs. However, after a concerted analytical programme involving a number of laboratories it was found that the unidentified compound was actually due to monophenyltin, probably resulting from the degradation of triphenyltin (TPhT), a widely used pesticide.

The most important requirements for interfacing the separation system to the ESD are that the analyte is quantitatively transferred from one to the other without loss or rearrangement. Figure 1 shows a schematic diagram of the on-line coupling of HPLC or GC to ICP-MS. Conventional ICP-MS operates on liquid samples that are introduced *via* nebulization at a flow rate of 0.1 to 1 mL min^{-1}. With liquid-based separations using HPLC, a suitable length of tubing can be used to couple the column to the nebulizer. Alternatively, for some elemental species HPLC can be hyphenated to ESD via VG (see Section 3.5). With the other separation systems (GC, CE, SFC) the interface has required development work to be carried out to accommodate the differences between the separation system and the requirements of the ICP-MS.

The main difficulties when coupling HPLC to ICP-MS involve eluents containing a high proportion of an organic modifier, because this can destabilize the plasma, necessitating a cooled spraychamber (−5 to −15 °C) or low flow conditions, to reduce the solvent load. Oxygen addition is required to eliminate the deposition of carbon on the sampling cone and maintain the transmission of ions through the cone orifice. To withstand the extra wear generated, a platinum tipped sample cone has to be used. The advent of low-flow and desolvating nebulizers has helped with coupling HPLC to ICP-MS and more recent applications have not used cooled spray chambers. This type of sample introduction system allows the use of gradient elution, which makes possible shorter chromatographic runs and more versatile separation systems. Recent developmental work has produced a sheathless interface using a microflow total consumption nebulizer, which facilitates the use of eluents containing 100% organic solvent, without spray chamber cooling or oxygen addition [45]. This makes the coupling of

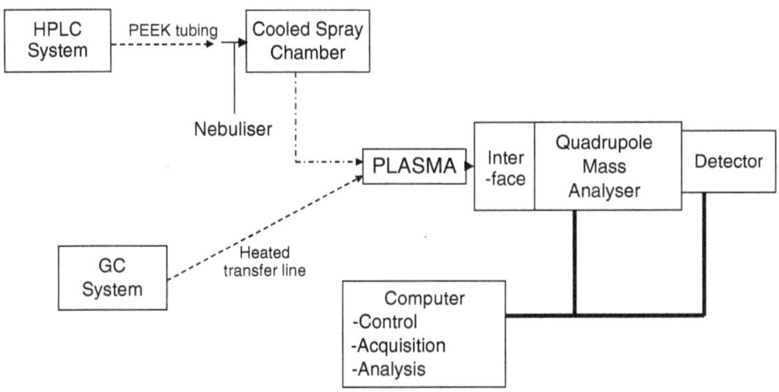

Figure 1. Schematic diagram of the coupling used for the hyphenation of GC or HPLC to ICP–MS.

low-flow capillary HPLC separations to ICP-MS possible and offers significant advantages over conventional columns because small sample volumes (nL) can be used, the chromatographic system provides enhanced peak resolution with a better signal-to-noise ratio and consequently a lower LOD.

Coupling GC to ICP-MS requires heating of the transfer line to a temperature higher than that used in the separation so as to prevent cold spots, which lead to peak broadening or complete retention of the analyte within the system. The first use of a heated transfer line was described in 1992 [46,47] and consisted of an aluminum bar with a slit, in which the capillary column was contained, before introduction into the central channel of the torch. The necessary argon make-up flow was heated in the GC oven prior to its introduction through a T-piece and sheathed the column, helping to avoid condensation in the transfer line. This interface was successfully applied to the analysis of high boiling point compounds such as Fe, Ni, and V containing porphyrins [48,49]. Another interface design in which a heated quartz transfer line was inserted through the torch to the base of the plasma has been developed commercially [50]. Recently the construction and evaluation of a low cost interface which could be adapted for use with most GC and ICP-MS instrumentation has been described [51].

The main advantage provided by using GC separations is that around 100% of the injected sample reaches the detector and because no liquid is introduced the plasma is not cooled. With HPLC only a few percent of the sample reaches the plasma due to the inefficiency of conventional nebulizer-spraychamber configurations and the wet aerosol cools the plasma, reducing the energy available to ionize the analyte. In general GC methods have better S/N ratio characteristics than HPLC methods, because of the sharp and narrow peak shapes generated. Another important characteristic of GC-ICP-MS is the ability to perform multi-elemental speciation studies, which is generally not possible with HPLC because of the limitations in chromatographic selectivity.

With GC separations the volatility of the analyte is the principle factor determining how long the analyte stays on the column, so as long as the chemical species are stable and volatile they can be separated regardless of the element. With HPLC separations other properties such as polarity determine how the chemical species behave, making it difficult to develop separations that accommodate the diverse range of OMC properties. Capillary GC separations also have the potential to deliver better compound resolution compared to HPLC. The main difference between the two approaches is that GC requires an extra step, so that the generally ionic, low volatility compounds are converted to a stable volatile form, with HPLC the target analytes are determined directly. The consequence of this extra derivatization step is that there is a significant chance the analyte could be lost or an artefact formed during the reaction.

Derivatization reactions, especially aqueous ethylation with sodium tetraethyl borate (STEB), used when GC separation is employed prior to detection of Hg compounds have been implicated in the formation of artefacts [52]. This derivatization step is inhibited by high concentrations of chloride ions [24]. The high stability of the MeHg chloro complex which is formed in high chloride-containing samples has been suggested as an explanation. The ability of halide ions to interfere with the ethylation reaction is of particular importance when MeHg extraction using HCl is employed and not just when seawater samples or other high chloride containing samples are analyzed [53]. Chloride and bromide ions have been reported to convert MeHg into Hg(0) and iodide promotes a disproportionation reaction of MeHg to produce both Hg(0) and Hg(II) [52]. The same study showed that derivatization using propylation did not cause this conversion.

The main advantage of HPLC compared to GC is that there is no need to derivatize the compounds prior to analysis. However, acidic or alkaline sample extracts do need pH adjustment when a silica-based column is used, otherwise the chromatographic medium could be damaged. This pH adjustment has been implicated in the artefactual formation of MeHg from Hg(II) [8]. Mercury compounds are notorious for exhibiting memory effects, i.e., adhering to internal components of HPLC instrumentation and various mobile phase additives have been used to try to reduce this. One very effective method to eliminate poor peak shapes, high blank values and non-eluting compounds, is to use polyetheretherketone (PEEK) instead of stainless steel components in the HPLC system and include 2-mercaptoethanol (2-ME) in the eluent [54]. Another sulfur-containing reagent used to reduce these effects is cysteine [25]. Other problems related to the analysis of Hg in biological and environmental samples have been encountered and these have been reviewed [55]. Figure 2a (see Section 3.3) shows a typical chromatogram obtained for the analysis of Hg species by using HPLC-ICP-MS when using 2-mercaptoethanol to reduce peak tailing.

SFC uses a liquefied gas as the eluent and programmed changes in pressure to facilitate separation, in a similar way to temperature programming in GC separations. Supercritical fluids have critical temperatures (temperature above which the fluid cannot be liquefied) below 200 °C and densities of the order 0.1–1 g L^{-1} at pressures of 1000–6000 psi. Carbon dioxide is the most common eluent for SFC analysis of metal(loid) species and in some applications has been doped with methanol. SFC-ICP-MS overcomes some of the limitations of HPLC and GC because it can be used to rapidly separate thermally labile, non-volatile, high molecular weight compounds and affords lower LODs. The interface between SFC and ICP-MS is commercially available and involves a restrictor to maintain the high pressure required for the separation system. However, only a few applications have used SFC-based methods and the majority of these have focused on the determination of OTCs in marine samples [56,57].

CE is a family of related techniques that employ narrow bore (20-200 μm in diameter) capillaries to perform high efficiency separations [58], facilitated by the application of a high voltage to the capillary, which generates electroosmotic and electrophoretic flow. The technique has been coupled to ICP-MS and ESI-MS [59] for the measurement of OMCs in biological and environmental samples. The initial difficulties in designing a suitable interface to couple CE separations with ICP-MS were centered on the high flow-rate requirements of conventional ICP nebulizers and the low-flow rate nature of CE. The suction generated with the conventional self-aspirating nebulizers, caused a loss in chromatographic resolution and the necessity to maintain an effective electrical connection to the end of the capillary posed problems. These difficulties were overcome by using a low-flow nebulizer and a small make-up buffer flow with an earth connection [60]. The main advantages of CE for speciation analysis include: minimal species interaction with separation media due to its absence from the capillary; potential to measure neutral, variably charged, and organometallic species in a single run; low sample consumption; and a high separation efficiency compared to other liquid chromatographic methods. However, because of the small sample size used it is difficult to detect the species present in real samples unless a low LOD detector is available.

3.3. Methods Based on Molecular Mass Spectrometry

Molecular mass spectrometry has been used in conjunction with some of the above mentioned chromatographic techniques for the analysis of OMCs. The most commonly used ionization techniques for HPLC and CE are atmospheric pressure ionization (API), of which there are two main variants, electrospray ionization (ESI) and chemical ionization (APCI). Traditional mass spectrometry using electron impact (EI) ion sources have been used with GC separations. The main characteristics of these molecular detection methods when used for the analysis of OMCs include: ionization specific to the analyte molecule; possibility for structural studies via tandem MS analysis; potential for high mass accuracy characterization; availability of a wide range of commercially available hyphenated instrumentation; wide m/z range analysis; and low LODs, although not as low as for ICP-MS.

The advantage of molecular detection is that it is possible to identify unknown chemical species in situations where standards may not be available and it offers the potential for structural elucidation. When using

API-MS for the analysis of environmental or biological samples it can suffer from significant matrix effects, so may require extensive sample clean-up procedures to be used, to eliminate the effect and reduce the formation of sodiated and potassiated ions. Matrix effects are still a difficult problem to contend with in API-MS analysis, where a "soft-ionization" process is used for ion generation. Unlike API-MS, ICP-MS is such a "hard-ionization" process that suppression of ion formation by the sample matrix is not considered a problem. Hence, the major shortcomings of ESI-MS compared to ICP-MS are the much poorer LOD and the adverse effect of the matrix present in biological and environmental samples.

The majority of methods using API-MS involve ESI-MS which was first developed in the mid-1980s [61,62] and used for the analysis of large molecular weight, non-volatile biomolecules and more recently for small polar metabolites [63]. In the case of organometallic analysis ESI was initially used for the determination of small polar or ionizable compounds such as tributyltin (TBT), or As species, but the greatest impact of ESI-MS has been made in the analysis of much larger molecules, particularly metalloproteins. The use of ESI-MS for the analysis of OMCs has been reviewed [64,65]. The complementary ionization source to ESI is APCI and this has found some limited use for the analysis of OMCs; Figure 2b shows the detection of mercury species by APCI-MS, after HPLC separation using 2-ME in the eluent and Figure 2c the APCI mass spectrum for the MeHg peak, corresponding to an adduct between MeHg and 2-ME and clearly shows the isotopic pattern for Hg.

The most important technical difference between ICP-MS and modern API instrumentation is the possibility to carry out tandem API-MS/MS experiments. The ions formed in the source are sampled in to the first quadrupole and then either the molecular ion or a fragment ion is isolated in a collision cell containing an inert gas with a collision voltage applied. Depending on the ion and the voltage the sampled ion is further broken down into different fragments. This approach, termed collision-induced dissociation (CID), results in highly specific analysis, provides the lowest LODs and the ability to investigate the structure of the molecule of interest. This technique has made a significant impact on our understanding of the biogeochemistry of As in the marine environment, where a range of As-containing sugar compounds are found. By using tandem MS, with an ESI source it is now possible to directly characterize these novel arsenicals directly after HPLC separation [66]. Until the advent of ESI-MS/MS these marine arsenicals were investigated using a natural products approach, whereby large quantities of material are extracted to isolate sufficient of the As compound for identification by NMR [3]. Electrospray principles and general applications were reviewed extensively in 2000 [67].

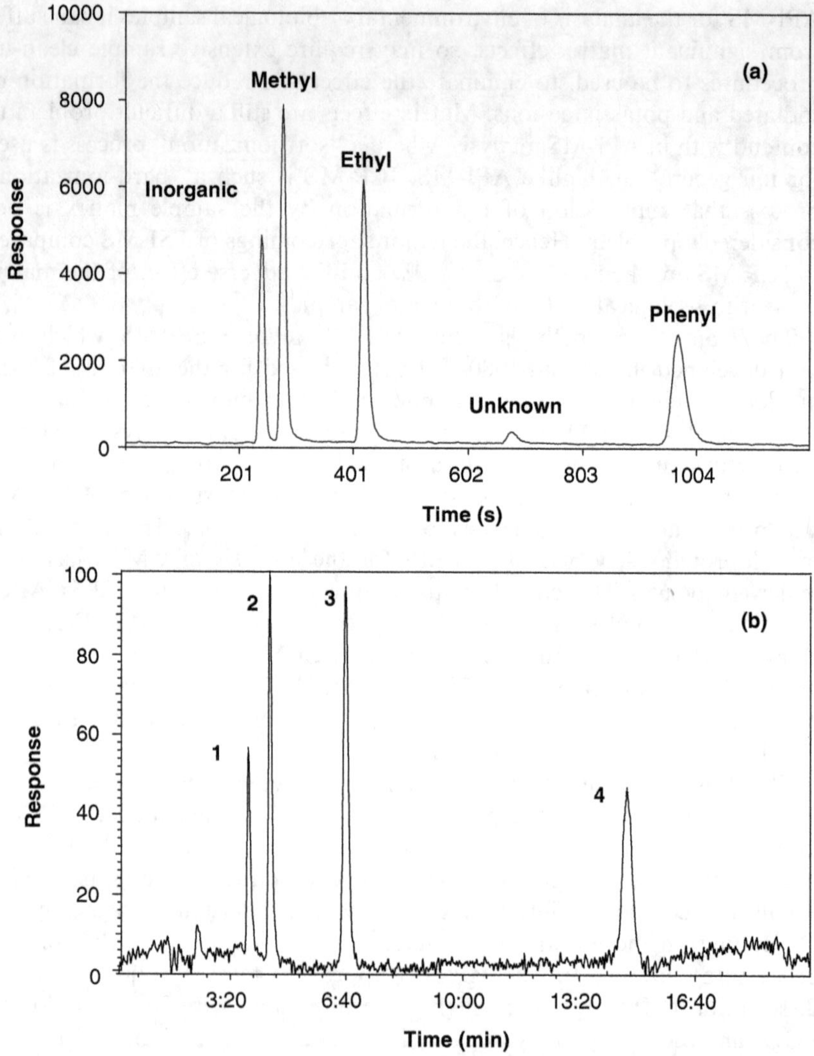

3.4. Complementary Mass Spectrometry Methods

Molecular detection via API-MS and ESD via ICP-MS can be considered as having ionization processes at opposite ends of the spectrum. Both techniques use sources at atmospheric pressure, however API is considered to be a soft-ionization technique, effectively converting the charged species present in the liquid phase into an ion in the gas phase, whereas ICP very effectively converts chemical species in the liquid phase into their constituent elemental ions.

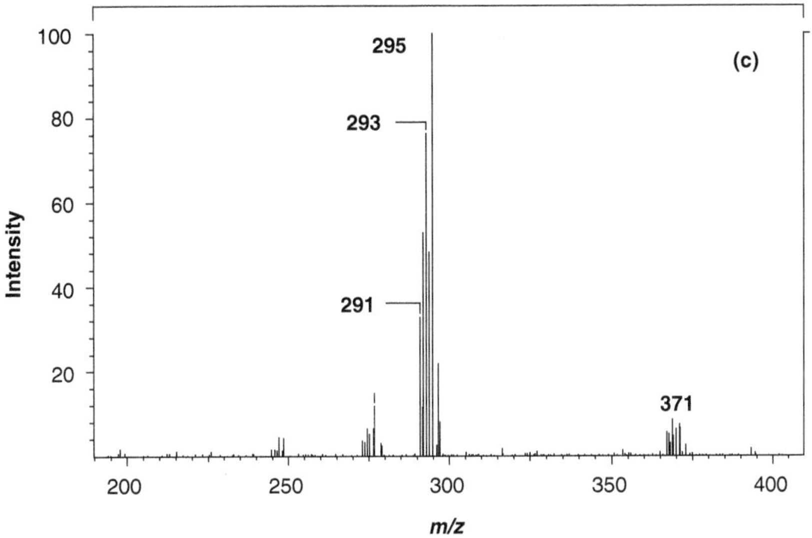

Figure 2. **(a)** Separation of different mono substituted mercury species by HPLC coupled to ICP-MS. The system used a reversed-phase column (250 × 4.6 mm i.d., 5 μm), an eluent of MeOH (50%), water (50%), containing 0.05% 2-mercaptoethanol (v/v) at a flow rate of 1 mL min^{-1}. The spraychamber was cooled to −10 °C and oxygen was added post-nebulization. The concentration of each component of the standard was 10 ng g^{-1}. **(b)** Separation of different mercury species by HPLC coupled to APCI-MS. The same HPLC conditions as in (a) were used. 1 = inorganic, 2 = methyl, 3 = ethyl, 4 = phenyl. Standard concentration was 10 ng g^{-1} for each component. **(c)** Mass spectrum for a 10 ng g^{-1} standard of methylmercury chloride. The most abundant ion at *m/z* 295 corresponds to a methylmercury/2-mercaptoethanol adduct, whereas the cluster at *m/z* 371 corresponds to a methylmercury/2-mercaptoethanol adduct containing two 2-mercaptoethanol groups and loss of two protons.

HPLC-API-MS provides structural information, but without an authentic standard, quantitation is not possible because ionization is molecule specific. HPLC-ICP-MS can give accurate and precise quantification with an elemental standard even at trace concentrations, but identification is only possible with a retention time standard and even in this situation mistakes can be made. By using these techniques in combination it is possible to generate a diverse range of information for a particular analytical problem. Figure 2 shows the results obtained for the speciation analysis of Hg using the same HPLC separation system, coupled to ICP-MS (Figure 2a) and APCI-MS (Figure 2 b,c). More recent work in this area has used the same column coupled in parallel to both detectors, which can provide quantitative and structural data simultaneously [68]. However, it is not possible to always achieve this aim because of the differences in sensitivity of the two detectors for some analyte-matrix combinations and often it is necessary to split the flow so that more reaches the API detector.

For GC separations there are more options because of the potential to use VG as an interface mechanism and so ICP-MS, microwave-induced plasma (MIP), and AFS can provide elemental analysis and conventional MS based on EI, in various mass analyzer configurations, can be used for structural analysis. CE applications have more niche applications and this chromatographic technique can be interfaced both to ICP-MS and ESI-MS, however a suitable device to enable coupling to both detectors simultaneously awaits development.

3.5. Methods Based on Vapor Generation

Vapor generation has been widely used as a gas-phase sample introduction technique for species of As, Hg, Sb, and Sn that can be readily converted into stable hydrides or the elemental form and Table 2 presents LODs for a selection of VG systems. There are several recent general reviews of speciation analysis by VG coupled to various detectors [69–74].

The basic design of a VG system has three or four stages: generation of the hydride or elemental form; vapor collection (optional); transfer of vapor to atomizer or spectroscopic excitation source; and atomization. Very high transport efficiencies, approaching 100%, can be achieved, whilst separating the analytes from undesirable matrix components. Because only a vapor is passed to the detector, chemical and spectral interferences are essentially eliminated, as is the need for a nebulizer, which improves transport efficiency. These factors help to lower the achievable LODs and VG is a technique that offers high sensitivity. Moreover, for VG operated in batch mode, relatively large sample volumes (e.g., 100 mL for batch versus 0.1 mL for HPLC flow) can be applied, further lowering the LODs achievable.

Hydride generation (HG) using sodium tetrahydroborate (STBH; $NaBH_4$) is by far the most common means of forming hydrides. The reaction for element E with an oxidation state $m+$ may be described:

$$NaBH_4 + 3H_2O + HCl \rightarrow H_3BO_3 + NaCl + 8H \tag{1.}$$

$$E^{m+} + 8H \rightarrow EH_m + H_2(\text{excess}) \tag{2.}$$

HG occurs very rapidly when an alkaline solution of STHB is mixed with an acidified sample solution. Post-reaction, hydrides, and other gases (mainly H_2) are transported via an inert carrier gas to a gas-liquid separator and then passed into the detector (e.g., AAS, AES, AFS or ICP).

Table 2. Selection of HG-based analytical systems with detection limits for determination of organometal(loid)s.

Analytical system	Sample	Organometal(oid) species (detection limit)[a]	References
HG pre-separation			
HG-CT-GC-ICP-MS (pH gradient HG)	Soil	As: MeAs (0.098)[b], Me_2As (0.011)[b], Me_3As (0.015)[b]	[112]
		Sb: MeSb (0.007)[b], Me_2Sb (0.005)[b], Me_3Sb (0.001)[b]	[112]
		Sn: MeSn (0.093)[b], Me_2Sn (0.07)[b], Me_3Sn (0.01)[b]	[112]
HG-SPME-GC-MS[a]	Sediments	Hg: MeHg (20 pg)	[113]
HG-CT-GC-AFS[d]	Sediments	Hg: mono-MeHg (0.03)[c], mono-EtHg (0.03)[c]	[114]
HG-CT-GC-ICP-MS[d]	Sediments	Hg: mono-MeHg (0.02)[c], mono-EtHg (0.01)[c]	[114]
HG post-separation			
HPLC-HG-ICP-MS (IP-RP column)	Spring water	As: MeAs (5.6), Me_2As (3.6)	[115]
HPLC-HG-AAS (IP-RP column)	Groundwater	As: MeAs (110), Me_2As (150)	[116]
HPLC-HG-ICP-AES (AEx column)	Spiked water	As: MeAs (380), Me_2As (2,130)	[117]
HPLC-UV-HG-AFS (AEx column)	Standards	As: MeAs (14), Me_2As (11), AB (15), AC (9), TMAO (17)	[118]
HPLC-HG-ETAAS (silica-based ion exchange)	Sediment, mussel tissue	Sn: MBT, DBT, TBT (135-942)	[119]
Flow-CE-HG-AFS	Human urine	As: MeAs (11,200)[c], Me_2As (8,900)[c]	[120]
	Lake & river water	Hg: MeHg (16,600)[c], EtHg (15,900)[c], PhHg (13,300)[c]	[121]

[a]Detection limits are given as pg of elemental form, unless otherwise stated.
[b]$\mu g\, kg^{-1}$ dry weight
[c]$ng\, L^{-1}$; HG was with TBH unless otherwise stated
[d]phenylation derivatization
CT, cryogenic trap; CE, capillary electrophoresis; ETAAS, electrothermal AAS

HG using STBH can be operated as a batch, continuous-flow or flow-injection system. Problems can occur through inadequate control of reaction conditions and separation of by-products, especially H_2, which then enters the atomizer. Such problems are mainly associated with batch systems, and are largely eliminated in flow systems. Transition and noble metals can cause severe signal suppression and such chemical interferences are considered to be the most serious form of interference in HG [71]. Considerable effort has been made to reduce or eliminate interferences through addition of chemical agents which complex the interfering metal ions, e.g., L-cysteine, L-histidine, EDTA, tartaric acid, KI [72,75]. For multi-elemental analysis a universal method for minimising chemical interferences has not been found because of the great variety of operating conditions of the HG reaction reported in the literature, although L-cysteine and thiourea are generally regarded as the most promising masking reagents for severe interference metals such as Co(II), Cr(III), Cu(II), Ni(II), and Fe(III).

The reaction between STBH and the analyte in solution is markedly dependent upon pH, which influences both the level of protonation of the analyte and the hydrolysis of STBH. Selective batch mode methods have been used to speciate inorganic and methylated forms of As in the absence of a chromatographic separation [73]. Sample pre-treatment, the dependency of HG on pH and control of STBH and HCl concentrations, allows the non-chromatographic determination of methylated As(III) species and methylated As(V) species [76]. Although selective HG in batch mode operation is a simple and inexpensive approach to As speciation, it is limited to inorganic and simple methylated species and has the disadvantage of long reaction times, slow sample throughput and reliance on strict control of reaction conditions. This approach to the speciation of As, Sb, Se, and Te has recently been reviewed [73].

For speciation analysis of organometal(loid)s a chromatographic separation is almost invariably required, although as described above, chemical parameters can be used. For example, Me_3SbCl_2 has a derivatization optimum near to neutral pH, while $MeAsO(OH)_2$ and MeAsO(OH) require acidic conditions for derivatization [77]. A pH gradient procedure designed to overcome differences in pH optima for derivatization of different methyl species has been used for As, Bi, and Sb in a single run [78]. This involved adjusting the pH from 7 to 1 using citrate buffer during the HG stage, with coupling to GC-ICP-MS [78]. Anderson et al. [79,80] incorporated mercaptoacetic acid into the STBH/HCl reaction mixture and reported similar response profiles for As(III), As(V), MMA, and DMA. Incorporation of L-cysteine into reaction mixtures as a pre-reductant has been used widely in HG As speciation analysis. Not only does it minimize interferences from transition metals, it also reduces the concentration of acid required and improves the stability of the hydrides [75,81]. A further consideration is that increased demethylation occurs with decreasing pH during HG of methylated forms of As and other elements, including Bi, Sb, and Hg. Hirner [82] has described the artefacts that arise in speciation analysis from the application of derivatization techniques. Various acids, buffers and redox media have been utilized successfully for HG speciation analysis of inorganic and methylated forms of As [71,73], although a universal HG method has not emerged.

Electrochemical VG in atomic spectrometry is an alternative sample-introduction technique to chemical VG. Several advantages of this approach have been reported, including: the use of similar reaction media for analysis of all HG elements; the possibility of reduced interference from related species; and independence of HG efficiency from oxidation state of analyte. Avoidance of STBH as a derivatizing agent is also an advantage because it is expensive, must be prepared daily and can introduce contaminants. Although electrochemical HG has been widely used for total element determination, there is as yet little information on its application in speciation analysis. Denkhaus et al. [83] present a detailed summary of mechanistic electrolytic HG-AAS for the determination of As, Sb, Se, and Sn. The fundamentals, interferences, and application of electrochemical HG have been recently reviewed [70].

Cryogenic trapping (CT) of volatile hydrides is a useful approach for the determination of methylated forms of metal(loid)s, including those of As, Sb, Bi, Hg, and Se. The approach has also been used for focusing the hydrides formed, leading to efficient species separation and improved LODs. Columns packed with glass beads, glass wool or a suitable chromatographic material are immersed in liquid nitrogen. Removal of the liquid nitrogen alone or combined with subsequent electrothermal heating, releases and separates the hydrides according to their boiling points, which are then detected [84]. Generally, traps filled with chromatographic material show improved separation and species recovery compared with glass bead or wool filled traps [73]. For analysis of environmental gases for methylmetal(loid) species, samples have been passed directly to a series of cryogenic traps by a vacuum pump, or collected into gas bags (Tedlar bags) prior to cryogenic trapping [85]. Low temperature GC-ICP-MS has been used to analyze loaded cryogenic gas traps, with thermal desorption within the temperature range −100 to 165 °C [85].

A major disadvantage of the VG approach is that it does not differentiate between species with the same level of methylation. For example, dimethylarsinic acid (DMA) and dimethylarsinous acid (DMA^{III}) both form dimethylarsine, so all three species present in a sample are indistinguishable. A further issue with pre-column derivatization is that demethylation and transalkylation can occur, which may give rise to several species from a single organometal(loid) analyte [82,86]. For As speciation, a fully automated flow-injection-HG-CT-AAS has been reported using a poly(tetrafluoroethane) (PTFE) trap heated by microwave radiation [87]. Duester et al. [78] used a multi-organometal(loid) standard for determination of methylated As, Sb, and Sn species in soils, by HG-CT-GC-ICP-MS. The multi-standard comprised: $MeAs(ONa)_2$, $Me_2AsO(OH)$, Me_3AsO,

$MeSnCl_3$, Me_2SnCl, Me_3SnCl, $(C_4H_9)SnCl_3$, and Me_3SbBr_2. Other workers have reported on improved LODs for As species with novel cryogenic traps, such as replacing a conventional glass U-trap with a chromatographic packed cold finger trap [88]. Such improvements have led to better performance in terms of species separation. Terlecka [71] has reviewed As speciation in water samples by hyphenated techniques, including those involving HG.

Continuous-flow and flow-injection HG systems are more widely used than batch systems as they offer the advantages of higher volatilization efficiency with STBH, more effective transport of analytes to the atomizer; improved detector sensitivity and precision, and increased tolerance to interferences. Because not all OMCs form stable hydrides, an on-line degradation stage, such as microwave digestion or UV photolysis, may be required for speciation analysis by flow HG. This applies particularly to As-containing compounds such as AB, arsenocholine (AC), arsenosugars, and the tetramethylarsonium ion that do not form stable hydrides under normal conditions. With such degradative treatment, the organic counter-ion species would be destroyed and only the methylmetal(loid) portions detected, so that full molecular speciation is not provided.

Most flow HG systems utilize HPLC as a liquid separation stage interfaced with an ESD: HPLC-HG-AAS; HPLC-HG-ICP-AES; HPLC-HG-AFS; HPLC-HG-ICP-MS. Figure 3 illustrates the sequential stages of a HPLC-UV-HG-detector system. Detection limits and sensitivity to interferences depend on the detector used (Table 2). AFS as a flow-through detector couples well with on-line HG and has been extensively used. Advantages of AFS include high sensitivity for most of the hydride forming elements, high sampling frequency, ease of operation, and low cost [89]. HG eliminates light scattering and background interferences from the matrix, resulting in increased sensitivity for AFS [89]. In continuous flow systems (e.g., HPLC-HG), separation of matrix components such as transition metal species prior to HG also helps to minimize interferences in environmental sample analysis; hyphenation of flow injection with HG-AFS has been reviewed [89,90].

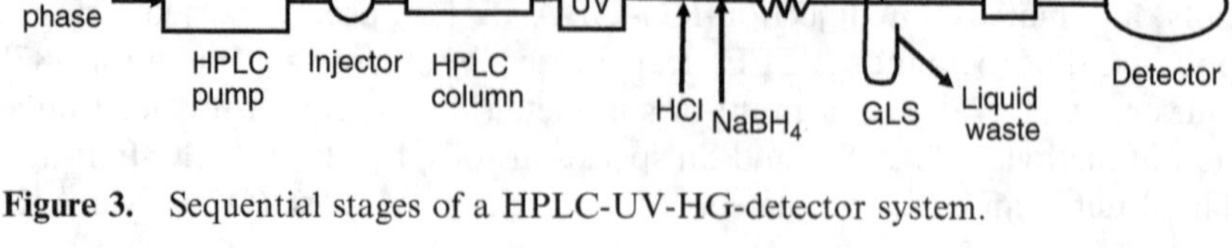

Figure 3. Sequential stages of a HPLC-UV-HG-detector system.

In As speciation studies, incorporation of HG between HPLC and ICP-AES has been shown to significantly reduce the severe spectral interference and enhance sensitivity [91]; HG hyphenated with different AES sources (e.g., ICP, MIP) has been reviewed [72]. Similarly, for As speciation using HPLC-ICP-MS, incorporation of HG eliminates spectral interferences that may occur due to the formation of ArCl ions and reduces the detection limit to around 1ng L^{-1} [71,72]. AAS offers high sensitivity, selectivity, and low LODs with different separation techniques, when combined with HG, e.g., HPLC-HG-AAS. The mechanism of hydride formation and atomization in HG-AAS has been reviewed [69]. The main advantages of HPLC-ICP-MS over HPLC-HG-AAS for speciation studies are the lower LODs and capability to detect non-hydride forming species without the requirement for an additional mineralization step.

3.6. Methods for Quantification

Molecular standards are not required for quantitation with ICP-MS detection because the argon plasma is such a good source of ions that the chemical species entering the plasma from the chromatographic separation are rapidly converted into their constituent elemental ions and this is essentially independent of the original molecule, although this needs to be assessed for the compound of interest. For identification purposes retention time standards are required. In most situations it is recommended that standard additions or the use of a suitable internal standard are used for calibration, so that matrix effects are mitigated.

A significant advantage of using mass spectrometry for organometallic analysis is the ability to carry out accurate and precise quantitation, which for the highest accuracy applications will involve calibration based on isotope dilution mass spectrometry. The basis of trace analysis using IDMS is the addition of an isotopically altered material known as the spike, to the sample containing the analyte. After allowing time for equilibrium, the resulting isotopic ratio between ions representative of the spike and the analyte are measured by MS. Provided the spike is present in an equilibrated and equivalent state to the analyte, it can perform the role of a "perfect" internal standard and enable exact compensation to be made for all stages of the analytical procedure, from the sample preparation steps to the final determination. IDMS using ESDs employs standards containing an enriched isotope of the metal of interest as the spike. Figure 4 shows the analysis of a harbor sediment reference material spiked with TBT enriched with ^{116}Sn by HPLC-ICP-MS. This shows how the isotopic ratio for DBT matches the natural ratio of ^{120}Sn to ^{116}Sn, but when TBT elutes the ratio changes considerably, due to the elution of the ^{116}Sn-enriched spike material. The

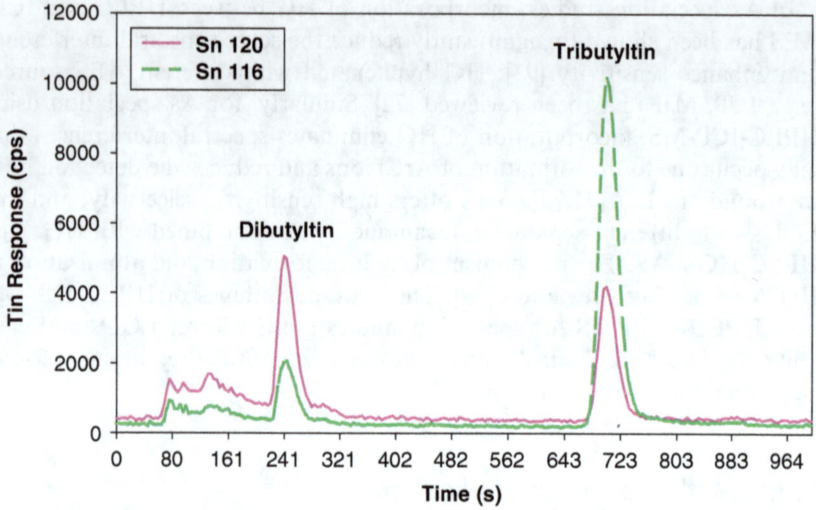

Figure 4. Analysis of the harbor sediment reference material PACS-1 using HPLC-ICP-ID-MS. The system used a reversed-phase column (150 × 2.1 mm i.d., 5 μm), an eluent of acetonitrile (65%), acetic acid (10%), water (25%) containing 0.05 % triethylamine (v/v) at a flow rate of 0.2 mL min^{-1}. The spraychamber was cooled to –10 °C and oxygen was added post-nebulization.

isotopic ratios rather than the response for a particular isotope are used to calculate the concentration of the analyte. When using IDMS with molecular MS, enriched stable isotopes of carbon or nitrogen are incorporated into an analogue of the analyte, which is then used as the spike material. In practice there are a few fundamental differences between molecular and elemental IDMS that result in different procedures and equations being used. More information on how to carry out both forms of IDMS and the differences between them are available [92]. Suffice to say, the correct use of either approach will provide high accuracy results with low measurement uncertainty.

A framework encompassing two different strategies for carrying out these measurements by ID-ICP-MS has been described [93]: *species-specific* spiking (ssIDMS), whereby the sample is spiked with an enriched metal(loid) containing analogue of the analyte at the beginning of the analytical procedure and *species-unspecific* spiking (suIDMS) where an enriched inorganic metal(loid) spike is added continuously to the eluent from the chromatographic column. In both approaches the isotope ratio between the spike and analyte isotope are measured. The former method requires that the structure of the chemical species in the sample is known and that a suitable isotopically enriched spike material is available, the latter method has been used where the OMC of interest is unidentified or an analogue of the analyte containing an enriched isotope is not available. ssIDMS is the superior method because any chemical or physical losses of the analyte during the analytical procedure will be corrected for in the final IDMS measurement, assuming that both the spike and the analyte reach chemical equilibrium prior to analysis. The real value of IDMS in speciation analysis was highlighted during the development of a GC-ICP-MS method for the analysis of MeHg in environmental water samples [52]. The methodology identified a systematic error during the derivatization step, which was completely corrected for using the ssIDMS approach. IDMS using ICP-MS for the measurement of OMCs in different materials has been reviewed [94].

Due to the monoisotopic nature of As [95] it is not possible to use IDMS in As speciation analysis. External calibration was compared with standard additions for the HG speciation of As in algal samples [28], with no significant differences (95% confidence level) between the calibration curve slopes for As(III), MMA, DMA, and As(V). Unlike As speciation analysis, Hg is amenable to IDMS as Hg comprises seven isotopes. As with all speciated IDMS methods, spike materials must be available and this is a limiting factor as few OMCs prepared with a suitably enriched isotope are. A commercially available enriched MeHg spike material was first offered as a certified reference material (CRM) in 2004 [96]. One advantage of the MeHg spike is that it has a certified concentration, enabling one way ssIDMS to be applied to MeHg determinations. The use of solid sampling electrothermal vaporization ICP-MS for the determination of both Hg(II) and MeHg in biological reference materials using suIDMS reduced the risk of forming artefacts attributable to analyte extraction because of the absence of an extraction step [97].

In ssIDMS, the added spike material and native analyte must achieve a state of equilibrium to ensure the quality of IDMS data [98]. If the added enriched spike and endogenous analyte behave differently at any stage of the sample processing or analysis then the results will be biased. Complete equilibration between an enriched MeHg spike and the MeHg found in the CRM DORM-2 was estimated to have been achieved within the 6 minutes following sample spiking [94]. This is in contrast to other published equilibration times, e.g., 14 hours has been used to ensure equilibrium between spike and naturally abundant analyte in 3 different biological materials [98]. Similarly, in the analysis of a fish CRM DOLT-3, equilibration was ensured by measurement of isotope amount ratios of spiked methanolic KOH extracts after two days with the measurement repeated two weeks later [37]. No significant differences were observed on extract storage; if equilibrium had not been reached after the initial two days then the repeated analysis two weeks later would have found larger mass fractions of MeHg. This is because the spike would be preferentially extracted over the analyte from the sample in the initial determination after two days. This very large range of MeHg spike equilibration times has been addressed in a recent review, which concluded that after 5 minutes of MAE with TMAH, MeHg from a biological sample and the spike MeHg had come into equilibrium [53].

4. QUALITY MANAGEMENT

In practice, one of the most important characteristics of organometallic analysis stems from the fact that the total concentration of the metal(loid) being studied can be measured very accurately using well validated instrumental methods. This gives a very powerful means to determine whether the method being used is providing reliable results because the combined concentration of all the individual species in an extract of the sample must be equivalent to the total concentration in that extract. Any significant difference between the two values is indicative of a systematic error in the analysis.

After some sustained work in protocol performance testing, most notably the Standards, Measurement and Testing (SMT) Programme of the European Commission, the pitfalls that can be encountered during this type of analysis are better understood and methods to evaluate and eliminate them are now well established for OMCs [99]. For the extraction step QA considerations mean the extraction efficiency needs to be validated and this can be done either by spike recovery experiments or by using a representative certified reference material. The main criticism of spike recovery experiments is that the spike is not bound in the sample matrix in the same way as the endogenous analyte being measured, however low recoveries would indicate an inadequate method. Other national bodies, including NRC Canada and NIST in USA have played an important role in improving the framework for generating valid and traceable speciation measurements by the provision of a range of CRMs. The CRMs available with values for some of the more important OMCs now includes sediments, fish, and shellfish tissue and human matrices such as hair and urine. However, real samples are rarely identical to the matrix CRM available, so care should be taken when comparing the data from each.

5. FUTURE DEVELOPMENTS

Legislation, the main driving force for analytical measurements is lacking for all but a few defined chemical species. International legislation concerning food safety, the environment and occupational health, is most often based on total metal(loid) concentrations. In most regulations only specific contaminants and "their compounds" are specified, but some guideline values, regulations, or action limits, have been assigned for OMCs. Examples which stipulate the measurement of chemical species include: MeHg in fish [100,101]; TBT and TPhT species in the marine environment related to antifouling paints [102]. More significantly perhaps, the EU Water Framework Directive sets objectives that should ensure that all water meets "good status" by the year 2015. As part of this legislation a list of priority hazardous substances has been established and this includes: Cd and its compounds, Hg and its compounds, Pb and its compounds, Ni and its compounds, and TBT (organotin) [103].

As the requirement for more risk-based information becomes accepted, the more likely government agencies and regulatory bodies will realize the importance of chemical speciation. This will result in a greater need for CRMs, better availability of proficiency testing schemes for routine laboratories, a greater range of isotopically enriched standards, suitably integrated separation and detection equipment, with associated software and improvements in sample preparation approaches.

ACKNOWLEDGEMENTS

We thank Dr. Peter Sutton for provision of the ^{116}Sn TBT enriched compound used in Figure 4.

ABBREVIATIONS AND DEFINITIONS

2-ME	2-mercaptoethanol
AAS	atomic absorption apectroscopy
AB	arsenobetaine = trimethylarsonioacetate
AC	arsenocholine
AEC	anion exchange chromatography
AES	atomic emission spectrometry
AEx	anion exchange
AFS	atomic fluorescence spectroscopy
APCI	atmospheric pressure chemical ionization
API	atmospheric pressure ionization
API-MS	atmospheric pressure ionization mass spectrometry
As(III)	arsenite
As(V)	arsenate
ASE	accelerated solvent extraction

BCR	European Community Bureau of Reference
C_{18}	octadecylsilane chromatographic phase
CE	capillary electrophoresis
CID	collision-induced dissociation
CRM	certified reference material
CT	cryogenic trapping
DBT	dibutyltin
DMA	dimethylarsinic acid
DMA^{III}	dimethylarsinous acid
DML	dimethyllead
DMT	dimethyltin
DOLT-3	dogfish liver certified material-3
DORM-2	dogfish muscle certified material-2
DPT	diphenyltin
EDTA	ethylenediamine-*N*,*N*,*N′*,*N′*-tetraacetic acid
EI	electron impact ionization
ESD	element-specific detector
ESI	electrospray ionization
ESI-MS	electrospray ionization-mass spectrometry
Et	ethyl group
ETAAS	electrothermal atomic absorption spectrometry
FPD	flame photometric detection
GC	gas chromatography
GC-AES	gas chromatography-atomic emission spectrometry
GC-FPD	gas chromatography-flame photometric detector
GC-ICP-MS	gas chromatography-inductively coupled plasma mass spectrometry
GC-MS	gas chromatography-mass spectrometry
GC-QF-AAS	gas chromatography-quartz furnace-atomic absorption spectrometry
GLS	gas-liquid separator
HG	hydride generation
HG-AAS	hydride generation-atomic absorption spectrometry
HG-CT-AAS	hydride generation-cryogenic trapping-atomic absorption spectrometry
HPLC	high performance liquid chromatography
HPLC-API-MS	high performance liquid chromatography-atmospheric pressure ionization mass spectrometry

HPLC-HG-AAS	high performance liquid chromatography-hydride generation-atomic absorption spectrometry
HPLC-HG-AFS	high performance liquid chromatography-hydride generation-atomic fluorescence spectrometry
HPLC-HG-ICP-AES	high performance liquid chromatography-hydride generation-inductively coupled plasma atomic emission spectrometry
HPLC-HG-ICP-MS	high performance liquid chromatography-hydride generation-inductively coupled plasma mass spectrometry
HPLC-ICP-MS	high performance liquid chromatography-inductively coupled plasma-mass spectrometry
ICP-MS	inductively coupled plasma-mass spectrometry
ICP-OES	inductively coupled plasma optical emission spectroscopy
IDMS	isotope dilution mass spectrometry
IP-RP	ion pair-reverse phase
IUPAC	International Union of Pure and Applied Chemistry
LC-ESI-MS	liquid chromatography-electrospray ionization-mass spectrometry
LC-MS/MS	liquid chromatography-tandem mass spectrometry
LOD	limit of detection
m/z	mass-to-charge ratio
MAE	microwave assisted extraction
MALDI-TOF-MS	matrix assisted laser desorption ionization-time of flight-mass spectrometry
MBT	monobutyltin
Me	methyl group
MeHg	methylmercury
MIP	microwave induced plasma
MMA	monomethylarsonic acid
MMT	monomethyltin
MS/MS	tandem MS analysis
NIES	National Institute for Environmental Sciences (Japan)
NIST	National Institute of Standards and Technology (USA)
NMR	nuclear magnetic resonance

NRC	National Research Council (Canada)
OMC	organometallic compounds
OTC	organotin compounds
PACS-1	marine sediment reference material (National Reserach Council of Canada)
PEEK	polyetheretherketone
Ph	phenyl group
pK_a	acid dissociation constant
PTFE	poly(tetrafluoroethene)
QA	quality assurance
QC	quality control
QF-AAS	quartz furnace atomic absorption spectroscopy
Se-Cys	selenocysteine
Se-Me-Cys	Se-methylselenocysteine
Se-Met	selenomethionine
SFC	supercritical fluid chromatography
SMT	Standards, Measurement and Testing Programme of the European Commission
SPE	solid phase extraction
SPME	solid phase micro-extraction
ssIDMS	species-specific isotope dilution mass spectrometry
STBH	sodium tetrahydroborate, $NaBH_4$
STEB	sodium tetraethylborate
suIDMS	species-unspecific isotope dilution mass spectrometry
TBT	tributyltin
TETRA	tetramethylarsonium ion
TFA	trifluoroacetic acid
TMAH	tetramethylammonium hydroxide
TMAO	trimethylarsine oxide
TML	trimethyllead
TMT	trimethyltin
TPT	triphenyltin
TPrT	tripropyltin
Tris	trishydroxymethylaminomethane
VG	vapor generation

REFERENCES

1. R. Cornelius, *NATO Advanced Study Institute on Metal Speciation in the Environment*, Cesme, Turkey, 1989, p. 30.

2. D. M. Templeton, F. Ariese, R. Cornelis, L. G. Danielsson, H. Muntau, H. P. Van Leeuwen and R. Lobinski, *Pure Appl. Chem.*, 2000, **72**, 1453–1470.
3. J. R. Cannon, J. S. Edmonds, K. A. Francesconi, C. L. Raston, J. B. Saunders, B. W. Skelton and A. H. White, *Austral. J. Chem.*, 1981, **34**, 787–798.
4. E. Sabbioni, M. Fischbach, G. Pozzi, R. Pietra, M. Gallorini and J. L. Piette, *Carcinogenesis*, 1991, **12**, 1287–1291.
5. A. H. Stern and A. E. Smith, *Environ. Health Persp.*, 2003, **111**, 1465–1470.
6. R. Renner, *Environ. Sci. Technol.*, 2004, **38**, 481A–481A.
7. J. Feldmann, V. W. M. Lai, W. R. Cullen, M. S. Ma, X. F. Lu and X. C. Le, *Clin. Chem.*, 1999, **45**, 1988–1997.
8. J. Qvarnstrom and W. Frech, *J. Anal. At. Spectrom.*, 2002, **17**, 1486–1491.
9. Q. Tu, J. Qian and W. Frech, *J. Anal. At. Spectrom.*, 2000, **15**, 1583–1588.
10. C. M. Tseng, A. DeDiego, F. M. Martin, D. Amouroux and O. F. X. Donard, *J. Anal. At. Spectrom.*, 1997, **12**, 743–750.
11. A. I. C. Ortiz, Y. M. Albarran and C. C. Rica, *J. Anal. At. Spectrom.*, 2002, **17**, 1595–1601.
12. P. Jitaru, H. G. Infante and F. C. Adams, *J. Anal. At. Spectrom.*, 2004, **19**, 867–875.
13. K. L. Ackley, C. B'Hymer, K. L. Sutton and J. A. Caruso, *5th Rio Symposium on Atomic Spectroscopy*, Cancun, Mexico, 1998, pp. 845–850.
14. S. C. Hight and M. T. Corcoran, *J. Assoc. Official Anal. Chemists*, 1987, **70**, 24–30.
15. M. J. Vazquez, M. Abuin, A. M. Carro, R. A. Lorenzo and R. Cela, *Workshop on Errors in Methylmercury Determinations During Sample Preparation, Derivatization and Detection*, Wiesbaden, Germany, 1998, pp. 1211–1224.
16. E. Bulska, D. C. Baxter and W. Frech, *Anal. Chim. Acta*, 1991, **249**, 545–554.
17. D. Sanchez-Rodas, A. Geiszinger, J. L. Gomez-Ariza and K. A. Francesconi, *Analyst*, 2002, **127**, 60–65.
18. M. J. Ruiz-Chancho, J. F. López-Sánchez, E. Schmeisser, W. Goessler, K. A. Francesconi and R. Rubio, *Chemosphere*, 2008, **71**, 1522–1530.
19. E. Sanz, R. Muñoz-Olivas, C. Cámara, M. K. Sengupta, S. Ahamed and J. Environ. Sci. Health, *Part A*, 2007, **42**, 1695–1705.
20. A. H. Ackerman, P. A. Creed, A. N. Parks, M. W. Fricke, C. A. Schwegel, J. T. Creed, D. T. Heitkemper and N. P. Vela, *Environ. Sci. Technol.*, 2005, **39**, 5241–5246.
21. I. Pizarro, M. Gómez, M. A. Palacios and C. Cámara, *Anal. Bioanal. Chem.*, 2003, **376**, 102–109.
22. S. Hirata and H. Toshimitsu, *Anal. Bioanal. Chem.*, 2005, **383**, 454–460.
23. R. Fischer, S. Rapsomanikis and M. O. Andreae, *Anal. Chem.*, 1993, **65**, 763–766.
24. N. Bloom, *Canadian J. Fisheries Aquat. Sci.*, 1989, **46**, 1131–1140.
25. R. Rai, W. Maher and F. Kirkowa, *J. Anal. At. Spectrom.*, 2002, **17**, 1560–1563.
26. S. Rattanachongkiat, G. E. Millward and M. E. Foulkes, *J. Environ. Monitoring*, 2004, **6**, 254–261.
27. G. I. Callum, M. M. Ferguson and J. M. A. Lenihan, *The Analyst*, 1981, **106**, 1009–1013.

28. S. García Salgado, M. A. Quijano Nieto and M. M. Bonilla Simón, *J. Chromatogr. A*, 2006, **1129**, 54–60.
29. J. A. Brisbin and J. A. Caruso, *The Analyst*, 2002, **127**, 921–929.
30. U. Kohlmeyer, J. Kuballa and E. Jantzen, *Rapid Comm. Mass Spectrom.*, 2002, **16**, 965–974.
31. E. H. Larsen, J. Engman, J. J. Sloth, M. Hansen and L. Jorhem, *Anal. Bioanal. Chem.*, 2005, **381**, 339–346.
32. C. Gerbersmann, M. Heisterkamp, F. C. Adams and J. A. C. Broekaert, *Anal. Chim. Acta*, 1997, **350**, 273–285.
33. P. A. Gallagher, J. A. Shoemaker, X. Wei, C. A. Brockhoff-Schwegel and J. T. Creed, *Fresenius' J. Anal. Chem.*, 2001, **369**, 71–80.
34. S. S. Kannamkumarath, K. Wrobel, K. Wrobel and J. A. Caruso, *J. Agricult. Food Chem.*, 2004, **52**, 1458–1463.
35. G. Centineo, E. B. Gonzalez and A. Sanz-Medel, *J. Chromatography A*, 2004, **1034**, 191–197.
36. L. R. Bravo-Sanchez, J. R. Encinar, J. I. F. Martinez and A. Sanz-Medel, *Spectrochim. Acta Part B-Atomic Spectroscopy*, 2004, **59**, 59–66.
37. L. Yang, Z. Mester and R. E. Sturgeon, *J. Anal. At. Spectrom.*, 2003, **18**, 431–436.
38. J. C. van Loon, *Anal. Chem.*, 1979, **51**, 1139A–1150A.
39. K. T. Susuki, *Anal. Biochem.*, 1980, **102**, 31–34.
40. L. Ebdon, S. Hill and R. W. Ward, *The Analyst*, 1986, **111**, 1113–1138.
41. L. Ebdon, S. Hill and R. W. Ward, *The Analyst*, 1987, **112**, 1–16.
42. N. P. Vela and J. A. Caruso, *J. Anal. At. Spectrom.*, 1993, **8**, 787–794.
43. X. F. Lu, L. L. Arnold, S. M. Cohen, W. R. Cullen and X. C. Le, *Anal. Chem.*, 2003, **75**, 6463–6468.
44. P. Quevauviller, R. Ritsema, R. Morabito, W. M. R. Dirkx, S. Chiavarini, J. M. Bayona and O. F. X. Donard, *Appl. Organomet. Chem.*, 1994, **8**, 541–549.
45. D. Schaumloffel, J. R. Encinar and R. Lobinski, *Anal. Chem.*, 2003, **75**, 6837–6842.
46. A. Kim, S. Hill, L. Ebdon and S. Rowland, *J. High Resol. Chromatogr.*, 1992, **15**, 665–668.
47. A. W. Kim, M. E. Foulkes, L. Ebdon, S. J. Hill, R. L. Patience, A. G. Barwise and S. J. Rowland, *J. Anal. At. Spectrom.*, 1992, **7**, 1147–1149.
48. W. G. Pretorius, L. Ebdon and S. J. Rowland, *J. Chromatogr.*, 1993, **646**, 369–375.
49. L. Ebdon, E. H. Evans, W. G. Pretorius and S. J. Rowland, *J. Anal. At. Spectrom.*, 1994, **9**, 939–943.
50. A. Prange and E. Jantzen, *J. Anal. At. Spectrom.*, 1995, **10**, 105–109.
51. M. J. Watts, A. W. Kim, D. S. Vidler, R. O. Jenkins, J. F. Hall and C. F. Harrington, *Instrumentation Science & Technology*, 2007, **35**, 15–31.
52. N. Demuth and K. G. Heumann, *Anal. Chem.*, 2001, **73**, 4020–4027.
53. K. G. Heumann, *Anal. Bioanal. Chem.*, 2004, **378**, 318–329.
54. C. F. Harrington and T. Catterick, *J. Anal. At. Spectrom.*, 1997, **12**, 1053–1056.

55. M. Leermakers, W. Baeyens, P. Quevauviller and M. Horvat, *Trends Anal. Chem.*, 2005, **24**, 383–393.
56. W. L. Shen, N. P. Vela, B. S. Sheppard and J. A. Caruso, *Anal. Chem.*, 1991, **63**, 1491–1496.
57. E. Blake, M. W. Raynor and D. Cornell, *J. Chromatogr. A*, 1994, **683**, 223–231.
58. V. Majidi, *Microchem. J.*, 2000, **66**, 3–16.
59. J. W. Olesik, J. A. Kinzer, E. J. Grunwald, K. K. Thaxton and S. V. Olesik, *Spectrochim. Acta, Part B-Atomic Spectroscopy*, 1998, **53**, 239–251.
60. J. W. Olesik, J. A. Kinzer and S. V. Olesik, *Anal. Chem.*, 1995, **67**, 1–12.
61. M. Yamashita and J. B. Fenn, *J. Phys. Chem.*, 1984, **88**, 4451–4459.
62. J. B. Fenn, M. Mann, C. K. Meng, S. F. Wong and C. M. Whitehouse, *Science*, 1989, **246**, 64–71.
63. W. Morelle and J. C. Michalski, *Curr. Anal. Chem.*, 2005, **1**, 29–57.
64. E. Rosenberg, *J. Chromatogr. A*, 2003, **1000**, 841–889.
65. I. I. Stewart, *Spectrochim. Acta, Part B-Atomic Spectroscopy*, 1999, **54**, 1649–1695.
66. V. Nischwitz and S. A. Pergantis, *Rapid Comm. Mass Spectrom.*, 2006, **20**, 3579–3585.
67. J. F. de la Mora, G. J. Van Berkel, C. G. Enke, R. B. Cole, M. Martinez-Sanchez and J. B. Fenn, *J. Mass Spectrom.*, 2000, **35**, 939–952.
68. J. Feldmann, *Anal. Bioanal. Chem.*, 2008, **390**, 1683–1684.
69. A. R. Kumar and P. Riyazuddin, *Anal. Sci.*, 2005, **21**, 1401–1410.
70. F. Laborda, E. Bolea and J. R. Castillo, *Anal. Bioanal. Chem.*, 2007, **388**, 743–751.
71. E. Terlecka, *Environmental Monitoring and Assessment*, 2005, **107**, 259–284.
72. P. Pohl, *Trends Anal. Chem.*, 2004, **23**, 87–101.
73. A. R. Kumar and P. Riyazuddin, *Int. J. Environ. Anal. Chem.*, 2007, **87**, 469–500.
74. J. L. Capelo, C. Fernandez, B. Pedras, P. Santos, P. Gonzalez and C. Vaz, *Talanta*, 2006, **68**, 1442–1447.
75. P. Carrero, A. Malave, J. L. Burguera, M. Burguera and C. Rondon, *Anal. Chim. Acta*, 2001, **438**, 195–204.
76. H. Hasegawa, Y. Sohrin, M. Matsui, M. Hojo and M. Kawashima, *Anal. Chem.*, 1994, **66**, 3247–3252.
77. R. A. Diaz-Bone, J. Felix and A. V. Hirner, *European Winter Plasma Conference on Plasma Spectrometry*, Garmisch-Partenkirchen, Germany, 2003, p. 296.
78. L. Duester, R. A. Diaz-Bone, J. Kosters and A. V. Hirner, *J. Environ. Monitor.*, 2005, **7**, 1186–1193.
79. R. K. Anderson, M. Thompson and E. Culbard, *Analyst*, 1986, **111**, 1153–1158.
80. R. K. Anderson, M. Thompson and E. Culbard, *Analyst*, 1986, **111**, 1143–1152.
81. D. L. Tsalev, A. Dulivo, L. Lampugnani, M. DiMarco and R. Zamboni, *J. Anal. At. Spectrom.*, 1996, **11**, 989–995.
82. A. V. Hirner, *Anal. Bioanal. Chem.*, 2006, **385**, 555–567.
83. E. Denkhaus, F. Beck, P. Bueschler, R. Gerhard and A. Golloch, *Fresenius J. Anal. Chem.*, 2001, **370**, 735–743.

84. P. J. Craig, S. N. Forster, R. O. Jenkins and D. Miller, *Analyst*, 1999, **124**, 1243–1248.
85. J. Feldmann, E. M. Krupp, D. Glindemann, A. V. Hirner and W. R. Cullen, *Appl. Organomet. Chem.*, 1999, **13**, 739–748.
86. I. Koch, J. Feldmann, J. Lintschinger, S. V. Serves, W. R. Cullen and K. J. Reimer, *Appl. Organomet. Chem.*, 1998, **12**, 129–136.
87. J. L. Burguera, M. Burguera, C. Rivas and P. Carrero, *Talanta*, 1998, **45**, 531–542.
88. T. M. Hsiung and J. M. Wang, *J. Anal. At. Spectrom.*, 2004, **19**, 923–928.
89. M. L. Chen, A. M. Zou, Y. L. Yu and R. H. He, *Talanta*, 2007, **73**, 599–605.
90. M. Leermakers, W. Baeyens, M. De Gieter, B. Smedts, C. Meert, H. C. De Bisschop, R. Morabito and P. Quevauviller, *Trends Anal. Chem.*, 2006, **25**, 1–10.
91. R. T. Gettar, R. N. Garavaglia, E. A. Gautier and D. A. Batistoni, *J. Chromatogr. A*, 2000, **884**, 211–221.
92. M. Sargent, R. Harte and C. F. Harrington, *in Guidelines for Achieving High Accuracy in Inorganic and Organic Analysis by Isotope Dilution Mass Spectrometry*, Royal Society of Chemistry, London, UK, 2002.
93. L. Rottmann and K. G. Heumann, *Fresenius J. Anal. Chem.*, 1994, **350**, 221–227.
94. R. Clough, J. Truscatt, S. T. Belt, E. H. Evans, B. Fairman and T. Catterick, *Appl. Spectrosc. Rev.*, 2003, **38**, 101–132.
95. C. Dietz, J. Sanz, E. Sanz, R. Muñoz-Olivas and C. Cámara, *J. Chromatogr. A*, 2007, **1153**, 114–129.
96. J. P. Snell, C. R. Quetel, L. Lambertsson and J. Qvarnstrom, *J. Anal. At. Spectrom.*, 2004, **19**, 1315–1324.
97. I. Gelaude, R. Dams, M. Resano, F. Vanhaecke and L. Moens, *Anal. Chem.*, 2002, **74**, 3833–3842.
98. R. C. R. Martin–Doimeadios, E. Krupp, D. Amouroux and O. F. X. Donard, *Anal. Chem.*, 2002, **74**, 2505–2512.
99. P. Quevauviller, *Method Performance Studies for Speciation Analysis*, Royal Society of Chemistry, Cambridge, UK, 1998.
100. (USEPA) US Environmental Protection Agency, *2001. Water Quality Criterion for the Protection of Human Health: Methylmercury*, US Environmental Protection Agency (EPA), Washington, DC, Report no. EPA-823-R-01-001 (19 December 2008).
101. National Research Council (NRC), *Toxicological Effects of Methylmercury*, National Academy Press, Washington, DC, 2000.
102. Anonymous, *J. Environ. Monitor.*, 1999, **1**, 11N.
103. European Commission, *Directive 2000/60/EC of the European Parliament and of the Council of 23 October 2000 Establishing a Framework for Community Action in the Field of Water Policy, Official Journal*, Vol. 327, 22 December 2000, pp. 1–73.
104. N. Poperechna and K. G. Heumann, *Anal. Chem.*, 2005, **77**, 511–516.
105. M. J. Moreno, J. Pacheco-Arjona, P. Rodriguez-Gonzalez, H. Preud'Homme, D. Amouroux and O. F. X. Donard, *J. Mass Spectrom.*, 2006, **41**, 1491–1497.
106. C. H. Tang and W. H. Wang, *Anal. Chim. Acta*, 2007, **581**, 370–376.

107. M. Siwek, A. B. Noubar, J. Bergmann, B. Niemeyer and B. Galunsky, *Anal. Bioanal. Chem.*, 2006, **384**, 244–249.
108. H. G. Infante, A. A. Borrego, E. Peachey, R. Hearn, G. O'Connor, T. G. Barrera and J. L. G. Ariza, *J. Agricult. Food Chem.*, 2009, **57**, 38–45.
109. I. De Gregori, W. Quiroz, H. Pinochet, F. Pannier and M. Potin-Gautier, *Talanta*, 2007, **73**, 458–465.
110. E. Sanz, R. Muñoz-Olivas and C. Cámara, *Anal. Chim. Acta*, 2005, **535**, 227–235.
111. P. A. Gallagher, J. A. Shoemaker, X. Wei, C. A. Brockhoff-Schwegel and J. T. Creed, *Fresenius' J. Anal. Chem.*, 2001, **369**, 71–80.
112. L. Duester, R. A. Diaz-Bone, J. Kosters and A. V. Hirner, *J. Environ. Monitor.*, 2005, **7**, 1186–1193.
113. S. Mishra, R. M. Tripathi, S. Bhalke, V. K. Shukla and V. D. Puranik, *Anal. Chim. Acta*, 2005, **551**, 192–198.
114. Y. X. Mao, G. L. Liu, G. Meichel, Y. Cai and G. B. Jiang, *Anal. Chem.*, 2008, **80**, 7163–7168.
115. T. Guerin, A. Astruc and M. Astruc, *Talanta*, 1999, **50**, 1–24.
116. S. L. Chen, S. J. Yeh, M. H. Yang and T. H. Lin, *Biol. Trace Elem. Res.*, 1995, **48**, 263–274.
117. G. Rauret, R. Rubio and A. Padro, *Fresenius J. Anal. Chemi.*, 1991, **340**, 157–160.
118. S. Simon, H. Tran, F. Pannier and M. Potin-Gautier, *J. Chromatogr. A*, 2004, **1024**, 105–113.
119. M. Grotti, P. Rivaro and R. Frache, *J. Anal. At. Spectrom.*, 2001, **16**, 270–274.
120. X. B. Yin, X. P. Yan, Y. Jiang and X. W. He, *Anal. Chem.*, 2002, **74**, 3720–3725.
121. X. P. Yan, X. B. Yin, D. Q. Jiang and X. W. He, *Anal. Chem.*, 2003, **75**, 1726–173.

Met. Ions Life Sci. **2010**, *7*, 71–110

3

Evidence for Organometallic Intermediates in Bacterial Methane Formation Involving the Nickel Coenzyme F_{430}

Mishtu Dey, Xianghui Li, Yuzhen Zhou, and Stephen W. Ragsdale

Department of Biological Chemistry, University of Michigan Medical School,
1150 W. Medical Center Dr., 5301 MSRB III, Ann Arbor MI 48109-0606, USA
< sragsdal@umich.edu >
(Current address of M.D.: Department of Chemistry, Massachusetts Institute of Technology,
77 Massachusetts Ave., Cambridge, MA 02139, USA)

Metal Ions in Life Sciences, Volume 7 Edited by Astrid Sigel, Helmut Sigel, and Roland K. O. Sigel

Published by the Royal Society of Chemistry, www.rsc.org DOI: 10.1039/9781849730822-00071

ABSTRACT: Bioorganometallic chemistry underlies the reaction mechanisms of metalloenzymes that catalyze key processes in the global carbon cycle. Metal ions that appear well suited for the formation of metal-carbon bonds are nickel, iron, and cobalt. The formation and reactivity of alkylcobalt species (methylcobalamin and adenosylcobalamin) at the active sites of B_{12}-dependent methyltransferases and isomerases have been well studied and serve as models to guide hypothesis for how organometallic reactions occur in other systems. This review focuses on methyl-coenzyme M reductase (MCR), which is responsible for all biologically produced methane on earth. At its active site, this enzyme contains a nickel corphin (F_{430}), which bears similarity to the cobalt corrin in cobalamin (B_{12}). Several mechanisms have been proposed for the

MCR-catalyzed reaction, and a methylnickel species is a central intermediate in all but one of these mechanisms. After introducing some important concepts of bioorganometallic chemistry and describing methanogenesis and some of the key properties of MCR, this review discusses research that has led to the generation and characterization of alkylnickel species in MCR and in model complexes related to F_{430}. Then, the focus shifts to the reactions that these alkylnickel species can undergo both in the enzyme and in bioinspired models: protonolysis to form alkanes and thiolysis to form thioethers, including methyl-SCoM (the natural methyl donor for MCR). Throughout, results are discussed in relation to the proposed models for the MCR mechanism.

KEYWORDS: carbon dioxide fixation · cobalamin · carbon monoxide dehydrogenase · hydrogenase · metallobiochemistry · methanogenesis · nickel · tetrapyrrole

1. INTRODUCTION

1.1. Development of Bioorganometallic Chemistry

Today the term "bioorganometallic chemistry" is broadly used to link organometallics with medicine and enzymology, thus, signifying the role of organometallic chemistry in biology. In 1985 Jaouen and Vessières first used the term bioorganometallic chemistry to describe the study of organometallic species of biological and medicinal interests and Halpern in 1986 first described mechanisms involved in bioorganometallic chemistry [1,2]. The term takes into account complexes formed using classical organometallic ligands such as CO, alkyls, and biologically active molecules such as enzymes, proteins, steroids, DNA or RNA nucleosides, which have in common a direct metal-carbon bond and are important in biological processes [3–7]. Several reviews covering various aspects of bioorganometallic chemistry have been reported and the historical perspective on the development of the field has also been well reviewed [8–11]. The placement of bioorganometallic chemistry and its great implication in the context of research are summarized in Figure 1. Precisely, bioorganometallic chemistry is defined as the study of biomolecules that contain a direct carbon-metal bond.

Bioorganometallic species are of great significance in biology as therapeutics, environmental toxins, and intermediates formed at the active sites of metalloenzymes. The use of organometallic complexes in medicine was studied primarily due to their unusual reactivity, which led to the discovery of the first organometallic drug "Salvarsan" [11]. This organoarsenic compound was used as an antimicrobial agent and was one of the first pharmaceuticals, for which Paul Ehrlich won the Nobel prize in 1908 [12–14]. Cisplatin complexes are well known for their antitumor activities since their

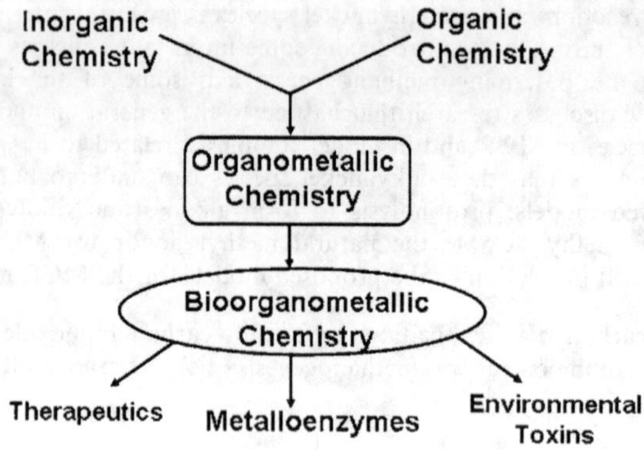

Figure 1. Origin and scope of bioorganometallic chemistry.

discovery in 1965 [15–17]. It was Köpf and Köpf-Maier, who in 1979 reported the antitumor activity of transition metal cyclopentadienyl complexes [18]. The organometallic iron complex, ferroquine, a novel antimalarial drug candidate is currently in development at Sanofi-Aventis [19]. Organometallic compounds can serve as biosensors, for example, a ferrocene complex is used to monitor glucose levels in diabetics [20].

The toxicity of organometallic compounds in the environment has been long recognized because they release volatile gases. In 1893, the Italian Physicist Bartolomeo Gosio first published that the toxic gas, alkylarsenic, was produced by the microbial conversion of arsenic [21]. Later in 1933, Challenger first identified this volatile, foul smelling "Gosio gas" as trimethylarsine [$(CH_3)_3As$] [22]. Subsequently, he reported that trimethylarsenic gas was produced by molds in a biological process involving *S*-adenosylmethionine, and hence the term "biological methylation" was coined to describe this process [23–26].

Another seminal development in the bioorganometallic field spans back to the middle of the twentieth century with the unexpected finding of metal-carbon bonds in the three biologically active forms of B_{12}: the vitamin (cyano), the coenzyme (adenosyl), and the methyl forms (below). Thus B_{12} occupies a preeminent place in the history of naturally occurring biorganometallic species [27–29].

In the 1980's, organometallic chemistry was invoked to explain the biological roles of the nickel-containing enzymes, methyl-coenzyme M reductase [30], and carbon monoxide dehydrogenase (CODH)/acetyl-CoA synthase (ACS) [31,32]. NiFe and FeFe hydrogenases also contain both Fe-CO and Fe-CN species that are important in their mechanisms and are biological examples of organometallic compounds containing an iron-carbon bond [33]. Stable iron-CO complexes of heme proteins are important in transcriptional activation and in inhibition of enzyme activity. The copper-containing ethylene receptor protein in plants appears to be another example of a naturally occurring organometallic species.

Although only a few roles of organometallic chemistry in nature have been so far uncovered, they have provided insights into novel roles of metals in biology. It is likely that novel bioorganometallic complexes are yet to be discovered.

1.2. Bioorganometallic Complexes in Enzymes

1.2.1. General Principles Exemplified by Cobalamin-Dependent Enzymes

Vitamin B_{12} (cyanocobalamin) was long considered to be the only naturally occurring species with a covalently linked cobalt-carbon bond. The observation that raw liver cures pernicious anemia led Folkers and coworkers to extract and crystallize the active component in 1948 [34] and Dorothy Hodgkin determined its structure in 1956, a time in which structural determination of biomolecules using X-ray crystallography was in its infancy [35]. The discoveries that the biologically active form of vitamin B_{12}, B_{12} coenzyme (5′-deoxyadenosylcobalamin, AdoCob) and the corresponding methylcobalamin (methylCob), are all organometallic compounds containing covalently linked cobalt-carbon bonds, opened up the area that is now known as bioorganometallic chemistry.

Vitamin B_{12} is a cobalt-containing corrin-like cofactor similar to the nickel coenzyme F_{430}, in which the central metal atom is ligated by four nitrogen atoms from the tetrapyrrole ring (Figure 2). In B_{12}, the cobalt center also axially ligates a dimethylbenzimidazole ligand. Depending on the type of carbon ligand at the upper axial site, the cofactor can exist in different forms. Thus, vitamin B_{12}, AdoCob (also called coenzyme B_{12}), and methylCob contain cyano, 5′-deoxyadenosyl, and methyl ligands, respectively, at the upper axial site.

Cleavage of the Co-C bond could occur by homolytic or by two types of heterolytic mechanisms (Figure 3). The homolytic and heterolytic metal-carbon bond cleavage reactions in the enzymatic mechanisms of AdoCob- and methylCob-dependent enzymes [36], respectively, will be briefly described as a prelude to the discussion of F_{430}-based enzymology because the B_{12}-dependent reactions provide well-characterized frameworks on which the F_{430} mechanisms are partly based.

The AdoCob-dependent enzymes catalyze 1,2-rearrangements in which substrate is converted to product via replacement of a hydrogen atom on one carbon with a substituent on an adjacent saturated carbon (Figure 4). The

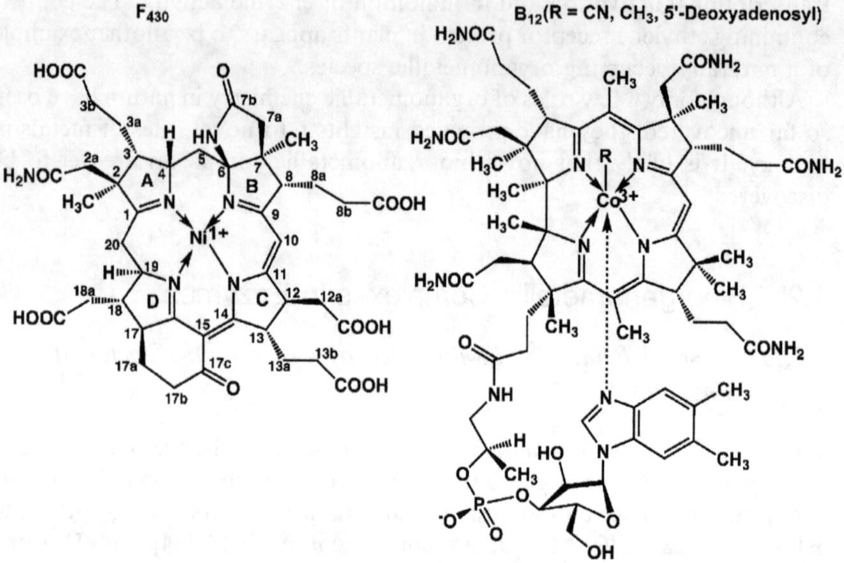

Figure 2. Structures of F_{430} and B_{12}.

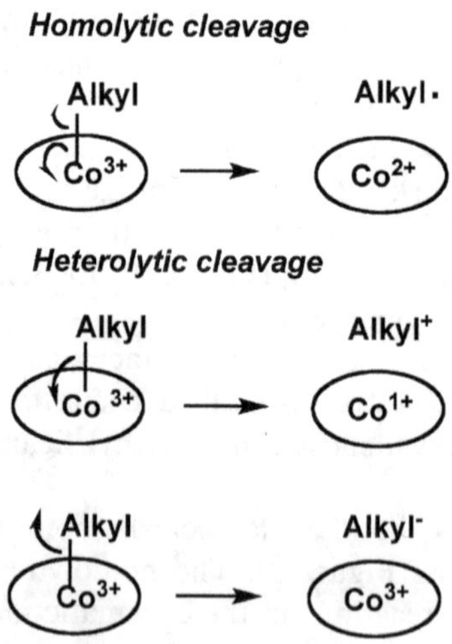

Figure 3. Mechanisms of Co-C bond cleavage.

key step in the overall reaction is the enzyme-induced homolytic cleavage of the cobalt-carbon bond leading to the formation of a 5′-deoxyadenosyl (dAdo) radical and the cob(II)alamin cofactor. The bond dissociation energy for homolytic cleavage of the cobalt-carbon bond of AdoCob is

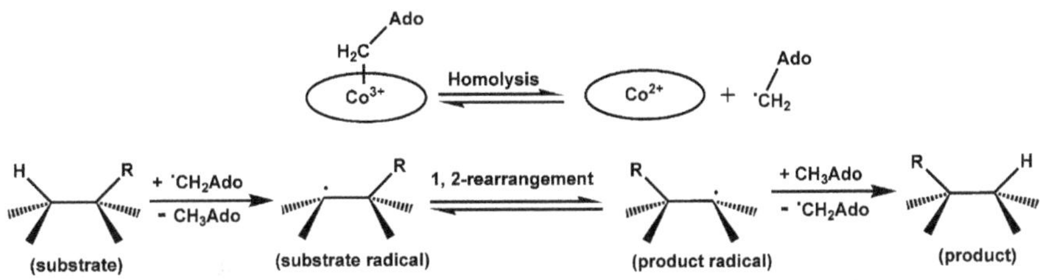

Figure 4. Coenzyme B_{12}-dependent 1,2-rearrangement.

$\sim 30\,\text{kcal}\,M^{-1}$. The dAdo carbon radical propagates to the substrate by abstracting a hydrogen atom to form the substrate radical and deoxyadenosine. This radical then undergoes a 1,2-rearrangement or isomerization forming the corresponding product radical, which subsequently reabstracts a hydrogen atom from 5′-deoxyadenosine to form product and regenerate the dAdo radical, which can undergo another round of catalysis or recombine with Co(II). B_{12}-dependent isomerases that follow this general scheme include mutases, e.g., lysine amino mutase and methylmalonyl-CoA mutase, and dehydratases, e.g., glycerol dehydratase and ethanolamine ammonia lyase. The B_{12}-dependent class II ribonucleotide reductases follow a variation of this mechanism in which homolysis of the cobalt-carbon bond is coupled to a hydrogen atom abstraction from a cysteine residue of the protein, and the resulting Cys radical propagates through the protein to finally abstract a hydrogen atom from substrate ribonucleotide, initiating the reduction of C-2 of ribose to deoxyribose [36].

The methylCob-dependent reactions, on the other hand, involve *heterolysis* of the cobalt-carbon (i.e., methyl) bond [36,37] followed by transfer of the methyl group as a carbocation. The methyl transfer reaction has been proposed to take place via two sequential S_N2 reactions. In the first step, the methyl group is first transferred from methyl tetrahydrofolate to an activated cob(I)alamin center, generating methylCob(III)alamin and tetrahydrofolate. In the second step, the methyl group of methylCob(III)alamin is transferred to homocysteine to yield methionine. The key steps in the methyltransferase mechanism include: (i) substrate binding and activation of the methyl group to enhance its reactivity toward nucleophilic attack; (ii) nucleophilic attack of Co(I) on the methyl group to generate methylCob(III); and (iii) methyl group transfer to the methyl group acceptor, forming product, which is then released. Due to its high reactivity, Cob(I)alamin has been described as a "supernucleophile".

Enzymes that undergo B_{12}-dependent methyl transfer include methionine synthase, and the anaerobic methyltransferases in methanogenic archaea and acetogenic bacteria that play an important role in making cell carbon

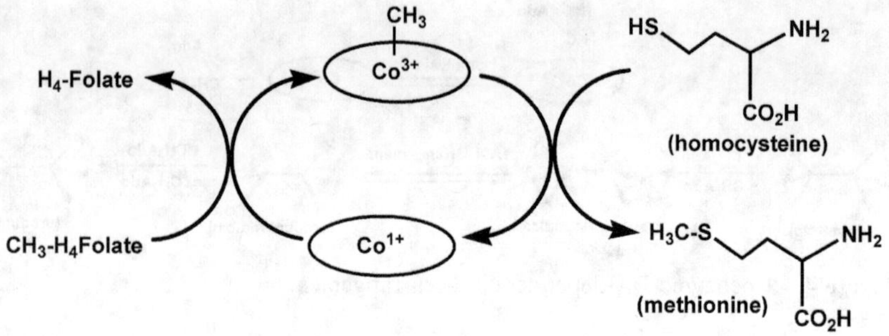

Figure 5. B_{12}-dependent methyl transferase reaction in methionine synthase.

[36]. A classic example of methyl transferase reaction involves methionine synthase, where the methylCob cofactor serves as an intermediate and catalyzes the transfer of the methyl cation from methyltetrahydrofolate (CH_3-H_4folate) to homocysteine to form methionine and tetrahydrofolate described in Figure 5 [38].

One can consider that the organometallic methylCob species is formed through an S_N2 mechanism, an oxidative addition mechanism, or an electron transfer mechanism [39]. In the S_N2 mechanism, the methyl group being transferred is partially bonded both to the incoming nucleophile (Co(I)) and to the departing leaving group (N5 of CH_3-H_4folate). In the mechanism involving oxidative addition, the cobalamin is proposed to form a three-centered bond with the CH_3-N5 moiety of CH_3-H_4folate. The distinction between S_N2 and oxidative addition mechanisms is the relative orientation of cobalamin versus the CH_3-N5 bond of CH_3-H_4folate. The oxidative addition mechanism requires that the C-N bond to be cleaved be parallel to the plane of the corrin ring. Thus, high-resolution structures of the methylCob-dependent metalloenzymes, especially bound to transition state inhibitors, may distinguish between these two mechanisms. In the proposed single electron transfer mechanism, one electron is transferred from Co(I) to CH_3-H_4folate to activate the methyl group (Figure 6).

What is the origin of the catalytic power of enzyme to form and cleave the organometallic bond? The rate of the Co-C bond cleavage is enhanced 10^9- to nearly 10^{14}-fold by AdoCob-dependent enzymes, relative to the rate of the uncatalyzed reaction [40,41]. In AdoCob-dependent enzymes, the homolysis of the Co-C bond of AdoCob to Co(II) and an Ado radical is a simple bond dissociation reaction with the same free energy of activation as the bond dissociation energy ($\sim 30\,\text{kcal}\,\text{mol}^{-1}$), and the reaction coordinate diagram is simply the portion of the Morse potential curve that raises with increasing distance. Since this reaction coordinate diagram has no maximum, there is no transition state, and the reaction can only be catalyzed by destabilizing

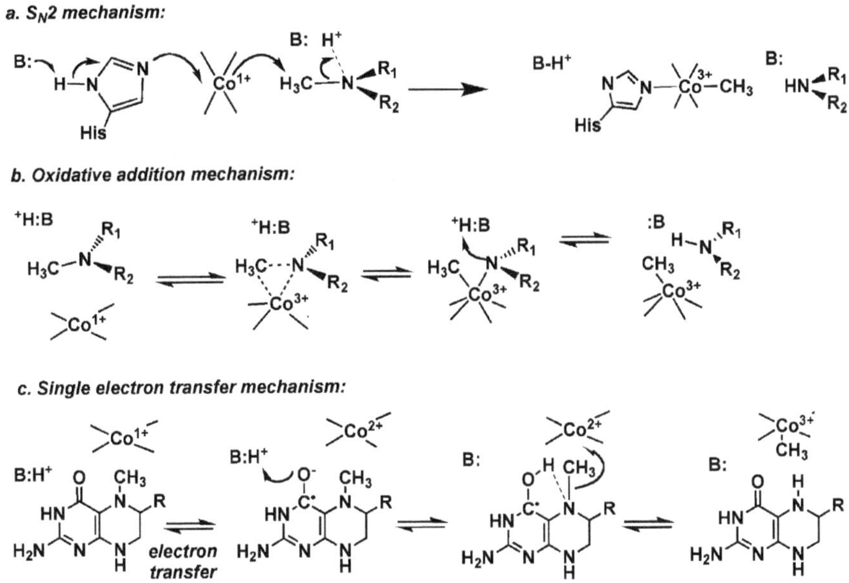

Figure 6. General mechanisms illustrating the formation and cleavage of organometallic species using B_{12} as an example. This figure is revised from [140].

the reactant, by stabilizing the products of the Co-C bond homolysis, or by a combination of these two effects. One possible explanation for the large rate enhancement is offered by the strain hypothesis [42], where it is assumed that the enzyme destabilizes the ground state of the reacting system and thus reduces the activation barrier for the chemical step. This catalytic effect has been attributed to reactant state destabilization (RSD) and, in particular to the distortion of the corrin ring in the mechanochemical trigger mechanism [42]. This could involve an "upward" fold of the corrin to sterically accelerate Co-C bond cleavage. Another possibility is that manipulation of the axial Co-N bond by the enzyme could stabilize the cob(II)alamin product state. However, theoretical and spectroscopic studies have indicated that the strain hypothesis is not justified [43]. Kinetic studies show that the entropy of AdoCob activation by AdoCob-dependent ribonucleotide reductase from *Lactobacillus leichmannii* is essentially the same as that for the non-enzymatic thermal homolysis of AdoCob, but the enthalpy of activation is 13 kcal mol^{-1} lower. Thus, in this case, catalysis of Co-C bond cleavage appears to be entirely enthalpic [44]. Theoretical studies by Warshel's group indicate that the electrostatic interaction between the ribose and the protein are responsible for the major catalytic contribution [45].

Thus, B_{12}-dependent enzymes provide classic examples of interfacing organometallic chemistry and biology as well as serving as paradigms that will be referred to in discussions of other organometallic reactions.

1.2.2. Organometallic Complexes in Carbon Monoxide Dehydrogenase and Acetyl-Coenzyme A Synthase

Before discussing MCR and coenzyme F_{430}, we will briefly discuss the bioorganometallic chemistry involving carbon monoxide dehydrogenase (CODH)/acetyl coenzyme A synthase (ACS), hydrogenase, and a Cu ethylene-sensing enzyme. CODH catalyzes the reversible reduction of atmospheric CO_2 to CO and ACS catalyzes the synthesis of acetyl coenzyme A from CO, the methyl group from methylCob (bound to a corrinoid iron-sulfur protein), and the thiolate from coenzyme A [46]. CODH can occur as a monofunctional enzyme or in association with ACS as a bifunctional CODH/ACS machine, which is central to the Wood-Ljungdahl pathway of anaerobic CO_2 fixation, a major component of the global carbon cycle that is found in various anaerobic microbes, including methanogens and acetogens (Figure 7).

The active site of the anaerobic CODH has been shown to contain a NiFeS cluster, known as the C-cluster, where CO_2 reduction to CO takes place [47]. The C-cluster is a cuboidal $NiFe_3S_4$ cluster tethered to an additional iron exo to the cube, which is known as the unique iron (or as ferrous component 2) (Figure 8). Each metal of the cuboidal cluster is ligated by a cysteine residue and three bridging sulfides. The unique iron is also ligated by a histidine residue. CO_2 reduction (CO oxidation) occurs through Ni-CO

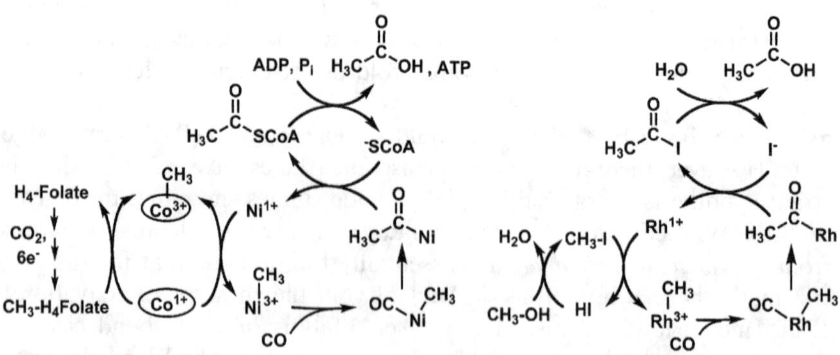

Figure 7. Left: Wood-Ljungdahl pathway for acetate synthesis; **right**: Monsanto industrial process for acetic acid synthesis.

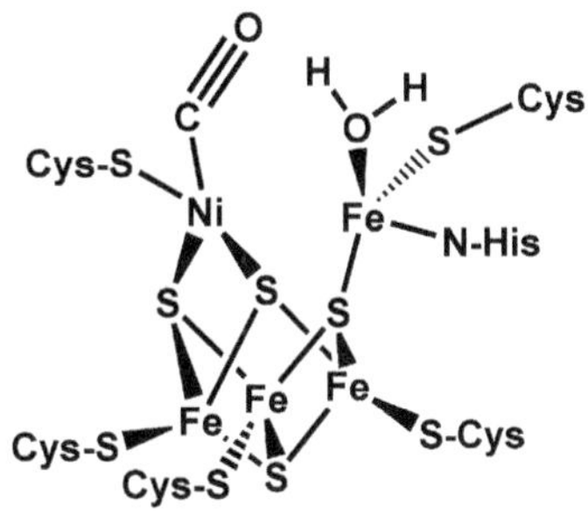

CODH$_{Mb}$

Figure 8. Active site of methanogenic CODH from *Methanosarcina barkeri* (CODH$_{Mb}$) based on work described in [48].

and Ni-COOH intermediates, both of which apparently have been trapped at the enzyme active site and observed in crystal structures [48,49].

ACS catalyzes acetyl coenzyme A synthesis at its active site A cluster, which consists of a [4Fe-4S] cluster that is bridged by a cysteine residue to a dinickel center containing the proximal nickel (Ni$_p$), which in turn is connected to a distal nickel (Ni$_d$) by two bridging cysteine sulfur atoms [47]. Ni$_p$ binds CO and the methyl group in random order [50], catalyzes C-C bond formation to form acetyl-Ni, then binds CoA, and catalyzes the thiolysis of the acetyl group to form acetyl-CoA. It is the proximal nickel site where CO is thought to bind after it travels through a gas channel from the C-cluster of CODH to the ACS active site. The mechanism of acetyl-CoA synthesis is still being debated. The reactions catalyzed by CODH-ACS in the Wood-Ljungdahl pathway were noted to exhibit similarities to those of the industrial Monsanto process for acetic acid synthesis (Figure 7), in that both involve metal-carbonyl, metal-methyl, and metal-acetyl bonds [32].

1.2.3. An Organometallic Active Site Containing Carbon Monoxide and Cyanide in Hydrogenases

[NiFe]-hydrogenases and [FeFe]-hydrogenases both require a CO and two CN$^-$ ligands bound to iron at their active site (Figure 9). The hydrogenases (or H$_2$ases) catalyze the reversible oxidation of molecular hydrogen into protons and electrons [33]. The active site of the [NiFe]-hydrogenase consists of a Ni subsite with two terminal cysteine ligands and two bridging cysteines to the Fe subsite [51], which contains the two cyanide and one CO ligand coordinated to the Fe center, as first identified by FTIR spectroscopy [52]. Using a combination of radioisotope labeling and mass spectrometry [53],

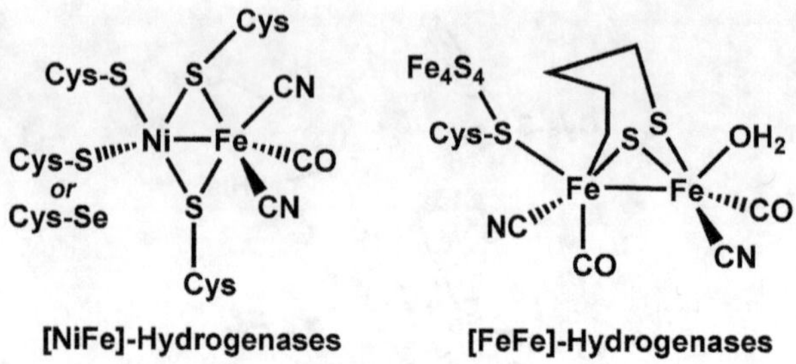

Figure 9. Active sites of [NiFe]- and [FeFe]-H_2ases.

Bock and coworkers demonstrated that the source of cyanide is an organic thiocyanate that is formed from carbamoyl phosphate by a several-step pathway. The [FeFe]-hydrogenase also contains CN^- and CO ligands. The function of the diatomic ligands is apparently to maintain the Fe centers in a low valent Fe^{2+} state. The catalytic mechanism [33] and the assembly of the metallocenters [54] of the hydrogenases have been recently reviewed.

Besides the organometallic complexes described above, CO and CN^- are known to bind and in some cases inactivate the metal centers at the active sites of various proteins, for example, hemoglobin and cytochrome oxidase. CO is also recognized to be a signal molecule that works by binding to metalloproteins, usually heme sites in various proteins, e.g., guanylate cyclase, and CooA, a transcriptional regulator that derepresses transcription of the CO oxidation system.

1.2.4. Formation of Organocopper Complexes in the Ethylene Receptor Protein

Similar to the gaseous signaling molecule CO that is sensed by heme-containing proteins in animals, nature has developed similar biosensors in plants. ETR1, an ethylene receptor in plants plays an important role in fruit ripening and influences growth and development. Theoretical studies in the 1960's indicating Cu(I) as a possible receptor in plants for ethylene [55,56] were followed two decades later by the characterization of the *Arabidopsis thaliana ETR1*, and demonstration that this protein requires copper ion for high-affinity ethylene binding [57]. Extended X-ray absorption fine structure (EXAFS) and resonance Raman characterization of sulfur-ligated Cu(I) ethylene complexes $[Cu([9]aneS_3)(C_2H_4)]^+$ and its CO analogue $[Cu([9]aneS_3)(CO)]^+$ provide evidence for a copper-carbon species that may resemble the proposed ethylene binding site in ETR1 (Figure 10) [58].

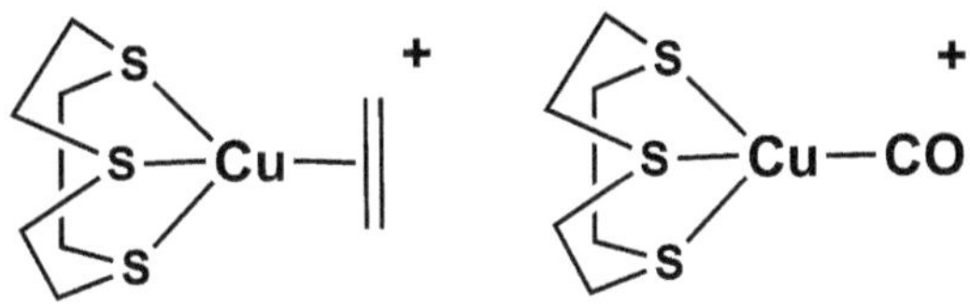

Figure 10. A bioorganometallic copper-carbon model complex of the proposed ethylene binding site of ETR1 Cu(I)-ethylene complex.

1.2.5. Bioorganometallic Chemistry and Methyl-Coenzyme M Reductase

A bioorganometallic Ni-CH_3 species has been invoked in the catalytic reactions involving both methane formation and anaerobic oxidation of methane [59]. The catalytic mechanism of methane synthesis by MCR is yet to be defined. However, two of the three published mechanisms propose methane formation by the intermediacy of an organometallic methylnickel species generated by the reduction of methyl-SCoM. Although a true methylnickel intermediate has thus far not been observed with the natural substrate methyl-SCoM, in recent studies bromo- and iodomethane has been shown to react with active MCR to generate a bioorganometallic methyl-Ni(III) species at the active site of MCR. The catalytic mechanism of MCR is the major subject of this chapter.

1.3. Detection and Characterization of Organometallic Species

Trapping and understanding of the organometallic species are important for unveiling the mystery of the enzymatic reactions mentioned above. Different spectroscopies [UV-visible, Fourier transform infrared (FTIR), electron paramagnetic resonance (EPR), electron nuclear double resonance (ENDOR), Mössbauer, and hyperfine sublevel correlation (HYSCORE), nuclear magnetic resonance (NMR) spectroscopy], theoretical computation, model complexes, and crystallography have been extensively used to detect and characterize these organometallic species. Direct evidence by EPR [60], Mössbauer [60], ENDOR [61], and FTIR [62] spectroscopies as well as indirect evidence from theoretical work [63], has led to the definition of the NiFeC site in ACS as $[Fe_4S_4]^{2+}$-Ni_p^{+}(CO)-Ni_d^{2+}. In the enzyme, only the NiFeC species has been directly observed, while evidence for the other intermediates (CH_3-Ni, and acetyl-Ni) in the catalytic cycle is indirect.

An important test of a mechanistic model is to use model complexes that serve as well-defined structural and functional mimics. For example, synthetic Ni complexes [64,65], after reductive activation, perform a nucleophilic attack on the methyl group carbon to form a methyl-Ni(II) species. A major value of the model complexes is that they can be studied by NMR spectroscopy and X-ray diffraction [66]; for example, NMR spectroscopy, particularly ^{2}H NMR, has shown to be a very useful technique for the characterization of a high-spin methyl-Ni(II) compound. The large high-field shift from the methyl group after *in situ* methylation of a derivative of F_{430} provides a direct proof for the presence of a carbon-nickel bond [66]. The characterization of a CH_3-F_{430}Ni(II), which was postulated as an intermediate in the formation of methane in the reaction of F_{430}Ni(I) and electrophilic methyl donors, provides indirect evidence for the methyl-Ni intermediate in the MCR reaction. Similarly, in the reaction of MCR with its activated substrate analogs, such as 3-bromopropanesulfonate (BPS) [67], brominated acid [68], and methyl iodide [69], a methyl-Ni intermediate has been characterized by UV-visible [67,68], EPR [67–69], and ENDOR, HYSCORE [69,70] spectroscopies. Besides these methods mentioned above, X-ray crystallography may eventually reveal the structure of an organometallic species at the heart of MCR.

Many methods have been developed to characterize the organometallic species, but so far few of the organometallic intermediates have been directly trapped and characterized in the catalytic reaction of enzyme. In order to unravel the mechanism of methanogenesis, the synergistic cooperation of biochemists, spectroscopists, crystallographers, and synthetic bioinorganic chemists is required.

2. A BRIEF INTRODUCTION TO METHANOGENESIS

2.1. The Impact of Methanogenesis on the Carbon Cycle, Energy, and the Environment

Before discussing coenzyme F_{430} and its role in the mechanism of methane formation, we will briefly describe the microbial basis of methanogenesis and its importance to energy and the environment. The first record of the observation of methanogenesis has been colorfully related by Wolfe [71]. Beginning in 1776, a series of letters between Father Carlo Campi and the Italian physicist Alessandro Volta described observations and experiments on the "combustible air" from marshy soil. Almost a century later, Bechamp provided the first evidence that methane can be formed by a microbial process [72]. In 1906, N. L. Sohngen demonstrated the natural cyclical process of microbial methane generation and its utilization as an energy and carbon source and, in 1910, put forth the equation for methane formation (eq 1) [73].

$$4H_2 + CO_2 \rightarrow CH_4 + 2H_2O \tag{1}$$

In 1933, Stephenson and Stickland inaugurated the modern era of methanogenesis with the isolation of the first pure methanogenic culture and by reporting the first examination of a methanogenic enzyme [74]. The pioneering of anaerobic aseptic techniques by Hungate [141] accelerated the pace of studies of the microbiology of methanogens and enabled mass culturing, which permitted initial biochemical studies.

It is now recognized that methanogens are obligate anaerobes that are responsible for all biological methane production on earth [75], synthesizing globally $\sim 10^9$ tons of methane per year [73]. Methanogens also have an evolutionary history of at least 3 billion years and have been classified within the third domain of life, as the founding members of the domain Archaea (from greek; ancient, primitive) [76,77]. Methanogens are widely distributed in anaerobic environments, including aquatic sediments (ponds, marshes, swamps, rice soils, lakes, and oceans), the intestinal tract of animals (including the intestines of humans and the rumen of herbivores), sewage digesters, landfills, heart wood of living trees, decomposing algal mats, oil wells, and mild-ocean ridges. Some methanogens are extremophiles, found in environments such as hot springs and submarine hydrothermal vents as well as in the "solid" rock of the earth's crust, kilometers below the surface [78].

Methanogenesis is the final step of energy conservation in methanogens and plays an important role in biomass biodegradation. In the carbon cycle (Figure 11), fermentative bacteria degrade natural polymers to H_2, CO_2, formate, and acetate (Figure 11, step 3). These one- and two-carbon compounds are then converted by methanogens to CH_4 (step 4). Methanogenesis has important beneficial effects on the global carbon cycle by depleting H_2 that is generated in anaerobic environments and inhibits the natural biodegradation of organic compounds (step 3). Some of the methane diffuses into the aerobic environment (step 5) to undergo oxidation to CO_2 by aerobic methanotrophic bacteria (step 6), while part of the methane undergoes anaerobic oxidation by a process called reverse methanogesis or anaerobic oxidation of methane (AOM) (step 7). AOM in marine sediments consumes more than 70 billion kilogram of methane annually [79] and is performed by microbial consortia, largely composed of archaea and sulfate-reducing bacteria (SRB) [80], which can couple methane oxidation to sulfate reduction. Environmental genomic analyses indicate that several different methanogen-related archaeal groups are involved in AOM and two groups

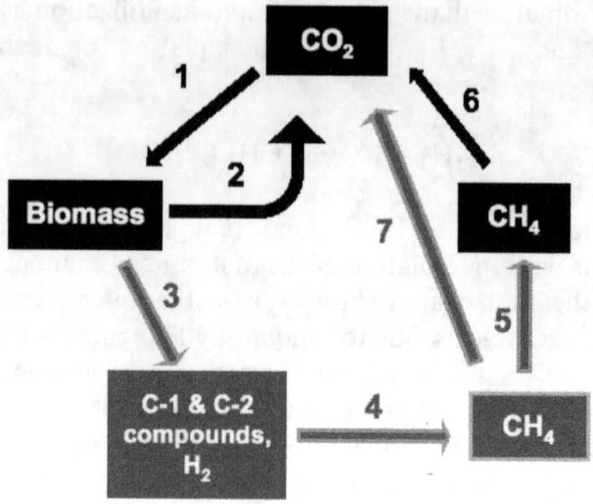

Figure 11. The global carbon cycle. Step **1** carbon dioxide fixation; **2**, aerobic degradation of biomass; **3**, anaerobic fermentation; **4**, methanogenesis by methanogens; **5**, diffusion of methane from anaerobic to aerobic environment; **6**, aerobic oxidation of methane; **7**, anaerobic oxidation of methane (reverse methanogenesis). The black and grey backgrounds indicate aerobic and anaerobic environments, respectively.

of putative anaerobic methane-oxidizing archaea (ANME-1 and ANME-2) and several SRB groups typically occur together in methane-rich marine sediments [79]. A MCR-like Ni-protein has been retrieved from habitats where methane-oxidizing microbial communities are abundant [81,82].

Because the sources and sinks of methane do not match, an increasing amount of methane has been escaping into the atmosphere, which is a source of concern because methane is a potent greenhouse gas that is 21 times more effective at trapping heat in the atmosphere than carbon dioxide [83]. Over the past two centuries, the atmospheric methane concentration has been increasing by about 1 ppb each year, and has more than doubled over the past two centuries, now accounting for 16% of global greenhouse gas emissions from human activities [84].

Besides being a greenhouse gas, methane is also a primary constituent of natural gas and an important energy source [85,86]. Approximately 22 percent of the energy consumption of the U.S. comes from natural gas, with slightly more than half of homes using natural gas as their heating fuel. Methane is considered a clean fuel because it emits less sulfur, carbon, and nitrogen than coal or oil, and leaves little ash. Thus, the U.S. government launched a methane-to-market partnership in November 2008 to promote the capture and use of methane as a clean energy source. Currently including 21 national governments and more than 200 organizations, the partnership has a goal of reducing annual methane emissions by 2015 by an amount equivalent to removing 33 million cars from the roadways for one year.

2.2. Methane on Mars and Titan

Living systems produce more than 90% of the earth's atmospheric methane [87] with the balance being generated by geochemical reactions. Recently, methane has been detected from Mars and Titan [88–90] and there is evidence that the methane is being continually produced [87]. The methane is of course a biomarker and could originate from living organisms on Mars and Titan, but the methane could also be abiotically produced. Either explanation would be fascinating in its own way, revealing either that life exists elsewhere in the universe or that both Mars and Titan harbor large underground bodies of water together with unexpected levels of geochemical/biological activity.

3. GENERAL PROPERTIES OF METHYL-COENZYME M REDUCTASE AND COENZYME F_{430}

3.1. Discovery of Methyl-Coenzyme M Reductase and Its Cofactor, Coenzyme F_{430}

In 1965, Bartha and Ordal first demonstrated a bacterial growth requirement for nickel when characterizing two strains of hydrogen-oxidizing bacteria [91]. This observation altered the long accepted concept that nickel is toxic/carcinogenic. Since then, eight nickel enzymes have been discovered and characterized: urease, hydrogenase, carbon monoxide dehydrogenase, acetyl-coenzyme A synthase, methyl-coenzyme M reductase, Ni-superoxidase, Ni-dependent glyoxylase, and *cis-trans* isomerase [92].

The first reported observation of F_{430} was in 1977 when LeGall discovered a non-fluorescent, yellow compound in cell extracts of *Methanothermobacter thermautotrophicus* ΔH (*M. thermautotrophicus* ΔH) and reported this finding to Wolfe [93]. F_{430} was named so due to its strong absorbance at 430 nm. At the time of its discovery, the significance of F_{430} was not known because adding the free cofactor to cell extracts neither inhibited nor stimulated methanogenesis [93]. Later, Wolfe and Thauer and their coworkers demonstrated that F_{430} binds nickel in a 1:1 (mol:mol) stoichiometry [94,95]. At about the same time Thauer's group also demonstrated that radiolabeled δ-[4-^{14}C] 5-aminolevulinic acid is incorporated into F_{430}, which provided evidence that F_{430} is a tetrapyrrolic compound [96]. Extensive ^{13}C and ^{1}H

NMR studies were performed to solve the structure of F_{430}, thereby confirming that it is indeed a tetrapyrrole coenzyme (Figure 2) [97]. It is the most reduced tetrapyrrole in nature, consisting of only five double bonds in the macrocycle, four of which are conjugated and one is isolated [73,75,98]. F_{430} is the first biologically occurring nickel tetrapyrrole described and appears to be unique to methanogens and methanotrophs [73].

3.2. Methyl-Coenzyme M Reductase Reaction and Structure

MCR is an essential and abundant protein (about 10% of the total protein) in all methanogenic archea, since it catalyzes the last step (eq 1) in methanogenesis, the process by which methanogens conserve energy. The MCR-catalyzed reaction has been reviewed [99] and involves the conversion of methyl-coenzyme M (CH_3-SCoM) and N-7-mercaptoheptanoylthreonine phosphate (CoBSH) to methane plus the mixed disulfide, CoBS-SCoM (eq 2), which is subsequently reduced by heterodisulfide reductase in an energy-generating step [100]. In the MCR-catalyzed reaction, the conversion of CoBSH to CoBS-SCoM yields two electrons that contribute to the reduction of methyl-SCoM to methane. As mentioned above, MCR also appears to catalyze the first step in AOM (reverse methanogenesis).

$$CH_3\text{-SCoM} + \text{CoB-SH} \rightarrow CH_4 + \text{CoBS-SCoM} \quad (2)$$

MCR catalysis requires the F_{430} cofactor. Based on the X-ray crystal structures of three EPR-silent and inactive Ni(II) states of this enzyme (MCR-silent, MCR_{ox1}-silent and MCR_{red1}-silent), F_{430} is tightly bound and deeply buried at the bottom of a 30 Å channel that connects to the surface [101–103]. This channel is sufficiently deep to accommodate the two substrates and apparently shields the reaction from solvent. The phosphate group of CoBSH binds at the upper lip of the well with its thiol group located 6–8.2 Å from the central Ni atom of F_{430} depending on the state of the enzyme (see below), as observed in the different crystal structures. The Ni atom coordinates the four planar tetrapyrrole nitrogens and a lower axial oxygen ligand contributed by the carbonyl oxygen of the side chain of Gln-α′147. In the Ni(II)-silent form of MCR, the upper axial nickel ligand is the sulfonate oxygen of CoBS-SCoM; whereas, in the Ni $(II)_{ox1}$-silent form, this site is occupied by the thiol(ate) group of CoM-S(H) (see Figure 12 in Section 3.3). A five-coordinate form of Ni(II)-MCR_{red1}-silent, lacking an upper axial ligand, has also been observed in the crystal structure.

All the structures show two equal independent active sites located 50 Å apart.

3.3. The Oxidation and Coordination States of Methyl-Coenzyme M Reductase

MCR can exist in several nickel oxidation and coordination states (Figure 12). The active Ni(I) state of MCR, called MCR_{red1} [103–105] is green (λ_{max} ~ 390 nm) and paramagnetic, exhibiting EPR spectra with g-values at 2.25, 2.07 and 2.06, which is typical of an approximately square planar Ni(I) system with an unpaired electron in the $d_{x^2-y^2}$ orbital [106,107]. The MCR_{red1} state is five-coordinate leaving an open upper axial coordination site available for interaction with CH_3-SCoM [108]. The MCR_{red1} state can be generated *in vivo* by bubbling cells with 100% H_2 for 30 min before harvesting [109]. Under these conditions, there is also an increase in the MCR_{red2} form, in which the Ni(I) center coordinates with the sulfur of the $^-SCH_2CH_2SO_3^{2-}$ ligand and one of the tetrapyrrole nitrogens is protonated. The MCR_{red2} form can be induced by incubating MCR_{red1} with HSCoM and CoBSH *in vitro* [110]. Because of

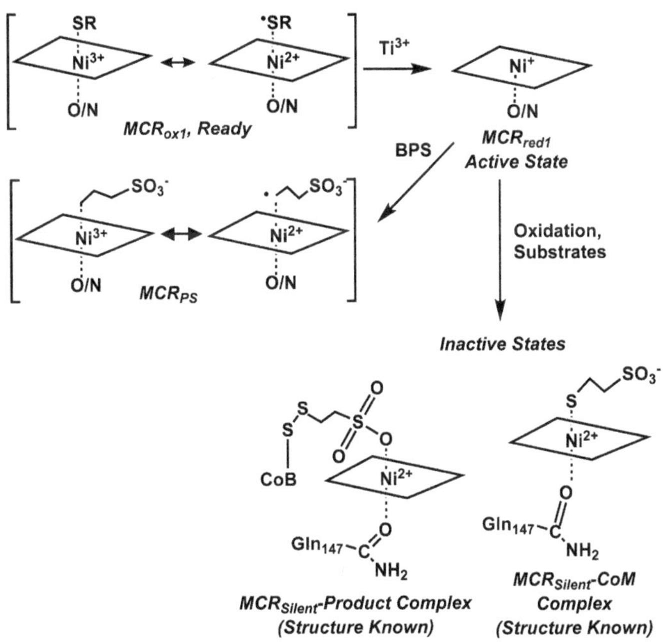

Figure 12. Various states of MCR based on work described in [139].

the low redox potential of the Ni(II)/(I) couple, great care must be taken to isolate and maintain the enzyme in the Ni(I) oxidation state; otherwise, it undergoes oxidative inactivation to Ni(II) (MCR_{ox1}-silent, see below), turning bright yellow (λ_{max} ~ 420 nm).

MCR_{ox1} is assigned as a high spin-Ni(II) coupled to a thiyl radical (Figure 12) based on an array of spectroscopic (XAS), UV-visible, EPR, pulsed-EPR (ENDOR and HYSCORE), MCD, and computational methods (TD-DFT). The catalytic inactive MCR_{ox1} state is relatively stable in the presence of oxygen and has been called the "ready" state because it can be converted *in vitro* to active MCR_{red1} [105] by incubation with the strong reductant, titanium(III) citrate [103]. MCR_{ox1} can be formed *in vivo* by switching the gas before harvesting from 80% H_2/20% CO_2 to 80% N_2/20% CO_2 [109] or by treating the growing cells with sodium sulfide just before harvest [111]. MCR_{red2} can also be converted into MCR_{ox1} by oxidation with polysulfide [105,112].

The MCR_{PS} (called MCR_{BPS} earlier) state is formed when MCR_{red1} reacts with the potent inhibitor, bromopropanesulfonate (BPS) [105]. This state is characterized by UV-visible spectra that are very similar to the Ni(II) protein, yet it has an EPR spectrum with g-values of 2.223, 2.115 (Figure 12). EPR, ENDOR, and HYSCORE spectroscopic studies have determined the electronic structure of the active site Ni center to be formally Ni(III) with a covalent methyl-Ni bond [69,70].

Recently, a Ni(III)-F_{430} hydride complex was detected by continuous wave and pulse EPR spectroscopy when mixing MCR_{red1} with HSCoM, CoBSH, or its analogue CH_3-SCoB, which has been shown to activate methane [113]. This Ni(III)-F_{430} hydride complex supports the involvement of MCR in reverse methanogenesis.

3.4. Activation of Methyl-Coenzyme M Reductase

It has been hypothesized that the activation of MCR involves a one-electron reduction of the Ni from the 2+ to the 1+ state, as well as a two-electron reduction of the tetrahydrocorphinoid ring system based on the marked shifts in the UV-visible and Raman spectra associated with the formation of MCR_{red1} [114]. However, electrochemical studies [115] followed by a variety of spectroscopic and computational results showed that reduction of F_{430} with Ti(III) citrate reduces Ni(II) to Ni(I), but the tetrapyrrole ring is intact [116]. A novel form of the coenzyme, called F_{330} is generated by reducing F_{430} with sodium borohydride ($NaBH_4$) and is named so because it exhibits a prominent absorption peak at 330 nm [116]. Spectroscopic (mass spectrometric, one- and two-dimensional NMR, resonance Raman, X-ray absorption, and magnetic circular dichroism) and computational studies revealed that F_{330} contains a low-spin Ni(II) and a C=O double bond reduction on the macrocycle (the carbonyl group at carbon 17c undergoes reduction to an alcohol) [116].

3.5. Proposed Mechanisms for Methane Formation

On the basis of kinetic, structural, and spectroscopic studies and computational analysis of the enzyme in its various states, insight into the enzyme mechanism is beginning to emerge. Two general mechanisms have been considered for the MCR-catalyzed reaction: Mechanism I involving an organometallic methyl-Ni(III) intermediate and mechanism II involving a methyl radical.

As shown in Figure 13, mechanism I is initiated with a nucleophilic attack by the Ni(I) center of MCR_{red1} on the methyl group of the methyl-SCoM forming a methyl-Ni(III) intermediate. The proton of CoBSH is transferred to the resulting $CoMS^-$ anion, resulting in the formation of HSCoM and a $CoBS^-$ anion [103]. In the subsequent step, HSCoM transfers an electron to the methyl-Ni(III) intermediate, forming methyl-Ni(II) and a thiyl radical on HSCoM. The methyl-Ni(II) species undergoes protolysis to form methane, then the CoM radical reacts with $CoBS^-$ forming the heterodisulfide (CoBS-SCoM)$^{\bullet -}$ radical anion. The heterodisulfide radical anion is highly reducing and transfers an electron to the Ni(II) to regenerate active Ni(I)-MCR_{red1} and the heterodisulfide product, CoBS-SCoM.

Although a true methyl-Ni intermediate has not been identified upon reaction of MCR_{red1} with the native substrate, methyl-SCoM, the relative positions of CoM, CoB, and F_{430} in the crystal structures is consistent with a nucleophilic attack of Ni(I) on CH_3-SCoM and formation of a Ni(III)-CH_3 intermediate. In addition, alkyl-Ni intermediates, formed by reaction of MCR_{red1} with BPS, have been characterized as a high-spin Ni(II)/alkyl

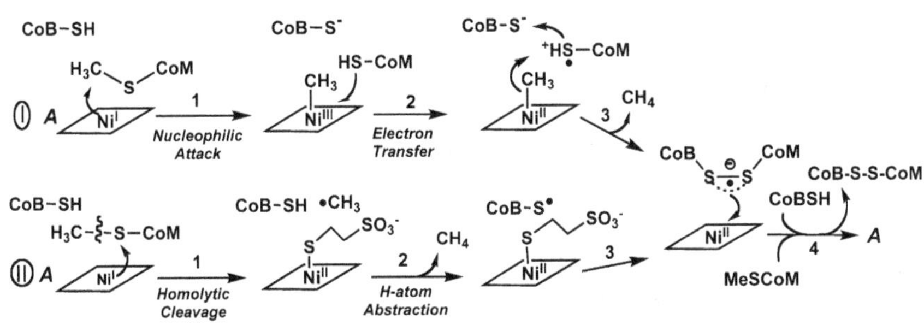

Figure 13. Proposed mechanisms of the MCR-catalyzed methane formation reaction.

radical species. This intermediate undergoes protonation to form the corresponding alkane or to react with various thiol groups (including CoM) to form the methylthioether (mimicking the reverse of the first step in methane formation or the final step in methane oxidation). Furthermore, a methyl-Ni(II) intermediate has been shown in the reduction of activated methyl sulfonium to methane by free reduced F_{430} pentamethyl ester [117,118], as described below.

Mechanism II, which is based on density functional theory computations by Siegbahn and Crabtree [119–121], avoids the methyl-Ni(III) species because cleavage of the strong methyl-S bond of methyl-SCoM to form a relatively weak methyl-Ni(III) species was determined to be extremely endothermic (45 kcal/mol). Therefore, mechanism II proposes attack of Ni(I) on the sulfur atom adjacent to the methyl group of methyl-SCoM, resulting in homolytic cleavage of the methyl-sulfur bond to generate a methyl radical and a Ni(III)-thiolate ↔ Ni(II)-thiol radical complex (MCR_{ox1}-like species) (Figure 13). The methyl radical then abstracts a hydrogen atom from CoBSH to generate methane and a CoBS radical. In the subsequent step, the CoBS radical reacts with bound CoM to generate a disulfide radical anion, which reduces Ni(II) to Ni(I) and forms the heterodisulfide product similar to that in mechanism I. An argument against mechanism II is that inversion of stereoconfiguration (as observed in the case of ethyl-coenzyme M) would require hydrogen abstraction by the intermediate methyl radical before it has time to rotate inside the active site.

Recently a new mechanism, which is also based on DFT calculations (Figure 14) has been proposed [113]. This catalytic cycle starts with the protonation of MCR, either on the Ni center or on the C-ring nitrogen of the corphin, followed by oxidative addition of CH_3-SCoM. The coordination around the center is substantially distorted, and the Ni adopts a position above the four nitrogen atoms of the corphin ring. The sulfur of the deprotonated CoBSH ($SCoB^-$) then interacts with the sulfur of the SCoM ligand and elimination of CH_3-S-SCoM, leaves a CH_3-substituted Ni.

4. ORGANONICKEL INTERMEDIATES ON METHYL-COENZYME M REDUCTASE

4.1. Alkylnickel Model Complexes Related to Coenzyme F_{430} and Their Reactions: Protonolysis, Thiolysis, Hydride Transfer

Two of the three proposed mechanisms of methane formation by MCR suggest the intermediacy of a methylnickel species generated by the

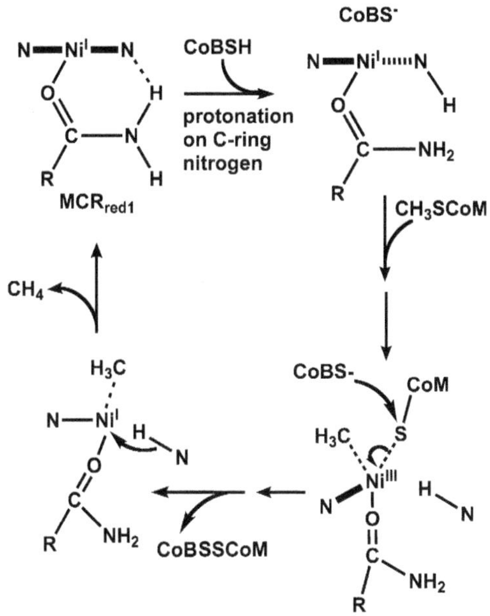

Figure 14. Proposed mechanism based on DFT computations based on work described in [121].

reduction of methyl-SCoM [103,117,122,123]. In order to gain insight into biological methane formation, most of the early mechanistic studies were performed using the pentamethyl ester of the free coenzyme F_{430} (F_{430M}) [118,122,124,125]. F_{430M} was used instead of F_{430} due to its higher stability, easier purification, and solubility in aprotic solvents compared to the pentaacid precursor, F_{430}. The first definitive evidence that F_{430} could undergo redox changes was provided by Jaun and Pfaltz with F_{430M}. The Ni(II)-F_{430M} state was shown by UV-visible and EPR spectroscopy to be efficiently reduced with sodium amalgam in THF to generate Ni(I)-F_{430M} [125], which is analogous to the catalytically active form of F_{430} in MCR.

In a seminal study, Jaun and Pfaltz investigated the reactivity of Ni(I)-F_{430M} towards compounds containing an activated methyl group bound to halogen, oxygen, or sulfur, and demonstrated methane formation from methyl iodide, methyl tosylate, and methyl sulfonium salts [122]. When Ni(I)-F_{430M} was used as a catalyst, methane formation from methyl tosylate was much slower than from methyl iodide. This finding was interesting because it demonstrated that reduction of iodomethane to methane proceeds via a methyl-Ni(II) (methyl-Ni(II)F_{430M}) intermediate.

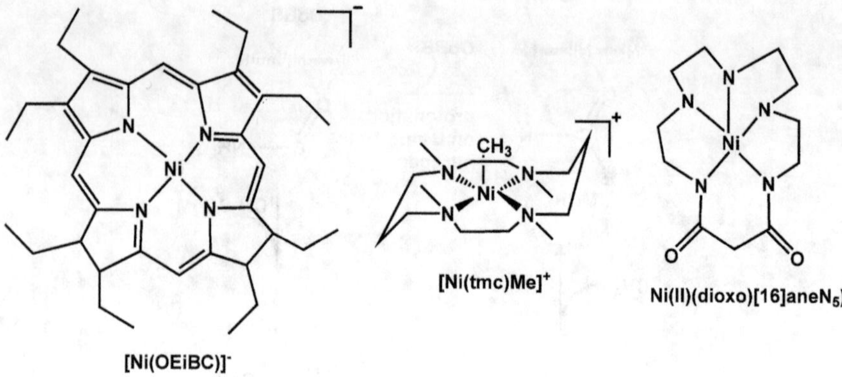

Figure 15. Model complexes of coenzyme F_{430}: **left**, Ni-octaethylisobacteriochlorin {[Ni(OEiBC)]$^-$}; **center**, [Ni(tmc)Me]$^+$; **right**, Ni(II) complex of 1,4,7,10,13-pentaazacyclohexadecane-14,16-dionate.

Concurrently, Stolzenberg and Stershic studied the reactivity of a nickel tetrapyrrole of octaethylisobacteriochlorin, Ni(I)-OEiBC (Figure 15, left), and demonstrated methane formation from its reaction with methyl iodide and methyl *p*-toluenesulfonate [126,127]. This study also provided evidence for a transient alkyl-Ni(III) intermediate that undergoes reduction to Ni(II) and protonolysis to yield methane and iodide. Subsequent studies of Ni(I)-OEiBC demonstrated that several alkyl halides react very rapidly with the Ni(I) center by an S_N2 reaction, leading to cleavage of the carbon-halogen bond and thus forming the alkylnickel complexes [128]. The carbon-nickel bond in the R-Ni^{II}(OEiBC)$^-$ complex can be cleaved through protonation, alkylation, and internal proton transfer if the alkyl group has a β-hydrogen, which could undergo β-hydride elimination as shown in equations (3) and (4) [129].

$$\text{R-Ni}^{\text{II}}(\text{OEiBC})^- + \text{RX} \rightarrow \text{Ni}^{\text{II}}(\text{OEiBC})^- + \text{R-R} \tag{3}$$

$$\text{R-Ni}^{\text{III}}(\text{OEiBC})^- \xrightarrow{e^-} \text{H-Ni}^{\text{III}}(\text{OEiBC})^- + \text{R}_{-\text{H}} \tag{4}$$

These reactions suggested reactivity of coenzyme F_{430} in MCR in reductive dehalogenation of a broad range of substrates, as discussed in a later section.

Because little is known about the binding and cleavage of methyl-SCoM at the enzyme active site, synthetic nickel macrocyclic complexes have been developed to gain insight into thioether ligation to the nickel center. A thioether binding to nickel in the +1 oxidation state is unprecedented and very few reports exist for thioether binding to Ni(II). A nickel complex showing methyl-SCoM binding was isolated by Riordan et al. using $Ni(tmc)^{2+}$ (tmc = 1,4,8,11-tetramethyl-1,4,8,11-tetraazacyclotetradecane) (Figure 15, center). Interestingly, NMR and IR characterization of the resulting complex reveals binding of a sulfonate oxygen of methyl-SCoM rather than the anticipated thioether ligation [130]. However, further reductivity of this methyl-SCoM bound-Ni(II) complex and the feasibility in liberating methane is not known.

Drain, Sable, and Corden demonstrated the unusual reactivity of a synthetic nickel(II) complex, [1,4,7,10,13-pentaazacyclohexadecane-14,16-dionato(2^-)]Ni(II), toward methyl-SCoM in water liberating methane and CoBS-SCoM disulfide [131,132]. The reaction is catalytic in the presence of oxidants such as I_2 and $NaClO_4$. The proposed mechanism includes thioether ligation to Ni and the oxidation of Ni(III) coupled with methane formation to generate a Ni(III)-CoM thiolate species. This result is highly significant as it is the only nickel complex reported to uniquely activate methyl-SCoM.

While the natural substrate methyl-SCoM was unreactive to Ni(I)-F_{430M}, interestingly, the more electrophilic methyl-sulfur bond of dialkyl(methyl) sulfonium ion is cleaved by Ni(I)-F_{430M} to produce methane *via* a methyl-Ni(II) intermediate [122]. Therefore, methane formation from the reaction between highly activated electrophilic methyl donors and Ni(I)-F_{430M} is described in Figure 16.

The reductive cleavage of sulfonium ions catalyzed by Ni(I)-F_{430M} and formation of a potential methyl-Ni(II)F_{430} intermediate were confirmed by ^2H NMR experiments [117,118]. Using the isotopically labeled organometallic reagent $(CD_3)_2Mg$, a CD_3-Ni(II)-F_{430M} derivative was characterized by NMR, which was shown to undergo protonolysis to methane. The NMR spectrum of CD_3-Ni(II)-F_{430}M resembled that of [CH_3-Ni(II)(tmc)][CF_3SO_3], which was then the only other isolated organometallic methylnickel synthetic complex whose molecular structure was known [133] and served as a structural model for the methylnickel intermediate of MCR.

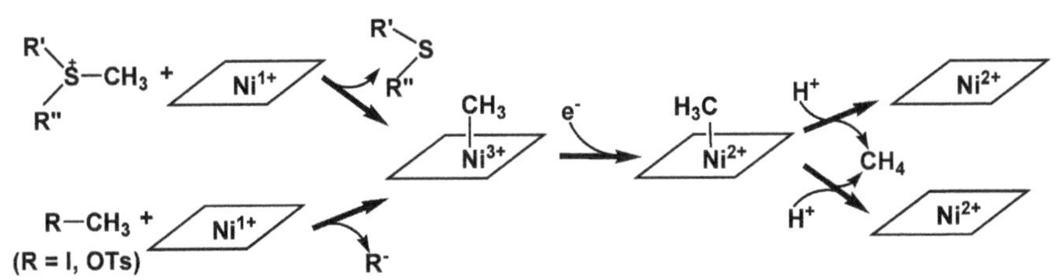

Figure 16. Methane formation from the reaction between Ni(I)-F_{430M} and activated methyl donors: methyl-sulfonium ions and iodomethane, Me-OTs.

4.2. Strategy for Trapping Intermediates at the Active Site of Methyl-Coenzyme M Reductase

No intermediate in the MCR-catalysed reaction have yet been trapped. To trap an intermediate (B, in Figure 17) in a reaction with the general scheme of eq (5), the ratio of k_1 to k_2, must be in the appropriate range to allow accumulation of a detectable amount of the intermediate. For example, if k_1 is much smaller than k_2, the intermediate can not be observed. In such a case, to solve this problem, one must either increase k_1 or decrease k_2, by perturbing the system, using substrate analogs and/or site directed mutagenesis. Of course, this strategy also works for more complicated reactions, like eq (6).

$$A \xrightarrow{k_1} B \xrightarrow{k_2} C \tag{5}$$

$$A \xrightarrow{k_1} B \xrightarrow{k_2} C \rightarrow\rightarrow\rightarrow D \tag{6}$$

This strategy was used in the study of the MCR mechanism. As described above, neither a methyl-Ni(III) intermediate nor a Ni(III)-SCoM species has been observed upon reaction of MCR_{red1} with the native substrate

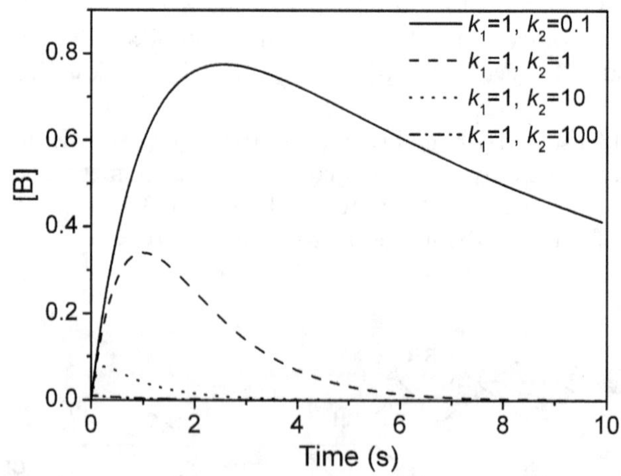

Figure 17. Kinetic control of experimental observation of reaction intermediates. The concentrations of the intermediates are a function of the relative values of k_1 and k_2. As the value of k_2 is increased from 0.1 to 100 s^{-1}, the maximum concentration of the intermediates during the reaction decreases from 78% to 1% of the total amount of initial substrate.

methyl-SCoM. In order to trap the intermediate, the strategy indicated above was used to react MCR with substrate analogs. To rapidly generate the alkyl-Ni species, we used highly activated methyl-SCoM analogs, methyl iodide, bromoalkyl sufonates and bromoalkyl carboxylates. Even in the absence of the second substrate, CoBSH, alkyl-Ni(III) species were obtained. To decrease k_2, CoBSH analogs with variations in the length of the carbon chain of CoBSH were used. A radical intermediate has been obtained in the reaction of MCR with methyl-SCoM and CoB_6SH (Dey et al., unpublished). The successful use of this strategy gives a new light into the mechanism of MCR.

4.3. Formation of Alkylnickel Intermediates at the Active Site of Methyl-Coenzyme M Reductase

The catalytic mechanism of MCR remains to be elucidated. Although MCR has wonderful spectroscopic handles for following redox changes at the active site, no spectral changes have been observed during catalysis apparently because the intermediates form and decay too rapidly to accumulate. Thus, we have resorted to the strategy of using different substrate analogs of methyl-SCoM and CoBSH to affect the elementary rate constants of the MCR mechanism and to trap and observe intermediates by various spectroscopic and kinetic methods. This work has led to the identification and characterization of different states of MCR [69,108,112,139], including several alkyl-Ni(III) species as well as organic radicals.

4.3.1. *Alkylnickel Species from Halogenated Alkyl Sulfonates and Alkyl Carboxylates*

Studies in the late 1980's using cell extracts of *M. marburgensis* demonstrated the potency of BPS to inhibit methanogenesis [134]. To date BPS remains the most potent inhibitor of methanogenesis with an apparent K_i of 50 nM. In 1992, Thauer and coworkers first observed that when active Ni(I)-MCR_{red1} was incubated with BPS, a unique EPR signal with g-values at 2.223 and 2.115 was observed [135,136], which we will call "MCR_{PS}". Because of its air-sensitivity and its similarity to the MCR_{red1} spectrum, this MCR_{BPS} signal, as it was called earlier, was assigned as a Ni(I) state [135]. Yet, the EPR signal does not exhibit measurable hyperfine interactions from the halogen, leading to its assignment as a high-spin Ni(II)/alkyl radical species [137]. Further analyses suggested that the bromide group of BPS is released to form a Ni-alkyl adduct that can be described as either a

Ni(III)-propylsulfonate or a high-spin Ni(II) attached to an alkylsulfonyl radical [112]. In 2006, it was recognized that MCR_{PS} has UV-visible and EPR spectral features resembling MCR_{ox1} and that protonolysis of this species leads to the formation of propanesulfonate, which is similar to the proposed formation of methane from methyl-Ni(III) described in mechanism I. Further, as described below, reaction of MCR_{PS} with thiols regenerates MCR_{red1} and forms a thioether [67]. HYSCORE-EPR experiments better defined the features of the MCR_{PS} species and provided strong evidence for the assignment as an organometallic Ni(III)-propylsulfonate species [106].

The description of MCR_{PS} as an alkyl-Ni(III) complex in resonance with an alkyl-Ni(II) radical is nearly identical to that of MCR_{ox1} except that the upper axial nickel ligand is a carbon in case of MCR_{PS} versus a thiolate sulfur for MCR_{ox1}. The most striking feature of MCR_{PS} is that, being an alkyl-Ni(III) complex, it is electronically and chemically similar to the first proposed intermediate in mechanism I. Furthermore, this alkyl-Ni(III) species is surprisingly stable in the enzyme active site, whereas it had been expected to be sufficiently oxidizing that it would undergo rapid reduction to the alkyl-Ni(II) state.

When MCR_{red1} is incubated with other structurally related sulfonates, an EPR signal nearly identical to MCR_{PS} is observed [67,112,136]. Even a series of brominated carboxylic acids of chain lengths varying from 4 to 16 methylene groups can react with active Ni(I)-MCR_{red1} to form related Ni(III)-alkanoic acids, and the EPR spectra of these adducts are nearly identical to those of Ni(III)-MCR_{PS} [68]. There is no detectable hyperfine splitting from the halogen (nuclear spins of Cl, Br $= 3/2$; I $= 5/2$) in any of the haloalkyl complexes described above, demonstrating that the halogen group is distant from the paramagnetic nickel center, thereby, suggesting that the halide undergoes elimination during the formation of the alkyl-Ni(III) complex. Thus, the reactions of halogenated alkane-sulfonates and -carboxylates with active Ni(I)-MCR presumably involve the nucleophilic attack of Ni(I)-MCR_{red1} on the terminal carbon adjacent to the halogen atom to eliminate halide and generate the EPR-active alkyl-Ni(III) species as outlined in eq (7) below. This generates a six-coordinate Ni(III) complex, with the alkyl group occupying the upper axial site. This reaction is analogous to the proposed reaction of active Ni(I)-MCR_{red1} with the natural substrate, methyl-SCoM, to generate a methyl-Ni(III) intermediate during biological methane synthesis. The alkyl-Ni(III) complexes formed from the halogenated alkane-sulfonic and -carboxylic acids that elicit the alkyl-Ni(III) signature are sensitive to oxygen and over time decay to an inactive Ni(II) state.

$$[\text{Ni(I)-MCR}_{\text{red1}}] + \text{RX} \rightarrow [\text{R-Ni(III)-MCR}]^{+} + \text{X}^{-} \quad (7)$$

The UV-visible absorption spectra of the alkyl-Ni(III) complexes resemble those of inactive Ni(II) forms of MCR with an absorption maximum at

420 nm in contrast to the active Ni(I)-MCR_{red1}, which absorbs at 385 nm. Because the halogenated compounds react rapidly with active-Ni(I) to form the alkyl-Ni(III) complex, UV-visible stopped-flow methods demonstrated that formation of the alkyl-Ni(III) complexes is faster than the rate of methanogenesis and is saturable with substrate concentration.

Surprisingly, all brominated acids ranging from the relatively small bromobutyric acid (Br4A) to the relatively large bromohexadecanoic acid (Br16A) can react with MCR_{red1} to form an EPR-active Ni(III)-MCR_{XA} species and have been categorized into two classes, based on their reactivity. The shorter brominated acids, Br4A-Br8A, react rapidly with MCR_{red1} to form the MCR_{XA} state, and are thought to mimic binding of methyl-SCoM, with their carboxylate groups interacting with side chain Arg120. The longer bromo acids, Br9A-Br16A, apparently mimic CoBSH and the heterodisulfide product. The relatively long brominted acids are proposed to bind with their carboxyl group interacting with the solvent and the positively charged residues at the upper lip of the active site channel with the bromoalkyl chain reaching toward the Ni(I) center, where it could react rapidly and form the MCR_{XA} complex.

On the basis of these studies, a model has been proposed that illustrates three modes of binding of various carboxylates of different chain lengths that can be classified as (a) methyl-SCoM-like/BPS-like, (b) CoBSH-like, and (c) heterodisulfide product-like. These studies reveal the unexpected reactivity and flexibility of the MCR active site to accommodate a broad range of substrates, provide a molecular ruler for the substrate channel in MCR, and may aid in the development of other substrate analogues and/or inhibitors of MCR.

4.3.2. Methylnickel Formation at the Methyl-Coenzyme M Reductase Active Site

Although an organometallic methyl-Ni(III) intermediate has been proposed to be a catalytic intermediate in methane synthesis [117–118,138], such an intermediate has never been trapped during the reaction of MCR with native substrates. However, methyliodide [69] and methylbromide [70] react with active Ni(I)-MCR_{red1} to form an organometallic methyl-Ni(III) (denoted MCR_{Me}) species, apparently by an oxidative addition reaction described in equation (8). The most striking feature of MCR_{Me} is that electronically and chemically it represents the proposed intermediate in the first step of mechanism I.

$$[\text{Ni(I)-MCR}_{red1}] + \text{CH}_3\text{I} \rightarrow [\text{CH}_3\text{-Ni(III)-MCR}]^+ + \text{I}^- \qquad (8)$$

The formation of the methyl-Ni(III) species was confirmed by EPR spectroscopy and the covalent linkage between the methyl group and the

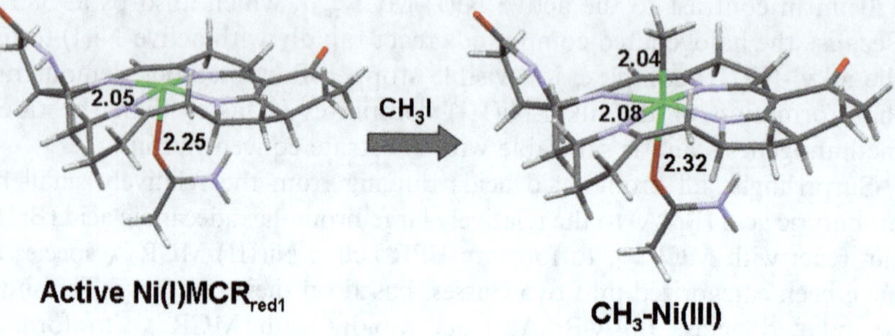

Figure 18. EXAFS structure of the methyl-Ni(III) bioorganometallic species at the MCR active site. Based on [108].

nickel center was confirmed by high resolution ENDOR and HYSCORE experiments using different isotopes of methyliodide [69,70]. X-ray absorption spectroscopy of the alkyl-Ni(III) state of MCR reveal a six coordinate Ni center with an upper axial Ni-C bond at 2.04 Å, four Ni-N bonds at 2.08 Å, and a lower axial Ni-O interaction at 2.32 Å, unambiguously establishing the organometallic nature of the methyl-Ni(III) species (Figure 18) [108].

As previously suggested, it was expected that the methyl-Ni(III) species formed during methanogenesis would be highly oxidizing and undergo immediate conversion to a methyl-Ni(II) state [103]; however, the methyl-Ni(III) species is relatively stable in the MCR active site. The rate at which active MCR_{red1} reacts with methyliodide to form the methyl-Ni(III) intermediate ($1900\,M^{-1}\,s^{-1}$ at 20 °C) is comparable to the maximum rate of methane formation with methyl-SCoM and CoBSH ($k_{cat} = 4.5\,s^{-1}$ at 20 °C; $k_{cat}/K_M = 930\,M^{-1}\,s^{-1}$ and $1.9 \times 10^4\,M^{-1}\,s^{-1}$ at 65 °C), which suggests the catalytic competence of the methylnickel species [69]. The catalytic intermediacy of the methyl-Ni(III) species is also indicated by its ability to regenerate active Ni(I)-MCR_{red1} and to form methane, as discussed below.

Presumably the reason that no observable spectroscopic changes are observed upon reaction of the natural methyl donor methyl-SCoM with MCR is because formation of the first intermediate requires activation in a process that requires CoBSH (the second substrate) and this kinetic coupling between the first and second steps makes k_1 much slower than k_2 (see Figure 17, the kinetic simulation), preventing accumulation of detectable amounts of the intermediate. On the other hand, the activated bromoalkyl substrate analogs rapidly react and form a stable intermediate in the absence of the second substrate. Of course, one must also worry about how closely these reactions with the substrate analog mimic the reaction with the natural substrate.

4.4. Reactions of the Organonickel Species at the Methyl-Coenzyme M Reductase Active Site

4.4.1. *Alkane Formation from Alkylnickel Species*

As an organometallic species, the alkylnickel bond can be cleaved homolytically or heterolytically. One heterolytic reaction that parallels the early steps in mechanism I (Figure 3) is protonolysis of the alkyl-Ni(III) complex on MCR to form alkanes. As described above, active Ni(I)-MCR_{red1} reacts with BPS to form an alkyl-Ni(III) MCR_{PS} complex that undergoes protonolysis upon acid quenching to yield the corresponding alkane, propanesulfonic acid, which was identified by NMR spectroscopy and high performance liquid chromatography (HPLC) analysis [67]. Single turnover experiments revealed that the rates for BPS decay and the product HPS formation are identical and equal the rates of Ni(III)-MCR_{PS} formation and Ni(I)-MCR_{red1} decay. These results indicated that the reaction of Ni(I)-MCR_{red1} with BPS parallels the early steps in mechanism I, as summarized by equations (9) and (10).

$$\text{Ni(I)-MCR}_{\text{red1}} + \text{BPS} \rightarrow \text{Ni(III)-MCR}_{\text{PS}} + \text{Br}^- \tag{9}$$

$$\text{Ni(III)-MCR}_{\text{PS}} + \text{H}^+ \rightarrow \text{HPS} + \text{Ni(II)-MCR} \tag{10}$$

Similarly, reaction of the methyl-Ni(III) species with the natural substrate, CoBSH, generates methane, although inactive Ni(II)-enzyme is generated (unpublished results). Mechanism I also indicates that protonolysis of alkyl-Ni leads to the formation of a transient Ni(II) species, which is reduced back to the active Ni(I) state by the CoBSSCoM radical anion. Perhaps in the absence of HSCoM, loss of the methyl group leads to a highly oxidizing Ni(III) species that rapidly captures an electron from the protein. Another possibility is that the Ni(II) is generated by homolytic cleavage of the methyl nickel bond, which directly or indirectly abstracts a hydrogen atom from CoBSH to generate methane, a CoBSH-based thiyl radical, and the inactive Ni(II) enzyme (Figure 19). In the absence of HSCoM, there would be no

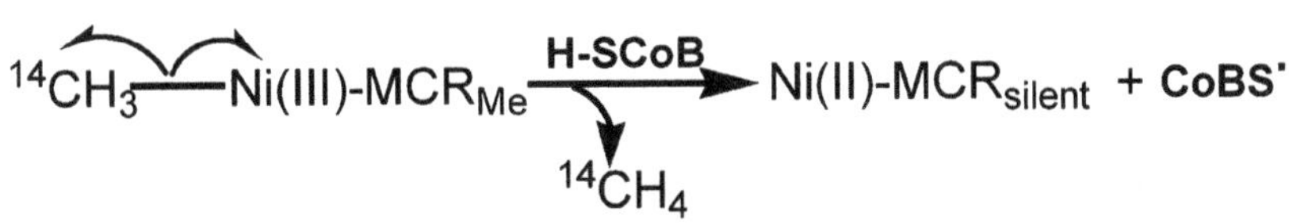

Figure 19. Homolytic cleavage of methyl-Ni(III) species to produce methane.

mechanism to reactivate the Ni center. However, there is no spectroscopic evidence for a CoBS thiyl radical.

The alkyl-Ni(III) adducts of brominated acids also appear to undergo alkanogenesis to liberate alkanoic acids, although, in this case, the product acids were not isolated and the suggestion for alkanoic acid formation was based on the yield and stability of the alkyl-Ni(III) complexes. Unlike the relatively stable MCR_{PS} and MCR_{Me} complexes, EPR signals from the organometallic adducts with the longer bromo acids (Br9A-Br16A), accumulate with a significantly lower yield. It was suggested that the relative instability of these alkyl-Ni(III) complexes results from homolytic cleavage of the nickel-carbon bond, giving Ni(II)-MCR_{silent} and the corresponding alkanoic acid radical, which abstracts a hydrogen atom from the environment of the protein to form the alkanoic acid [68].

4.4.2. Formation of Thioethers and Esters from Alkyl-Ni(III) Species

As described above, the anaerobic oxidation of methane may occur by a reversal of methanogenesis. Thus, according to mechanism I (Figure 13), the final step in AOM would be the reaction of methyl-Ni(III) with HSCoM to generate methyl-SCoM. Surprisingly, the alkyl-Ni(III) species generated at the MCR active site reacts with thiols to form active Ni(I)-MCR and a thioether product, as first discovered in the reaction of the replacement of the characteristic UV-visible and EPR signals of MCR_{PS} with those of MCR_{red1} [67]. The thioether product CoMS-PS was identified by mass spectrometric analysis [139]. The rate of conversion of the MCR_{PS} to Ni(I)-MCR_{red1} is dependent on the concentration of HSCoM. Besides demonstrating that the MCR_{PS} complex can be converted to regenerate the active enzyme, these results demonstrate that BPS is not an irreversible inhibitor, as thought, but a reversible redox inactivator.

As described above, MCR_{red1} also forms alkyl-Ni(III) adducts with a variety of alkanesulfonates and the resulting MCR_{XA} complexes (where $X = 5$–8) react with HSCoM to form thioether products and regenerate the active Ni(I)-MCR_{red1}. However, the alkyl-Ni(III) complexes from longer brominated acids (9–16 carbons) do not appear to react with HSCoM, perhaps because they block the channel in the enzyme and prevent access of HSCoM to the active site [68].

The HSCoM-dependent conversion of the alkyl-Ni(III) complexes of sulfonates and carboxylates to active MCR_{red1} with HSCoM occur rather slowly. For instance, the second order rate constant of the MCR_{PS} conversion to MCR_{red1} with HSCoM is approximately 60,000-fold slower than the second order rate constant for MCR_{PS} formation ($1.6 \times 10^5\, M^{-1}s^{-1}$).

On the other hand, MCR_{PS} reacts with a number of thiols to form the thioether product and regenerating the active Ni(I) state of the enzyme [67,139], including mercaptoethanol ($0.65\,s^{-1}$), cysteine ($9\,s^{-1}$), and Na_2S ($14\,s^{-1}$). The two-electron reductant, sodium borohydride also reacts with MCR_{PS} and reduces it to the active Ni(I) state; however, the low potential one-electron reductant Ti(III) citrate reacts poorly, if at all, with MCR_{PS} [139]. On the other hand, the reaction of the methyl-Ni(III) species at the MCR active site reacts with Ti(III) citrate to regenerate active Ni(I)-MCR_{red1} and to form methane (k_{cat} of $0.011\,s^{-1}$), similar to reactions reported for derivatives of F_{430} in solution (above).

A surprising reaction was discovered when MCR_{red1} is reacted with 4-bromobutyrate (Br4A). First, one observes the formation of the alkyl-Ni(III) complex (MCR_{4A}) ($k_{max} = 15\,s^{-1}$), followed by a "self-reactivation" that occurs in the absence of any reductant to regenerate MCR_{red1} and an ester product, which has been identified by mass spectrometry as 4-(4-bromobutanoyloxy)butanoic acid.

5. PERSPECTIVE AND PROSPECTIVE

This review has focused mainly on the organometallic aspect of MCR-based catalysis, however, one must step back and recognize that the alkylnickel species has not yet been observed as an intermediate with the natural methyl donor methyl-SCoM. Furthermore, as described briefly above, on the basis of density functional theory calculations, it was proposed [119] that such an intermediate is not feasible because conversion of methyl-SCoM to methyl-Ni would be thermodynamically unfavorable (endothermic by 45 kcal/mol). Mechanism 2, described above, which has a methyl radical, instead of an organometallic intermediate, as the hallmark was less objectionable. On the other hand, it has been pointed out [99] that transfer of the methyl group from methyltetrahydrofolate to Co(I) to form methylCob in the B_{12}-dependent methyltransferases like methionine synthase is similar in many respects to the transfer of a methyl group from methyl-SCoM to Ni(I) as proposed in mechanism 1 for MCR. The key to the cobalamin-dependent reaction is activation of the methyl group by protonation of the nitrogen to which it is attached; similarly, if a methylnickel intermediate is formed during MCR catalysis, an activation step would be necessary. Regardless, the enzyme-bound MCR cofactor can undergo alkylation (including methylation) by various activated alkyl group donors and the resulting alkyl-Ni(III) species can undergo biologically relevant reactions: protonolysis to form the alkane (such as methane) and thiolysis to form thioethers, including methyl-SCoM (the natural substrate) when methyl-Ni(III) is reacted with HSCoM.

The various proposed mechanisms are hypothesis, frameworks to guide experiments. One might consider mechanisms that could find common ground between mechanisms 1 (methylnickel) and 2 (methyl radical). One can look forward to experiments that probe how the C-S bond of methyl-SCoM is labilized and/or activated. The use of substrate analogs may be expanded to finally be able to trap the initial intermediates in the MCR mechanism. Mutagenesis experiments that target the active site may interrupt the mechanism at different points and perhaps even enable direct structural characterization of bound intermediates and mutations that target distant residues may provide information on protein dynamics that may be key to catalysis. It will be interesting to complete the biosynthetic pathway for F_{430} and to characterize these enzymes; furthermore, the enzymes responsible for the posttranslational modifications of MCR have yet to be identified. In addition, the transport proteins, molecular chaperones, and metallochaperones involved in maturation of MCR have yet to be identified. We also do not yet know how cells activate MCR. Genetic tools are now available for studies of methanogens and a true multidisciplinary effort is now possible to unravel many of the remaining questions about how this highly interesting nickel metalloenzyme catalyzes the formation of methane, a clean-burning energy-rich gas with major environmental implications.

ACKNOWLEDGMENTS

We are grateful to DOE (DE-FG02-08ER15931) for supporting our research on methanogenesis.

ABBREVIATIONS AND DEFINITIONS

ACS	acetyl coenzyme A synthase
AdoCob	adenosyl cobalamin
AOM	anaerobic oxidation of methane
BPS	3-bromopropanesulfonate
Br16A	bromohexadecanoic acid
Br4A	4-bromobutyric acid
CH_3-H_4folate	methyltetrahydrofolate
CH_3-SCoM	methyl-coenzyme M
CoBSH	coenzyme B, mercaptoheptanoyl threonine phosphate
CODH	carbon monoxide dehydrogenase
CooA	product of the cooA gene
Cys	cysteine

dAdo	deoxyadenosyl
DFT	density functional theory
ENDOR	electron nuclear double resonance
EPR	electron paramagnetic resonance
EXAFS	extended X-ray absorption fine structure
F_{430M}	pentamethylester of F_{430}
FTIR	Fourier transform infrared spectroscopy
H_2ases	hydrogenases
HPLC	high performance liquid chromatography
HPS	propane sulfonate
HSCoM	coenzyme M
HYSCORE	hyperfine sublevel correlation
MCR	methyl-coenzyme M reductase
methylCob	methylcobalamin
Ni_d	distal nickel
Ni_p	proximal nickel
NMR	nuclear magnetic resonance
OEiBC	octaethylisobacteriochlorin
RSD	reactant state destabilization
SRB	sulfate-reducing bacteria
TD-DFT	time dependent density functional theory
THF	tetrahydrofuran
tmc	1,4,8,11-tetramethyl-1,4,8,11-tetraazacyclotetradecane
XAS	X-ray absorption spectroscopy

REFERENCES

1. J. Halpern, *Pure Appl. Chem.*, 1986, **58**, 575–584.
2. G. Jaouen and A. Vessières, *Pure Appl. Chem.*, 1985, **57**, 1865–1874.
3. R. H. Fish and G. Jaouen, *Organometallics*, 2003, **22**, 2166–2177.
4. G. Jaouen, A. Vessières and I. S. Butler, *Acc. Chem. Res.*, 1993, **26**, 361–369.
5. J. C. Fontecilla-Camps and S. W. Ragsdale, in *Advances in Inorganic Chemistry*, Academic Press, San Diego, CA, 1999, **47**, 283–333.
6. S. W. Ragsdale, *Curr. Opin. Chem. Biol.*, 1998, **2**, 208–215.
7. L. D. Slep and F. Neese, *Angew. Chem. Int. Ed. Engl.*, 2003, **42**, 2942–2945.
8. J. Halpern, *Pure Appl. Chem.*, 2001, **73**, 209–220.
9. C. G. Hartinger and P. J. Dyson, *Chem. Soc. Rev.*, 2009, **38**, 391–401.
10. C. G. Riordan, in *Comprehensive Coordination Chemistry II*, ed. J. McCleverty and T. J. Meyer, Elsevier, Oxford, 2003, **8**, Chapter 26.
11. C. S. Allardyce, A. Dorcier, C. Scolaro and P. J. Dyson, *Appl. Organometal. Chem.*, 2005, **19**, 1–10.
12. C. E. Carraher Jr. and C. U. Pittman Jr., *Macromolecules Containing Metal and Metal-Like Elements*, 2004, **3**, 1–18.

13. R. Mukhopadhyay, B. P. Rosen, L. T. Phung and S. Silver, *FEMS Microbiol. Rev.*, 2002, **26**, 311–325.
14. W. R. Cullen and K. J. Reimer, *Chem. Rev.* 1989, **89**, 713–764.
15. B. Rosenberg, L. V. Camp and T. Krigas, *Nature*, 1965, **205**, 698.
16. P. Yang and M. Guo, *Coord. Chem. Rev.*, 1999, **185–186**, 189–211.
17. J. Ruiz, M. D. Villa, N. Cutillas, G. L. Pez, C. d. Haro, D. Bautista, V. Moreno and L. Valencia, *Inorg. Chem.*, 2008, **47**, 4490–4505.
18. H. Köpf and P. Köpf-Maier, *Angew. Chem. Int. Ed. Engl.*, 1979, **91**, 509.
19. F. Dubar, J. Khalife, J. Brocard, D. Dive and C. Biot, *Molecules*, 2008, **13**, 2900–2907.
20. K. D. Gleria, H. A. O. Hill and C. J. McNeil, *Anal. Chem.*, 1986, **58**, 1203–1205.
21. B. Gosio, *Arch. Ital. Biol.*, 1893, **18**(253), 298.
22. F. Challenger, C. Higginbottom and L. Ellis, *J. Chem. Soc.*, 1933, 95–101.
23. F. Challenger, *Q. Rev. Chem. Soc.*, 1955, **9**, 255–286.
24. F. Challenger, *Adv. Enzymol.*, 1951, **12**, 429–491.
25. F. Challenger, *Chem. Rev.*, 1945, **36**, 315–361.
26. D. J. Thomas, S. B. Waters and M. Styblo, *Toxicol. Appl. Pharmacol.*, 2004, **198**, 319–326.
27. D. C. Hodgkin, *Proc. R. Soc. London, Ser. A*, 1965, **300**, 294.
28. D. C. Hodgkin, *Science*, 1965, **50**, 979–988.
29. D. Dolphin (Ed.), in *B_{12}, Vol. 1 and 2, Wiley Interscience, New York*, 1982, p. 671 and 505, respectively.
30. B. Jaun, *Helv. Chim. Acta*, 1990, **73**, 2209–2217.
31. S. W. Ragsdale, L. G. Ljungdahl and D. V. DerVartanian, *Biochem. Biophys. Res. Commun.*, 1983, **115**, 658–665.
32. S. W. Ragsdale and H. G. Wood, *J. Biol. Chem.*, 1985, **60**, 3970–3977.
33. W. Lubitz, M. V. Gastel and W. Gärtner, in *Nickel and Its Surprising Impact in Nature*, Vol. 2 of *Metal Ions in Life Sciences*, Ed. A. Sigel, H. Sigel and R. K. O. Sigel, John Wiley and Sons, Chichester, UK, 2007, 279–322.
34. E. L. Rickes, N. G. Brink, F. R. Koniuszy, T. R. Wood and K. Folkers, *Science*, 1948, **107**, 396–339.
35. D. C. Hodgkin, J. Kamper, M. Mackay, J. Pickworth, K. N. Trueblood and J. G. White, *Nature*, 1956, **178**, 64–66.
36. R. Banerjee and S. W. Ragsdale, *Ann. Rev. Biochem.*, 2003, **72**, 209–247.
37. B. Kräutler, in *Vitamin B_{12} and B_{12}-Proteins*, Ed. B. Kräutler, D. Arigoni and B. T. Golding, Wiley-VCH, Weinheim, 1996, 201–216.
38. C. L. Drennan, M. M. Dixon, D. M. Hoover, J. T. Jarrett, C. W. Goulding, R. G. Matthews and M. L. Ludwig, in *Vitamin B_{12} and B_{12}-Proteins*, Ed. B. Kräutler, D. Arigoni and B. T. Golding, Wiley-VCH, Weinheim, 1996, 133–156.
39. S. W. Ragsdale, in *Vitamins and Hormones*, Ed. G. Litwack, Elsevier, Amsterdam, 2008, 293–324.
40. T. Toraya, *Chem. Rec.*, 2002, **2**, 352–366.
41. S. Chowdhury and R. Banerjee, *Biochemistry*, 2000, **39**, 7998–8006.
42. K. L. Brown, *Chem. Rev.*, 2005, **105**, 2075–2149.
43. J. M. Sirovatka, A. K. Rappé and R. G. Finke, *Inorg. Chim. Acta*, 2000, **300**, 545–555.

44. K. L. Brown and J. Li, *J. Am. Chem. Soc.*, 1998, **120**, 9466.
45. P. K. Sharma, Z. T. Chu, M. H. Olsson and A. Warshel, *Proc. Natl. Acad. Sci. USA*, 2007, **104**, 9661–9666.
46. P. A. Lindahl and D. E. Graham, in *Nickel and Its Surprising Impact in Nature*, Vol. 2 of *Metal Ions in Life Sciences*, Ed. A. Sigel, H. Sigel and R. K. O. Sigel, John Wiley and Sons, Chichester, UK, 2007, 357–416.
47. J. Seravalli and S. W. Ragsdale, *Biochemistry*, 2008, **47**, 6770–6781.
48. W. Gong, B. Hao, Z. Wei, D. J. J. Ferguson, T. Tallant, J. A. Krzycki and M. K. Chan, *Proc. Natl. Acad. Sci. USA*, 2008, **105**, 9558–9563.
49. J. H. Jeoung and H. Dobbek, *Science*, 2007, **318**, 1461–1464.
50. J. Seravalli and S. W. Ragsdale, *J. Biol. Chem.*, 2008, **283**, 8384–8394.
51. A. Volbeda, M. H. Charon, C. Piras, E. C. Hatchikian, M. Frey and J. C. Fontecilla-Camps, *Nature*, 1995, **373**, 556–557.
52. K. A. Bagley, E. C. Duin, W. Roseboom, S. P. J. Albracht and W. H. Woodruff, *Biochemistry*, 1995, **34**, 5527.
53. S. Reissmann, E. Hochleitner, H. Wang, A. Paschos, F. Lottspeich, R. S. Glass and A. Bock, *Science*, 2003, **299**, 1067.
54. P. M. Vignais and B. Billoud, *Chem. Rev.*, 2007, **107**, 4206–4272.
55. S. P. Burg and E. A. Burg, *Science*, 1965, **148**, 1190.
56. E. M. Beyer Jr., *Plant Physiol.*, 1976, **58**, 268.
57. G. E. Schaller and A. B. Bleecker, *Science*, 1995, **270**, 1809–1811.
58. J. Hirsch, S. D. George, E. I. Solomon, B. Hedman, K. O. Hodgson and J. N. Burstyn, *Inorg. Chem.*, 2001, **40**, 2439–2441.
59. M. Krüger, A. Meyerdlerks, F. O. Glockner, R. Amann, F. Widdel, M. Kube, R. Reinhardt, J. Kahnt, R. Bocher, R. K. Thauer and S. Shima, *Nature*, 2003, **426**, 878–881.
60. J. Xia, Z. Hu, C. V. Popescu, P. A. Lindahl and E. Munck, *J. Am. Chem. Soc.*, 1997, **119**, 8301–8312.
61. C. L. Fan, C. M. Gorst, S. W. Ragsdale and B. M. Hoffman, *Biochemistry*, 1991, **30**, 431–435.
62. J. Chen, S. Huang, J. Seravalli, H. Gutzman Jr., D. J. Swartz, S. W. Ragsdale and K. A. Bagley, *Biochemistry*, 2003, **42**, 14822–14830.
63. R. P. Schenker and T. C. Brunold, *J. Am. Chem. Soc.*, 2003, **125**, 13962–13963.
64. R. C. Linck, C. W. Spahn, T. B. Rauchfuss and S. R. Wilson, *J. Am. Chem. Soc.*, 2003, **125**, 8700–8701.
65. R. Krishnan and C. G. Riordan, *J. Am. Chem. Soc.*, 2004, **126**, 4484–4485.
66. W. G. Dougherty, K. Rangan, M. J. O'Hagan, G. P. Yap and C. G. Riordan, *J. Am. Chem. Soc.*, 2008, **130**, 13510–13511.
67. R. C. Kunz, Y. C. Horng and S. W. Ragsdale, *J. Biol. Chem.*, 2006, **281**, 34663–34676.
68. M. Dey, R. C. Kunz, D. M. Lyons and S. W. Ragsdale, *Biochemistry*, 2007, **46**, 11969–11978.
69. M. Dey, J. Telser, R. C. Kunz, N. S. Lees, S. W. Ragsdale and B. M. Hoffman, *J. Am. Chem. Soc.*, 2007, **129**, 11030–11032.
70. N. Yang, M. Reiher, M. Wang, J. Harmer and E. C. Duin, *J. Am. Chem. Soc.*, 2007, **129**, 11028–11029.

71. R. S. Wolfe, *Am. Soc. Microbiol. News*, 1996, **62**, 529–534.
72. H. A. Barker, in *Bacterial Fermentations*, Ed. H. A. Barker, Wiley, New York, 1956, 1–27.
73. R. K. Thauer, *Microbiology*, 1998, **144**, 2377–2406.
74. M. Stephenson and L. H. Stickland, *Biochem. J.*, 1933, **27**, 1517–1527.
75. U. Deppenmeier, *Prog. Nucleic Acid Res. Mol. Biol.*, 2002, **71**, 223–283.
76. G. E. Fox, E. Stackebrandt, R. B. Hespell, J. Gibson, J. Maniloff, T. A. Dyer, R. S. Wolfe, W. E. Balch, R. S. Tanner, L. J. Magrum, L. B. Zablen, R. Blakemore, R. Gupta, L. Bonen, B. J. Lewis, D. A. Stahl, K. R. Luehrse, K. N. Chen and C. R. Woese, *Science*, 1980, **209**, 457–463.
77. C. R. Woese, O. Kandler and M. L. Wheelis, *Proc. Nat. Acad. Sci. USA*, 1990, **87**, 4576–4579.
78. J. L. Garcia, B. K. Patel and O. B., *Anaerobe*, 2000, **6**, 205–226.
79. S. J. Hallam, N. Putnam, C. M. Preston, J. C. Detter, D. Rokhsar, P. M. Richardson and E. F. DeLong, *Science*, 2004, **305**, 1457–1462.
80. A. J. Zehnder and K. Wuhrmann, *Science*, 1976, **194**, 1165–1166.
81. S. J. Hallam, P. R. Girguis, C. M. Preston, P. M. Richardson and E. F. DeLong, *Appl. Environ. Microbiol.*, 2003, **69**, 5483–5491.
82. M. Kruger, A. Meyerdierks, F. O. Glockner, R. Amann, F. Widdel, M. Kube, R. Reinhardt, J. Kahnt, R. Bocher, R. K. Thauer and S. Shima, *Nature*, 2003, **426**, 878–881.
83. J. Schimel, *Nature*, 2000, **403**, 375–377.
84. R. G. Kallen and W. P. Jencks, *J. Biol. Chem.*, 1966, **241**, 5851–5863.
85. E. F. DeLong, *Nature*, 2000, **407**, 577–579.
86. D. R. Blake and F. Sherwood Rowland, *Science*, 1988, **239**, 1129–1131.
87. S. K. Atreya, P. R. Mahaffy and A. Wong, *Planet. Space Sci.*, 2007, **55**, 358–369.
88. M. J. Mumma, G. L. Villanueva, R. E. Novak, T. Hewagama, B. P. Bonev, M. A. Disanti, A. M. Mandell and M. D. Smith, *Science*, 2009, **323**, 1041–1045.
89. T. Owen and H. B. Niemann, *Philos. Transact. A Math. Phys. Eng. Sci.*, 2009, **367**, 607–615.
90. T. C. Onstott, D. McGown, J. Kessler, B. S. Lollar, K. K. Lehmann and S. M. Clifford, *Astrobiology*, 2006, **6**, 377–395.
91. R. Bartha and E. J. Ordal, *J. Bacteriol.*, 1965, **89**, 1015–1019.
92. S. W. Ragsdale, *J. Inorg. Biochem.*, 2007, **101**, 1657–1666.
93. R. P. Gunsalus and R. S. Wolfe, *FEMS Microbiol. Lett.*, 1978, **3**, 191–193.
94. G. Diekert, B. Klee and R. K. Thauer, *Arch. Microbiol.*, 1980, **124**, 103–106.
95. W. B. Whitman and R. S. Wolfe, *Biochem. Biophys. Res. Commun.*, 1980, **92**, 1196–1201.
96. G. Diekert, R. Jaenchen and R. K. Thauer, *FEBS Lett.*, 1980, **119**, 118–120.
97. A. Pfaltz, B. Jaun, A. Fassler, A. Eschenmoser, R. Jaenchen, H. H. Gilles, G. Diekert and R. K. Thauer, *Helv. Chim. Acta.*, 1982, **65**, 828–865.
98. A. A. DiMarco, T. A. Bobik and R. S. Wolfe, *Annu. Rev. Biochem.*, 1990, **59**, 355–394.
99. S. W. Ragsdale, in *The Porphyrin Handbook*, Ed. K. M. Kadish, K. M. Smith and R. Guilard, Academic Press, New York, 2003, pp. 205–228.

100. T. Ide, S. Baumer and U. Deppenmeier, *J. Bacteriol.*, 1999, **181**, 4076–4080.
101. W. G. Grabarse, F. Mahlert, E. C. Duin, M. Goubeaud, S. Shima, R. K. Thauer, V. Lamzin and U. Ermler, *J. Mol. Biol.*, 2001, **309**, 315–330.
102. W. G. Grabarse, F. Mahlert, S. Shima, R. K. Thauer and U. Ermler, *J. Mol. Biol.*, 2000, **303**, 329–344.
103. U. Ermler, W. Grabarse, S. Shima, M. Goubeaud and R. K. Thauer, *Science*, 1997, **278**, 1457–1462.
104. S. Rospert, R. Bocher, S. P. Albracht and R. K. Thauer, *FEBS Lett.*, 1991, **291**, 371–375.
105. R. C. Kunz, Y.-C. Horng and S. W. Ragsdale, *J. Biol. Chem.*, 2006, **281**, 34663–34676.
106. D. Hinderberger, R. P. Piskorski, M. Goenrich, R. K. Thauer, A. Schweiger, J. Harmer and B. Jaun, *Angew. Chem. Int. Ed. Engl.*, 2006, **45**, 3602–3607.
107. A. H. Maki, N. Edelstein, A. Davison and R. H. Holm, *J. Am. Chem. Soc.*, 1964, **86**, 4580–4587.
108. R. Sarangi, M. Dey and S. W. Ragsdale, *Biochemistry*, 2009, **48**, 3146–3156.
109. S. P. J. Albracht, D. Ankel-Fuchs, R. Böcher, J. Ellermann, J. Moll, J. W. van der Zwann and R. K. Thauer, *Biochim. Biophys. Acta*, 1988, **955**, 86–102.
110. F. Mahlert, W. Grabarse, J. Kahnt, R. K. Thauer and E. C. Duin, *J. Biol. Inorg. Chem.*, 2002, **7**, 101–112.
111. D. F. Becker and S. W. Ragsdale, *Biochemistry*, 1998, **37**, 2639–2647.
112. M. Goenrich, F. Mahlert, E. C. Duin, C. Bauer, B. Jaun and R. K. Thauer, *J. Biol. Inorg. Chem.*, 2004, **9**, 691–705.
113. J. Harmer, C. Finazzo, R. Piskorski, S. Ebner, E. C. Duin, M. Goenrich, R. K. Thauer, M. Reiher, A. Schweiger, D. Hinderberger and B. Jaun, *J. Am. Chem. Soc.*, 2008, **130**, 10907–10920.
114. Q. Tang, P. E. Carrington, Y.-C. Horng, M. J. Maroney, S. W. Ragsdale and D. F. Bocian, *J. Am. Chem. Soc.*, 2002, **124**, 13242.
115. R. Piskorski and B. Jaun, *J. Am. Chem. Soc.*, 2003, **125**, 13120–13125.
116. M. Dey, R. Kunz, K. M. V. Heuvelen, J. L. Craft, Y.-C. Horng, Q. Tang, D. F. Bocian, S. J. George, T. C. Brunold and S. W. Ragsdale, *Biochemistry*, 2006, **45**, 11915–11933.
117. S.-K. Lin and B. Jaun, *Helv. Chim. Acta*, 1991, **74**, 1725–1738.
118. S.-K. Lin and B. Jaun, *Helv. Chim. Acta*, 1992, **75**, 1478–1490.
119. V. Pelmenschikov, M. R. A. Blomberg, P. E. M. Siegbahn and R. H. Crabtree, *J. Am. Chem. Soc.*, 2002, **124**, 4039–4049.
120. V. Pelmenschikov and P. E. Siegbahn, *J. Biol. Inorg. Chem.*, 2003, **8**, 653–662.
121. E. C. Duin and M. L. McKee, *J. Phys. Chem. B*, 2008.
122. B. Jaun and A. Pfaltz, *J. Chem. Soc.Chem. Comm.*, 1988, 293–294.
123. A. Berkessel, *Bioorg. Chem.*, 1991, **19**, 101–115.
124. S. P. J. Albracht, D. Ankel-Fuchs, R. Böcher, J. Ellermann, J. Moll, J. W. van der Zwann and R. K. Thauer, *Biochim. Biophys. Acta*, 1988, **955**, 86–102.
125. B. Jaun and A. Pfaltz, *J. Chem. Soc. Chem.Comm.*, 1986, 1327–1329.
126. A. M. Stolzenberg and M. T. Stershic, *J. Am. Chem.Soc.*, 1988, **110**, 5397.
127. A. M. Stolzenberg and M. T. Stershic, *Inorg. Chem.*, 1987, **26**, 3082.
128. M. C. Helvenston and C. E. Castro, *J. Am. Chem. Soc.*, 1992, **114**, 8490.

129. G. K. Lahiri, L. J. Schussel and A. M. Stolzenberg, *Inorg. Chem.*, 1992, **31**, 4991–5000.
130. M. S. Ram, C. G. Riordan, R. Ostrander and A. L. Rheingold, *Inorg. Chem.*, 1995, **34**, 5884–5892.
131. C. M. Drain, D. B. Sable and B. B. Corden, *Inorg. Chem.*, 1988, **27**, 2396–2398.
132. C. M. Drain, D. B. Sable and B. B. Corden, *Inorg. Chem.*, 1990, **29**, 1428–1433.
133. M. J. D'Aniello, Jr. and E. K. Barefield, *J. Am. Chem. Soc.*, 1976, **98**, 1610–1611.
134. J. Ellermann, S. Rospert, R. K. Thauer, M. Bokranz, A. Klein, M. Voges and A. Berkessel, *Eur. J. Biochem.*, 1989, **184**, 63–68.
135. S. Rospert, D. Linder, J. Ellermann and R. K. Thauer, *Eur. J. Biochem.*, 1990, **194**, 871–877.
136. S. Rospert, M. Voges, A. Berkessel, S. P. Albracht and R. K. Thauer, *Eur. J. Biochem.*, 1992, **210**, 101–107.
137. J. Ellermann, R. Hedderich, R. Bocher and R. K. Thauer, *Eur. J. Biochem.*, 1988, **172**, 669–677.
138. G. K. Lahiri and A. M. Stolzenberg, *Inorg. Chem.*, 1993, **32**, 4409–4413.
139. R. C. Kunz, M. Dey and S. W. Ragsdale, *Biochemistry*, 2008, **47**, 2661–2667.
140. R. G. Matthews, *Acc. Chem. Res.*, 2001, **34**, 681–689.
141. R. E. Hungate, *Bacteriol. Rev.*, 1950, **14**, 1–49.

Met. Ions Life Sci. **2010**, *7*, 111–151

4

Organotins. Formation, Use, Speciation, and Toxicology

Tamás Gajda and Attila Jancsó

Department of Inorganic and Analytical Chemistry, University of Szeged,
P.O. Box 440, H-6701 Szeged, Hungary
⟨tamas.gajda@chem.u-szeged.hu⟩
⟨jancso@chem.u-szeged.hu⟩

Metal Ions in Life Sciences, Volume 7 Edited by Astrid Sigel, Helmut Sigel, and Roland K. O. Sigel

Published by the Royal Society of Chemistry, www.rsc.org DOI: 10.1039/9781849730822-00111

ABSTRACT: The speciation of organotin(IV) cations in natural waters, in sewage or in biofluids is strongly influenced by the complex formation with the available metal-binding compounds, i.e., both high and low molecular weight ligands of biological and environmental interest. The primary intention of this chapter is to discuss the aquatic solution chemistry of organotin cations and their complexes formed with low and high molecular weight bioligands. Besides, some synthetic aspects, applications and sources of organotin pollution, their destinations in the environment, and toxicology will be also shortly discussed.

KEYWORDS: accumulation of organotin compounds in the environment · bioinorganic speciation · organotin(IV) · organotin pollution · tributyltin(IV)

1. INTRODUCTION

Since the beginning of the bronze age tin and its alloys have been important to mankind, but organotin compounds have been known only in the past 150 years. Today more than 800 organotins are known and tin has a larger number of organometallic derivatives in commercial use than any other element. The first industrial application dates back to 1940, and the worldwide production of organotin chemicals increased drastically in the past sixty years. In 1996 the annual world production of organotins was roughly estimated to be 50,000 tons [1]. After 1992 the production slowly decreased due to the legislative restrictions in developed countries. However, the consumption of organotins in developing countries still increased in the last decade.

Due to its effect on the aquatic life, tributyltin(IV) (TBT) is one of the most toxic compounds that man has ever introduced in the environment on purpose. Therefore, TBT and other organotins represent a very high risk for the aquatic and terrestrial ecosystem.

Davies' recent monograph gives an impressive overview of organotin chemistry, which concentrates mainly on the preparative and structural aspects [2]. Besides, many excellent books and reviews appeared in the last decade dealing with organotin chemistry in general [3], and in more specialized topics, such as asymmetric synthesis [4,5] and coordination chemistry [6–8] focusing on the solid state complexes. The readers are kindly directed to these publications for a more general view on organotin chemistry.

The speciation of organotin(IV) cations in natural waters, in sewage or in biofluids is strongly influenced by complex formation with the available metal-binding compounds, i.e., both high and low molecular weight ligands of biological and environmental interest. The primary intention of this chapter is to discuss the aquatic solution chemistry of organotin cations and their complexes formed with low and high molecular weight bioligands. To the best of our knowledge, no review devoted to this topic has been published so far. Besides, some synthetic aspects, applications, and sources of organotin pollution, their destinations in the environment, and toxicology will also shortly be discussed.

2. SYNTHETIC ASPECTS

The first report on the preparation of organotin compounds dates back to the middle of the 19th century when Frankland managed to produce diethyltin diiodide (Et_2SnI_2) from the reaction of ethyl iodide and tin [9]. A few years later an alternative route to the direct method was published which described the reaction of diethyl zinc and tin tetrachloride to form tetraethyltin as the final product [10]. A major break-through in the synthetic methods for the preparation of organotin compounds was brought by Grignard's organomagnesium halides at the very beginning of the 20th century. The use of Grignard's reagents for building the carbon-tin bond is still one of the key reactions in synthetic organotin chemistry. In spite of the above cited early reports on the synthesis of these new types of organometallic substances, approximately 100 years passed before organotin compounds attracted wider interest due to their discovered possible practical applications.

Indeed, there are four major routes for creating new carbon-tin bonds that are summarized by the following reactions (1)–(4) [2]:

(1) The oldest method uses the reaction of metallic tin or tin(II) halide with an organic halide:

$$Sn + 2\,RX = R_2SnX_2 \qquad (1)$$

(2) The most frequent way is the reaction of organometallic reagents of lithium, magnesium or aluminium (including also Grignard's reagents) with tin(II) or tin(IV) halides:

$$SnX_4 + 4\,RMgX = R_4Sn + 4\,MgX_2 \qquad (2)$$

(3) The addition of trialkyltin hydrides to alkenes or alkynes produces the fourth carbon-tin bond around the central tin:

$$R_3SnH + \rangle C{=}C\langle \;=\; R_3Sn{-}\overset{|}{\underset{|}{C}}{-}\overset{|}{\underset{|}{C}}{-}H \qquad (3)$$

(4) Metallic (e.g., lithium) derivatives of triorganotin with alkyl halides give tetraorganotin compounds:

$$R_3SnM + R'X = R_3SnR' + MX \qquad (4)$$

Next to Davies' comprehensive book [2], there are many books and reviews discussing the various aspects and modifications of these principal reactions, together with several other alternatives for the formation of the carbon-tin bond (see for example [3,11–14]). During the previous decades a huge number of publications appeared on the synthesis of new organotin compounds, formed with a large variety of ligands and their structural investigations, mostly in the solid state but sometimes also in solution. Within the frame of this review it is not possible to provide even an overview about these achievements, nevertheless we try to summarize the most important methods for building new carbon-tin bonds and the synthetic aspects of a selected range of compounds by keeping the usual classification that is based on the number of carbon-tin bonds present in the substances. This chapter focuses on organotin(IV) compounds. Divalent organotin compounds are generally unstable and polymerize with the formation of Sn-Sn bonds. Lower valence state organotin materials have been discussed elsewhere in excellent books and reviews [2,15–19].

2.1. Tetraorganotins

The route used most often for the preparation of tetraorganotins is based on the reaction of the appropriate Grignard reagent (applied generally in excess), or other organometallic reagents (RM or R_2M', M = Na, Li, M' = Zn) with a tin(IV) halide ($SnCl_4$) (see [14] and references therein). This method results in high yields (more than 90%) for the preparation of tetravinyl, tetraallyl, tetraalkyl, and tetraaryl tins, however, for the preparation of tetraorganotins with longer alkyl groups than butyl other methods provide better results [14]. Alkyl- and vinyltin compounds can be prepared by hydrostannation of alkenes and alkynes with an R_3SnH reagent [20,21]. Thoonen et al. described in detail (with references) several refined methods to obtain various symmetric tetraorganotins and asymmetric, $R_2R'_2Sn$- and $R_3R'Sn$-type derivatives [14]. For the preparation of $R_2R'R''Sn$-type compounds, a dialkyltin halide (R_2SnX_2) is converted first to a mixed tetraorganotin ($R_2R'_2Sn$) by the use of R′MgX. One of the organic groups of $R_2R'_2Sn$ is selectively cleaved by the addition of one equivalent of a halogen. The final product is then obtained by adding the second Grignard reagent (R″MgX) [22]. The preparation of racemic and optically active tetraorganotins (RR′R″R‴Sn) was described by Gielen [23]. From Me_4Sn as a starting material three methyl groups were replaced by cyclohexyl, isopropyl, and ethyl substituents in alternating steps of methyl group cleavage by bromine and alkylation by the appropriate Grignard reagents containing the desired organic groups.

Monostannacycloalkanes ($R_2Sn(CH_2)_n$) form a special class of tetraorganotins with tin being part of the cycloalkane ring [2]. Cyclic organotin compounds with a coordinating heteroatom, having in many cases penta- or hexacoordinated structures, can be isolated by using C,Y-type chelating ligands (Y = a heteroatom-containing substituent) [24]. A subclass of the above tetraorganotins, called diptych or triptych compounds, containing trigonal-bipyramidal tin centers and two or three cycles were discussed by Tzschach and Jurkschat, focusing mostly on nitrogen-containing derivatives [25].

Tetraorganotins are starting material for the synthesis of organotin derivatives with less carbon-tin bonds, i.e., organotin(IV) halides by the Kocheshkov redistribution reaction (5) [14] (see Section 2.2 below), organotin compounds with tin-oxygen (R_3SnO_2CR', Et_3SnOPh) or tin-sulfur (R_3SnSR') bonds from tetraalkyltins by cleaving an alkyl group by the proper carboxylic acid (R′COOH), phenol (PhOH) or mercaptane (R′SH), respectively [13].

Tetraorganotins are important as mediators in synthetic organic chemistry. The use of the Stille cross-coupling reaction, a palladium-catalyzed coupling of organic electrophiles and (tetra)organostannanes is a well established way for the selective formation of new carbon-carbon bonds [26,27]. The above mentioned allylstannanes are important reagents in asymmetric synthesis [4,5]. Transmetallation reactions between allyltin compounds and other Lewis acid metal halides have been used to prepare allylic derivatives of several other elements, e.g., boron, phosphorus, arsenic, copper, and other metals [2].

2.2. Triorganotins

The usual way to prepare triorganotin compounds is to use the Kocheshkov redistribution reaction (5), resulting in triorganotin halides from tetraorganotins and a tin tetrahalide [13,14]. (For the preparation of triorganotins(IV) halides, x = 3 in the reaction below).

$$x\,R_4Sn + (4-x)SnX_4 \xrightarrow{\Delta} 4\,R_xSnX_{4-x} \quad (5)$$

Instead of tin(IV) tetrahalides, tin(II) dihalides may also be used for the dealkylation of tetraalkyltins [28]. Cleavage of the carbon-tin bond can be achieved in other ways, i.e., by the use of different halogens (preferably bromine) [13,29] or HX reagents (resulting in the formation of alkanes as side products) [13].

Triorganotin halides, e.g., R_3SnCl, serve as starting basis for preparing various other triorganotin substances. The replacement of the chlorine substituent by a nucleophile (e.g., $X = OH$, $OCOR'$, OR', NR_2, SR', etc.) leads to the appropriate R_3SnX derivative [2].

Triorganotin(IV) hydrides can be produced by the use of a metal hydride, as nucleophile (e.g., $LiAlH_4$). These hydrides are important starting materials for the preparation of metallic derivatives of triorganotin (R_3SnM) (with significance in organic synthesis), alkyl- and vinyltin compounds, and they can also be converted to symmetric ditins (R_3SnSnR_3) by using palladium catalysts [30]. They can react with various substrates in addition and substitution reactions following different homolytic or heterolytic mechanisms [2].

The alkaline hydrolysis of triorganotin(IV) chlorides leads to the corresponding hydroxides (R_3SnOH) or oxides ($[R_3Sn]_2O$) [31]. The formation and structural features of a large number of organotin assemblies containing Sn-O bonds (including tri-, di-, and monoorganotin compounds) have been reviewed by Chandrasekhar et al. in recent reviews [32,33].

2.3. Diorganotins

The oldest method for the preparation of organotin compounds is the reaction of metallic tin with an alkyl halide producing a diorganotin(IV) dihalide [9]. Similarly to triorganotins, the simplest way for the preparation of diorganotin compounds is based on the Kocheshkov redistribution reaction (5) [13,14]. Diorganotin(IV) dihalides can also be synthesized by the reaction between tetraorganotins and HCl [34] or by the exchange reaction (6) between two diorganotin(IV) dihalides, leading to a mixed dihalide derivative [35]:

$$R_2SnX_2 + R_2SnY_2 \rightarrow 2\,R_2SnXY \quad (6)$$

Instead of cleaving the Sn-C bond of tetraorganotins, selective dialkylation of $SnCl_4$ is also a way to form dialkyltin(IV) dichlorides by using alkylaluminium reagents [36].

Diorganotin(IV) dihalides go through a hydrolysis pathway amongst aqueous conditions which results in oligomeric/polymeric diorganotin(IV) oxides ($[R_2SnO]_n$), after the formation of various intermediates [2]. Generally, the first products that can be isolated are the tetraorganodistannoxanes ($XR_2SnOSnR_2X$). The chemistry and structure of these compounds is discussed in a complete section of "Tin Chemistry" by Jurkschat [37]. Distannoxanes (e.g., $ClR_2SnOSnR_2Cl$) have, with special exceptions, a dimeric structure with a SnOSnO central core [38] with peripheral alkyl groups that causes an excellent solubility in non-polar solvents. The X ligands in the dimeric structure can often form bridges between the central and terminal tin atoms, resulting in fused rings with 5-coordinate tin atoms. The synthesis and structural aspects of diorganotin compounds containing the four-membered $[Sn(\mu\text{-}OH)]_2$ units are discussed in detail by Chandrasekhar's group [39]. Distannoxanes deserve interest due to their useful properties as catalysts of organic reactions [2], e.g., in transesterifications, as shown by Otera [40].

2.4. Monoorganotins

The use of the Kocheshkov redistribution reaction (5) for the synthesis of monoorganotin halides is limited for R = vinyl, phenyl, mesityl, allyl, and acryl ester substituents [14]. In the case of alkyl substituents, the third step of the overall process (between R_2SnX_2 and SnX_4 to give selectively $RSnX_3$) fails and thus the practical way to prepare monoalkyltin(IV) trihalides is to lead the reaction until the mixture contains R_2SnX_2 and $RSnX_3$ which can then be separated by distillation. Nevertheless, suitable catalysts for the problematic step have been found and high yields and selectivity for different monoalkyltin(IV) trihalides, (e.g., *n*-$HexSnCl_3$, $MeSnCl_3$, *n*-$BuSnCl_3$) have been achieved [41]. Reaction (7) between tin(II) dihalides and organic halides, in the presence of different catalysts, gave good results for the synthesis of monoorganotin(IV) tribromides [42] or allyltin(IV) trichlorides [43].

$$SnX_2 + RX \rightarrow RSnX_3 \tag{7}$$

The alkaline hydrolysis of different monoorganotin trihalides [44] or alkyltin trialkoxides may lead, in many cases, to complex cluster structures (e.g., $[(BuSn)_{12}O_{14}(OH)_6](Cl)_2 \cdot 2H_2O$ or $[(BuSn)_{12}O_{14}(OH)_6](OH)_2)$ [45]) that might be interesting as possible catalysts [46].

Monoorganotin compounds also have great potentials in organic synthesis, e.g., in coupling reactions with secondary alkyl bromides in the presence of nickel catalysts [47], and these achievements have been reviewed recently by Echavarren [48].

3. APPLICATIONS AND SOURCES OF ORGANOTIN POLLUTION

In spite of the early discovery of organotin compounds, their widespread use started only in the 1940s due to the expansion of polyvinyl chloride (PVC) production. It was found that the addition of organotin derivatives can prevent the decomposition of heated PVC caused by HCl elimination from the polymer backbone [49]. Ever since organotin chemicals have found various practical applications and their annual production was already around 50,000 tons in the mid 1990s [1]. The practical uses of organotins are more or less limited to tri-, di-, and monoorganotins (Table 1), nevertheless, tetraorganotins are crucially important starting materials or intermediates in the synthesis of these derivatives (see Section 2) and have a great potential in organic synthesis as reagents or mediators in organic reactions. A few examples for tetraorganotin derivatives having insecticidal effects have also been documented [50,51].

3.1. Mono- and Diorganotin Compounds

The most important and oldest application of mono- and diorganotin compounds is their use as stabilizers in the PVC industry. The advantageous properties of these compounds on preventing the heat- and photo-induced decomposition of PVC were discovered in the 1940s by Yngve [52]. Recently, PVC stabilizers have been estimated to make up approximately 60–70% of the annual organotin consumption [53]. One of the problems that rise in the production of PVC is that it loses its stability around 180–200 °C and elimination of HCl from the polymer backbone starts to occur, resulting in the color change of the material through yellow and red to black and also the embrittlement of the polymer. The addition of organotin compounds (e.g., DBT dithiolates) in a quantity of 5–20 g/kg PVC [2] can prevent these problems by (i) scavenging the released HCl – that would otherwise catalyze further eliminations – and by (ii) stabilizing the unstable allylic chloride sites [53].

There are various applications of organotin-stabilized PVCs that involve pipes for drinking, sewage, and drainage water, foils (e.g., in packaging [54]),

Table 1. Practical applications of organotin compounds.

Organotin Derivatives	(Industrial) Applications
R_4Sn	Insecticides
R_3SnX	
$(Bu_3Sn)_2O$, Ph_3SnX, Bu_3SnX, $(CH_2CHMeCO_2SnBu_3)_n$	Antifouling paints-biocides
Ph_3SnX, Bu_3SnX, $(c\text{-Hex})_3SnX$	Agricultural-fungicides, acaricides, insecticides, antifeedants
Bu_3SnX, Bu_3Sn(naphthenate)	Wood preservatives-fungicides, insecticides
Bu_3SnX	Stone, leather, paper protection
Ph_3SnX	Impregnation of textile-fungicide, antifeedant
$(Bu_3Sn)_2O$, $Bu_3SnOCOPh$	Disinfectants
R_2SnX_2	
R_2SnX_2 (R = Me, Bu, Oct; X = isooctyl mercaptoacetate, laurate)	Stabilizers for PVC
Me_2SnX_2	Glass coating
	Homogenous catalysts for
Bu_2SnX_2 (X = octanoate, laurate)	-polyurethane foam formation
Bu_2SnX_2 (X = octanoate, laurate)	-room temperature vulcanization of silicone
Bu_2SnX_2 (X = laurate)	Antihelminthics in poultry farming
$RSnX_3$	
$RSnX_3$ (R = Me, Bu, Oct; X = isooctyl mercaptoacetate)	Stabilizers for PVC
$(BuSnO_2H)_n$, $BuSn(OH)_2Cl$	Homogenous catalysts
$BuSnCl_3$	Glass coating

Compiled from [2,49,55].

window frame sidings and fittings, etc. The possible sources of organotin pollution to the environment have been summarized by Cima, Craig, and Harrington [55], including di- and monoorganotin derivatives originating directly from stabilized PVC materials [56]. In a thorough study, samples of raw, treated, and tap water from houses located on freshly installed PVC pipelines in Canada, were analyzed for organotin derivatives [57]. No organotin compounds were detected in raw or treated water, however,

MMT and DMT derivatives in a concentration range of 0.5–257 ng Sn/L and 0.5–6.5 ng Sn/L were found in about half of the tap water samples, suggesting that the contamination originated from the water distribution system. MBT and DBT were also shown to be leached from chlorinated PVC pipes designed for high temperature water distribution systems [58]. Mono- and diorganotin stabilizers from PVC materials can be addressed as the origin of organotin chemicals found in municipal wastewater [56].

The landfill disposal of organotin-stabilized PVC materials, in general, is also a notable source of organotin pollution to the environment [55,56]. In a study by Takahashi et al. several plastic products, including baking parchments, were analyzed and a very significant amount of DBT and MBT (up to 130000–140000 ng/g) were detected in some of the samples [59]. Furthermore, they found that a fraction of organotins could partially transfer to the foodstuff placed in the baking parchments and prepared in an oven at 170 °C (720 ng/g DBT) and a decent amount of total butyltin (63000 ng/g) still remained in the baking parchments after cooking [59].

Mono- and diorganotin derivatives, mostly MBT, are precursors in glass coating. SnO_2 films are deposited on various hot glass surfaces to strengthen the material and to allow the use of lighter and cheaper glassware [2,53]. A very recent study has described the covalent functionalization and solubilization of metal oxide nanostructures (e.g., TiO_2 and ZnO) and multi-walled carbon nanotubes by organotin reagents [60] that might become a useful way for the preparation of nanostructure dispersions used in composites [60]. Mono- and diorganotin compounds have important uses in homogenous catalysis, especially in transesterification reactions, urethane coatings/polyurethane foam formation or silicone vulcanization at room temperature [2,53]. The most common catalysts that are used in the polyurethane synthesis are the dibuthyltin(IV) dioctanoate and dibutyltin(IV) dilaurate [2,55].

In spite of the above mentioned applications of mono- and diorganotin compounds, their presence in the environment originates mainly from the degradations of trisubstituted organotin substances (e.g., TBT) [49,55,56,61–63] (Figure 1). A significant level of MBT, DBT, MMT, and DMT, as well mono- and diphenyltin (mostly in soil) have been detected in the environment, e.g., in various seawater and freshwater sites [49,63–66], sediments [49,63–67], soils [49,68] or municipal wastewater and sewage sludge [49,56].

3.2. Triorganotin Compounds

Triorganotin chemicals were used worldwide as biocides in the production of antifouling paints which was the most important application of these

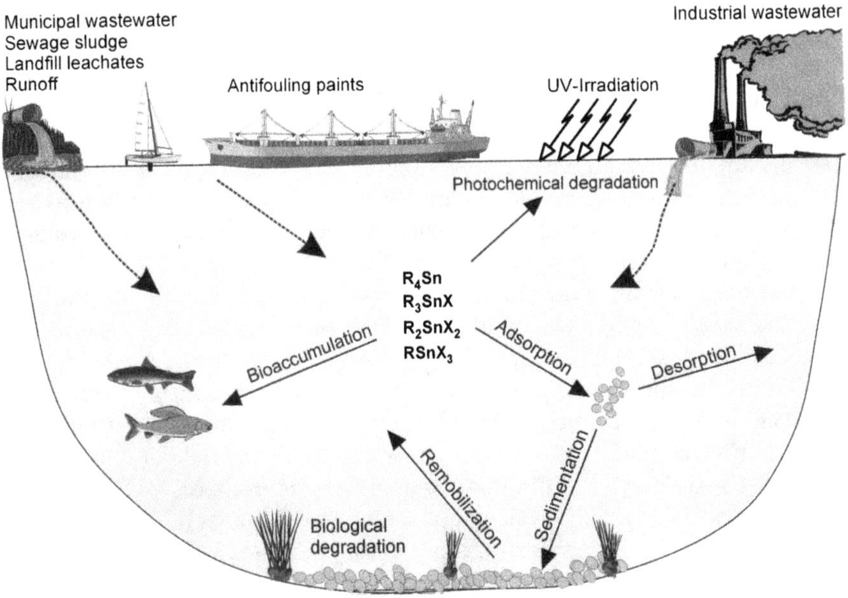

Figure 1. Distribution and fate of organotins and their general routes into the aquatic environment. Reproduced from [49] by permission from Elsevier, copyright (2001).

derivatives until the beginning of this decade. According to the AFS 2001 Convention (International Convention on the Control of Harmful Antifouling Systems on Ships), adopted by the International Maritime Organization (IMO) on October 5, 2001, and which entered into force on September 17, 2008, the use of these compounds in antifouling paints is banned [69]. However, it seems to be unavoidable to give an overview on this organotin application due to the significant impacts it has had and still has on the environment.

Fouling of the vessel hulls by aquatic organisms (e.g., algae, barnacles, weeds) results in the increase of vessel weight and roughness. It causes a notable increase in fuel consumption – a 6% increase for every 100 μm increase in average hull roughness [70] – and also the frequent need of cleaning in drydocks, thus the increase of costs. TBT derivatives, having biocidal properties in contrast to mono- or diorganotin chemicals, started to be in use from the early 1970s when they began to replace Cu_2O in antifouling paints [49]. In the first period, tributyltin oxide was physically dispersed in the paint matrix, forming a free association paint [49,53], however, the release of the biocide was uncontrolled and fast that limited the lifetime of such antifouling covers to 1.5–2 years [53]. In modernized self-polishing copolymer-type antifouling paints the biocide was part of an acrylic copolymer (methyl methacrylate with tributyltin methacrylate) [70], that could provide a constant and controlled biocide level around the immersed vessel structures preventing the settling of aquatic organisms, and it also had a significant increase of lifetime (~5 years) [70]. The release of the biocide from such antifouling paints occurs through a hydrolysis reaction as seawater interacts with it and cleaves the TBT from the copolymer, causing an erosion of the paint [70].

Already from the 1980s on the use of TBT-containing biocides in antifouling paints started to be regulated, due to the observed negative effects of the released TBT on the environment. The most reflective case of TBT pollution, having a dramatic effect on oyster growth and reproduction in Arcachon Bay in France from 1975 to 1982 [71], initiated international attention, which later led to regulations and finally to the complete ban of TBT derivatives from antifouling paints. The above cited IMO convention has been ratified already by 36 countries (status of convention as of January 31, 2009 (http://www.imo.org)), representing more than fifty percent of the world's merchant shipping tonnage [69].

Nevertheless, the extensive use of TBT biocides in the previous decades resulted in the accumulation of TBT derivatives in the aquatic environment. Evidently, areas with strong ship traffic (e.g., harbors) and shipyards, where the reparation and cleaning of vessel hulls take place, are the most affected [64–66,70,72]. Prior to strong legislations the concentration of TBT in the polluted zones was in the range of 1–2000 ng Sn/L [55] which is very significant considering that TBT concentration around 1 ng/L is believed to cause imposex in female snails [49]. Due to the legislations, the TBT level in water should show a decreasing tendency [55].

A very serious and presumably long-lasting problem is the contamination of sediments where the decomposition of organotin derivatives is much slower than in seawater (especially close to the surface), the estimated half-life of TBT being in the range of several years [2,49,55,73,74]. The level of TBT contaminations detected in sediments of highly polluted zones can be as high as a few thousand ng Sn/g dry weight [49,55]. The organotin contaminants in the upper layer of the sediment are available to various organisms and can be remobilized, too [49]. The sources of organotin contaminations and their fate in the aquatic environment are summarized in Figure 1. The best available techniques for the removal of TBT from the shipyard wastes and from contaminated sediments are highlighted in a very recent review [75].

The biocidal properties of triorganotins have been discovered in the 1950s by van der Kerk and Luijten [76] and this important discovery opened the way for their agricultural uses as pesticides. They are widely used as fungicides, bactericides, herbicides, acaricides, insecticides or antifeedants [77,78]. The most common derivatives are the triphenyltin (TPT) and tricyclohexyltin (TCHT) compounds [53], but besides, TBT derivatives also have applications for similar purposes. TPT compounds are applied generally as fungicides on potatoes, sugar beets, pecans, peanuts, coffee, cocoa, rice, sunflower, tomato, onion, etc. [49,53,77] while TCHTs are extremely efficient as acaricides for several fruits (e.g., apple, pear, grape, citrus fruit), tea and wine [49,53,78]. Due to the direct use of these chemicals on plants, they can easily penetrate the soil where they can be adsorbed [68] and later desorbed, opening the way also to the aquatic environment by leaching and run off [49,79]. Triorganotins can also appear in wastewater and in sewage sludge [56,80], thus the dumping of wastewater or sludge to seas or the disposal of sewage sludge on landfills must also be considered as sources of (tri)organotin pollution [56].

TBT compounds, like tributyltin(IV) oxide or tributyltin(IV) naphthenate, having fungicidal properties, are used as wood preservatives [2,49,63]. For the impregnation of wood, a double-vacuum process, performed in a special chamber, is the most efficient technique used in timber industry [2]. The preservative stays safely in the wood impregnated by this method, and leaching is considered to be negligible [49].

A number of tri- and diorganotin compounds have been reported to possess cytotoxic or anticancer activities *in vitro* and in a few cases, also *in vivo* [81–85]. However, the mechanism of the antitumor activity of organotin compounds has not yet been explored [85]. Whether organotin compounds can become competitive anticancer therapeutic drugs in the future is still an open question.

4. (BIO)INORGANIC SPECIATION IN THE AQUATIC ENVIRONMENT

4.1. Aqueous Complexes with Hydroxide Ion and Other Inorganic Ligands

The equilibrium speciation of organotin(IV) cations in aqueous environments is fundamentally determined by their strong Lewis acid character, i.e., their ability to form stable coordination compounds. Although the Lewis acidity of mono-, di-, and triorganotin(IV) cations is characterized by different hardness, all of them show a strong tendency to hydrolyze in aqueous solutions. Therefore, hydroxide ion is by far the most important inorganic ligand for these cations. After the pioneering work of Tobias et al. [86,87], the hydrolysis of different organotin(IV) cations have been studied in several

Table 2. Hydrolysis constants of $(CH_3)_xSn^{(4-x)+}$ cations at $I = 0$ M and $T = 298$ K.

	$-\log^*\beta^{\#}_{pq}$		
species (p,q)[a]	$(CH_3)Sn^{3+}$	$(CH_3)_2Sn^{2+}$	$(CH_3)_3Sn^{+}$
1,1	1.5	2.86	6.14
1,2	3.46	8.16	18.88
1,3	9.09	19.35	
1,4	20.47		
2,2		4.99	
2,3		9.06	
2,5	7.69		

[a]p and q stand for the stoichiometric numbers in Mp(OH)q species
Adapted from [95].

laboratories (see for example [88–91]). Systematic studies on the ionic strength and temperature dependence of the hydrolysis constants for mono-, di-, and trialkyltin(IV) cations have been published only recently [92–97].

The propensity for hydrolysis follows the trend $RSn^{3+} > R_2Sn^{2+} > R_3Sn^{+}$ (Table 2), according to the hardness of organotin(IV) cations [95]. Aside from mononuclear hydroxo complexes, hydroxo-bridged dinuclear complexes are also formed, but the stability of dinuclear species strongly decreases with increasing number of alkyl-substituent on tin(IV) (Figure 2). Some papers [86,89] reported the formation of higher oligomers at high concentration of the metal ion ($[(CH_3)_2Sn^{2+}] > 20$ mM), too, but these species are not relevant from an environmental point of view.

A very important feature of the organotin(IV) hydroxo complexes is their high solubility, which is more or less the same as those of the aqua ions. This surprising fact has fundamental impact on their speciation in the aquatic environment.

The hydrolysis constants of the different $R_xSn^{(4-x)+}$ cations do not show a clear dependence on the nature of the alkyl(aryl) groups [96], which provides the possibility to deduce the coordination ability of the most used but rather insoluble butyl- and phenyltin(IV) derivatives, from the studies performed with methyl- or ethyltin(IV) cations.

The dependence of the hydrolysis constants of $R_xSn^{(4-x)+}$ cations in different media ($NaNO_3$, NaCl, Na_2SO_4, Na(Cl/F), $Na(Cl/CO_3)$) can be explained by the formation of ion pairs between the aqua/hydroxo complexes and the above listed inorganic ions, which was taken into account both in terms of stability constants and of the specific ion interaction theory using the Pitzer equations [92–97]. The formation of both parent and hydroxo mixed ligand complexes has been detected, with relatively high stability. The presence of the above listed anions in seawater significantly

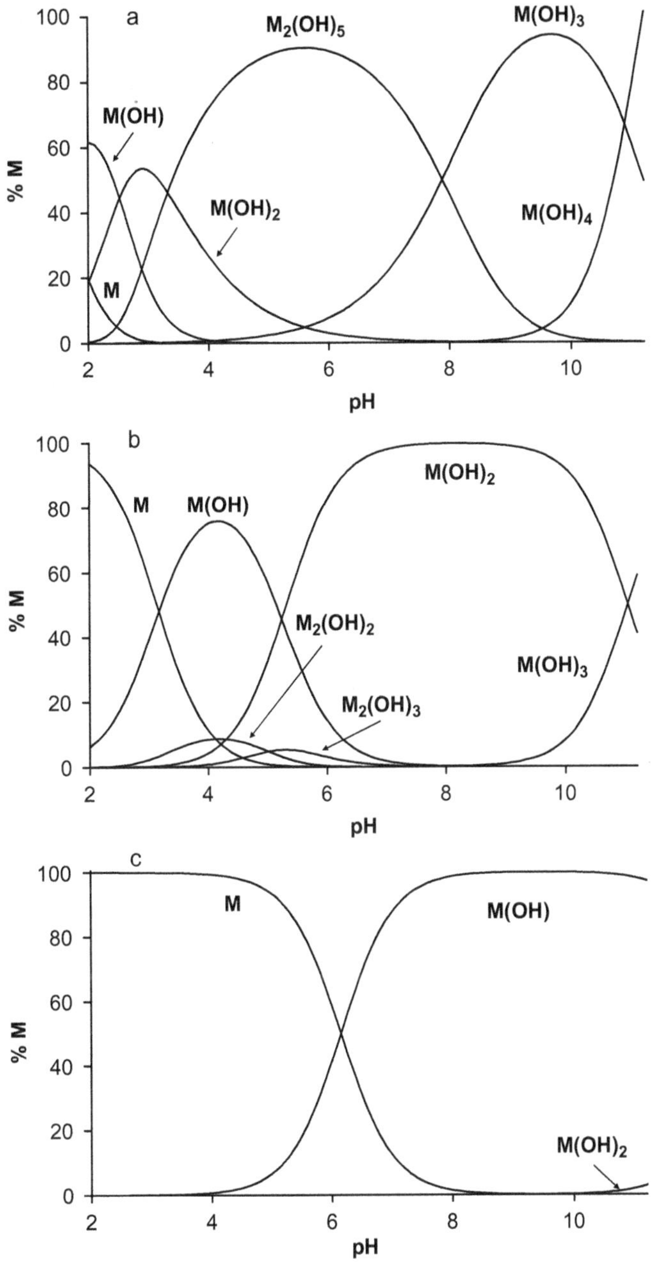

Figure 2. Species distribution curves for the hydrolysis of $(CH_3)_xSn^{(4-x)+}$ cations ($M = (CH_3)Sn^{3+}$ (a), $(CH_3)_2Sn^{2+}$ (b), $(CH_3)_3Sn^+$ (c), [M] = 0.003 M, $I = 0$ M). Calculated with equilibrium constants given in [95].

influences the formation of hydrolytic species of RSn^{3+}, while their effect is moderate and negligible in the cases of R_2Sn^{2+} and R_3Sn^+, respectively.

Only a few data are available for the ortho- and pyrophosphate [98] and tripolyphosphate [99] complexes of organotin(IV) cations, indicating relatively strong interactions, especially in the acidic pH range.

4.2. Aqueous Complexes with Naturally Occurring Small Organic Ligands

The speciation of organotin(IV) cations in natural waters, in sewage or in biofluids is strongly influenced by the complex formation with the available metal-binding compounds. In both high and low molecular weight ligands of biological and environmental interest, the carboxylate group is one of the most important metal-binding sites.

Organotin(IV) cations form rather stable complexes even with acetate (log $K_{ML} = 2.81$, $I = 0.1$ M $NaNO_3$, M = $(CH_3)_2Sn^{2+}$ [100]), comparable to the first row transition metal ions, but due to their strong tendency to hydrolyze the percentage of the acetate-complexed organotins is rather low in the acidic pH range. Obviously, dicarboxylic acids (e.g., malonic or succinic acids) form more stable complexes with organotins. Similarly to the hydroxo species, the stability of organotin(IV) complexes of these ligands significantly decreases with decreasing cation charge (e.g., log $K_{ML} = 8.6$, 5.43 and 2.74, for the MMT, DMT, and TMT complexes of malonic acid, respectively, at $I = 0$ M [101]). However, the ligand and the hydroxide ion are in strong competition for the metal ion, therefore, the formation of malonato complexes does not correlate with the above listed stability order (Figure 3). Though at pH 4 the concentration of malonato complexes follows the order MMT > DMT > TMT, at neutral pH only the TMT complexes are present in the solution in considerable amount (Figure 3). Although only a few comparative studies are available on the different $R_xSn^{(4-x)+}$ complexes [101], the above mentioned behavior can be generalized for most of the hard base ligands.

The presence of additional donors in the ligands may considerably increase the stability of the formed complexes. Figure 4 compares the speciation of the DET-succinic acid (SA), and -malic acid (MA) systems. The additional stabilization of the -OH group can be clearly seen from the basicity-corrected stability constants of the complexes ML (log K^*_{ML} = log β_{ML} − log β_{H_2L}). Log K^*_{ML} is nearly two orders of magnitude lower in the case of SA than in that of MA (log $K^*_{ML} = -4.56$ and −2.93, respectively [91]), indicating the additional stabilization provided by the coordinated OH group. The presence (or absence) of the hydroxyl groups governs the

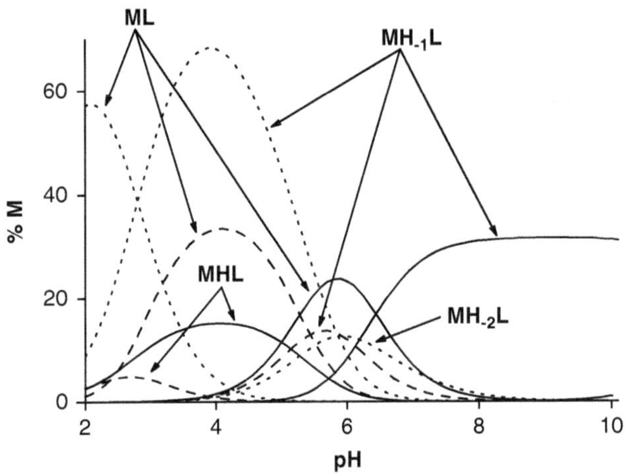

Figure 3. Species distribution curves of the $(CH_3)_xSn^{(4-x)+}$-malonic acid systems ($M = (CH_3)Sn^{3+}$ (dotted lines), $(CH_3)_2Sn^{2+}$ (broken lines), $(CH_3)_3Sn^+$ (full lines), $I = 0$ M, $2[M] = [L] = 0.002$ M). Calculated with equilibrium constants given in [101]; the distribution curves of the hydrolytic species are not shown for the sake of clarity.

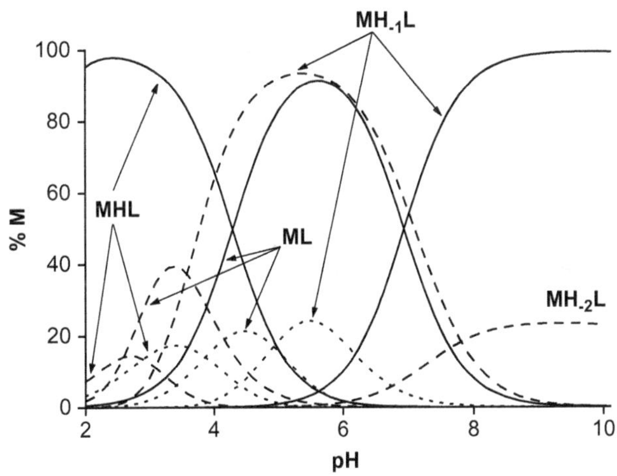

Figure 4. Species distribution curves of the $(C_2H_5)_2Sn^{2+}$-succinic acid (dotted lines), -malic acid (dashed lines) and -mercaptosuccinic acid (full lines) systems ($M = (CH_3)_2Sn^{2+}$, $I = 0.1$ M, $2[M] = [L] = 0.002$ M). Calculated with equilibrium constants given in [91]; the distribution curves of the hydrolytic species are not shown for the sake of clarity.

successive deprotonation processes, too. The pK value for the reaction $ML = MH_{-1}L + H^+$ is much higher for SA than for MA (pK=4.92 and 3.58, respectively [91]). In the case of SA a mixed hydroxo species is formed in the above process, while metal-promoted deprotonation of the hydroxyl group takes place in the case of MA [91]. A similar stability enhancement has been reported for the succinic/tartaric acid [91] and tricarballylic/citric acid pairs [101].

Based on the equilibrium study of ten different carboxylates with MMT, DMT, and TMT cations, Sammartano et al. formulated an empirical correlation between complex stability and some simple structural parameters [101],

$$\log\beta(I = 0) = -6.0 + 1.63n_{\text{carb}} + 1.4n_{\text{OH}} + 4.58r + 3.9z_{\text{cat}} \quad (8)$$

where n_{carb} and n_{OH} are the number of carboxylic and alcoholic groups in the ligand, respectively, r is the stoichiometric coefficient of H^+ (+) or OH^- (–) in the given complex, and z_{cat} is the charge of the methyltin cations $(CH_3)_xSn^{(4-x)+}$. This correlation indicates mainly electrostatic interactions between organotin(IV) cations and O-donor ligands, which is also supported by the fact that the major contribution to the stability of these complexes is the entropic term [102].

Interestingly enough, the replacement of OH group(s) by thiol group(s) in hydroxycarboxylic (lactic, malic or tartaric) acids results in a fundamental stability increase of the formed complexes [91]. This is in sharp contrast with the hard Lewis acid behavior of organotin(IV) cations concluded above from the interaction with O-donor ligands, and indicates the exceptional coordination ability of these cations. Indeed, in the DMT-2-mercaptopropionic (MPA), -mercaptosuccinic (MSA), and -dimercaptosuccinic (DMSA) acid systems, between pH 2–11 the metal ion is completely transformed into thiolate-bound species (Figure 4). In the neutral pH range trigonal bipyramidal $\{COO^-,S^-\}$ and $\{COO^-,S^-,OH^-\}$ coordinated complexes are in equilibrium in the case of MPA and MSA, while an exceptionally stable, octahedral $\{2COO^-,2S^-\}$ coordinated dimer is present in solution in the case of DMSA [91].

Although the hydroxyl group is considered as a hard base, the coordination affinity of polyhydroxylated ligands toward organotin(IV) cations largely depends on the steric arrangement of the OH groups and on the availability of other donor(s) in chelating position(s). Most monosaccharides are able to coordinate to DMT only in the alkaline pH range, above pH 8–9 [103,104]. However, fructose in excess over DMT may compete with the hydroxide ion even in the neutral pH range, due to the favorable *ax-eq-ax* arrangement of the OH groups in this ligand [103]. The presence of carboxylate(s) in open chain polyhydroxy derivatives (such as gluconic acid or in *N*-D-gluconylamino acids) results in a considerably higher stability of the diorganotin(IV) complexes [105,106], suppressing completely the hydrolysis, but the effect is less pronounced in the cases of the cyclic ascorbic [107] and glucuronic acids [108].

Phosphomonoesters of monosaccharides also show an enhanced affinity toward DMT as compared to the parent sugars themselves [109]. In the acidic pH range the phosphate group is the primary binding site with possible participation of the non-deprotonated sugar OH groups. In the neutral pH range $DMT(OH)_2$ is the dominating species, while at $pH > 10$ alcoholate(s) of the sugar moiety become potent competitor(s) of hydroxide ion. Mononucleotides behave in a similar manner with DMT [104,109,110], but are able to partially suppress the hydrolysis of MMT and TMT in the neutral pH range [110]. The coordination of the base nitrogen(s) was not reported at any pH [104,109]. Due to the presence of the triphosphate unit, nucleoside 5'-triphosphates have an increased binding affinity toward DMT in the acidic pH range, but hydrolytic species dominate in the neutral pH range, too [104,111].

Obviously, the increasing number of phosphomonoester units results in a higher stability of the complexes formed. Phytic acid (myo-inositol hexakisphosphate), a widely distributed ligand in plants with high sequestration ability, forms very stable mono-, di-, and trinuclear complexes with DMT [112].

Only a few studies are available on the equilibrium speciation of organtin(IV)-amino acid complexes [90,98,113]. Amino acids with non-coordinating side chains form MHL, ML, and $MH_{-1}L$ complexes with DMT [90,113]. The protonated species is monodentate $\{COO^-\}$ coordinated. The comparison of amino acids having different basicity and different size of chelate rings formed during complexation revealed $\{COO^-,OH^-\}$ type coordination in ML [90], although bidentate $\{COO^-,NH_2\}$ type binding was also assumed [113]. In the neutral pH range mixed hydroxo complexes are present, and the DMT-binding ability follows the order Gly < Ala < Phe < Val [90,113]. The imidazole side chain of histidine does not coordinate to DMT, since the stability of histidine and glycine complexes is similar [90]. On the contrary, the presence of a sulfur atom in a chelating position considerably enhances the stability of the formed complexes [114,115]. Equilibrium studies on the DET- and DMT-cysteine systems [114,115] revealed similar speciation and stabilities of the complexes. With increasing pH highly stable $\{COO^-,S^-\}$, $\{COO^-,S^-,NH_2\}$ and $\{COO^-,S^-,NH_2,OH^-\}$ coordinated complexes dominate in solution at pH = 3,5, ~6, and 10, respectively (Figure 5), suppressing completely the hydrolysis of DET. Similarly to thiocarboxylic acids [91], the high stability is due to the favored thiolate coordination. Comparison with *N*-acetyl cysteine (Figure 5) proves the coordination and additional stabilization of the amino group above pH 6 in the case of cysteine. *S*-methylcysteine forms more stable complexes than glycine, also indicating the coordination of the thioether group [114].

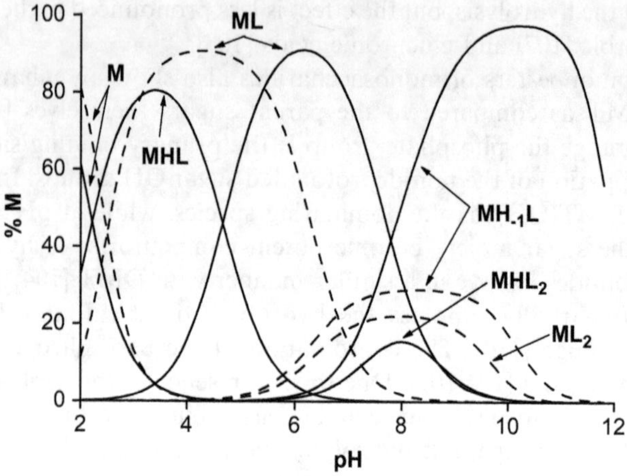

Figure 5. Species distribution curves of the $(C_2H_5)_2Sn^{2+}$-*N*-acetyl-cysteine (dashed lines) and -cysteine (full lines) systems ($M = (C_2H_5)_2Sn^{2+}$, $I = 0.1$ M, $2[M] = [L] = 0.002$ M). Calculated with equilibrium constants given in [114]; the distribution curves of the hydrolytic species are not shown for the sake of clarity.

Peptides are efficient metal ion binders in biology and form stable complexes with organotin(IV) cations. Although the X-ray diffraction study of some crystalline organotin(IV)-peptide complexes provided definite evidence of the formation of an Sn-amide bond [7], diorganotin(IV)-induced amide deprotonation in aqueous solution has been reported recently at surprisingly low pH (4–5) [90,105,116–118]. Amide coordination is essential for the strong metal ion binding of oligopeptides at physiological pH. It is known for many metal ions that the presence of a suitable anchoring donor is of crucial importance to promote amide deprotonation [119]. In contrast with most other metal ions, the C-terminal COO^-, and not the N-terminal NH_2, is the primary anchor for DMT in its complexes with several Gly-X and X-Gly peptides [90,116]. The deprotonation of ML leading to the amide-coordinated $MH_{-1}L$ can be attributed to the cooperative proton loss of the amino and amide nitrogens followed by a water release from the coordination sphere of the cation (Figure 6). The amide-coordinated trigonal bipyramidal $MH_{-1}L$ complex is very stable, and the side-chain donor groups (imidazole, carboxylate, etc.) do not influence its stability and structure.

The replacement of the terminal amino group by a thiol group in mercaptopropionyl-glycine results in a considerably enhanced stability and a different primary binding site [118]. The thiolate is coordinated to the metal ion already at pH 2, therefore it takes over the anchoring role in the amide deprotonation. The speciation of different DMT-(pseudo)dipeptide $MH_{-1}L$

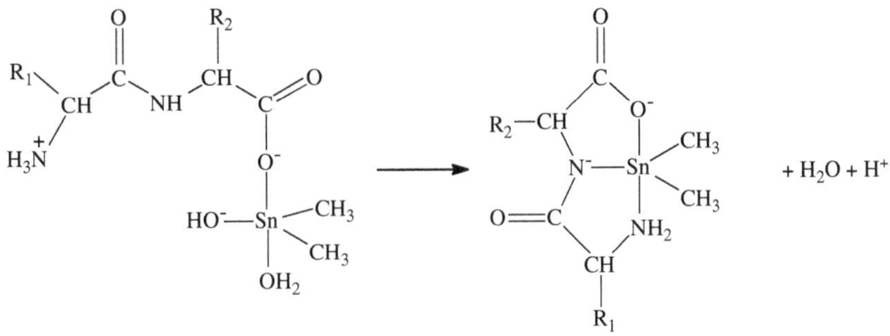

Figure 6. Schematic structure showing the cooperative deprotonations of amide and amino nitrogens in DMT-peptide complexes.

complexes (Figure 7) clearly shows the following donor set preference: $\{NH_2,N^-,COO^-\}<\{O^-,N^-,COO^-\} \ll \{S^-,N^-,COO^-\}$.

In the case of reduced glutathione the coordination of thiolate is the governing factor in their $(CH_3)_xSn^{(4-x)+}$-complexes, and the deprotonation of amide nitrogen(s) was not observed [120]. Recently a mitochondrial membrane protein named stannin has been identified that sensitizes neuronal cells to TMT intoxication. A nonapeptide fragment of stannin containing the putative metal-binding Cys-Xaa-Cys motif has favored preference for diorganotins,

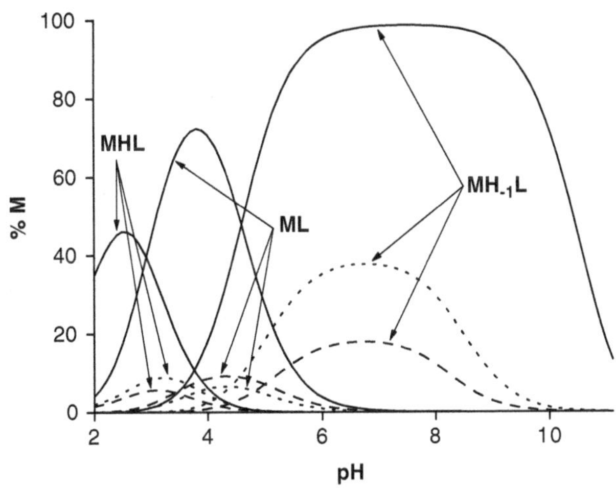

Figure 7. Species distribution curves of the $(CH_3)_2Sn^{2+}$-Ala-Gly (dashed lines), -salicyl-glycine (dotted lines) and -mercaptopropionyl-glycine (full lines) systems ($M=(CH_3)_2Sn^{2+}$, $I=0.1$ M, $2[M]=[L]=0.002$ M). Calculated with equilibrium constants given in [117] and [118]; the distribution curves of the hydrolytic species are not shown for the sake of clarity.

Table 3. Formation constants of some selected dimethyltin(IV) (DMT) and copper(II) complexes ($I = 0.1$ M, $T = 298$ K).

Ligand	Species	log β(DMT)	log β(Cu^{2+})
Acetic acid	ML	2.81 [100]	1.73
Malic acid	ML	4.65 [91]	3.67
Gluconic acid	ML	3.42 [106]	2.51
Citric acid	MHL	10.83 [99]	9.55
	$M_2H_{-1}L$	6.65	4.92
5'-GMP	MHL, log K_{M+HL}	4.68 [104]	3.9
Glycine	ML	7.99 [90]	8.20
Gly-Gly	ML	6.61 [90]	5.55
	$MH_{-1}L$	1.80	1.56
Ala-Gly	ML	6.80 [118]	5.34
	$MH_{-1}L$	1.81	1.66
Gly-Asp	ML	7.51 [116]	6.61
	$MH_{-1}L$	2.30	1.85
Mercaptopropionylglycine	ML	9.52 [118]	7.6
	$MH_{-1}L$	4.93	1.4
Oxydiacetic acid	ML	5.18 [122]	3.97
	log $K_{ML}{}^{a}$	−1.56	
Iminodiacetic acid	ML	9.41 [122]	10.57
	log $K_{ML}{}^{a}$	−4.14	
N-Methyliminodiacetic acid	ML	9.62 [123]	11.04
NTA	ML	10.38 [99]	12.94
EDDA	ML	12.41 [123]	16.2

[a]basicity-corrected stability constants (log $\beta_{ML} - \sum pK$)
The values for copper(II)were taken from [189].

which induces dealkylation of TMT, i.e., the formation of a {$2S^-$}-coordinated DMT-peptide complex and the release of methane [121].

Only few reports have been published on the interaction of DMT with amino-polycarboxylates [99,122,123]. Although IDA, MIDA, and NTA (see Table 3) form stable ML complexes with DMT around pH 4, they are not able to prevent metal ion hydrolysis in the neutral pH range [99,122,123]. The ML complex of NTA is only slightly more stable than that of MIDA, thus the third carboxylate of NTA is weakly bound or not at all [99,123]. The sequestering capacity of the studied aminopolycarboxylates at pH 7 follows the order 2,6-pyridinedicarboxylic acid ≥ EDDA > EDTA > NTA > IDA ~ MIDA. In contrast to most metal ions, EDDA forms more stable complexes with DMT than EDTA, due to the steric effect of the two tin-bound methyl groups, which destabilizes the ML complex, and promotes the formation of M_2L [123].

It is noteworthy that $(CH_3)_xSn^{(4-x)+}$ cations form more stable complexes with (poly)carboxylic acids, (poly)hydroxycarboxylic acids, nucleotides, and peptides than the most commonly studied cations with identical charges. Table 3 compares the formation constants of some representative DMT and copper(II) complexes. Although, the coordination modes are not necessarily identical, only the DMT complexes of ligand with amino groups are less stable than those of copper(II), except the peptide complexes. For example, the DMT complexes of citric acid are more stable, while its EDDA complex is less stable than the corresponding copper(II) species (Table 3). The higher stability of the DMT-peptide complexes is probably due to the favored formation of a covalent metal-amide bond. The preference of DMT for O-donors over an amino group is clearly seen from the basicity-corrected stability constants of IDA and ODA (see Table 3). The available data clearly show the $N < O < S$ donor preference of organotin(IV) cations, which does not fit into the hard-soft classification.

Indeed, there are conflicting reports in the literature concerning the interaction of organotin(IV) cations with polyamines. Complexation has not been observed in the DMT-histamine [90] and TMT-bipyridyl [98] systems, while others reported strong complex formation [124]. Clearly, further studies are needed to establish the organotin(IV) binding ability of polyamines in aqueous environment.

4.3. Interaction with Biological Macromolecules

Humic substances of biological origin in natural waters and in sediments have a high metal ion sequestering ability due to their carboxylate and phenolate functions and therefore, they considerably alter the distribution of many inorganic pollutants in environmental matrices. Organotin(IV) binding to insoluble and soluble humic acids may provide a mean for the transport of these compounds from contaminated sediments to the overlying water [125]. The conditional stability constant of humic acid-organotin(IV) (MBT, DBT, TBT, tripropyltin, TET, TPT) complexes, determined by dialysis techniques, are between log $K = 4.6$–6.1, suggesting that humic acids have a significant affect on the fate and transport of organotin(IV) compounds in low salinity lacustrine sediments [125].

In spite of the high toxicity of organotin compounds, the literature on their binding to biological macromolecules at the molecular level is rather scarce. Trialkyltin(IV) derivatives have been reported to interact with thiolate and imidazole side chains of native cat and rat hemoglobin in a trigonal bipyramidal environment [126,127].

Mitochondrion-dependent apoptosis of rat liver induced, by selective interaction of TBT with two proximal thiol groups of an adenine nucleotide translocator, opening of the permeability transition pore, thereby decreasing membrane potential and releasing cytochrome *c* from mitochondria [128]. As mentioned above, in 1992 a small mitochondrial membrane protein named stannin has been identified that sensitizes neuronal cells to TMT intoxication [129]. This protein is largely expressed, in a direct correlation with TMT toxicity, in multiple tissues such as spleen, brain, lymph, or liver. Stannin has two conserved vicinal cysteines (C32 and C34) that may constitute an organotin binding site [130]. The model peptide of this binding site has been shown to dealkylate TMT to DMT via the CXC sequence [121], suggesting that stannin may carry out a dealkylation reaction resembling that of the bacterial protein organomercurial lyase. The coordination of TMT/DMT may induce substantial structural and/or dynamical changes of stannin, recruiting other binding partners to initiate the apoptotic cascade [131].

Based on some similar observations [132,133], thiol groups seem to be the main protein targets for organotin(IV), especially when vicinal thiols are available. However, most thiol groups are present in the hydrophobic core of the globular proteins and are not accessible to the thiol reagents [134]. Due to their high hydrophobic properties, neutral organotin(IV) compounds, such as TBT(OH), are able to interact with both surface and internal thiol groups, which might induce irreversible inactivation of many proteins/enzymes [132].

A different mechanism of interaction has been reported to exist between TBT and F_1F_0 ATP synthase. TBT interacts with the selectivity filter of the ion channel of subunit ‘a’ of ATP synthase through non-covalent interactions without any explicit involvement of the thiols in the coordination of the tin atom. This interaction prevents Na^+ ions from passing through the channel, which can be suppressed by high sodium ion concentration, indicating competition between inhibitor and Na^+ binding [135].

Organotin binding to DNAs seems to be less preferred than to proteins. Among MMT, DMT, and TMT, only MMT interacts with calf thymus DNA under physiological conditions [136]. An increase of the DNA melting point was observed on increasing TMT concentration, indicating an interaction with the phosphodiester groups. At pH 7.4 DMT and TMT are present mainly in neutral hydrolyzed form, which prevents electrostatic interaction with DNA [136]. These species are able to interact with DNA only in their cationic forms at acidic pH, which is consistent with earlier findings in the DMT-5’-d(CGCGCG)$_2$ system [104].

5. CONCENTRATION AND DESTINATION IN THE ENVIRONMENT

The environmental appearance of organotin compounds originates mostly from anthropogenic sources. These compounds are present in the aquatic environment, in seas close to the shores or even in deep sea, in sediments, in rivers and lakes and in mainland soil. The concentration and distribution of the organotin derivatives are influenced by several factors, like the solubility of the species in aqueous medium, adsorption to solid particles in water or to the soil, degradation/transformation processes that all influence the persistence and accumulation of the contaminants in the ecosystem.

5.1. Solubility, Stability, Transformation, and Degradation

The solubility of organotin compounds ($R_{(4-n)}SnX_n$ with $n = 0–3$) is strongly dependent on the quality of the R and X groups and also on their relative number [55]. Obviously, the increasing number and length/hydrophobicity of the R substituents decrease the solubility in general but the relation with the number of R groups is not always straightforward [137]. Definitely, triorganotin compounds in general have a low solubility; depending on the circumstances [pH (5–7), temperature (10–25 °C), salt content] it falls in the range from 0.1 to ca. 50–70 mg/L [137–139]. Di- and monomethyltin(IV) chlorides are dissolved in water extremely well, the corresponding data falling in the 10^4–10^5 mg/L range [49,137].

As it was hinted above, the solubility of species highly depends on the various circumstances, including temperature, pH, ionic strength of the solution, and on the quality and quantity of the inorganic and organic ligands that may be present in the solution. In a model study, the applied artificial seawater conditions were shown to decrease the solubility of four selected organotin derivatives by a factor of 2–30 [138]. In the absence of coordinating ligands organotins are present in solution as cations or as different hydrolysis species, depending on pH. The pK_a values of TBT and TPT cations were found to be 6.25 and 5.20, respectively [140]. Accordingly, the dominant species in neutral conditions are the neutral, monohydroxo species. Schwarzenbach et al. studied the 1-octanol-water [140] and later, the liposome-water [141] partitioning of TBT and TPT and determined D_{ow} values (overall distribution ratio) as a function of pH. The profiles followed the hydrolysis of the cations and increased and levelled off in parallel with the formation of the hydroxo species in 1-octanol-water [140], however, a slightly decreasing tendency with increasing pH was seen in liposome-water with both compounds [141]. It was suggested that the sorption of the cationic species by the phosphatidylcholine liposomes was governed by complex formation with the phosphate groups and not just by electrostatic interactions [141].

Amongst environmental conditions a very important factor determining the distribution and fate of species is the adsorption (and desorption) of organotins to solid particles (e.g., to the sea sediments), characterized by K_d

values. The adsorption behavior of organotin contaminants can be characterized in general by cation exchange processes on the negatively charged metal oxide or clay mineral surfaces, however, beside the sediment composition, there are many factors, including the molecular structure of the organotins, complexation processes with negatively charged ligands, salinity, and pH, that influence substantially the adsorption and desorption processes [49]. Adsorption and desorption of organotins is considered to be reversible, however, TBT and TPT derivatives were shown to remain in the sediments of harbors for a long time [142], consequently their slow release process may have long-term ecotoxicological consequences by influencing the bioavailability of organotin contaminants [139,143].

Organotin compounds can be considered as stable materials, regarding the stability of the carbon-tin bond (dissociation energy is ~190–220 kJ/mol) since it is stable to heat (up to ~200 °C), atmospheric conditions (O_2), and water [55]. Nevertheless, amongst environmental conditions, there are several types of degradation processes that provide routes for their transformations to other organotin derivatives or finally, to inorganic tin species (Figure 8).

The loss of organic substituents can be described by the following simple pathway:

$$R_4Sn \rightarrow R_3SnX \rightarrow R_2SnX_2 \rightarrow RSnX_3 \rightarrow SnX_4$$

and the processes can occur by biological cleavage (aerobic or anaerobic) and by abiotic mechanisms, like UV radiation or chemical cleavage [49,55,139]. In addition, in a recent work, a nine amino acid-peptide with a CXC motif, corresponding to the putative TMT binding site of the membrane protein stannin has been synthesized and studies have revealed a strong dealkylating property of the peptide for trisubstituted organotins having 1–3 carbons in the R groups [121].

Regarding the kinetic aspects, it seems that photolysis can be a relatively fast route in water until limited depth or in the very top layer of soil. It has probably very minor significance in sediments or in the deeper soil layers [49]. TPT and TCHT were found to degrade fast by UV radiation, however, the measured half-lives for TBT compounds are much longer and fall in the range of a few weeks to a few months [49,55,74,144]. The increasing salinity and humic acid concentration were shown to decrease remarkably the UV degradation rates of methyltins (especially TMT) at laboratory conditions [145].

Biological degradation processes are probably the most important degradation routes of organotin compounds, at least for TBT derivatives [61]. Collected half-lives of various organotin compounds in different conditions reflect that the dealkylation of TBT to DBT and MBT is a rather

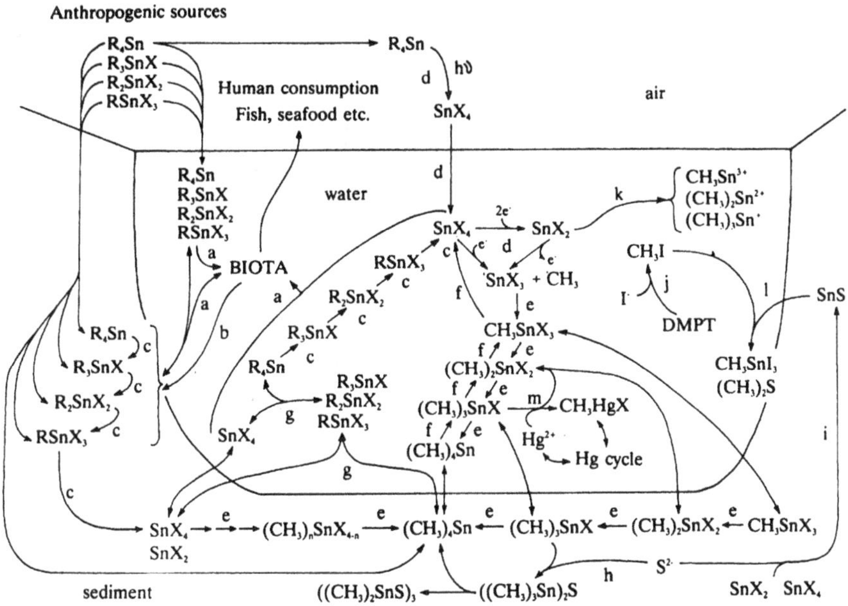

Figure 8. A model for the biogeochemical cycling of organotins. The main reactions detailed are: (a) bioaccumulation; (b) deposition or release from biota on death or other processes; (c) biotic and abiotic degradation; (d) photolytic degradation and resultant free radical production; (e) biomethylation; (f) demethylation; (g) disproportionation reactions; (h) sulfide-mediated disproportionation reactions; (i) SnS formation; (j) formation of methyl iodide by reaction of dimethyl β-propiothetin (DMPT) with aqueous iodide; (k) CH_3I methylation of SnX_2; (l) oxidative methylation of SnS by CH_3I to form methyltin triiodide; and (m) transmethylation reactions between organotins and mercury. Reproduced from [62] by permission from Elsevier, copyright (2000).

slow process in sediments [49,55]; the estimated half-lives vary between a few months to several years. High concentration of TBT was found to inhibit the microbial degradation process by having adverse effects on the development of the microorganisms [146,147]. A review from 1999 by White, Tobin, and Cooney gives an overview on the interaction of microorganisms with organotins, including the mechanisms of toxicity, uptake, resistance, and biotransformations of the organotin derivatives [63]. In a more recent review, Dubey and Roy focus on the biodegradation of TBT derivatives by various organisms, especially bacteria, and discuss the biochemical and genetic basis of organotin resistance [61]. They claim that further efforts to explore the exact mechanism of biodegradation and the genes that are involved in the process could allow the use of bacteria for the remediation of organotin-polluted sites [61]. Indeed, a TBT-resistant bacterium, *Aeromonas veronii,* has been isolated lately and the authors claim that it degrades and utilizes TBT as a carbon source [148].

Similarly to sediments, microbial degradation of organotin compounds may be the most relevant pathway of organotin dealkylation in soil. The bacterial decomposition of triphenyltin(IV) acetate to di- and mono-phenyltin and inorganic tin was observed in a soil sample with a half-life of about 140 days, nevertheless, decomposition did not occur in sterile soil [149]. Other authors reported shorter half-lives [150], however, these data are strongly dependent on the conditions, including sunlight, soil type (affecting the adsorption and thus the bioavailability), moisture content, and the actual microbial activity [150]. Due to the same reasons, half-life values for TBT also vary in a wide range, between 1 day and 4 years [151]. Nevertheless, TBT is much more persistent then TPT, and its degradation products, DBT and MBT are also persistent [68,151,152].

Beside degradation processes biomethylation also influences the available forms of organotins in the environment. Methyltin derivatives may be formed by biomethylation processes representing the only non-anthropogenic origin of organotin in the environment [49,55,62]. Methylcobalamin is believed to be the main methylating agent for tin compounds [62]. Methyltin formation in anaerobic sediments has been associated with sulfate-reducing bacteria, e.g., *Desulfovibrio* sp. [62]. Other methyl donors, e.g., methyliodide, produced by certain algae and seaweeds can also be involved in the methylation of inorganic tin(II) salts in aqueous medium (tin(IV) compounds do not react) [49] which was also supported by laboratory model experiments [153]. Besides, transmethylation of methyltins by other heavy metals also has significance [49,153]. TBT and its degradation products can also be methylated, owing to the observed dibutyldimethyltin and tributylmethyltin species in contaminated sediments [154].

5.2. Bioaccumulation

Organotin compounds, especially in the aquatic environment, are available for uptake for organisms at various levels of the food web. Organisms may take up organotins from the water or sediment phase via the body surface (bioconcentration) or via the food chain (biomagnification) [139]. Concentration and speciation of the available forms of organotins either in the aqueous or solid phase and the excretion and/or degradation processes of the organism influence the bioaccumulation of contaminants [139]. The uptake of organotins is influenced by the lipophilic character of the compounds (e.g., the fraction of the neutral forms), however, this factor might not be as important as could be postulated from octanol-water partitioning model studies [155]. The microbial uptake is generally considered to be a biphasic process. The first step is biosorption when metal ions can bind to the predominantly anionic cell surfaces by various interactions (to hydroxyl, phosphate or carboxylate functions of the cell wall polymers) and the second step is a metabolism-dependent transport of the metal across the membrane [63].

Bioaccumulation of organotins has been reported in a wide range of organisms. The bacterium *Pseudomonas* sp. was shown to accumulate a very high amount of TBT, up to 2% of its cellular dry weight without any significant biotransformation [156]. Avery, Codd, and Gadd reported the biosorption of various tri-substituted organotin compounds; the uptake increased with increasing molecular mass of the organotins (TPT > TBT > tripropyltin $\geq$ TMT $\geq$ triethyltin) [157]. They observed a weak effect of pH, a strong inhibitory effect of salinity on TBT uptake and a TBT-concentration dependence [157]. The bioaccumulation of various organotins was investigated in algae and in some cases, significant bioconcentration factors (BCF) were determined (for *S. obliquus* BCF $> 3.32 \times 10^5$ (TBT) and 1.4×10^5 (TPT)) [158]. Some of the studied algae showed toxicity resistance for TBT and they metabolized TBT to the less toxic DBT [158]. Significant amounts of butyltins and phenyltins (up to ~90 and 210 ng/g dry weight, respectively) were found in sediment samples and deep sea organisms (gastropods, sea cucumbers, galatheid crabs, and bivalves) taken from the Nankai Trough, Japan (~3000 m water depth) [159]. Organotin contaminants can get into animals being at higher levels of the food chain, e.g., vertebrates [160–163] or humans [160,164].

Butyltin residues were analyzed in the sediment and in some vertebrates at the Polish Coast by Kannan and Falandysz who reported high concentrations of butyltins in some fishes (14–455 ng/g wet weight) and birds (35–870 ng/g wet weight) and a very high level was found in the liver of a long-tailed duck (4600 ng/g wet weight). The published data suggest the trophic transfer of the studied compounds through the aquatic food chains [160]. Butyltin levels in human liver in the range of 2.4–11 ng/g (wet weight) was reported by Kannan and Falandysz [160] and in the range of 0.8–28.3 ng/g (wet weight) by Nielsen and Strand [164]. These concentrations appear to be smaller, compared to animal samples taken from the same area [160], suggesting a relatively fast excretion or metabolic mechanism for organotins operating in humans [160].

Finally, accumulation of TBT was shown in the roots of willow trees [165]. The observed very small translocation to the higher aerial plant parts was believed to reduce the risk of spreading TBT contamination along the terrestrial food chain.

6. TOXICITY

The toxicity of organotin compounds is very broad and complex. Organotin compounds cause neurotoxicity in animals and humans, and they are known to have detrimental effects on the immune response. Polarity plays an important role in the uptake and accumulation rates of a compound by an organism and therefore strongly determines the toxicity, which is therefore directly linked to the number and nature of the organic moieties. Tri- and disubstituted organotins are known to be the most toxic, and their toxicity decreases with increasing alkyl chain length independent of the counter ions. However, there is also much difference between organisms. TET is the most toxic compound of all organotins to mammals, TMT and TBT show the highest toxicity for insects and marine species, respectively. Furthermore, alkyltin compounds are generally more toxic than aryltins.

Unlike other organometals, organotin compounds are very selective toxins, targeting specific organs in mammals. For example, triorganotins with alkyl chains of intermediate length (TBT and TPT salts), are primarily immunotoxic, while compounds with short alkyl groups (TET and TMT) exhibit neurotoxic activity [166]. On the other hand, TMT and TET behave differently, inducing selective damage to distinct regions of the central nervous system. TMT-induced toxicity is localized within the hippocampus and neocortex of the brain, while TET predominately affects regions of the spinal cord. The higher trialkyltin homologs, such as trioctyltins, were found to be only slightly toxic, however, their metabolitic conversion may produce immunotoxic dialkyltins, too [167].

Although diorganotin compounds are less toxic than triorganotins, they manifest teratogenic, immuno- and developmental toxicity.

Mono- and tetraorganotins are much less toxic, the first because they are too polar, the latter because they are practically not polar at all. But it should be kept in mind that organotin compounds can be converted into each other. The presence of non-toxic mono- or tetraorganotin compounds can lead to a dangerous situation when conversion (bioalkylation, degradation) becomes possible.

Triorganotin compounds affect a variety of biochemical and physiological systems and their action may vary with compound and dose, but the effect strongly depends on the species and route of administration. Consequently, it is almost impossible to give a short overview of all the different effects on different species. Therefore only a few specific cases will be discussed.

6.1. Effects on Aquatic Life

In marine and freshwater ecosystems TBT is the most common contaminant of exceeding acute and chronic toxicity levels. Some aquatic organisms display a remarkable ability to accumulate TBT. For example, in oyster samples collected along the Essex coast (UK) prior to TBT regulations, 3.5–8.6 mg/kg (wet weight) TBT was detected [168]. TBT presents the highest toxicity by disturbing the function of mitochondria, and has been demonstrated to cause impairments in growth, development, reproduction, and survival of many marine species [169]. For example, the 48h or 72h lethal concentrations (LC_{50}, lowest concentration to cause 50% lethality in the test population) of TBT for marine invertebrates range between 50–5000 ngL^{-1} [170], a concentration reached in harbor areas. In fact, growth impairment is a much more sensitive response to TBT exposure than mortality.

Of particular concern has been the decline of marine molluscs in costal areas due to imposex. Imposex occurs when male sex characteristics are superimposed on normal female gastropods. In studies with intertidal mud snails, the imposex condition was linked to pollution in marinas and mainly to TBT [171]. This is because gastropods bioaccumulate TBT and its endocrine disruptive effects result in an elevated testosterone level that promotes development of male sex characteristics [172]. Imposex results in impaired reproductive fitness or sterility in the affected animals and is one of the clearest examples of environmental endocrine disruption. It remains an open question whether *in vivo* organotins act primarily as protein and enzyme inhibitors, or rather mediate their endocrine disrupting effects at the transcriptional level. Accordingly, the induction mechanism of imposex was attributed to the direct inhibition of the testosterone processing P450 aromatase enzyme by TBT [173]. On the other hand, recent research has shown that aromatase mRNA levels can be downregulated in human ovarian granulosa cells by treatment with organotins or ligands for the nuclear hormone receptor retinoid X receptors (RXRs) [174]. Organotins (both TBT and TPT) bound to RXRs with high affinity, inducing downstream of the RXR cascade and the development of imposex, namely the differentiation and growth of male type genital organs in female gastropods [175].

Based on laboratory and field observations, Gibbs and Bryan [176] proposed a relationship between TBT exposure of tin in water and morphological modifications of the genital tract in gastropods, as follows: 0–0.5 ngL^{-1} normal breeding; 1–2 ngL^{-1} breeding capacity retained by some females, others sterilized by blockage of oviduct as indicated by presence of aborted capsule masses; 3–5 ngL^{-1} virtually all females sterilized, oogenesis apparently normal; 10 ngL^{-1} oogenesis suppressed, spermatogenesis initiated; 20 ngL^{-1} testis developed to variable extent, vesicula seminalis with ripe sperm in most-affected animals; 100 ngL^{-1} sperm-ingesting gland undeveloped in some individuals [176]. These observations reflect that gastropods are hypersensitive to TBT exposure, and they are affected at concentrations which are possible even in the open sea, far away from costal regions [49].

TBT exposure leads to masculinization of several fish species, too. TBT exposure at an environmentally relevant level (0.1 ngL^{-1}) on zebrafish from hatching to 70 days resulted in a male-biased population [177]. The sperm motility of fishes exposed to TBT for 70 days at concentration of 10 ngL^{-1} significantly decreased, and all sperm lacked flagella [177].

6.2. Risks to Mammals and Human Health

Obviously, marine mammals are the species most exposed to organotin compounds, especially TBT. In contrast with several aquatic invertebrates, these animals, particularly cetaceans, have a low capacity to degrade organotin compounds [178]. Therefore, they accumulate organotins mostly in liver, kidney, and brain. The highest level of total butyltin concentration (MBT + DBT + TBT = 10 mg/g wet weight) in cetaceans was found in the liver of a dead finless porpoise from the Seto Inland Sea, Japan [179].

Acute oral toxicity for several organotin compounds to rat has been determined, and showed a toxicity order TET > TMT > DMT > DBT > TBT [49]. TBT-oxide, DBT, and dioctyltin compounds are potent thymolytic and immunotoxic agents in rats [180]. It has been reported that up to 5 ppm tributyltin oxide in the rat diet produced immunotoxicity in a 2-year feeding study, and 50 ppm increased the incidence of tumors of endocrine origin.

Administration of TMT to adult animals causes neuronal degeneration in the hippocampus, amygdala, pyriform cortex, and neocortex [181], while exposure to TMT during development impairs later learning and memory [182].

The consumption of contaminated drinking water (PVC water pipes), beverages, or in particular marine food has been reported as an important route of human exposure. Indeed, in untreated wastewater of the city of Zürich (Switzerland), approximately 1 μg/L mono-, di-, and tributyltin have been determined [183].

Human exposure to high doses of TMT resulted in memory deficits, seizures, altered affect, hearing loss, disorientation, and in some instances death [184]. In a recent accidental poisoning by high doses of DMT and

TMT motor ataxia, memory loss, disorientation, and speech difficulty have been reported even after the urinary alkyltin level returned to the normal range. The patient showed severe hypokalemia, which suggests that TMT induces acute renal leakage of K^+. After treatment with 2,3-dimercapto-propanol the patient recovered from coma [185].

In 1954 a widespread accidental poisoning occured in France, caused by triethyltin iodide. Of the ~1000 persons affected at least 100 deaths and more than 200 intoxications occurred [186]. Among others, visual disturbance, cardiac and respiratory failures have been reported. Most of these symptoms were due to the formation of a cerebral edema. Of all the intoxicated people only ten recovered completely.

Due to their high toxicity, TMT and TET have not been implemented in industrial or agricultural applications, yet traces of TMT have been documented in the urine of humans not exposed directly to TMT [187], leading to concerns about possible environmental exposure to these toxins and/or methylation of other tin species *in vivo*.

Imposex has already been documented for as many as 150 species. It is obvious that TBT and other organotins have adverse hormonal effects on many organisms. Although humans may be exposed to relatively high doses of organotins, little is known concerning the long term effects (chronic toxicity) of these compounds in humans [170]. According to the WHO there is no direct danger for human health, not even for heavy fish consumers [168]. But this remains a point of discussion [188].

7. CONCLUDING REMARKS

Organotin compounds are of high toxicological relevance, and their effect is mostly related to aquatic environments. Consequently, the aquatic chemistry of organotin compounds is of crucial importance. Further studies dealing with the interaction of organotin(IV) cations with different naturally occurring ligands may furnish essential details on their transport processes, biospeciation, and bioavailability.

Although exponentially increasing data are available on the toxicity of organotins to invertebrates, still little is known concerning the long term effects (chronic toxicity) and mode of action of these compounds in humans.

ACKNOWLEDGMENT

This work was supported by the Hungarian Research Foundation (OTKA NI61786).

ABBREVIATIONS

BCF	bioconcentration factor
Bu	butyl group
c-Hex	cyclohexyl
DBT	dibutyltin(IV)
DET	diethyltin(IV)
DMPT	dimethyl β-propiothetin
DMSA	dimercaptosuccinic acid
DMT	dimethyltin(IV)
EDDA	ethylenediamine-*N,N'*-diacetic acid
EDTA	ethylenediamine-*N,N,N',N'*-tetraacetic acid
Et	ethyl group
IDA	iminodiacetic acid
IMO	International Maritime Organization
MA	malic acid
MBT	monobutyltin(IV)
Me	methyl group
MIDA	*N*-methylimino-diacetic acid
MMT	monomethyltin
MPA	2-mercaptopropionic acid
MSA	mercaptosuccinic acid
n-Hex	normal-hexyl
NTA	nitrilotriacetic acid
Oct	octyl group
ODA	oxydiacetic acid
Ph	phenyl group
PVC	polyvinyl chloride
SA	succinic acid
TBT	tributyltin(IV)
TCHT	tricyclohexyltin(IV)
TET	triethyltin(IV)
TMT	trimethyltin(IV)
TPT	triphenyltin(IV)
WHO	World Health Organization

REFERENCES

1. K. Fent, *Crit. Rev. Toxicol.*, 1996, **26**, 1–117.
2. A. G. Davies, *Organotin Chemistry*, 2nd edn., Wiley-VCH, Weinheim, 2004, pp. 1–426.

3. M. Gielen, A. G. Davies, K. Pannell and E. Tiekink (Ed.), *Tin Chemistry: Fundamentals, Frontiers, and Applications*, John Wiley and Sons, Ltd., Chichester, UK, 2008.
4. J. A. Marshall, *Chem. Rev.*, 1996, **96**, 31–48.
5. Y. Yamamoto and N. Asao, *Chem. Rev.*, 1993, **93**, 2207–2293.
6. R. Barbieri, L. Pellerito, G. Ruisi, A. Silvestri, A. Barbieri-Paulsen, G. Barone, S. Posante and M. Rossi, in *Chemical Processes in Marine Environments*, Ed. A. Gianguzza, E. Pelizetti and S. Sammartano, Springer, Berlin, 2000.
7. M. Nath, S. Pokharia and R. Yadav, *Coord. Chem. Rev.*, 2001, **215**, 99–149.
8. L. Pellerito and L. Nagy, *Coord. Chem. Rev.*, 2002, **224**, 111–150.
9. (a) E. Frankland, *Justus Liebigs Ann. Chem.*, 1849, **71**, 171–213.
(b) E. Frankland, *Q. J. Chem. Soc.*, 1850, **2**, 263–296.
10. G. B. Buckton, *Phil. Trans. R. Soc. London*, 1859, **149**, 417–435.
11. J. G. A. Luijten and G. J. M. van der Kerk, in *Organometallic Compounds of the Group IV Elements*, Vol. 1, Part II, Ed. A. G. MacDiarmid, Marcel Dekker, New York, 1968, 91–189.
12. M. Gielen and J. Nasielski, in *Organotin Compounds*, **Vol. 3**, Ed. A. K. Sawyer, Marcel Dekker, New York, 1972, pp. 625–822.
13. R. K. Ingham, S. D. Rosenberg and H. Gilman, *Chem. Rev.*, 1960, **60**, 459–539.
14. S. H. L. Thoonen, B. -J. Deelman and G. van Koten, *J. Organomet. Chem.*, 2004, **689**, 2145–2157.
15. P. D. Lickiss, in *Chemistry of Tin*, 2nd edn., Ed. P. J. Smith, Blackie, London, 1998, pp. 176–202.
16. M. Weidenbruch, *J. Organomet. Chem.*, 2002, **646**, 39–52.
17. P. Jutzi and N. Burford, *Chem. Rev.*, 1999, **99**, 969–990.
18. M. Weidenbruch, *Eur. J. Inorg. Chem.*, 1999, 373–381.
19. M. Driess and H. Grützmacher, *Angew. Chem. Int. Ed. Engl.*, 1996, **35**, 828–856.
20. H. G. Kuivila, *Adv. Organomet. Chem.*, 1964, **1**, 47–87.
21. H. G. Kuivila, *Acc. Chem. Res.*, 1968, **1**, 299–305.
22. I. L. Marr, D. Rosales and J. L. Wardell, *J. Organomet. Chem.*, 1988, **349**, 65–74.
23. M. Gielen, *Acc. Chem. Res.*, 1973, **6**, 198–202.
24. J. T. B. H. Jastrzebski and G. van Koten, *Adv. Organomet. Chem.*, 1993, **35**, 241–294.
25. A. Tzschach and K. Jurkschat, *Pure Appl. Chem.*, 1986, **58**, 639–646.
26. J. K. Stille, *Angew. Chem. Int. Ed. Engl.*, 1986, **25**, 508–524.
27. P. Espinet and A. M. Echavarren, *Angew. Chem. Int. Ed.*, 2004, **43**, 4704–4734.
28. H. H. Anderson, *Inorg. Chem.*, 1962, **1**, 647–650.
29. S. D. Rosenberg, E. Debreczeni and E. L. Weinberg, *J. Am. Chem. Soc.*, 1959, **81**, 972–975.
30. T. N. Mitchell, *J. Organomet. Chem.*, 1986, **304**, 1–16.
31. J. M. Brown, A. C. Chapman, R. Harper, D. J. Mowthorpe, A. G. Davies and P. J. Smith, *J. Chem. Soc., Dalton Trans.*, 1972, 338–341.
32. V. Chandrasekhar, S. Nagendran and V. Baskar, *Coord. Chem. Rev.*, 2002, **235**, 1–52.
33. V. Chandrasekhar, K. Gopal, P. Sasikumar and R. Thirumoorthi, *Coord. Chem. Rev.*, 2005, **249**, 1745–1765.

34. V. Bade and F. Huber, *J. Organomet. Chem.*, 1970, **24**, 387–397.
35. D. A. Armitage and A. Tarassoli, *Inorg. Chem.*, 1975, **14**, 1210–1211.
36. W. P. Neumann, *Liebigs Ann. Chem.*, 1962, **653**, 157–163.
37. K. Jurkschat, in *Tin Chemistry: Fundamentals, Frontiers, and Applications*, Ed. M. Gielen, A. G. Davies, K. Pannell and E. Tiekink, John Wiley and Sons, Ltd., Chichester, UK, 2008, pp. 201–230.
38. M. Veith, D. Agustin and V. Huch, *J. Organomet. Chem.*, 2002, **646**, 138–145.
39. V. Chandrasekhar, P. Singh and K. Gopal, *Appl. Organometal. Chem.*, 2007, **21**, 483–503.
40. J. Otera, *Chem. Rev.*, 1993, **93**, 1449–1470.
41. S. Thoonen, B. -J. Deelman and G. van Koten, *Chem. Commun.*, 2001, 1840–1841.
42. E. J. Bulten, *J. Organomet. Chem.*, 1975, **97**, 167–172.
43. S. Thoonen, B. -J. Deelman and G. van Koten, *Tetrahedron*, 2003, **59**, 10261–10268.
44. V. Chandrasekhar and P. Singh, *Organometallics*, 2009, **28**, 42–44.
45. F. Banse, F. Ribot, P. Toledano, J. Maquet and C. Sanchez, *Inorg. Chem.*, 1995, **34**, 6371–6379.
46. S. Durand, K. Sakamoto, T. Fukuyama, A. Orita, J. Otera, A. Duthie, D. Dakternieks, M. Schulte and K. Jurkschat, *Organometallics*, 2000, **19**, 3220–3223.
47. D. A. Powell, T. Maki and G. C. Fu, *J. Am. Chem. Soc.*, 2005, **127**, 510–511.
48. A. M. Echavarren, *Angew. Chem. Int. Ed.*, 2005, **44**, 3962–3965.
49. M. Hoch, *Appl. Geochem.*, 2001, **16**, 719–743.
50. P. A. T. Hoye and R. G. Hargreaves, US Pat. 3914431, Oct. 21, 1975.
51. M. H. Gitlitz, J. E. Engelhart and D. A. Russo, US Pat. 4316853, Feb. 23, 1982.
52. V. Yngve, US Pat. 2219463, Oct. 29, 1940.
53. J. M. Batt, *The World of Organotin Chemicals: Applications, Substitutes, and the Environment*, Organotin Environmental Programme Association (ORTEPA). www.ortepa.org/WorldofOrganotinChemicals.pdf, 2000.
54. K. Figge, *Packag. Technol. Sci.*, 1989, **2**, 215–226.
55. F. Cima, P. J. Craig and C. Harrington, in *Organometalic Compounds in the Environment*, 2nd edn., Ed. P. J. Craig, John Wiley and Sons, Ltd., Chichester, UK, 2003, pp. 101–149.
56. K. Fent, *Sci. Total Environ.*, 1996, **185**, 151–159.
57. A.-I. Sadiki, D. T. Williams, R. Carrier and B. Thomas, *Chemosphere*, 1996, **32**, 2389–2398.
58. D. S. Forsyth and B. Jay, *Appl. Organometal. Chem.*, 1997, **11**, 551–558.
59. S. Takahashi, H. Mukai, S. Tanabe, K. Sakayama, T. Miyazaki and H. Masuno, *Environ. Pollut.*, 1999, **106**, 213–218.
60. A. Gomathi, S. Jafar Hoseini and C. N. R. Rao, *J. Mater. Chem.*, 2009, **19**, 988–995.
61. S. K. Dubey and U. Roy, *Appl. Organometal. Chem.*, 2003, **17**, 3–8.
62. G. M. Gadd, *Sci. Total Environ.*, 2000, **258**, 119–127.
63. J. S. White, J. M. Tobin and J. J. Cooney, *Can. J. Microbiol.*, 1996, **45**, 541–554.
64. D. Cao, G. Jiang, Q. Zhou and R. Yang, *J. Environ. Managem.*, 2009, **90**, S16–S24.

65. X. Wang, H. Hong, D. Zhao and L. Hong, *Mar. Poll. Bull.*, 2008, **57**, 419–424.
66. I. Rumengan, M. Ohji, T. Arai, H. Harino, Z. Arifin and N. Miyazaki, *Coast. Mar. Sci.*, 2008, **32**, 116–126.
67. J. Scancar, T. Zuliani, T. Turk and R. Milacic, *Environ. Monit. Assess.*, 2007, **127**, 271–282.
68. J.-H. Huang and E. Matzner, *J. Plant Nutr. Soil Sci.*, 2004, **167**, 33–38.
69. IMO News Magazine, Ed. L. Adamson, Issue 4, p. 10, 2008. http://www.imo.org.
70. T. Arai, H. Harino, M. Ohji and W. J. Langston (Ed.), *Ecotoxicology of Antifouling Biocides*, Springer, 2009, pp. i–437.
71. C. Alzieu, *Mar. Environ. Res.*, 1991, **32**, 7–17.
72. M. A. Fernandez, A. de Luca Rebello Wagener, A. M. Limaverde, A. L. Scofield, F. M. Pinheiro and E. Rodrigues, *Mar. Environ. Res.*, 2005, **59**, 435–452.
73. WHO, *Mono- and disubstituted methyltin, butyltin, and octyltin compounds*, Concise International Chemical Assessment Document 73, *www.who.int/ipcs/publications/cicad/cicad73.pdf*, 2006.
74. R. J. Maguire, *Water Qual. Res. J. Canada*, 2000, **35**, 633–679.
75. A. Kotrikla, *J. Environ. Manage.*, 2009, **90**, S77–S85.
76. G. J. M. van der Kerk and J. G. A. Luijten, *J. Appl. Chem.*, 1954, **4**, 314–319.
77. A. J. Crowe, *Appl. Organometal. Chem.*, 1987, **1**, 143–155.
78. A. J. Crowe, *Appl. Organometal. Chem.*, 1987, **1**, 331–346.
79. K. Kannan and R. F. Lee, *Environ. Toxicol. Chem.*, 1996, **15**, 1492–1499.
80. K. Fent and M. D. Müller, *Environ. Sci. Technol.*, 1991, **25**, 489–493.
81. M. Gielen, *Coord. Chem. Rev.*, 1996, **151**, 41–51.
82. M. Gielen, *Appl. Organometal. Chem.*, 2002, **16**, 481–494.
83. M. Gielen, M. Biesemans and R. Willem, *Appl. Organometal. Chem.*, 2005, **19**, 440–450.
84. S. Tabassum and C. Pettinari, *J. Organom. Chem.*, 2006, **691**, 1761–1766.
85. S. K. Hadjikakou and N. Hadjiliadis, *Coord. Chem. Rev.*, 2009, **253**, 235–249.
86. R. S. Tobias, I. Ogrins and B. A. Nevett, *Inorg. Chem.*, 1962, **1**, 638–646.
87. R. S. Tobias, H. N. Farrer, M. B. Hughes and B. A. Nevett, *Inorg. Chem.*, 1966, **5**, 2052–2055.
88. G. Arena, R. Purrello, E. Rizzarelli, A. Gianguzza and L. Pellerito, *J. Chem. Soc., Dalton Trans.*, 1989, 773–777.
89. T. Natsume, S. Aizawa, K. Hatano and S. Funahashi, *J. Chem. Soc., Dalton Trans.*, 1994, 2749–2753.
90. P. Surdy, P. Rubini, N. Buzas, B. Henry, L. Pellerito and T. Gajda, *Inorg. Chem.*, 1999, **38**, 346–352.
91. K. Gajda-Schrantz, L. Nagy, T. Fiore, L. Pellerito and T. Gajda, *J. Chem. Soc., Dalton Trans.*, 2002, 152–158.
92. C. De Stefano, C. Foti, A. Gianguzza, M. Martino, L. Pellerito and S. Sammartano, *J. Chem. Eng. Data*, 1996, **41**, 511–515.
93. C. De Stefano, C. Foti, A. Gianguzza, F. Marrone and S. Sammartano, *Appl. Organomet. Chem.*, 1999, **13**, 805–811.
94. C. De Stefano, C. Foti, A. Gianguzza, F. J. Millero and S. Sammartano, *J. Solution Chem.*, 1999, **28**, 959–972.

95. C. Foti, A. Gianguzza, D. Piazzese and G. Trifiletti, *Chem. Speciat. Bioavailab.*, 2000, **12**, 41–52.
96. C. Foti, A. Gianguzza, D. Milea and A. Sammartano, *Appl. Organomet. Chem.*, 2002, **16**, 34–43.
97. C. Foti, A. Gianguzza, D. Milea, F. J. Millero and A. Sammartano, *Marine Chem.*, 2004, **85**, 157–167.
98. M. J. Hynes and M. O'Dowd, *J. Chem. Soc. Dalton Trans.*, 1987, 563–566.
99. G. Arena, A. Contino, S. Musumeci and R. Purello, *J. Chem. Soc. Dalton Trans.*, 1990, 3383–3387.
100. G. Arena, A. Gianguzza, L. Pellerito, S. Musumeci, R. Purello and E. Rizzarelli, *J. Chem. Soc. Dalton Trans.*, 1990, 3383–3387.
101. A. De Robertis, A. Gianguzza, O. Giuffre, A. Pettignano and S. Sammartano, *Appl. Organometal. Chem.*, 2006, **20**, 89–98.
102. A. De Robertis, A. Gianguzza, O. Giuffre, A. Pettignano and S. Sammartano, *Appl. Organometal. Chem.*, 2007, **22**, 30–38.
103. N. Buzás, T. Gajda, L. Nagy, E. Kuzmann, A. Vértes and K. Burger, *Inorg. Chim. Acta*, 1998, **274**, 167–176.
104. A. Jancsó, L. Nagy, E. Moldrheim and E. Sletten, *J. Chem. Soc., Dalton Trans.*, 1999, 1587–1594.
105. B. Gyurcsik, N. Buzás, T. Gajda, L. Nagy, E. Kuzmann, A. Vértes and K. Burger, *Z. Naturforsch.*, 1995, **50b**, 515–523.
106. A. Szorcsik, L. Nagy, B. Gyurcsik, G. y. Vankó, R. Krämer, A. Vértes, T. Yamaguchi and K. Yoshida, *J. Radioanal. Nucl. Chem.*, 2004, **260**, 459–469.
107. M. Nath, R. Jairath, G. Eng, X. Song and A. Kumar, *Spectrochim. Acta Part A*, 2005, **61**, 77–86.
108. T. Fiore, C. Foti, A. Gianguzza, S. Orecchio and L. Pellerito, *Appl. Organomet. Chem.*, 2002, **16**, 294–301.
109. H. Jankovics, L. Nagy, N. Buzás, L. Pellerito and R. Barbieri, *J. Inorg. Biochem.*, 2002, **92**, 55–64.
110. C. De Stefano, A. Gianguzza, O. Giuffre, D. Piazzese, S. Oreccio and S. Sammartano, *Appl. Organomet. Chem.*, 2004, **18**, 653–661.
111. G. Arena, R. Cali, A. Contino, N. Loretta, S. Musumeci and R. Purello, *J. Chem. Soc., Dalton Trans.*, 1992, 2039–2043.
112. C. De Stefano, D. Milea and S. Sammartano, *Biophys. Chem.*, 2005, **116**, 111–120.
113. G. Arena, R. Cali, A. Contino, A. Musumeci, S. Musumeci and R. Purello, *Inorg. Chim. Acta*, 1995, **237**, 187–191.
114. N. Buzás, T. Gajda, E. Kuzmann, L. Nagy, A. Vértes and K. Burger, *Main Group Metal Chem.*, 1995, **18**, 641–649.
115. P. Cardiano, C. De Stefano, O. Giuffre and S. Sammartano, *Biophys. Chem.*, 2008, **133**, 19–27.
116. A. Jancsó, B. Henry, P. Rubini, G. Vankó and T. Gajda, *J. Chem. Soc., Dalton Trans.*, 2000, 1941–1947.
117. A. Jancsó, T. Gajda, A. Szorcsik, T. Kiss, B. Henry, G. Vankó and P. Rubini, *J. Inorg. Biochem.*, 2001, **83**, 187–192.
118. K. Gajda-Schrantz, A. Jancsó, C. Pettinari and T. Gajda, *Dalton Trans.*, 2003, 1–6.

119. I. Sóvágó, in *Biocoordination Chemistry*, ed. K. Burger, Ellis Horwood, New York, 1990, pp. 135–184.
120. F. Capolongo, A. M. Giuliani, M. Giomini and U. Russo, *J. Inorg. Biochem.*, 1993, **49**, 275–293.
121. B. A. Buck-Koehntop, F. Porcelli, J. L. Lewin, C. J. Cramer and G. Veglia, *J. Organomet. Chem.*, 2006, **691**, 1748–1755.
122. V. Cucinotta, A. Gianguzza, G. Maccarrone, L. Pellerito, R. Purello and E. Rizzarelli, *J. Chem. Soc., Dalton Trans.*, 1992, 2299–2303.
123. S.-I. Aizawa, T. Natsume, K. Hatano and S. Funahashi, *Inorg. Chim. Acta*, 1996, **248**, 215–224.
124. M. M. Shoukry, *Talanta*, 1996, **43**, 177–183.
125. E. J. O'Loughlin, S. J. Triana and Y.-P. Chin, *Environ. Toxicol. Chem.*, 2000, **19**, 2015–2021.
126. K. R. Siebenlist and F. Taketa, *Biochem. J.*, 1986, **233**, 471–477.
127. R. Barbieri, A. Silvestri, M. T. Lo Giudice, G. Ruisi and M. T. Musumeci, *J. Chem. Soc., Dalton Trans.*, 1989, 519–525.
128. A. Nishikimi, Y. Kira, E. Kasahara, E. F. Sato, T. Kanno, K. Utsumi and M. Inoue, *Biochem. J.*, 2001, **356**, 621–626.
129. S. M. Toggas, J. K. Krady and M. L. Billingsley, *Mol. Pharmacol.*, 1992, **42**, 44–49.
130. C. M. Patanow, J. R. Day and M. L. Billingsley, *Neuroscience*, 1997, **76**, 187–192.
131. B. A. Buck-Koehntop, A. Mascioni, J. J. Buffy and G. Veglia, *J. Mol. Biol.*, 2005, **354**, 652–665.
132. R. Gabbianelli, G. Falcioni and G. Lupidi, *Appl. Organometal. Chem.*, 2002, **16**, 3–8.
133. A. G. Atanasov, L. G. Nashev, S. Tam, M. E. Baker and A. Odermatt, *Environ. Health Persp.*, 2005, **113**, 1600–1606.
134. G. Nagel-Starczynowska and K. Kaletha, *Biochim. Biophys. Acta*, 1993, **261**, 1164–1169.
135. C. von Ballmoos, J. Brunner and P. Dimroth, *Proc. Natl. Acad. Sci.*, 2004, **101**, 11239–11244.
136. G. Barone, R. Barbieri, G. La Manna and M. H. J. Koch, *Appl. Organometal. Chem.*, 2000, **14**, 189–196.
137. RPA, *Revised assessment of the risks to health and the environment associated with the use of organostannic compounds (excluding use in antifouling paints)*, Final Report, prepared for the European Commission, http://www.rpaltd.-co.uk/documents/J456Organotins.pdf, 2003.
138. K. Inaba, H. Shiraishi and Y. Soma, *Wat. Res.*, 1995, **29**, 1415–1417.
139. H. Rüdel, *Ecotoxicol. Environ. Saf.*, 2003, **56**, 180–189.
140. C. G. Arnold, A. Weidenhaupt, M. M. David, S. R. Müller, S. B. Haderlein and R. P. Schwarzenbach, *Environ. Sci. Technol.*, 1997, **31**, 2596–2602.
141. R. W. Hunziker, B. I. Escher and R. P. Schwarzenbach, *Environ. Sci. Technol.*, 2001, **35**, 3899–3904.
142. K. Fent, *Toxicology*, 2004, **205**, 223–240.
143. W. J. Langston and N. D. Pope, *Mar. Poll. Bull.*, 1995, **31**, 32–43.

144. R. J. Maguire, J. H. Carey and E. J. Hale, *J. Agric. Food Chem.*, 1983, **31**, 1060–1065.
145. G. Zhai, J. Liu, B. He, J. Zhang, Q. Zhou and G. Jiang, *Chemosphere*, 2008, **72**, 389–399.
146. E. A. Clark, R. M. Sterritt and J. N. Lester, *Environ. Sci. Technol.*, 1988, **22**, 600–604.
147. O. Errecalde, M. Astruc, G. Mauryt and R. Pinel, *Appl. Organomet. Chem.*, 1995, **9**, 23–28.
148. A. Cruz, T. Caetano, S. Suzuki and S. Mendo, *Mar. Environ. Res.*, 2007, **64**, 639–650.
149. R. D. Barnes, A. T. Bull and R. C. Poller, *Pest. Sci.*, 1973, **4**, 305–317.
150. G. I. Paton, W. Cheewasedtham, I. L. Marr and J. J. C. Dawson, *Environ. Pollut.*, 2006, **144**, 746–751.
151. S. Dubascoux, G. Lespes, L. Denaix and M. Potin Gautier, *Appl. Organometal. Chem.*, 2008, **22**, 481–487.
152. J. Heroult, Y. Nia, L. Denaix, M. Bueno and G. Lespes, *Chemosphere*, 2008, **72**, 940–946.
153. V. Minganti, R. Capelli, G. Drava and R. De Pellegrini, *Chemosphere*, 2007, **67**, 1018–1024.
154. R. J. Maguire, *Environ. Sci. Technol.*, 1984, **18**, 291–294.
155. P. W. Looser, S. Bertschi and K. Fent, *Appl. Organometal. Chem.*, 1998, **12**, 601–611.
156. W. R. Blair, G. J. Olson, F. E. Brinckman and W. P. Iverson, *Microb. Ecol.*, 1982, **8**, 241–251.
157. S. V. Avery, G. A. Codd and G. M. Gadd, *Appl. Microbiol. Biotechnol.*, 1993, **39**, 812–817.
158. G. Huang, Z. Bai, S. Dai and Q. Xiet, *Appl. Organometal. Chem.*, 1993, **7**, 373–380.
159. H. Harino, N. Iwasaki, T. Arai, M. Ohji and N. Miyazaki, *Arch. Environ. Contam. Toxicol.*, 2005, **49**, 497–503.
160. K. Kannan and J. Falandysz, *Mar. Pollut. Bull.*, 1997, **34**, 203–207.
161. T. Ciesielski, A. Wasik, I. Kuklik, K. Skra, J. Namienik and P. Szefer, *Environ. Sci. Technol.*, 2004, **38**, 1415–1420.
162. H. Harino, M. Ohji, G. Wattayakorn, K. Adulyanukosol, T. Arai and N. Miyazaki, *Arch. Environ. Contam. Toxicol.*, 2007, **53**, 119–125.
163. J. Strand and J. A. Jacobsen, *Sci. Total Environ.*, 2005, **350**, 72–85.
164. J. B. Nielsen and J. Strand, *Environ. Res. Section A*, 2002, **88**, 129–133.
165. G. Ciucanim, H. Mosbaek and S. Trapp, *Environ. Sci. Pollut. Res.*, 2004, **11**, 267–272.
166. N. J. Snoeij, A. A. van Iersel, A. H. Penninks and W. Seinen, *Toxicol. Appl. Pharmacol.*, 1985, **81**, 274–286.
167. N. J. Snoeij, A. H. Penninks and W. Seinen, *Int. J. Immunopharmacol.*, 1988, **10**, 891–899.
168. C. Alzieu, *Ecotoxicology*, 2000, **9**, 71–76.
169. J. A. Haggera, M. H. Depledge and T. S. Galloway, *Mar. Pollut. Bull.*, 2005, **51**, 811–816.

170. B. Antizar-Ladislao, *Environ. Int.*, 2008, **34**, 292–308.
171. B. S. Smith, *J. Appl. Toxicol.*, 1981, **1**, 141–144.
172. T. Horiguchi, H. Shiraishi, M. Shimizu and M. Morita, *Appl. Organometal. Chem.*, 1997, **11**, 451–455.
173. M. M. Santos, L. F. Castro and M. N. Vieira, *Comp. Biochem. Physiol. C Toxicol. Pharmacol.*, 2005, **141**, 101–109.
174. M. Saitoh, T. Yanase, H. Morinaga, M. Tanabe, Y. M. Mu, Y. Nishi, M. Nomura, T. Okabe, K. Goto, R. Takayanagi and H. Nawata, *Biochem. Biophys. Res. Commun.*, 2001, **289**, 198–204.
175. J. Nishikawa, S. Mamiya and T. Kanayama, *Environ. Sci. Technol.*, 2004, **38**, 6271–6276.
176. P. E. Gibbs and G. W. Bryan, in *Tributyltin: Case Study of an Environmental Contaminant*, Ed. S. J. de Mora, *Cambridge Environmental Chemistry Series*, 1996, **8**, 212–236.
177. B. G. McAllister and D. E. Kime, *Aquat. Toxicol.*, 2003, **65**, 309–316.
178. S. Tanabe, *Mar. Pollut. Bull.*, 1999, **39**, 62–72.
179. S. Tanabe, M. Prudente, T. Mizuno, J. Hasegawa, H. Iwata and N. Miyazaki, *Environ. Sci. Technol.*, 1998, **32**, 193–198.
180. I. J. Boyer, *Toxicology*, 1989, **55**, 253–258.
181. A. W. Brown, W. N. Aldridge, B. W. Street and R. D. Verschoyle, *Am. J. Pathol.*, 1979, **97**, 59–82.
182. M. E. Stanton, K. F. Jensen and C. V. Pickens, *Neurotoxicol. Teratol.*, 1991, **13**, 525–530.
183. K. Fent, *Organotins in Municipal Wastewater and Sewage Sludge*, in *Organotin*, Ed M. A. Champ and P. F. Seligman, Chapman & Hall, London, 1996, p. 581.
184. J. Gui-bin, Z. Qun-fang and H. Bin, *Bull. Environ. Contam. Toxicol.*, 2000, **65**, 277–284.
185. C. I. Yoo, Y. Kim, K. S. Joeng, C. S. Sim, N. Choy, J. Kim, J. B. Eum, Y. Nakajima, Y. Endo and Y. J. Kim, *J. Occup. Health*, 2007, **49**, 305–310.
186. A. K. Saxena, *Appl. Organometel. Chem.*, 1987, **1**, 39–56.
187. R. S. Braman and M. A. Tompkins, *Anal. Chem.*, 1978, **51**, 12–19.
188. A. C. Belfroid, M. Purperhart and F. Ariese, *Mar. Pollut. Bull.*, 2000, **3**, 226–232.
189. D. Pettit and K. Powell, *IUPAC Stability Constants Database*, Academic Software, Otley, UK, 1997.

Met. Ions Life Sci. **2010**, *7*, 153–164

5

Alkyllead Compounds and Their Environmental Toxicology

Henry G. Abadin and Hana R. Pohl

Agency for Toxic Substances and Disease Registry, U.S. Department of Health and Human Services, Atlanta, GA 30333, USA

<hrp1@cdc.gov>

ABSTRACT: Alkyllead compounds are man-made compounds in which a carbon atom of one or more organic molecules is bound to a lead atom. Tetraethyllead and tetramethyllead are the most common alkyllead compounds that were used primarily as gasoline additives for many years. Consequently, auto emissions have accounted for a major part of lead environmental pollution. Alkyllead compounds can readily enter living organisms as they are well absorbed via all major routes of entry. Because of their lipid solubility, the alkylleads can also readily cross the blood-brain barrier. The toxicokinetic information on organic lead can be used as biomarkers of exposure for monitoring exposed individuals. The organic alkyllead compounds are more toxic than the

Metal Ions in Life Sciences, Volume 7 Edited by Astrid Sigel, Helmut Sigel, and Roland K. O. Sigel

Published by the Royal Society of Chemistry, www.rsc.org DOI: 10.1039/9781849730822-00153

inorganic forms of lead. Neurotoxicity is the predominant effect of lead (both for organic and inorganic forms), although lead affects almost every organ of the body. The use of alkyllead compounds has declined over the last 20 years, due to the world-wide effort to eliminate the use of leaded gasoline. This achievement can be viewed as a great accomplishment of public health preventive measures.

KEYWORDS: alkyllead · gasoline additives · neurotoxicity · pollution decrease

1. INTRODUCTION

Lead is a naturally occurring metal found in the Earth's crust at concentrations of about 15–20 mg/kg. Lead rarely occurs in its elemental state but, rather, in its +2 oxidation state in various ores throughout the Earth. Alkyllead compounds, on the other hand, are man-made compounds in which a carbon atom of one or more organic molecules is bound to a lead atom.

Alkyllead compounds are classified as tetraalkylleads, trialkylleads, or dialkylleads. Of these, the tetraalkyllead compounds, tetraethyllead (TEL), and tetramethyllead (TML), are the most common [1]. TEL and TML have been primarily used in the past as gasoline additives. Although use has been significantly reduced, the use of these alkyllead compounds does continue in some countries, and previous use has resulted in the widespread dispersal of lead compounds in the environment.

2. FORMATION OF ALKYLLEAD COMPOUNDS

Alkyllead is produced through several methods, including the electrolysis of an ethyl Grignard reagent or alkylation of a lead-sodium alloy. Alkyllead is used as a fuel additive to reduce "knock" in combustion engines. TEL was first distributed as an additive to automobile fuel in 1923; TML was introduced in 1960. These alkyllead compounds also help to lubricate internal engine components and protect intake and exhaust valves against recession [1].

Exposure is most likely to occur in occupational settings during production, distribution, and handling of alkylleads and in high-traffic areas. However, the compound's use in gasoline has widely dispersed inorganic lead forms in the environment, resulting in non-occupational exposures. Worldwide, there has been a decreasing trend in the allowable amount of lead additives in gasoline; however, many countries still allow lead in gasoline. Inevitably, workers engaged in the manufacture of these compounds are exposed to both inorganic and alkyllead. Some exposure also occurs at the petroleum refineries where TEL and TML are blended into gasoline [2].

3. RELEASES TO THE ENVIRONMENT

The primary source of lead in the environment has historically been anthropogenic emissions to the atmosphere. The U.S. Environmental Protection Agency (EPA) began a phaseout of the use of alkyllead in gasoline in 1973. By 1990, auto emissions accounted for 33% of all anthropogenic lead emissions, compared to 90% in 1984 [3,4]. Production of leaded gasoline decreased from 77.5 billion gallons in 1967 to 3.1 billion gallons in 1991 [1]. EPA totally banned the use of lead additives in motor fuels after December 31, 1995, except for aviation, race car, and other off-road vehicle fuels [1,5].

Between 1970 and 2006, air emissions of organic and inorganic lead compounds decreased by two orders of magnitude (Table 1). The greatest decrease between 1970 and 1985 can be attributed mostly to the reduction in leaded gasoline. In 2001, EPA estimated that 78% of emissions were from industrial processes, 12% from transportation, and 10% from fuel combustion [6].

Table 1. Historic Levels of Lead Emissions to the Atmosphere in the United States.

Short Tons of Lead Emitted Annually								
1970	1975	1980	1985	1990	1995	2000	2005	2006
220,000	160,000	75,000	23,000	5,000	4,000	2,000	3,000	4,000

Compiled from [26].
1 short ton = 907,185 kg

Worldwide, the use of leaded gasoline is slowly being reduced; however, it still accounts for a large proportion of air emissions in many cities where leaded gasoline is still used [7]. Consequently, preventing exposure to lead (e.g., elimination of lead in gasoline) is the primary prevention strategy for eliminating exposure [8]. Reductions in blood lead levels have been observed in the United States (Figure 1) and in other countries that have eliminated the use of leaded gasoline (e.g., Greece, India) [9–12]. Most recently, in countries such as Indonesia, where the phaseout of leaded gasoline began in 2001, and Lebanon, where it was banned in 2003, children's blood lead levels are expected to rapidly decline [13,14].

4. ENVIRONMENTAL FATE

Alkyllead is not significantly released during the combustion of leaded gasoline. Rather, lead is emitted as lead halides (mostly PbBrCl) and as double salts with ammonium halides (e.g., $2PbBrCl \cdot NH_4Cl$, $Pb_3(PO_4)_2$ and

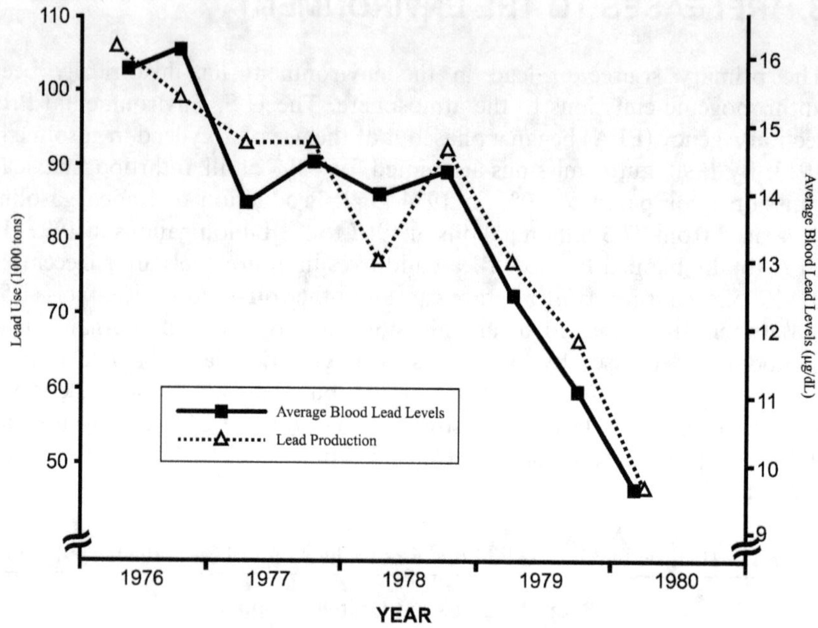

Figure 1. Leaded gasoline production and blood lead levels in the United States (1 short ton = 907,185 kg). Adapted from [65].

$PbSO_4$ [15,16]). After 18 hours, approximately 75% of the bromine and 30%–40% of the chlorine disappear, and lead carbonates, oxycarbonates, and oxides are produced. These lead oxides are subject to further weathering to form additional carbonates and sulfates [17].

Because of the decrease in production, alkyllead compounds are no longer present in significant quantities in the air. However, their degradation products are still present. TEL and TML exist almost entirely in the vapor phase in the atmosphere [18]. When exposed to sunlight, they decompose rapidly to trialkyl- and dialkyllead compounds, which are more stable in the atmosphere, decomposing eventually to inorganic lead oxides by a combination of direct photolysis, reaction with hydroxyl radicals, and reaction with ozone. The half-life of TEL in summer atmospheres is approximately 2 hours, and the half-life of TML is about 9 hours. In the winter, both compounds have half-lives of up to several days [19]. Trialkyl compounds occur almost entirely in the vapor phase, and dialkyl compounds occur almost entirely in particulate form.

Lead that is released into the environment ultimately deposits onto land or onto sediment in the case of a release to surface water. In the atmosphere, particulate lead is dispersed and eventually removed from the atmosphere by wet or dry deposition. Airborne lead particles can remain airborne for days and, therefore, may be transported far from the original source.

The fate of lead in soil is dependent upon the characteristics of the soil, such as pH, soil type (e.g., sandy, clay), particle size, organic matter content, presence of inorganic colloids, and the cation exchange capacity of the soil [20,21]. Lead may be immobilized by ion exchange with hydrous oxides or clays or by chelation with humic or fulvic acids in the soil [17]. Lead is likely to be retained in soils when the pH ≥ 5 and organic content of the soil is greater than 5%. Because of their insolubility, tetraalkyl lead compounds are not expected to leach in soil. However, dealkylation to the water soluble trialkyls in soils has been shown to occur and may result in leaching into groundwater. In addition, tetraethyl lead can be transported through a soil column when it is present in a migrating plume of gasoline [22,23].

In water, tetraalkyllead compounds are first degraded to their respective ionic trialkyllead species and are eventually mineralized to inorganic lead by biological and chemical degradation processes [24]. The amount of soluble lead in surface waters depends upon the existing chemistry of the water (e.g., pH and dissolved salt content). Most of the lead in water is in an undissolved form consisting of colloidal particles or particles of lead carbonate, lead oxide, lead hydroxide, or other lead compounds.

5. HEALTH EFFECTS

Alkyllead compounds are more toxic than inorganic forms. The tetraalkyllead compounds, in turn, are more toxic than trialkyllead compounds, and ethyl forms are more toxic than the methyl forms [25]. Neurotoxicity is the predominant effect of lead (organic and inorganic), although lead affects almost every organ of the body. In many aspects, the intoxication with organic lead is similar to intoxication with inorganic lead.

There are a number of mechanisms of lead toxicity. One of the most important is the ability of lead to mimic calcium in the body, leading to a disruption of physiologic processes. In addition, lead affects heme synthesis, which can result in hematological, neurological, renal, and hepatic effects [26].

Urinary lead increase is an important marker of exposure to organic lead [27]. In humans, urinary lead levels $> 200\ \mu g/L$ are associated with poisoning and levels $> 1{,}000\ \mu g/L$ with fatalities.

5.1. Studies in Humans

The onset of poisoning in humans may start with non-specific symptoms. When examined, the patients often present with pallor, tremor, increased tendon reflexes, and decreased blood pressure. There is a clear correlation between the time of onset and the severity of intoxication; the shorter the onset, the more severe poisoning is manifested. Symptoms of alkyllead poisoning include anorexia, insomnia, tremors, weakness, fatigue, nausea, vomiting, mood shifts, and impairment of memory. These can progress to mania, convulsions, coma, and death [27]. Brain edema and neuron death in the cerebral and cerebellar cortex, reticular formation, and basal ganglia are the prominent pathological findings. Coarse muscular tremors are one of the most often seen effects.

Among 222 current lead workers (air-lead concentrations: inorganic, 4–119 $\mu g/m^3$, and organic, 1–56 $\mu g/m^3$; blood lead weighted average: 240 μg/L), manual dexterity, verbal memory, and learning were related to exposures [28]. Workers with the highest exposures averaged scores 5%–22% lower in the neuropsychological tests than the control group. A self-referred subgroup of the workers underwent further clinical examination [29]. Neurobehavioral abnormalities (18 of 39 workers) and sensorimotor polyneuropathies (11 of 31 workers) were reported.

In a study of former (< 16 years latency) organolead workers – a cohort of more than 500 individuals – a negative correlation was found between tibia lead levels and performance in neuropsychological tests [30]. The mean tibia lead levels were 22.6 + 16.5 μg/g (up to 98.7 μg/g). Verbal memory and learning, visual memory, executive memory, and manual dexterity were tested to determine the relative contribution of past lead exposure and normal aging on cognitive function. Results indicated a progressive decline in cognitive function resulting from previous occupational exposure to lead.

The decline in cognitive function was explained by the occurrence of persistent brain lesions associated with an increased cumulative lead dose [31]. A total of 36% of former workers had a white matter lesion (WML) grade of 1 to 7 (0–9 scale) on an MRI examination. The adjusted odds ratio for a 1 μg/g increase in tibia lead was 1.042 for a grade of 5+ on the WML grading scale.

A major confounder in these organolead occupational studies is co-exposure to inorganic lead during the manufacturing process. Whether the effects can be attributed to organic lead, inorganic lead, or both is uncertain [32].

Other effects on the former workers in this cohort included increased blood pressure [33]. Mean blood lead levels were 4.6 μg/dL (±2.6 μg/dL); tibia lead levels averaged 19.3 μg/g (±9.4 μg/g). Lead levels were associated with an increase in systolic blood pressure of 0.64 mmHg and 0.73 mmHg, respectively, for blood and bone lead. Similar effects were found in another study that investigated a younger cohort [34].

Lead interferes with heme synthesis by altering the activities of δ-aminolevulinic acid dehydratase (ALAD) and ferrochelatase. As a consequence of these changes, heme biosynthesis is decreased and the activity of the rate-limiting enzyme of the pathway, δ-aminolevulinic synthetase (ALAS), which is feedback inhibited by heme, is subsequently increased. ALAD activity was significantly decreased in the blood of men occupationally exposed to alkyllead [35]. A mean of 220 and 677 units of enzyme activity were found in the exposed and control groups, respectively. The mean blood lead levels were 42.5 μg/dL in the exposed and 15 μg/dL in controls.

Several case studies reported on exposure to TEL in gasoline sniffers [36–38]. The studies noted that the initial acute phase of intoxication can probably be attributed to various volatile organic compounds (VOCs) in gasoline and the later phase can be attributed to the lead itself. However, the symptoms overlap, and the studies can be used only as supporting information.

As always, epidemiologic studies must account for confounding factors. For example, recent exposures to organic lead were positively correlated with increased blood lead levels in exposed workers [39]. Similarly, age and cigarette smoking were positively correlated with blood lead levels in the cohort. However, increased alcohol consumption was associated with lower blood lead levels. This finding is in contrast to results obtained in cohorts exposed to inorganic lead. The data suggest possible differences in enzyme-mediated metabolism of organic lead.

The treatment for organic lead intoxication is symptomatic. Alkyllead compounds are chelated to a much lesser degree than inorganic lead. Although chelation may slightly increase the excretion of lead, the recovery of the patient is not usually affected [27]. In support of this observation, Stewart et al. [40] reported that an increase in chelatable lead in organic lead workers mainly reflected the body burden of inorganic lead.

5.2. Studies in Animals

The lethal dose in rats is about 11 mg/kg for TEL and about 83 mg/kg for TML. When groups of rats were exposed to TEL at concentrations ranging from 12 to 46 mg/m^3 and TML at concentrations from 12 to 63 mg/m^3, rats that inhaled TML survived two or three times longer than those exposed to tetraethyl lead [41]. Dogs proved to be more sensitive than rats to the toxicity of both chemicals and to TML, in particular. The interspecies differences were unclear but were possibly due to toxicokinetic differences between rats and dogs.

The ability of TEL and lead acetate (both of equivalent lead content: 27.3 mg Pb/kg) to induce cochlear dysfunction was tested in guinea pigs following a single intraperitoneal injection [42]. The cochlear toxicity of TEL, as measured by electrophysiological measurements, was detected at doses that did not induce any damage by lead acetate.

6. TOXICOKINETICS

Alkyllead compounds are lipophilic and, therefore, well absorbed through the skin. Rapid and extensive dermal absorption of tetraalkyl lead compounds has been shown in rabbits and rats [43,44]. *In vitro* experiments have shown the rank order of absorption rates through excised skin from humans and guinea pigs as follows: tetrabutyllead > lead nuolate (lead linoleic and oleic acid complex) > lead naphthanate > lead acetate > lead oxide (non-detectable) [45].

Following inhalation exposure, TEL and TML are both rapidly absorbed. In a study of human volunteers exposed to ^{203}Pb labeled TEL for 1–2 minutes, 37% of the inhaled ^{203}Pb was initially absorbed in the respiratory tract, 50% of the ^{203}Pb was associated with the liver, and the remaining burden was widely distributed throughout the body; 20% was exhaled in the subsequent 48 hours [46]. In a similar experiment conducted with (^{203}Pb) tetramethyllead, 51% of the inhaled ^{203}Pb dose was initially deposited in the respiratory tract, of which approximately 40% was exhaled in 48 hours. The distribution of ^{203}Pb 1 hour after the exposure was similar to that observed following exposure to tetraethyllead.

The kinetics of ^{203}Pb in the blood of these subjects showed an initial declining phase during the first 4 hours (TML) or 10 hours (TEL) after the exposure, followed by a phase of gradual increase in blood lead that lasted for up to 500 hours after the exposure. Radioactive lead in blood was highly volatile immediately after the exposure and transitioned to a non-volatile state thereafter. These observations may reflect an early distribution of organic lead from the respiratory tract, followed by a redistribution of de-alkylated lead compounds.

Because of their lipid solubility, the alkylleads can also readily cross the blood-brain barrier. Due to the relatively high content of lipids, organic lead has a high affinity for the nervous system. Similarly, in the blood, about three times as much of alkylleads are found in the lipid fraction as compared to inorganic lead [47].

Alkyllead compounds are actively metabolized in the liver by oxidative dealkylation catalyzed by cytochrome P_{450}. The metabolites include trialkyllead (which is water-soluble) and inorganic lead [47]. Relatively few studies that address the metabolism of alkyllead compounds in humans have been reported. Occupational monitoring studies of workers who were exposed to TEL have shown that TEL is excreted in the urine as diethyllead, ethyllead, and inorganic lead [48–50]. Trialkyllead metabolites were found in the liver, kidney, and brain following exposure to the tetraalkyl compounds in workers. These metabolites have also been detected in brain tissue of non-occupational subjects [51,52]. In volunteers exposed by inhalation to 0.64 and 0.78 mg lead/m^3 of ^{203}Pb-labeled TEL and TML, respectively, lead was cleared from the blood within 10 hours, followed by a reappearance of radioactivity in the blood after approximately 20 hours [46]. The high level of radioactivity initially in the plasma indicates the presence of tetraalkyl/trialkyllead. The subsequent rise in blood radioactivity, however, probably represents water-soluble inorganic lead and trialkyl- and dialkyllead compounds that were formed from the metabolic conversion of the volatile parent compounds [46].

Independent of the route of exposure, absorbed lead is excreted primarily in urine and feces; sweat, saliva, hair and nails, and breast milk are minor routes of excretion [53–58].

The toxicokinetic data on organic lead can be used as biomarkers of exposure for monitoring exposed individuals. Increased blood lead levels were reported in workers exposed to organic lead [28,59]. Both the organic lead and its metabolite inorganic lead were found in the blood of these workers. Organic lead exposure results in a significant increase in lead concentration in urine as well [27]. In fact, a disproportionally high concentration of lead in urine, as compared to the expected concentration on the basis of the blood lead, is a marker of alkyllead exposure [60]. Lead deposited in teeth and bones can reflect chronic exposures. For example, lead levels in bones were used as biomarkers of lead exposure in gasoline sniffers [61] and exposed workers [62–64].

7. CONCLUDING REMARKS

The use of alkyllead compounds has declined over the last 20 years, due primarily to the worldwide effort to eliminate the use of leaded gasoline. Unlike exposure to inorganic lead, alkyllead exposure is mostly confined to occupational settings or the handling of gasoline. In addition, whereas oral exposure is the primary route for inorganic lead, inhalation and dermal exposure are the major exposure routes for the alkylleads. However, the resulting distribution of lead in the environment through the combustion of leaded gasoline in motor vehicles poses risks to the general population from exposure to inorganic lead. Decreases in population blood lead levels have been observed in the United States and in other countries that have eliminated the use of leaded gasoline.

ABBREVIATIONS

ALAD	δ-aminolevulinic acid dehydratase
ALAS	δ-aminolevulinic acid synthetase
ATSDR	Agency for Toxic Substances and Disease Registry
EPA	Environmental Protection Agency
MRI	magnetic resonance imaging
TEL	tetraethyl lead
TML	tetramethyl lead
VOC	volatile organic compound
WHO	World Health Organization
WML	white matter lesions

REFERENCES

1. (EPA) U.S. Environmental Protection Agency, *PBT national action plan for alkyllead*, Washington, DC, 2002, pp. 7–15.
2. (WHO) World Health Organization, *Environmental health criteria for lead*, http://www.inchem.org/documents/ehc/ehc/ehc003.htm#SubSectionNumber: 3.3.2., Geneva, Switzerland, 1977.
3. (EPA) U.S. Environmental Protection Agency, *National air quality and emissions trends report* 1995, Washington, DC, 1996, pp. 14–17.
4. (USDOI) U.S. Department of the Interior, *Minerals yearbook for 1990*, vol. 1, Government Printing Office, Washington, DC, 1991, pp. 657–684.
5. (EPA) U.S.,Environmental Protection Agency, *National air quality and emissions trends report, 2003 Special studies edition*, EPA454R03005, Research Triangle Park, NC, 2003, pp. 13–16.
6. (EPA) U.S. Environmental Protection Agency, *Air quality and emissions-progress continues in 2006*, http://www.epa.gov/airtrends.
7. M. Lovei, *Eliminating a silent threat: World Bank support for the global phaseout of lead from gasoline*, Lead poisoning prevention and treatment: implementing a national program in developing countries, Bangalore, India, The George Foundation, 1999, pp. 169–180.
8. P. A. Meyer, M. J. Brown and H. Falk, *Mutat Res*, 2008, **659**, 166–175.
9. S. Hernberg, *Am. J. Ind. Med.*, 2000, **38**, 244–254.
10. V. Nichani, W. I. Li, M. A. Smith, G. Noonan, M. Kulkarni, M. Kodavor and L. P. Naeher, *Sci. Total Environ.*, 2006, **363**, 95–106.
11. J. L. Pirkle, R. B. Kaufmann, D. J. Brody, T. Hickman, E. W. Gunter and D. C. Paschal, *Environ. Health Persp.*, 1998, **106**, 745–750.

12. V. M. Thomas, R. H. Socolow, J. Fanelli and T. G. Spiro, *Environ. Sci. Technol.*, 1999, **33**, 3942–3948.
13. R. Albalak, G. Noonan, S. Buchanan, W. D. Flanders, C. Gotway-Crawford, D. Kim, R. L. Jones, R. Sulaiman, W. Blumenthal, R. Tan, G. Curtis and M. A. McGeehin, *Sci. Total Environ.*, 2003, **301**, 75–85.
14. I. Nuwayhid, M. Nabulsi, S. Muwakkit, S. Kouzi, G. Salem, M. Mikati and M. Ariss, *Environ. Health*, 2003, **2**, 1–9.
15. P. D. E. Biggins and R. M. Harrison, *Environ. Sci. Technol.*, 1979, **13**, 558–565.
16. G. L. Ter Haar and M. A. Bayard, *Nature*, 1971, **232**, 553–554.
17. K. W. Olson and R. K. Skogerboe, *Environ. Sci. Technol.*, 1975, **9**, 227–230.
18. S. J. Eisenreich, B. B. Looney and J. D. Thornton, *Environ. Sci. Technol.*, 1981, **15**, 30–38.
19. W. R. A. De Jonghe and F. C. Adams, *Adv. Environ. Sci. Technol.*, 1986, **17**, 561–594.
20. (NSF) National Science Foundation, *Transport and distribution in a watershed ecosystem*, NSFRA770214, Washington, DC, 1977, pp. 105–133.
21. K. J. Reddy, L. Wang and S. P. Gloss, *Plant and Soil*, 1995, **171**, 53–58.
22. (EPA) U.S. Environmental Protection Agency, *Air Quality Criteria Document for Lead*, 1986, pp. 6/1–6/51.
23. (USAF) U.S. Air Force, *The fate and behavior of lead alkyls in the subsurface environment*, AL/EQ-TR-1994-0026, Tyndall AFB, FL, 1995.
24. L.-T. Ou, W. Jing and J. E. Thomas, *Environ. Toxicol. Chem.*, 1995, **14**, 545–551.
25. R. Eisler, *Lead hazards to fish, wildlife, and invertebrates: a synoptic review, Biol. Report 85 (1–14)*, Fish and Wildlife Service, U.S. Department of the Interior, Laurel, MD, 1988, pp. 1–14.
26. ATSDR, *Toxicological profile for lead*, Agency for Toxic Substances and Disease Registry, Department of Health and Human Services, Atlanta, GA, 2008, pp. 202–312.
27. P. Grandjean and T. Nielsen, *Residue Rev.*, 1979, **72**, 97–148.
28. B. S. Schwartz, K. I. Bolla, W. Stewart, D. P. Ford, J. Agnew and H. Frumkin, *Am. J. Epidemiol.*, 1993, **137**, 1006–1021.
29. C. S. Mitchell, M. S. Shear, K. I. Bolla and B. S. Schwartz, *J. Occup. Environ. Med.*, 1996, **38**, 372–378.
30. B. S. Schwartz, W. F. Stewart, K. I. Bolla, P. D. Simon, K. Bandeen-Roche, P. B. Gordon, J. M. Links and A. C. Todd, *Neurology*, 2000, **55**, 1144–1150.
31. W. F. Stewart, B. S. Schwartz, C. Davatzikos, D. Shen, D. Liu, X. Wu, A. C. Todd, W. Shi, S. Bassett and D. Youssem, *Neurology*, 2006, **66**, 1476–1484.
32. W. F. Stewart and B. S. Schwartz, *Am. J. Ind. Med.*, 2007, **50**, 729–739.
33. B. S. Glenn, W. F. Stewart, J. M. Links, A. C. Todd and B. S. Schwartz, *Epidemiology*, 2003, **14**, 30–6.
34. B. S. Glenn, K. Bandeen–Roche, B. K. Lee, V. M. Weaver, A. C. Todd and B. S. Schwartz, *Epidemiology*, 2006, **17**, 538–544.
35. J. A. Millar, G. G. Thompson, A. Goldberg, P. S. Barry and E. H. Lowe, *Br. J. Ind. Med.*, 1972, **29**, 317–320.
36. R. L. Boeckx, B. Postl and F. J. Coodin, *Pediatrics*, 1977, **60**, 140–145.

37. S. S. Seshia, K. R. Rjani, R. L. Boeckx and P. N. Chow, *Dev. Med. Child Neurol.*, 1978, **20**, 323–334.
38. S. C. Edminster and M. J. Bayer, *J. Emerg. Med.*, 1985, **3**, 365–470.
39. M. P. McGrail, W. Stewart and B. S. Schwartz, *J. Occup. Environ. Med.*, 1995, **37**, 1224–9.
40. W. F. Stewart, B. S. Schwartz, D. Simon, K. I. Bolla, A. C. Todd and J. Links, *Neurology*, 1999, **52**, 1610–1617.
41. R. K. Davis, A. W. Horton, E. E. Larson and K. L. Stemmer, *Arch. Environ. Health*, 1963, **6**, 473–479.
42. U. Tuncel, W. J. Clerici and R. O. Jones, *Hear. Res.*, 2002, **166**, 113–123.
43. E. P. Laug and F. M. Kunze, *J. Ind. Hyg. Toxicol.*, 1948, **30**, 256–259.
44. F. Thamann and R. A. Kehoe, *Am. J. Hyg.*, 1931, **13**, 478–498.
45. W. C. Bress and J. H. Bidanset, *Vet. Hum. Toxicol.*, 1991, **33**, 212–214.
46. M. J. Heard, A. C. Wells and D. Newton, "Human Uptake and Metabolism of Tetraethyl and Tetramethyl Lead Vapour Labelled with ^{203}Pb", *Int. Conf. on Management and Control of Heavy Metals in the Environment*, London, 1979, pp. 103–108.
47. J. E. Cremer, *Occup. Health Rev.*, 1965, **17**, 14–19.
48. Z. Turlakiewicz and J. Chmielnicka, *Br. J. Ind. Med.*, 1985, **42**, 682–685.
49. N. Vural and Y. Duydu, *Sci. Total Environ.*, 1995, **171**, 183–187.
50. G. G. Zhang, H. Z. He and H. M. Bolt, *Int. Arch. Occup. Environ. Health*, 1994, **65**, 395–399.
51. W. Bolanowska, J. Piotrowski and H. Garczynski, *Arch. Toxikol.*, 1967, **22**, 278–282.
52. T. Nielsen, K. A. Jensen and P. Grandjean, *Nature*, 1978, **274**, 602–603.
53. A. Chamberlain, C. Heard and M. J. Little, *Philos. Trans. R. Soc. London A*, 1978, pp. 557–589.
54. J. B. Hursh and J. Suomela, *Acta Radiol. Ther. Phys. Biol.*, 1968, **7**, 108–120.
55. J. B. Hursh, A. Schraub, E. L. Sattler and H. P. Hofmann, *Health Phys.*, 1969, **16**, 257–267.
56. R. A. Kehoe, *Food Chem. Toxicol.*, 1987, **26**, i–53.
57. M. B. Rabinowitz, G. W. Wetherill and J. D. Kopple, *J. Clin. Invest.*, 1976, **58**, 260–270.
58. J. L. Stauber, T. M. Florence, B. L. Gulson and L. S. Dale, *Sci. Total Environ.*, 1994, **145**(1–2), 55–70.
59. J. Schwartz, *Arch. Environ. Health*, 1995, **50**, 31–37.
60. I. A. Bergdahl, A. Schutz, L. Gerhardsson, A. Jensen and S. Skerfving, *Scand. J. Work Environ. Health*, 1997, **23**, 359–363.
61. H. D. Eastwell, *Med. J. Aust.*, 1985, **143**(Suppl no 9), S63–S64.
62. I. Tell, L. J. Somervaille, U. Nilsson, I. Bensryd, A. Schutz, D. R. Chettle, M. C. Scott and S. Skerfving, *Scand. J. Work Environ. Health*, 1992, **18**, 113–119.
63. B. S. Schwartz and W. F. Stewart, *Arch. Environ. Health*, 2000, **55**, 85–92.
64. B. S. Schwartz, W. Stewart and H. Hu, *Occup. Environ. Med.*, 2002, **59**, 648–649.
65. (EPA) U.S. Environmental Protection Agency, *Costs and benefits of reducing lead in gasoline: Final regulatory impact analysis*, Office of Policy, Planning, and Evaluation,Washington, DC, 1995, pp. E-5.

Met. Ions Life Sci. **2010**, *7*, 165–229

6

Organoarsenicals. Distribution and Transformation in the Environment

Kenneth J. Reimer, [a] ***Iris Koch,*** [a] ***and William R. Cullen*** [b]

[a]Environmental Sciences Group, Royal Military College of Canada, Kingston, Ontario, K7K 7B4, Canada
<reimer-k@rmc.ca>
<koch-i@rmc.ca>

[b]Chemistry Department, University of British Columbia, Vancouver, British Columbia, V6T 1Z1, Canada
<wrc@chem.ubc.ca>

Metal Ions in Life Sciences, Volume 7 Edited by Astrid Sigel, Helmut Sigel, and Roland K. O. Sigel

Published by the Royal Society of Chemistry, www.rsc.org DOI: 10.1039/9781849730822-00165

ABSTRACT: The widespread distribution of organoarsenic compounds has been reviewed in terms of the five kingdoms of life. Over 50 organoarsenicals are described. Pathways for their formation are discussed and significant data gaps have been identified.

KEYWORDS: arsenic · arsenobetaine · Challenger · freshwater · marine · speciation · terrestrial

1. INTRODUCTION

1.1. Background

Some 20 years ago we wrote a review, *Arsenic in the Environment* [1], in which we attempted to provide a summary of existing knowledge sufficiently complete to be used as a base for future work. Our hopes have been fulfilled in that the review is still widely referenced. Our expectations for this chapter are more limited because there has been an enormous increase in the number of publications dealing with arsenic speciation so that a comprehensive review would take far more space than we have available (for reviews see [2–7]). There are a number of reasons for this situation, the principal one being a response to the realization that the toxic effects of arsenic compounds are not limited to the results of chronic ingestion of arsenic trioxide, a favorite tool for homicide so lovingly chronicled by Agatha Christie and her colleagues [8]. Thus it became necessary to study the chronic and acute toxicity of all available arsenic species. Arsenic compounds can be divided into organoarsenicals, which possess an arsenic-carbon bond, and inorganic species, which do not. The structures of the main organoarsenicals found in the environment, together with the abbreviations that will be used in this chapter, are provided in Figures 1 and 2 (see below).

1.2. Analytical Considerations

The second reason for the increase in the number of publications is that the search for arsenic species has been enormously aided by a dramatic increase in our ability to isolate and identify the arsenicals found in most

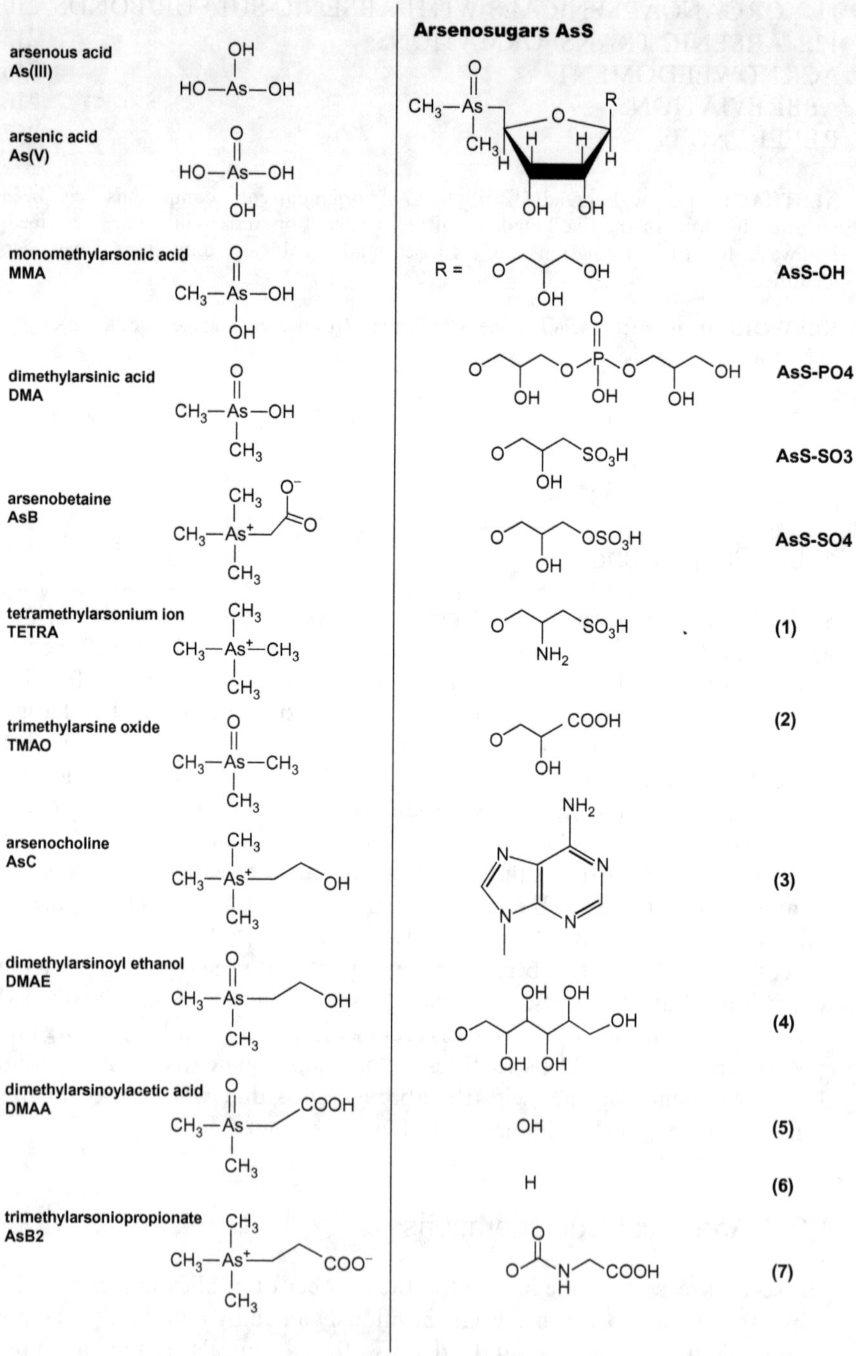

Figure 1. Non-volatile arsenic compounds found in the environment. The less common species are identified by numbers rather than letters. Some such as **5**, **6**, **14**, and **15** are believed to be metabolites of arsenosugars.

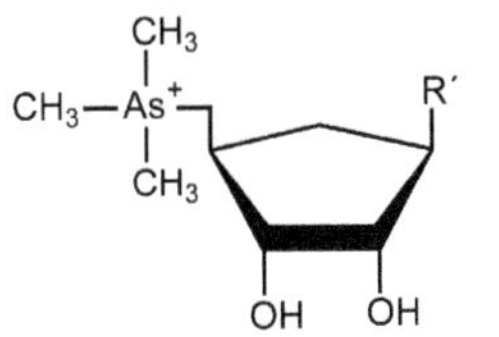

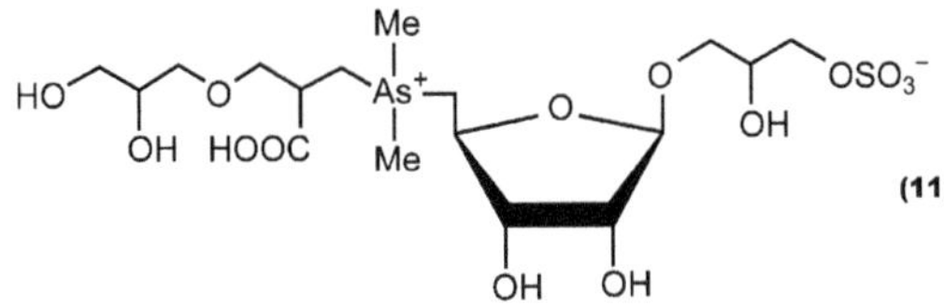

(11)

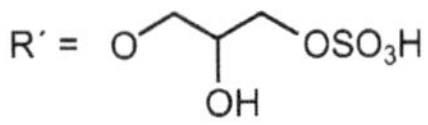

(8)

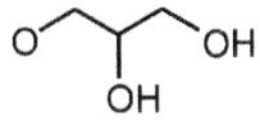

(9)

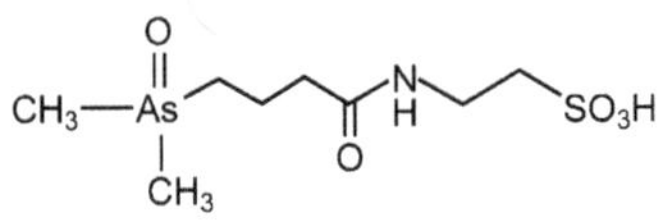

(12)

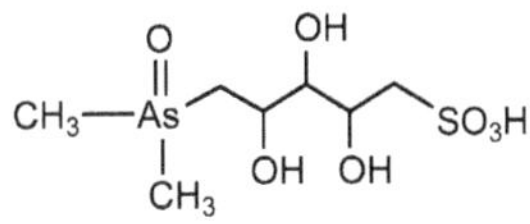

(13)

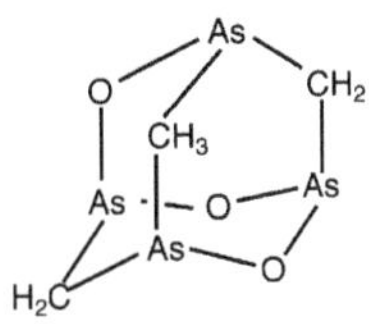

Arsenicin A

(10)

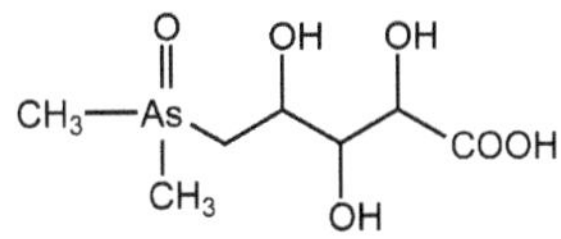

(14)

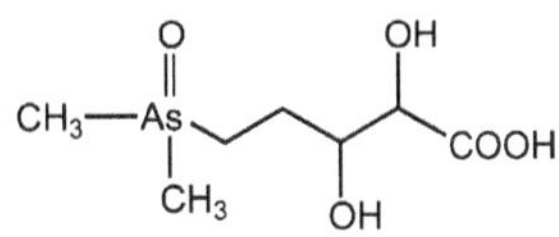

(15)

Figure 1. Continued.

environmental compartments. Analytical methods are described in detail in Chapter 2 of this volume but, given the fact that the arsenic composition of a sample is operationally defined by the analytical method (i.e., compounds can only be 'seen' if a method is capable of 'looking for' them), it is instructive to review some key factors.

Element-specific detection, in the form of inductively coupled plasma mass spectrometry (ICPMS) was just coming to the fore around in the 1980s so that the analytical method of choice for arsenic speciation became high performance liquid chromatography (HPLC) coupled to ICPMS; however, this had limitations because of the requirement for known standards. More recently the development of mass spectrometric ionization techniques compatible with HPLC effluents (e.g., electrospray ionization MS, ESI-MS) has allowed molecule specific detection (e.g., [9–11]). Even so, methods involving pH selective hydride generation and separation of derived arsines are still used in toxicology work or when inorganic and simple arsenic

Arsenolipids

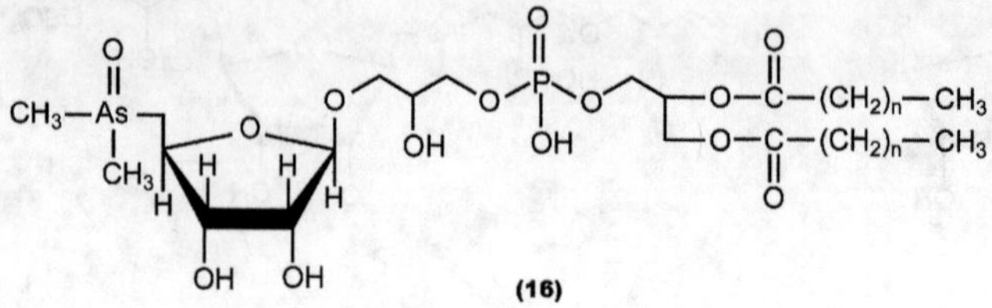

(16)

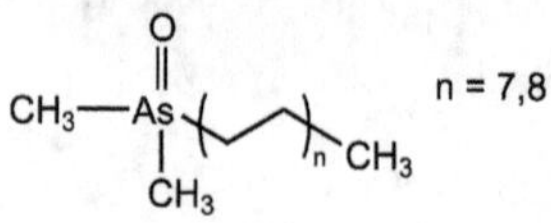

(17)

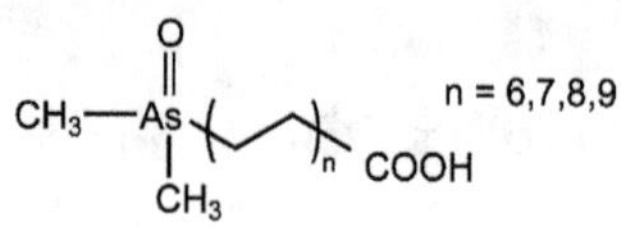

(19)

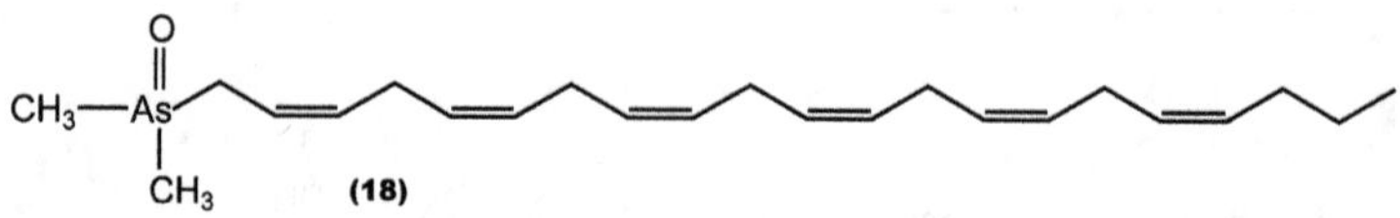

(18)

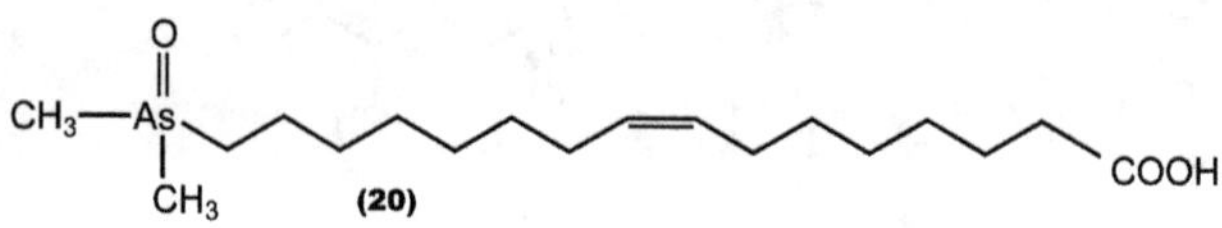

(20)

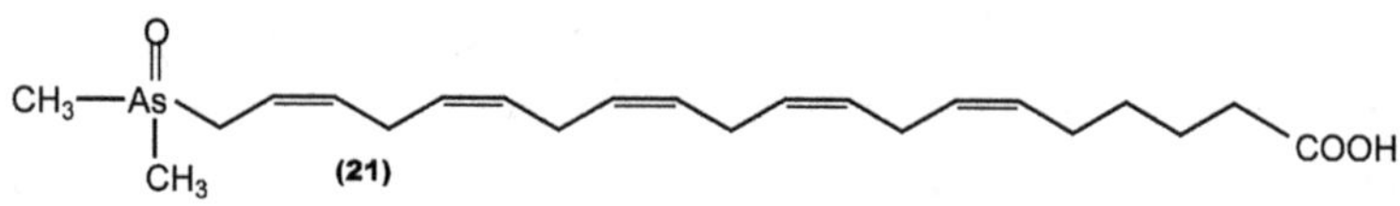

(21)

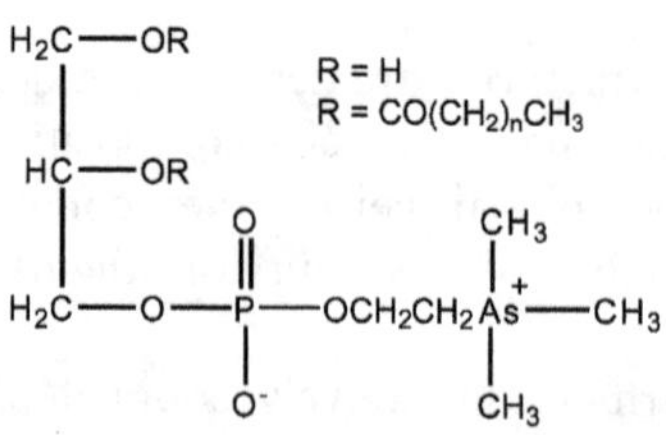

(22)

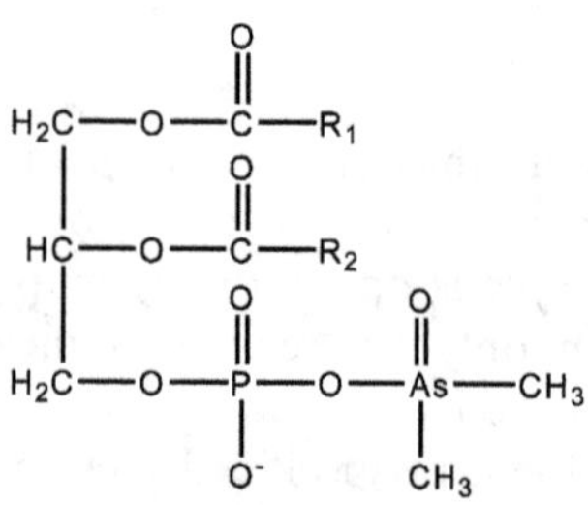

(23)

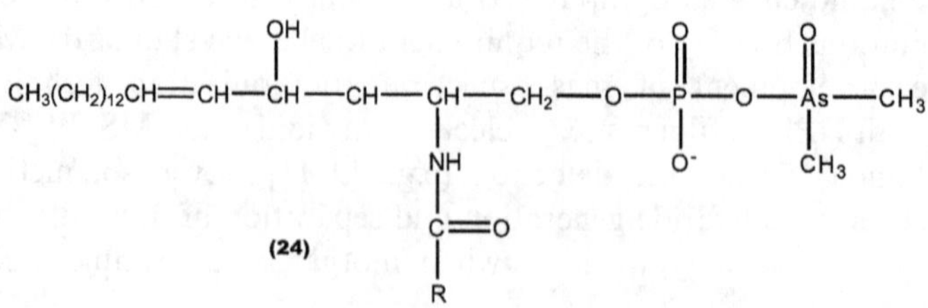

(24)

Figure 1. Continued.

compounds are targeted (e.g., [12]). Hydride species can be generated from more complex arsenicals previously thought to be inert to hydride generation, specifically arsenosugars, but under extreme conditions [13,14]; consequently, this has limited analytical utility. Essentially all of these speciation methods depend on being able to get the arsenicals into solution, something that is much easier for marine samples, with extraction efficiencies sometimes nearing 100%, than for terrestrial plants, with extraction efficiencies often less than 50%. Techniques such as X-ray absorption spectroscopy (XAS) – even more sophisticated (and costly) – are now providing information about these insoluble species.

It is interesting to note that even when extraction efficiencies are less than 100%, recent studies suggest that the most popular solvents used (methanol/water combinations) are actually quite efficient at extracting the organoarsenicals (but not necessarily the inorganic species). In one study, solvents (methanol/water) with increasing aqueous content extracted more inorganic arsenic whereas monomethylarsonic acid (MMA) and dimethylarsinic acid (DMA) extraction remained relatively constant [15]; higher methanol content extracts polar species more efficiently [16]. Sequential methods demonstrated that after maximum extraction of organoarsenicals by using aqueous methanol, a second slightly acidic extraction yielded mostly inorganic arsenic from terrestrial plants and marine algae [17–20], as well as additional MMA and DMA from marine animals [18].

An important methodological aspect of conventional arsenic speciation analysis is the potential change of species from the *in situ* forms to forms that can be detected using the selected instrumentation. It is not surprising that changes to inorganic arsenic species occur during harvesting and sample preparation [21,22], but sample preparation may also affect extraction of organoarsenicals. This was seen when extractable trimethylarsine oxide (TMAO) and DMA decreased in freeze-dried plant and soil samples (compared with fresh, air and nitrogen dried samples) [23]. Storage (even at –20 °C) of spruce needles, and fish and chicken extracts resulted in loss of arsenobetaine (AsB) [23,24]. (See also microbial decomposition in Section 4.6 and thioarsenicals in Section 11).

As mentioned previously, XAS is very helpful in providing information about unidentified arsenic, as well as *in situ* (unaltered) samples, since no sample preparation is needed. In a large number of studies using this technique, inorganic arsenic is found to predominate in whole (not previously extracted) samples (e.g., in soil [25], plants [26], and earthworms [27]; one exception is the mushroom *Agaricus bisporus* [28]). Notably, the inorganic arsenic in unaltered samples is often bound to sulfur (As(III)-S) (e.g., [26,27]). Likewise it appears that unextracted arsenic is also predominantly As(III)-S [12,17,29,30], confirming what others have proposed.

In some samples lipid-bound arsenic may also account for the residual arsenic [31].

The result of this analytical activity is that we now know far more about the arsenic species (around 50 to date) found in a wide range of microorganisms, algae, plants, and animals than we did in 1989. Some species are present in such low abundance that they were only revealed by using improved analytical methods. However, we are not much closer to understanding the biological processes that produce this bewildering array of species. There are a few highlights in the positive direction such as the Challenger pathway (Section 3) is operative in the marine alga *Polyphyas peniculus* [32]; methylarsenic(III) derivatives, putative intermediates in the Challenger pathway, are produced by the freshwater alga *Closterium aciculare* [33]; mussels living in seawater containing labeled DMA and MMA accumulate labeled arsenobetaine [34].

The distribution and formation of the compounds shown in Figure 1 is the focus of this chapter but it should be noted that nature may yet reveal novel arsenicals. For example, a polyarsenic compound (arsenicin A; Figure 1) was isolated from a marine sponge and it exhibits antibacterial activity [35]. Arsenicin A is the most unusual arsenic compound to be isolated from any environmental compartment. This structure does not fit any pattern related to the Challenger pathway and seems to be derived from $(HO)_2As\text{-}CH_2\text{-}As(OH)\text{-}CH_2\text{-}As(OH)\text{-}CH_2\text{-}As(OH)_2$. Although unique at this moment, other related species might be found because the compound was isolated from dichloromethane extracts, rather than the usual aqueous methanol mixture.

It is also worth noting that organoarsenicals have been found in petroleum products and coal. Natural gas samples from the Southern USA contain up to 63 $\mu g\,dm^{-3}$ as mostly trimethylarsine, but surprisingly, the other species found include ethyl derivatives such as ethyldimethylarsine, diethylmethylarsine, and triethylarsine [36]. Trimethylarsine sulfide and probably the oxide are present as solid deposits in the pipelines. An aqueous extract of oil had trimethylated arsenic (520 $ng\,cm^{-3}$), along with monomethylated arsenic (104 $ng\,cm^{-3}$) [37]. Organoarsenicals were found in coal from Slovenia and the Czech Republic, with tetramethylarsonium ion (TETRA) being predominant in coals with lower total arsenic concentrations (2.3–14.3 $mg\,kg^{-1}$); MMA and As(V) were also found [38]. The arsenic concentration in one sample was substantially higher than in the other samples at 142 $mg\,kg^{-1}$, but the extractable arsenic contained only traces of organoarsenicals and was mostly As(V). The majority of samples had at least trace concentrations of AsB (up to 37 $\mu g\,kg^{-1}$). In oil shale, conventional extraction techniques revealed the presence of phenylarsonic acid and MMA [39,40]; XAS with curve fitting of the unaltered samples also suggested the presence of phenylarsonic acid [41].

1.3. Toxicity of Organoarsenicals

The toxicity and carcinogenicity of organoarsenicals is dealt with in detail in Chapter 7 of this volume but it is important to note that three major changes in our thinking about the toxicity of arsenic species have occurred: (1) there is now general recognition of what was stated in 1989 that the methylarsenic(III) species are more toxic in a number of assays than the inorganic species (e.g., [42–44]) reversing the generally held opinion that the methylation of arsenic via the Challenger pathway is a detoxification process; (2) some thioarsenicals such as dimethylthioarsinic acid are more toxic in some assays than their oxy analogues [45]; and (3) trimethylarsine has a very low acute toxicity [46]. These findings have contributed to our understanding of arsenic transformations, and drive the search for new compounds such as thioarsenicals.

1.4. Organization

In this chapter we will examine the organoarsenicals found in the environment: in non-living compartments (natural waters, sediments, and the atmosphere), and in the five kingdoms of life: Prokaryotae (bacteria and cyanobacteria), Protoctista (including microalgae, and brown, red, and green algae), Fungi, Plantae (freshwater and terrestrial), and Animalia (parazoa or sponges; worms; molluscs; arthropods including insects, arachnids, and crustaceans; fish; amphibians; reptiles; birds; and mammals). Planktonic organisms that are at the bottom of the food chain and are a major source of food in the marine environment will be considered separately (after the Protoctista) since they span all the kingdoms. Humans will not be considered and the reader is directed to Chapter 14 of this book which deals with methylated metal(oids) in the human body. New developments in the isolation of arsenolipids and thioarsenicals are described. Lastly, we will examine the pathways giving rise to key organoarsenicals with a goal of determining if the presence of a particular compound is a consequence of biotransformation within (or by) an organism, accumulation through diet, or both.

2. ORGANOARSENICALS IN NATURAL WATERS AND SEDIMENTS

2.1. Water

Rivers and lakes have a range of arsenic concentrations that reflect the natural geology of the drainage area as well as anthropogenic inputs [7,47].

Although values in excess of 500 μg dm^{-3} have been found in surface waters as a consequence of arsenic-rich minerals and mining activity [48], the natural background is about 0.1 to 1.7 μg dm^{-3} [47]. Seawater has a relatively uniform natural arsenic content of 1 to 4 μg dm^{-3}, with an estimated median value of 3.7 μg dm^{-3} [47].

The presence of arsenate, arsenite, MMA, and DMA in both fresh- and seawater has been known for some time [1] but analytical improvements have extended this inventory. Hasegawa et al. [49] made use of the reagent diethylammonium diethyldithiocarbamate to selectively extract methylated arsenic(III) from the water of Lake Biwa, Japan. This then allowed the determination of methylated trivalent and pentavalent species in the same sample by using hydride generation methods: at 2 m depth the major organoarsenical was DMA(V), with no MMA(V) detected. Both MMA(III) and DMA(III) were detected in low amounts (maximum 1.3%).

Similar studies in seawater revealed that in one site in Uranouchi Inlet (Japan) the sum of the methylarsenicals comprised 10–82% of the total dissolved arsenic. The concentration of methylated arsenic(III) species was generally low and independent of that of the methylated arsenic(V) species [49,50]. Around the same time Bright et al. [51] revealed that dimethylarsenic(III) species, possibly thiols, could be produced by microbial action on Canadian lake sediments. These studies were the first to show that the methylated arsenic(III) compounds that are intermediates in the Challenger pathway (see Section 3 and Figure 2 there) can be released into the environment.

Howard and Comber [52] found that seawater contained arsenicals that were not detected by using conventional hydride generation methods. These became known as hidden arsenic species. They showed that the hidden species could be made hydride active by controlled UV irradiation of the sample and reported that on average hidden species comprised 25% of the total arsenic. The same phenomenon is found in fresh water systems.

Hasegawa et al. [53] classified the hidden arsenic species as UV-As and DMA-UV, which were species that released respectively inorganic arsenic and DMA on controlled UV irradiation. They looked at the dissolved <0.45 μm fraction, the colloidal 10 kDa–0.45 μm fraction, and the truly dissolved (<19 kDa) fraction, and found that the hidden species in Lake Kiba (Japan) are distributed mainly in the particulate fraction. The origin of these hidden species and the hydride active methylarsenicals is still uncertain; for example, the DMA concentration does not correlate with chlorophyll-a concentration. It was suggested that the species could be arsenobetaine and/or arsenosugars but Khokiattiwong et al. [54] found that, of 11 arsenicals introduced (in solution) to microbially enriched seawater, AsB and arsenocholine (AsC) were completely degraded, whereas the others underwent little or no change. AsB was transformed within hours to dimethylarsinoylacetic acid (DMAA) and then to DMA; AsC behaved similarly but at a slower rate. This relatively high rate of AsB and AsC degradation by microbes in seawater suggests that the likelihood of finding these species in seawater is not high.

2.2. Sediments

Ellwood and Maher [55] found that anoxic sediments from the marine Lake Macquarie, NSW Australia, contain high concentrations of As(III) and two arsenosugars AsS-SO4 and AsS-SO3 (see Fig. 1). Extraction, handling, and preservation influenced the extraction of the arsenicals, with phosphoric acid proving to be the best extractant for oxic sediments, and hydrochloric acid and sodium hydroxide proving to be marginally better for anoxic sediments.

The pore water from mine impacted lake sediment from Yellowknife (Canada) contains a variety of organoarsenicals amounting to about 10% of the total arsenic [48]. The main organoarsenic(V) species is DMA as determined by hydride generation at pH 1. There are also a number of arsenicals that afford hydrides at pH 6 and these are tentatively assigned to the class of thiols $(CH_3)_nAs(SR)_{3-n}$: model compounds (HSR = cysteine, glutathionine) do produce hydrides at pH 6. Non-hydride active arsenic species are also present. The authors postulate that the arsenic(III) species may have been produced by chemical reduction of bacterially derived arsenic(V) species by thiols present in the sediment [56]; however, there is also the chance that they may be bacterial metabolites.

Anaerobic enrichment cultures have been isolated from arsenic-contaminated lake sediment. Sulfate-reducing cultures produced the highest concentrations of methylarsenicals in both oxidation states. These same species are found in the pore water that was the source of the bacteria, supporting the possibility that the MMA(III), DMA(III) and TMAO are metabolites [51].

Takeuchi et al. [57] show that AsB is the dominant organoarsenical (up to 0.5% of total arsenic) in the surface of marine sediments sampled in Otsuchi Bay, Japan. Other prominent arsenic species were DMA and an unknown. The arsenicals were attributed to contributions from plankton and marine animals.

3. ORGANOARSENICALS IN THE ATMOSPHERE

In this section we will examine the release of arsenic compounds into the atmosphere. According to Matschullat [47] the atmosphere stores around

1.74×10^6 kg of arsenic, which is split between the Northern hemisphere, where most of the industrial activity takes place (1.48×10^6 kg), and the Southern hemisphere (0.86×10^6 kg). The total arsenic input into the atmosphere is around 3–8×10^7 kg per year with the bulk of this coming from volcanoes and anthropogenic sources such as copper smelting and coal combustion. One estimate of the release from bioproductivity from soil is 0.016–2.6×10^7 kg per year [58]: their extreme rate (26,000 tonnes per year) is unreasonably high and would account for around 50% of the total efflux. Frankenberger [59] suggests that bioproductivity could account for 35% of the total efflux, but again this seems too high.

These biovolatilization processes are part of a natural arsenic cycle: organoarsenicals that reach the atmosphere are not very stable and are mostly returned to soil as inorganic species. One study [60] concludes that a small amount of arsine in air is decomposed within four hours and that trimethylarsine is 30% decomposed in nine days: the rate of decomposition increases in the presence of water. Some biovolatilized species remain long enough to be returned in the rain. For example rain samples from Wolfsburg, Austria, contain 5.8 $\mu g\,dm^{-3}$ arsenic, consisting of arsenate (5.4 $\mu g\,dm^{-3}$) and DMA (0.2 $\mu g\,dm^{-3}$) [61]. The methylarsenic compounds in airborne particulate matter vary seasonally. In summer a high concentration of dimethyl and trimethyl forms of arsenic is observed, while in winter the levels are very low [62].

Biovolatilization of arsenic has been recognized for many years. In the late 1800s Bartolomeo Gosio, working in Rome, discovered that a number of fungi metabolized inorganic arsenic compounds, arsenites, and arsenates, to an arsenical gas with a garlic odor. This gas became known as Gosio gas and seemed to be a metabolic product of a number of fungi [63] and possibly bacteria [64].

The gas remained unidentified chemically until 1933 when Fredrick Challenger and his students at the University of Leeds, UK, established its identity as trimethylarsine $(CH_3)_3As$. Subsequent studies by the Leeds group led to the proposal of what we now refer to as the Challenger pathway for biomethylation shown in Figure 2 [63,65–67].

The methyl donor is *S*-adenosylmethionine (SAM) (Figure 3) and the reducing power probably comes from SH groups such as those in glutathione or more complex reductases. The arsenic(III) intermediates with one and two methyl groups are written along the middle line of Figure 2 as oxy species for convenience, and may not have any existence as an isolable species. Only their arsenic(V) analogues on the top line have been isolated from cultures [32,68].

So far it appears that volatilization is limited to the two kingdoms, Prokaryotae and Fungi, and details are given in the respective sections.

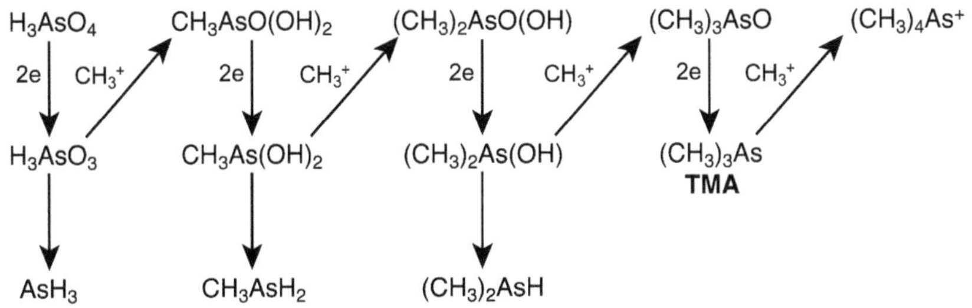

Figure 2. A modified Challenger pathway for the biomethylation of arsenic. The first two lines show how yeasts, fungi, and bacteria produce trimethylarsine (TMA) from inorganic arsenic species. The third line indicates how bacteria probably use the same route to produce arsine, methylarsine, and dimethylarsine. The figure was modified from [8].

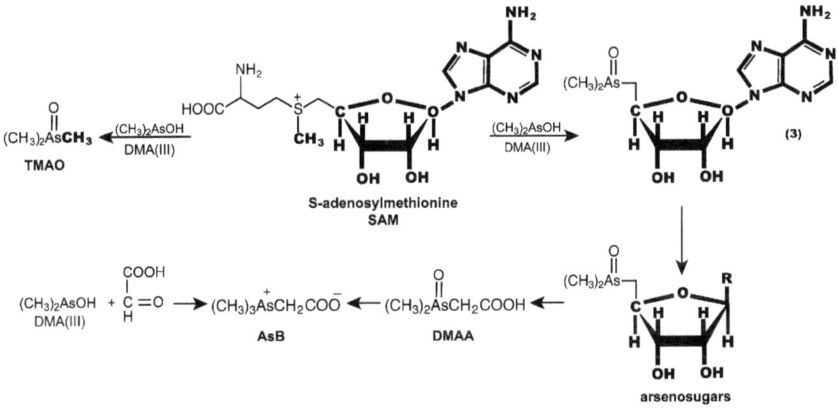

Figure 3. *S*-adenosylmethionine (SAM) as a source of methyl groups for the production of TMAO as in the Challenger pathway (Figure 2) and as a source of adenosyl groups for the production of arsenosugars. Two suggested routes to arsenobetaine (AsB) are also shown: one via DMAA derived from AsS, the other via glyoxylate.

4. PROKARYOTAE

4.1. Bacterial Transformations

In 1917 Puntoni [64] observed that the breath of patients being treated with sodium dimethylarsinate – believed to cure a variety of illnesses – had a strong garlic odor. He isolated *Bacillus subtilis* and *B. mesentericus ruber* from the feces of patients and claimed these produced Gosio gas on treatment with the "drug". This work could not be repeated [65].

The first substantiated report of the biovolatilization of arsenicals by bacteria appeared in 1971. McBride and Wolfe [69] discovered that a volatile arsenical was produced from arsenate by an anaerobic bacterium named *Methanobacterium* strain M.o.H. A gas was also produced from cell extracts of the same bacterium and with the help of radio labeling this was identified as dimethylarsine. The authors then assumed that the gas produced by the living bacterium was also dimethylarsine, but given the results described below this was probably a mistake: the gas is most likely trimethylarsine. McBride and Wolfe noted that gas production required the methyl donor methylcobalamin leading them to conclude that in the living cells metabolizing arsenate the methyl group was transferred from cobalt. This conclusion slowed the further development of the subject since a lot of effort was put into attempts to show this was the case [8].

An early report (1977) of the methylation of arsenic by lake sediments and bacterial isolates from the sediments such as *Aeromonas* sp. and *Flavobacterium* as well as by *E. coli* with occasional production of trimethylarsine has been generally overlooked [70].

Michalke and coworkers [71] confirmed that anaerobic bacteria, typically those found in sewage digesters, are capable of methylating arsenic, and they report that trimethylarsine is the main product from the methogenic archaea *Methanobacterium formicicum*, *Methanosarcina barkeri*, and *Methanobacterium thermoautotrophicum*; the sulfate reducers *Desulfovibrio vulgaris* and *D. gigas*; and the peptolytic bacterium *Clostridium collagenovorans*. When *Methanobacterium formicicum*, the most efficient gas producer in this group (both in quantity and number of products) was exposed to 0.3 mM arsenate, the head space contained almost equal amounts of arsine, methylarsine, and an unknown, and slightly lower amounts of dimethylarsine and trimethylarsine. In the same study *C. collagenovorans*, *D. vulgaris*, and *D. gigas* produced only small amounts of AsH_3 [67,71].

Aerobic cultivation of bacteria from the human gastrointestinal tract (isolated from feces) with AsB showed that after 7 days incubation the AsB had been degraded to DMA, DMAA, and TMAO but after 30 days AsB reappeared in the samples, possibly due to the deterioration/lysis of microbial cells and release of bound AsB, or alternatively the enzymatic formation of AsB from DMAA. No change in AsB was observed for the anaerobic system [72]. It was noted that most of ingested AsB is excreted unchanged in urine but this work indicates the potential for the involvement of human commensal bacteria in processing an important dietary source of arsenic.

Lysed cell extracts of *Pseudomonas fluorescence* A NCIMB 13944, isolated from *Mytilus edulis*, transformed 17% of arsenic provided as DMAA to AsB (maximum transformation was obtained with added SAM) [73]. The same bacterium degraded AsB to DMAA [74].

Microflora isolated from the tails and hepatopancreas of the freshwater crayfish *Procambarus clarkii* degraded AsB to DMA and MMA, as well as to an unknown species. The same microflora transformed (oxidized) AsC to AsB such that AsC was consumed completely; after 24 days the AsB concentration decreased which could not be accounted for by the authors who suggested possible volatilization [75].

Bacteria in anaerobic sediment convert arsenosugars in kelp to dimethylarsinoyl ethanol (DMAE) and DMA [76,77]. This observation was the inspiration for the proposal that arsenosugars are precursors to AsB as indicated in Figure 3. Almost the reverse process of AsB to DMAA to DMA takes place in seawater enriched with bacteria (originating from crabs) [54].

4.2. Sewage Sludge and Landfills

In one of the first reports of volatile species from municipal waste deposits Hirner et al. [78] used ICPMS to reveal that sewage gas contained arsenic in the range 16.1–30.4 $\mu g\, dm^{-3}$ and landfill gas contained arsenic concentrations that were an order of magnitude higher. There was evidence for the presence of arsine, dimethylarsine, trimethylarsine, and ethyldimethylarsine in both types of gases and additionally methylarsine in sewage gas. TMA predominated in landfill gas [79,80].

The sludge from a German municipal waste water treatment facility contained 15.2 $mg\, kg^{-1}$ arsenic. The volatile arsenicals detected in the headspace of this digester sludge after anaerobic digestion ($ng\, dm^{-3}$ quantities) comprised mostly trimethylarsine, with arsine, methylarsine, and dimethylarsine also present [71]. The authors believe the laboratory conditions were close to those established in the bulk facility because the composition of volatile As, Sb, Bi, Se, and Sn species produced in the laboratory experiment resembled that in the gas released from the sewage treatment plant.

Gas production is influenced, both positively and negatively, by the presence of antibiotics [81]. According to Michalke and Hensel [81], studies with pure cultures such as those described above allow limited insight into the productivity of the respective strain within its original habitat. Many variables play important roles, including pH, temperature, metal species, concentration, redox potential, etc. They generalize to state that "the responsible organisms of the metal(loid)-metabolizing biosphere and the underlying molecular process of the biotransformation of inorganic metal(loids) to their volatile derivatives are largely unknown."

4.3. Compost

In one commercial compost source containing 1.8 mg kg^{-1} of arsenic, trimethylarsine within the compost gas was measured at 400 ng m^{-3}. Garden compost contained a similar volatile concentration of trimethylarsine at 657 ng m^{-3} [82]. Diaz-Bone et al. [82] write, "the biomethylation potential was surprising as composting is a predominantly aerobic process. (Most biological waste facilities are aerobic with ca 10% anaerobic). Methylation may be restricted to the micro-anaerobic compartments within the compost, but it is unlikely that such a high biomethylation is caused by only this fraction of the compost". Maillefer et al. [80] found only methyl iodide in the gas from a municipal leaf composting operation.

4.4. Soil

The first incidence of arsenic volatilization from soil was observed during studies concerned with the stability of arsenical pesticides and herbicides in soil. Under aerobic conditions ^{14}C-labeled DMA lost 35% of its activity to the air over a 24 week period. Demethylation also took place producing arsenate and labeled CO_2. Under anaerobic conditions (flooded soil) the volatilization increased to 61% and a garlic odor was detected. Di- and trimethylarsine have been detected above lawns and fields treated with arsenate [83,84].

Cheng and Focht [85] isolated a *Pseudomonas* sp and an *Alcaligenes* sp from soil. They found that in flooded soil (anaerobic conditions) with added glucose and urea *Pseudomonas* sp afforded arsine, whose presence was confirmed by the use of mass spectrometry.

Isolates of *Corynebacterium* sp, *E. coli*, *Flavobacterium* sp, *Proteus* sp and *Pseudomonas* sp acclimated to growth with sodium arsenate for 6 months produced dimethylarsine from arsenate. Six bacteria species including *Nocardia* sp and *Pseudomonas* produced both mono- and dimethylarsine from methylarsonate. The former also produced trimethylarsine [86].

Turpeinen et al. [87] studied arsenic-contaminated soil from a CCA wood preservative plant where the arsenic concentration was in the range 212–632 mg kg^{-1} and the water extractable arsenic amounted to around 0.3%. Trimethylarsine was found in the soil gas and the maximum concentration was encountered at 30 cm depth.

Anaerobic incubation of an alluvial soil that contained 8.9 mg kg^{-1} arsenic gave trimethylarsine as the dominant species, along with arsine, methylarsine, and dimethylarsine; two unknown volatile arsenicals were produced in significantly lower concentrations. A number of other species including trimethylantimony and dimethylselenium were produced.

Trimethylarsine evolution did not start until after the production phase of the selenium derivative (20 days) [88]. Anaerobic cultures of a bacterium (named ASI-1) isolated from this soil biotransformed arsenate to the four usual arsines with methylarsine as the major product. One of the unknown arsenicals was also produced. ASI-1's relative, *Clostridium glycolicum*, was not able to biovolatilize arsenate (or antimony or bismuth).

ASI-1 appears to be the dominant member of the metal(loid) volatilizing population in the soil, but because the distribution of the volatile species from soil is different from the distribution in sewage gas, Michalke et al. [71] suggest the microbial populations in the two sources are different.

Islam et al. [89] concerned themselves with the possibility of biovolatilization of arsenic from soil that has been irrigated with arsenic-rich (8 to 61 $mg\,dm^{-3}$) water. They estimated the arsenic mobilization by bacteria in a range of soils, by measuring the actual production of volatile arsines by the soil under anaerobic conditions and in media designed to promote the growth of methanogens. These numbers were used to calculate the natural gasification potential which varied from soil to soil but maximized at 0.014 μg arsenic per kg soil per day: under enhanced conditions this increased to 0.68 μg As $kg^{-1}\,day^{-1}$. In soil column tests they found $<0.3\%$ of the arsenic in the soil is volatilized in 100 days.

4.5. Hot Springs and Fumeroles

A recent study from Yellowstone National Park (USA) found that the total volatile arsenic measured at the surface of geothermal features was in the range 0.5 to 200 $mg\,m^{-3}$ (average 36 $mg\,m^{-3}$), higher than any previously reported source. The air arsenic concentration dropped off rapidly with distance from the source and was below the detection limit, 0.030 $mg\,m^{-3}$, beyond 1–2 meters [90]. Samples were collected by using SPME fibers from numerous sites and chlorodimethylarsine was found at many of these, with trimethylarsine less abundant. Dichloromethylarsine and dimethyl(methylmercapto)arsine ($(CH_3)_2AsSCH_3$) were also identified as gas phase species. Quantification of the individual arsenicals proved to be impossible. Production of these unusual species could be biotic but it seems that an abiotic process must be partly involved.

An extremophilic eukaryotic alga of the order *Cyandiales* in a Yellowstone hotspring was isolated and found to both undergo redox reactions with inorganic arsenic and to produce DMA and TMAO. Methytransferase genes were cloned into *E. coli*, which was then able to methylate arsenic to the same compounds and to TMA [91].

4.6. Arsenic-Carbon Bond Cleavage

4.6.1. Demethylation. Pure Cultures

The bacterium *Mycobacterium neoaurum* that was isolated from sheep skin mattresses demethylates both methylarsonic acid and methylarsonous acid to mixtures of arsenate and arsenite. The demethylation occurs rapidly during the growth and stationary phases of the bacterium, and probably follows a reductive demethylation pathway, that is, the reverse of the oxidative addition methylation pathway of Figure 3 [92]. The same arsenical is demethylated by two isolates belonging to *Pseudomonas putida* strains isolated from the soil of Ohkunoshima Island (Japan), the site of chemical warfare agent production during the 1930s and 40s. The arsenic concentration in the soil ranged from 7 mg kg^{-1} to 12.5% and both aryl and alkyl arsenicals were present [93]. As mentioned previously (Section 4.1) *Pseudomonas fluorescens* A NCIMB 13944 degrades AsB to DMA via DMAA [74].

4.6.2. Demethylation. Mixed Communities

In a series of papers, Hanaoka and coworkers were able to demonstrate that demethylation of all organoarsenic species occurred in sediments under a variety of conditions. They suggested that these processes form a part of the marine cycle originating with inorganic arsenic, As(inorg): As(inorg) → AsB/TMAO/TETRA → TMAO → DMA → As(inorg) [94–96]. In addition, they note, as have others, AsC → AsB.

Bacterial cell densities of DMA-decomposing bacteria that use the arsenical as a carbon source are 1700 cells mL^{-1} in Lake Kahokugata and 330 cells mL^{-1} in Lake Kibagata (Japan). Fourteen isolates from Lake Kahokugata included two dominant types related to the genus *Pseudomonas*. The types were unique to each lake suggesting that DMA-decomposing bacteria are specific for the aquatic environment. Both MMA and inorganic arsenic are metabolites [97].

4.6.3. Dearylation

Arylarsenicals are found in the environment mostly as the result of anthropogenic input. The one exception appears to be phenylarsonic acid, which was identified in shale as mentioned earlier. Arsenicals containing unsubstituted aryl rings such as phenylarsonic acid, diphenylarsonic acid, and diphenylarsenic oxide, produced by hydrolysis and oxidation reaction of chemical agents such as dichlorophenylarsine and cyanodiphenylarsine, are particularly resistant to microbial degradation (e.g., [93]).

Arsenicals containing substituted aryl rings are now introduced into the environment through their use in animal medicine. The best known example is Roxarsone (3-nitro-4 hydroxyphenylarsonic acid), which is being used in many countries to control coccidosis and related diseases in chickens. Most of the arsenical is found unchanged in the chicken litter with some reduced to 3-amino-4-hydroxyarsonic acid [98]; the aryl ring is lost on composting so that inorganic arsenic is the major product [99]. Recently anaerobic cultures of *Clostridium* sp and *Alkaliphilus oremlandii* sp. were reported to reduce the 3-nitro to the 3-amino compound, with *Clostridium* sp taking the process to arsenate (30% of the arsenic added, with 3-amino at 60%) [100,101]. But there seems to be some problems with the identification of the *Clostridium* sp, and the authors express doubts about whether As(inorg) is produced by a metabolic process. Unlike the situation found for the demethylation of MMA, the cleavage of the As-C(aryl) bond is unlikely to take place by the reverse of the Challenger pathway (i.e., loss of $C_6H_5^+$) and probably takes place after the ring has been broken down.

5. PROTOCTISTA

5.1. Euglena

Euglena is a protist that has animal and plant characteristics. *Euglena gracilis* is an unusual example that can live in the low pH and high arsenic environment of acid mine drainage. Cells of *E. gracilis* grown in 200 mg dm^{-3} As(III) contain 315 mg kg^{-1} As (dry weight). However, Miot et al. [102] point out that if the water content of the cells is around 90% the arsenic concentration in the cell is not in excess of 31 mg kg^{-1}, which is seven times lower than the arsenic concentration in the growth medium. The XANES spectra of the arsenic-loaded cells indicate the presence of arsenic-sulfur species similar to the arsenic(III)-glutathione complex, $As(GS)_3$, as well as species containing As-C bonds amounting to as much as 28% of the total arsenic.

5.2. Freshwater Algae

Arsenic concentrations in freshwater algae are generally lower than in marine species, but some can accumulate the element to even higher levels. For example, the unicellular alga *Chlorella vulgaris* accumulates 2739 mg kg^{-1} in the cells when grown in 1000 mg dm^{-3} As(V) (bioconcentration factor 2.7). The cells contained mainly inorganic arsenic with some AsS-PO4. A ten times lower arsenic concentration in the growth solution resulted in a bioconcentration factor of 1.5, with As(V), As(III), DMA and AsS-PO4; AsB was absent. Cells grown without added arsenic contained only traces of the AsS-PO4. The extraction efficiencies were very low [103].

Maeda and coworkers extensively studied arsenic uptake by *C. vulgaris* exposed to inorganic arsenic (e.g., [104,105]), but this work employed alkaline hydrolysis followed by hydride generation to identify arsenic species in the algae and in the media. These indirect methods gave results that could be generated from a number of starting compounds in the cells, including TMAO, arsenobetaine (from trimethylarsine detection), DMA, and arsenosugars (from dimethylarsine detection).

Algae in natural waters reduce and methylate As(V) with the end product being either As(III) or methylated arsenicals. As(III) is produced during the log growth (fast) phase, with the peak concentration preceding or coincident with the algal bloom [106].

Hasegawa et al. [33] identified methylarsenic(III) species in the medium of the freshwater green alga *Closterium aciculare* collected from Lake Biwa (Japan) and grown under axenic conditions. The concentrations of the methylarsenic species accounted for up to 35% of the total methylarsenicals and the concentration of the reduced species in culture are of the same order as found in Lake Biwa, 0.1–0.2 nM, during natural phytoplankton blooms. These experiments show for the first time that methylarsenic(III) species, postulated intermediates in the Challenger biomethylation pathway, can be excreted by cells.

Green algae (unidentified) from the Danube River from a presumably uncontaminated area contained predominantly AsS-OH, with some AsS-PO4 and As(inorg), but no arsenosugars were present in dried dead samples from the shore [107]. The total sugar concentration in the living sample (3.2 mg kg^{-1}) was in the range of arsenosugar concentrations (0.3–4 mg kg^{-1}) found in freshwater algae from a hotspring [108] and from Yellowknife [109]; in the latter studies total arsenic ranged up to 250 mg kg^{-1} [108] but extraction efficiencies were 2–41% with predominantly inorganic arsenic extracted.

Although the cyanobacteria (also known as blue green algae) are in the kingdom Prokaryotae, they will be included in this section because they are commonly treated as a variant of algae. *Nostoc* is a genus of fresh water cyanobacteria that can be found in lakes, rivers and even moist rocks but is rarely found in marine habitats. Extracts of commercial samples of *Nostoc flagelliforme* from China contained AsS-OH as 93% of the extracted arsenic although extraction efficiency was low at 34% [110]. Microbial mats from hotsprings, which consist primarily of cyanobacteria and other bacteria, had small quantities (up to 4% of total arsenic) of arsenosugars (AsS-OH and AsS-PO4) [108].

5.3. Marine Algae

Much of what we know about arsenosugars comes from investigations on macroalgae and clam kidneys (clams are discussed in Section 9.6.2); details of the algal studies are available in a number of reviews [2–6,111]. These reviews describe the predominance of arsenosugars as the water-soluble arsenic species in marine macroalgae. The generally accepted route to their formation is shown in Figure 3, and involves the transfer of the adenosyl group from SAM to DMA(III). The product **3** has been isolated from clam kidneys.

Much less is known about unicellular and microalgae. The unicellular alga, *Polyphysa peniculus*, was grown axenically in artificial seawater in the presence of As(V), As(III), MMA and DMA in separate experiments [32,68]. DMA was not metabolized but was the major metabolic product from the other arsenicals in both the cells and the medium. Studies with CD_3-labeled methionine showed transfer of the label to arsenic, as would be expected from the Challenger pathway. Significant amounts of more complex arsenic species, such as arsenosugars, were not observed in the cells or the medium. However, these experiments were carried out at high arsenic concentrations ($>0.9\,mg\,kg^{-1}$) and there is the possibility that other metabolic processes may have been overwhelmed.

Foster et al. [19] studied axenic cultures of the microalgae *Dunaliella tertiolecta* and the diatom *Phaeodactylum tricornutum*. These were grown at arsenic concentrations typically found in seawater ($2\,\mu g\,dm^{-3}$) under different phosphorus concentrations. Although *D. tertiolecta* accumulated more arsenic ($13.7\,mg\,kg^{-1}$) than *P. tricornutum* ($1.9\,mg\,kg^{-1}$), media phosphorus concentrations (0.6–$3\,mg\,dm^{-3}$) had little influence on microalgae growth rates or arsenic accumulation. Lipid arsenic comprised a substantial amount of the total, up to 38%, and on hydrolysis gave mostly AsS-OH. Water-soluble species of microalgae *D. tertiolecta* contained mainly inorganic arsenic (54–86%) and lesser amounts of DMA and arsenosugars. *P. tricornutum* contained a different distribution with DMA and AsS-PO4 predominating.

What causes the accumulation of high concentrations of arsenosugars in macroalgae remains one of the unsolved mysteries of arsenic chemistry. Specifically, do the macroalga manufacture their own arsenosugars, or do they get them from other sources, such as epiphytes or symbiotic microorganisms? Examination of arsenic speciation of macroalgae with respect to taxonomic position has not given us the answer, since clear patterns do not emerge; for example, the distribution of inorganic arsenic and DMA appears to span many different orders of algae [20,112,113].

The arsenic species in the brown alga *Fucus gardneri* are AsS-OH, AsS-SO3 and AsS-SO4 but their concentration is seasonally dependent and the speciation is also different in the tips from the rest of the alga [114]. Similar differences in arsenosugar disposition were observed in *Fucus vesiculosus*, with AsS-SO4 at 0.95 mg kg^{-1} in the vesicles but only 0.09 mg kg^{-1} in the remainder of the frond [115].

In an attempt to understand the underlying mechanism of formation of the sugars, Granchinho et al. [116] grew whole young *Fucus* under axenic conditions. The first surprising result was that the alga lost about 73% of its original arsenosugars content, mostly as AsS-SO3, during the laboratory acclimation period. (Other samples showed a less dramatic response that was independent of the phosphate concentration [116]: the arsenosugars are detectable in the seawater media [117]). When the *Fucus* was exposed to arsenate (500 μg dm^{-3}) for 14 days there were increases in the concentration of As(III), DMA, and As(V), which were not detected in the control, and in AsS-OH (other arsenosugar species decreased). At the same time the concentration of the arsenate in the medium dropped to zero accompanied by the appearance of small amounts of As(III) and larger amounts of DMA. It is significant that DMA appeared within a few days whereas the As(III) appeared later. Although the Challenger pathway was clearly operative, it is not evident that sugars were produced at these high arsenic concentrations. Inorganic arsenic predominated in algae (*Fucus* sp.) collected from a contaminated area suggesting that metabolic pathways to arsenosugars may have been saturated, since arsenic in control samples from an uncontaminated area had more usual arsenic speciation [12].

A fungus grew with some *Fucus* samples in artificial seawater pH 7.7 under axenic conditions. This was identified as *Fusarium oxysporum melonis* and was studied in case it was the source of the arsenosugars. It did make DMA from As(V) but in very small amounts [118].

Another *Fucus* species, *Fucus serratus*, grown in aquaria with seawater amended with arsenate (0–100 μg dm^{-3}) also showed variation in species with time but the concentration of the major arsenical, AsS-SO3, was little changed [119]. A lack of additional arsenosugar formation with increasing concentrations was attributed to a toxic concentration being reached at 100 μg dm^{-3}, hindering metabolic pathways. Although the cultures were not axenic the alga probably was responsible for some of the formation of AsS-SO3; however, the authors optimistically interpreted these results as in favor of the alga being able to convert arsenate to arsenosugars.

Facile loss of the arsenosugars from *Laminaria digitata* was observed by Pengprecha et al. [77] who were repeating experiments first reported by Edmonds and Francesconi [120]. During the first 10 days of the experiment that involved the use of a mesocosm packed with kelp, anoxic sediment, and seawater, the arsenic in the aqueous phase was in the form of arsenosugars. DMAE was produced later along with DMA. The arsenicals in the aqueous phase after 106 days were As(III), As(V), MMA, and DMA (AsB and

AsC were absent). The formation of DMAE was taken by Edmonds and Francesconi [120] as support for their proposal that arsenobetaine was derived from arsenosugars. The absence of AsB from the products in this more recent experiment does not refute the argument because any AsB would be easily degraded under the anaerobic conditions. The more recent study seems to have overlooked the possibility of the formation of thioarsenosugars (Section 11).

The common arsenosugars discussed so far are not always the predominant arsenicals in algae. In one species of Antarctic algae, *Gigartina skottbergii*, 67% of the total arsenic was 5-dimethylarsinoyl-β-ribofuranose, **6** (see Fig. 1), identified by ESI-ITMS [121]. Some algal species are known to contain larger than usual proportions of inorganic arsenic (e.g., *Hijiki fusiforme, Sargassum fulvellum* [122], and *Laminaria* [123]). This is also the case for some recently reported algae species including representatives of brown algae (*Lobophora* sp), red algae (*Martensia fragilus, Laurencia* sp, *Champia viridis*) and green algae (*Ulva lactuta*), where 29–63% of the arsenic is As(V) [112]. DMA has also been found to be a major organoarsenical (16–41%) in *Ulva lactuta* (green), *Codium lucasii* (a green alga), *Amphirao anceps* (a red alga), and *Laurencia* sp [112].

Recent studies have reported the presence, for the first time, of arsenobetaine in extracts of marine algae [20,124,125], comprising up to 17% of extractable arsenic in four samples of red alga *Phyllophora antarctica* from Antarctica [126]. In most of the reports the authors expressed the possibility that the AsB originated from marine mesofauna adhered to the algae [20,124,125]. In the case of *P. antarctica*, great care was taken to remove the epiphytes (polychaetes) and these were found to contain much lower arsenic concentrations than the cleaned algae [126]. Low concentrations ($\mu g\,kg^{-1}$) of DMAA and the possible AsB precursor DMAE were identified in marine algae (*Ascophyllum nodosum* and *Fucus vesiculosus*) [9]. It seems safe to conclude that some algae contain AsB but the origin of this arsenical is still unclear.

6. PLANKTON

Plankton are a group of drifting organisms (from the Greek "planktos", meaning "wanderer" or "drifter") that are carried by ocean currents. Many planktonic organisms belong to lower trophic levels in the marine food web, although the tropic position of plankton as a whole is not straightforward. Japanese workers [127] studied speciation in marine zooplankton and phytoplankton that generally consisted of species that they believed belong to lower trophic levels in the marine food web. Their samples of zooplankton were collected from the ocean (600 m to surface) and phytoplankton came from laboratory cultures. The zooplankton contained most of their arsenic as AsB together with smaller amounts of arsenosugars, especially AsS-OH and AsS-SO4. In contrast, the phytoplankton did not contain detectable AsB but arsenosugars were present in species-specific concentrations; e.g., AsS-PO4 predominated in *Heterosigma* and AsS-SO4 in *Skeletonema costatum.* The authors suggest the speciation reflects their feeding habits, with carnivores accumulating AsB and herbivores accumulating arsenosugars. The arsonium sugar **9** was occasionally found in *S. costatum* but the authors argue that this arsenical is probably not the source of AsB in zooplankton and other marine animals as had been suggested [2].

In the same study, unidentified arsenic species were seen in relatively high concentrations in the zooplankton [127]. Unknowns also made up 30% of the arsenic species isolated from the photosynthetic protist *Chaetoceros concavicornis* [128] grown axenically in artificial seawater containing a low arsenic concentration (ca 1 $\mu g\,dm^{-3}$). AsS-SO4, normally the dominant arsenical in *Chaetoceros*, was present at 60%. A crustacean (copepod) *Gladioferens imparipes* fed these axenically grown *Chaetoceros* had a lower proportion of AsS-SO4 (20%) and TMAO appeared (70% of extracted arsenic), along with unknown compounds [128]. In normal seawater AsS-SO4 was 90% of extracted arsenic in the diatom and 70% in the copepod with 10% TMAO; in seawater containing elevated arsenic AsS-SO4 increased to >99% in the diatom but decreased to 20% in the copepod with 25% TMAO; and in seawater containing reduced arsenic AsS-SO4 was 60% and 20% (70% TMAO). The authors suggested that this increase in arsenosugar proportions in the diatom with increasing arsenic in the culture might be indicative of detoxification [128]. On the other hand, no clear pattern emerges for the copepod uptake of AsS-SO4 from its diet, although it is interesting that the maximum AsS-SO4 proportion was obtained in normal seawater, that is, in conditions most representative of the natural environment. However, the copepod appears to methylate As(V) presumed to be present in its culture conditions to TMAO, but does not synthesize AsB from arsenosugars. More recent unpublished work from a research group in Graz (K.A. Francesconi, personal communication, 2009) has found AsB, as well as arsenosugars, in copepods from the natural environment.

These important studies with copepods have been generally overlooked and are unique. The distribution of copepods in the marine environment, where they are the main source of protein, is nearly ubiquitous. They could also be the major source of arsenicals.

Takeuchi et al. [57] report that AsB is a major species in undifferentiated plankton collected from Otsuchi Bay (Japan). The plankton fraction greater than 100 μm contains 535 $\mu g\,kg^{-1}$ AsB (31% of the total arsenic) and the fraction greater than 350 contained 2272 $\mu g\,kg^{-1}$ AsB (53% of the total).

7. FUNGI

7.1. General

In this section we will discuss three types of fungi or fungi-containing organisms: those that are microscopic or mold-forming, those that produce mushrooms (fleshy, macroscopic fruiting bodies that contain spores for reproduction), and lichens, which are fungus symbionts with algae or cyanobacteria.

7.2. Microscopic and Mold-Forming Fungi

The production of Gosio Gas, trimethylarsine, by fungi was described above (Section 3). The best known of the fungi that can produce trimethylarsine, identified by Gosio as *Penicillium brevicaule* but now known as *Scopulariopsis brevicaulis,* was isolated from a moldy carrot. *S. brevicaulis* is abundant in nature, in soil, in stored grain and forage, and in slowly decaying semidry vegetables.

The odor threshold of Gosio gas in solution is less than $1\,\mu g\,dm^{-3}$, allowing as little as $1 \times 10^{-6}\,g$ of As_2O_3 in 1 g of sample to be detected by smell [129]. The following fungi were judged to have the capacity to produce an arsenical gas under the right conditions, on the basis of their ability to produce a garlic-smelling gas: *Aspergillus glaucus, A. virens, A. fischeri, A. sydowi, Mucor mucedo, M. ramosus, Penicillium previcaule* (now known as *Scopulariopsis brevicaulis*), *Cephalothecium roseum, Sterigmatocystis ochracea, Cryptococcus humanicus, Fusarium* sp., and *Paecilomyces* sp. It is important to note that Gosio found that some of the organisms such as *Penicillium notatum* do not produce trimethylarsine from arsenite but do so from dimethylarsinate [67]. Some of these early identifications may be in error or need refinement to the strain level. For example, *Mucor mucedo* obtained from the American Type Culture Collection is not a gas producer (unpublished results).

Challenger et al. [65] examined four different strains of *S. brevicaulis* and all were gas producers; however, the yield of trimethylarsine is low and production is slow. For example, after 105 days, a 2.12% yield of the arsine was obtained from arsenite (0.2%) on bread crumbs. Under different conditions, such as the addition of glucose to the media, the yield was increased to 5.3% after 77 days [130].

Merrill and French [131] found that only two of a large number of available wood rotting fungi were able to produce Gosio gas: *Lenzites trabea* and *Lenzites saepiaria.* The identification was based only on odor. Likewise the fungus responsible for athletes' foot and other similar afflictions,

Trichophyton rubrum, released a garlic odor from inorganic arsenic. This was said to be arsine but is more likely to be trimethylarsine [46,132].

Cox and Alexander [133,134] isolated *Candida humicola, Gliocladium roseum*, and a *Penicillium* sp from sewage. They all produce trimethylarsine, but only *C. humicola* produced it from inorganic arsenic. *C. humicola* gas production, which was at a maximum at pH 5.0, is inhibited by 0.10% phosphate. This investigation was the first to make use of instrumental methods, specifically GC-MS, for the identification of the arsenical. If the arsenic concentration is less than 1 mg dm^{-3} in the media, Gosio gas is not produced, but instead the end product is TMAO, the precursor to trimethylarsine in Figure 2.

Frankenberger and coworkers [59,135] isolated a *Penicillium* sp. from agricultural evaporation water. The fungus did not produce trimethylarsine from inorganic arsenic species but did so readily from MMA. The production maximum was seen at 100 mg dm^{-3}, pH 5–6, 20 °C and 0.1 to 50 mM phosphate. DMA was not metabolized to the same extent. Production of the arsine was suppressed by carbohydrates and sugar acids and many amino acids in the medium; however, phenylalanine promoted growth. Gas production was influenced by the presence of trace elements. In particular high concentrations (1000 μM) of Cu, Zn, and Fe are completely inhibitory.

It was not until 1994 that a definitive study was conducted on the extracellular metabolites of molds and fungi capable of generating Gosio gas [68]. Challenger had assumed that the whole pathway from arsenic uptake to gas elimination took place within the cells; however, *Apotricum humicola* (originally known as *Candida humicola*) rapidly reduced arsenate (1 mg dm^{-3}) and arsenite appears in the medium to be replaced by TMAO along with lesser amounts of DMA. Trimethylarsine is not produced at these low arsenate concentrations and the cells did not accumulate arsenic. A model that incorporates these results is shown in Figure 4. This is based on the finding that the diffusion coefficient of MMA is much lower than that of DMA, so that only DMA and TMAO are excreted into the media, and the observation that there may be a pathway involving the transfer of two methyl groups to MMA without going through a DMA intermediate is incorporated [68,136]. Labeling studies confirmed that the methyl group is transferred from *S*-adenosylmethionine [137].

During most of the 20th century Gosio gas was believed to be toxic and its evolution from moldy wall paper was claimed to be responsible for many human health problems including death. However, these associations have no foundation because trimethylarsine is not particularly toxic [8,46], although the gas is a potent genotoxin *in vivo* [138].

Lehr et al. isolated three fungi from sheep skin bedding that were able to methylate arsenic compounds [92]. Of these three (*Scopulariopsis koningii, Fomitopsis pinicola,* and *Pennicillium gladioli*) only the last produced trace

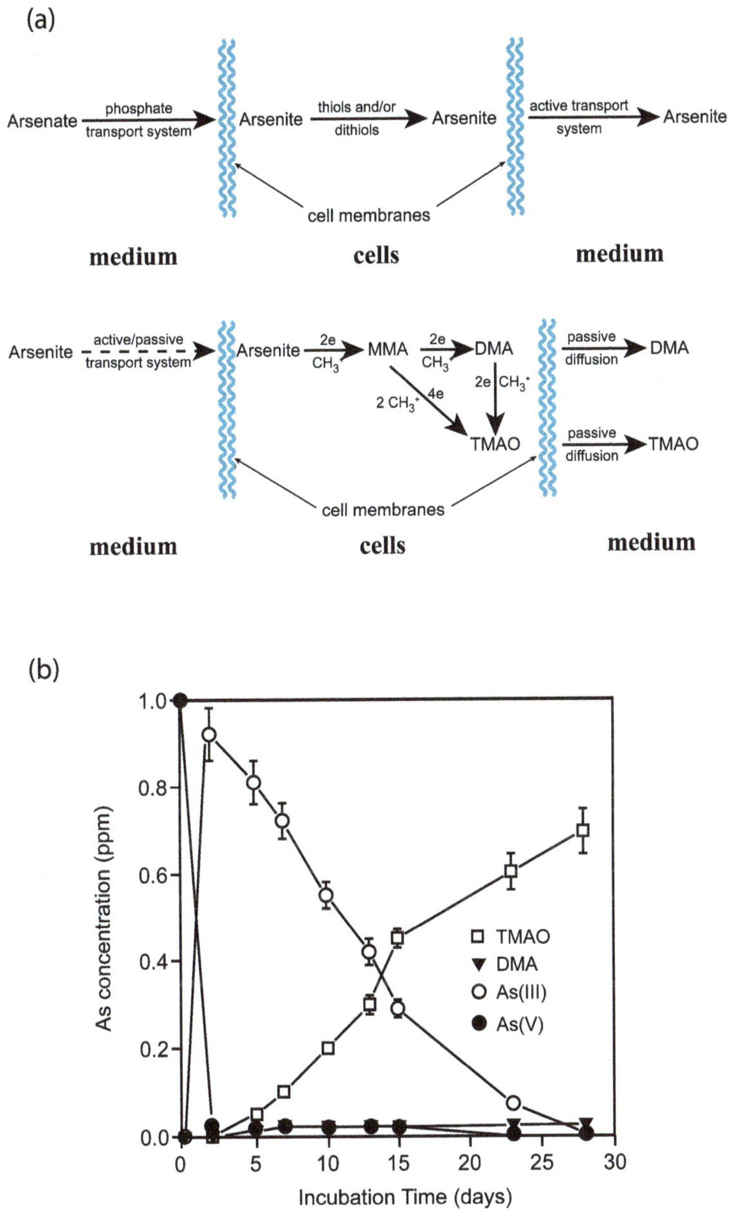

Figure 4. (**a**) A model proposed to account for the uptake of arsenate and the appearance of DMA and TMAO in the culture medium. (**b**) The metabolism of arsenate by *Apotricum humicola* (also known as *Candida humicola* or *Cryptococcus humicolus*). In the medium, As(V) is rapidly reduced to As(III) which in turn is converted to DMA and TMAO.

amounts of trimethylarsine and then only from MMA. *S. koningii* was able to efficiently methylate As(III), As(V), MMA, and DMA (each 500 $\mu g\,dm^{-3}$) to produce mainly TMAO in the medium and in the cells.

Estimates of the number of arsenic-tolerant fungi in arsenic-rich soil reveal that the number is greatest in heavily polluted soils (arsenic concentration greater than 400 $mg\,kg^{-1}$) under aerobic conditions [139]. Those capable of producing an arsenical gas, as judged by a nonspecific chemical test, were strains of *Aspergillus*. Only one strain of *Scopulariopsis* was isolated suggesting that it does not become predominant in soil polluted by arsenic.

In recent years there has been interest in mycorrhizal fungus, especially arsenic tolerant species. Inoculation of sunflower roots reduces toxicity of arsenic and improved plant growth, and the mycorrhizal roots colonized by the fungus are involved with DMA formation (no attempt was made to determine if DMA(III) or DMA(V) was formed, since HG was used), with indigenous soil microorganisms involved with promoting DMA to TMAO (no TMAO in sterile conditions) [140,141], although the sunflower itself is claimed to methylate *de novo* [142].

7.3. Mushrooms

Since our last review [1], investigation of the speciation of arsenic in mushrooms has revealed the presence of a surprising number of arsenic compounds including AsB, AsC, arsenosugars, TETRA, TMAO, DMA, MMA as well as inorganic arsenic. Extensive reviews are available [7,143] and not many additional higher fungi species have been studied since. Of the fungus species surveyed, nearly all have at least trace amounts of AsB in them and AsB was the major extracted compound in all species of Agaricaceae tested. DMA is also common in all fungi surveyed. AsC was found as the predominant species in a single fungus species (*Sparassis crispa*), but minor occurrences of this compound were observed in several other fungi. Likewise, TETRA occurred in a number of fungi, as did unknowns, but arsenosugars and TMAO occurred less frequently or rarely [7].

The Agaricaceae family, with the prevalence of AsB in all species studied to date, has been targeted for studying arsenic speciation and in particular the formation of AsB. The arsenical was not produced in early pure culture experiments with *Agaricus placomyces* [144] amended with inorganic arsenic. More recently *Agaricus bisporus*, as the most commonly cultivated form of the Agaricaceae family, has been used a convenient model species. Two controlled laboratory studies have been able to replicate the production of AsB in the fruiting bodies of *Agaricus bisporus*. In one study the amount produced was lower than that in a control (i.e., no arsenic amendment) experiment [145], whereas in the other study that used lower concentrations of added arsenic, AsB formation was significant [28]. In the latter study, a pasteurized control treatment not inoculated with the fungus did not have AsB in the compost, indicating that the AsB was produced by the fungus, or by organisms associated with the fungus. However, methylated species (up to TMAO) were detected in the control uninoculated compost (inoculated compost could not be separated from the mycelium and was thus not analyzed), indicating that some organisms capable of methylation survived the pasteurization process. These studies did not reveal the exact compartment in which the AsB is produced, but if microorganisms associated with the fungus are involved, this could be a potentially significant finding, if such organisms were commonly found in all environments, including those of marine origin.

7.4. Lichens

Lichens are associations of fungi and green algae or cyanobacteria and are popular atmospheric bioindicators of contamination. In recent years, work on arsenic species in lichens has expanded on past studies [108,146,147]. Organoarsenic compounds in *Hypogymnia physodes* (L.) Nyl. and *Cladonia rei Schaer* collected from the environment included MMA, DMA, AsB (more in *Cladonia* sp. than *Hypogymnia* sp.), TMAO, and AsS-OH, as well as AsS-PO4 in *H. physodes*. (Inorganic species predominate in both lichens, however). Low extraction efficiencies of this type of sample are thought to be attributable to soil content in the lichen [148] and application of soil extraction techniques improve extraction but the additional extracted species appear to be inorganic [148]. The organoarsenicals in transplanted *Parmelia caperata* L. Ach. were MMA and DMA only (inorganic species predominated) [149,150]. Exposure of *Hypogymnia physodes* (L.) Nyl. thalli (the lichen body) to an inorganic arsenic-containing solution resulted in a less complex species content (MMA and DMA) [151] than the *in situ* specimens described above [148].

Thus it appears that fungi and fungal communities (including lichens) are major contributors of AsB to the terrestrial environment, but the origin of this arsenical is still unknown.

8. PLANTAE

Plants contain mostly inorganic arsenic (e.g., [7,152]), and only exceptions to this general trend are reported here. Small amounts of organoarsenicals have been reported, including AsB and TETRA in soil, soil-like substrates, and soil porewaters (e.g., [23,153]).

DMA was the only organoarsenical in three species of angiosperms, but in the seagrass *Posidonia australis* up to 24% of water soluble arsenic (9% of total arsenic) was found as AsB in one sample, and in another sample 71% of extracted arsenic (35% of total arsenic) was a mixture of DMA, AsC, AsB, and three arsenosugars including the glycerol trimethylated arsenosugar **9** (the latter was 13% of extracted arsenic, or 6% of total arsenic) [154]. The presence of the organoarsenicals (other than DMA) were likely attributable to epiphytes that could not be washed off prior to analysis. In submergent plants from the Moira watershed, organoarsenic compounds (at trace levels) included MMA, DMA, TMAO, TETRA and possibly arsenosugars, but no AsB or AsC [155].

Epiphytes are less likely to be a problem for terrestrial plants, especially in above-ground parts that have been thoroughly washed. MMA, DMA, and TMAO, and TETRA have recently been reported in terrestrial plants from mine sites, where larger proportions of organoarsenicals (with respect to extracted arsenic) were attributed to the higher soil arsenic concentrations, although soil characteristics or habitat details were not considered, and the number of plants was small. Organoarsenicals, mostly DMA, reached a maximum of 25% of total arsenic in boxtree leaves from the most contaminated site [156].

Some examples of other plants in which higher proportions of organoarsenic species have recently been reported include bamboo, pepper plants, carrots, and rice [15,157–159]. Up to 29% of the total arsenic in bamboo shoots was DMA, which was found in all bamboo samples studied (MMA and TMAO appeared less frequently); total arsenic was less than $100\,\mu g\,kg^{-1}$ [157]. In pepper plants grown on arsenic-containing soil, 40% of total arsenic was DMA in fruits, and 4% was MMA in roots [15]. In four out of five carrot samples that had been archived from the 1980s, MMA was found to be the predominant compound, with other organoarsenicals including MMA(III), thioMMA (MMA with O replaced with S, Section 11) and traces of DMA; the presence of MMA was probably reflective of agricultural practices at the time of sample collection [158,160]. DMA is one of the dominant arsenic compounds found in American rice, and increases with increasing arsenic concentration (i.e., sum of species extracted, where EEs were >80%), whereas inorganic arsenic remained constant [159]. American rice was concluded to be less of a health hazard than Asian and European rice, which contain predominantly inorganic arsenic [159,161,162]. On the basis of earlier findings of inorganic arsenic in rice, the risks associated with rice consumption, especially by infants, were greatly overstated but widely disseminated [8,163,164], and therefore it is reassuring that a larger data set is now available. Differences in arsenic speciation were thought to be related to genetic differences in the rice types' abilities to methylate arsenic [159].

The speciation in the sunflower, a plant that has been extensively used to study As(III)-phytochelatin complexes, also includes a MMA(III)-phytochelatin complex (up to 13% of identified species), MMA(V), and DMA(V) (less than 1% methylation overall) [142,165]. In these studies the authors believe the methylated forms are synthesized "*de novo*" (although the plants were not cultured axenically), and that the possibility of methylation by microbial contamination of the hydroponic/Perlite solutions used is unlikely.

Axenic cell suspension cultures of the Madagascar periwinkle *Catharanthus roseus* are able to take up As(V) and excrete As(III) into the medium. Uptake of MMA ($2\,mg\,kg^{-1}$ As) is also facile. Limited methylation (4%) to DMA occurs, as well as demethylation (1%) to inorganic arsenic (1%) – this is the only study to date that has shown methylation and demethylation by the plant cells alone. DMA is the least toxic arsenical to the cells and it undergoes some demethylation (12%) [166].

9. ANIMALIA

Marine animals consistently contain arsenobetaine in their tissues, and this has been reviewed a number of times [2–6,111,167].

9.1. Porifera: Sponges

A single freshwater sponge *Ephydatia fluviatilis* from the Danube River, at a location used as fishing grounds (i.e., not extremely contaminated), has been analyzed and contained predominantly inorganic arsenic: AsS-OH along with some DMA were the only organoarsenicals, and AsB was absent [107]. On the other hand, AsB is commonly found in marine sponges [168–170] in proportions within the wide range 9–87% of water-soluble arsenic. When AsB did not predominate, arsenosugars usually did (the exceptions were *Acanthella* sp. and *Biemna fortis*, in which "other compounds" were dominant) [170]. While AsS-OH was ubiquitous among the marine sponges studied, its maximum proportion was only 48% in *Phyllospongia* sp., whereas AsS-PO4 accounted for up to 76% of water soluble arsenic in *Halichondria okadai*, but was absent in several other species [170].

It was noted earlier (Section 1.2) that sponges can contain unusual arsenic compounds such as arsenicin A (Fig. 1) [35].

9.2. Worms

9.2.1. *Terrestrial*

Most of the available arsenic speciation information on terrestrial earthworms comes from specimens collected from the natural environment, and inorganic arsenic predominates; in particular, As(III) bound to sulfur has been identified by XAS techniques [27,171]. Earthworms also contain AsB at low levels [27,172,173]. Notably, earthworms resistant to arsenic (acclimatized) contain proportionally more AsB [27] (although resistance is thought to be related to As(III)-S complexation), and higher proportions of AsB are seen in worms containing less arsenic and exposed to lower concentrations of arsenic [173,174]. The location of AsB (cautiously identified with the XAS method used) [171] was postulated to be the chloragogenous tissue of the earthworm, but no AsB was seen in whole earthworm, posterior, or body wall. Other organoarsenicals recently detected in earthworms are DMA, MMA, AsS-OH, -PO4, and -SO4 [173], concurring with an earlier study that showed the occurrence of DMA, AsS-OH, and -PO4, in addition to the aforementioned AsB [172,175].

The formation of ^{14}C-DMA was reported in a study using ^{14}C-labelled SAM, arsenite, and cytosol extracted from earthworms (*Lumbricus terrestris*), but no quantitative information was given [176]. These results may indicate that earthworms have the capacity to methylate As(inorg).

9.2.2. *Marine*

Polychaetes are worms habituating mostly marine environments and the arsenic speciation in their tissues depends on their ecology [177]. Two reviews are available [177,178].

The worms are remarkable in their ability to take up arsenic. For example, *Sabella spallanzanii* from the Mediterranean accumulates around 1036 mg kg^{-1} arsenic in the crown but only 48 mg kg^{-1} in the body tissues. The same animal in Australian waters accumulates around 713 mg kg^{-1} in the crown and 15 mg kg^{-1} in the body. The reverse situation is seen in *Serpula vermicularis*, also from the Mediterranean, with the crowns around 5 mg kg^{-1} and the body 52 mg kg^{-1} [178].

Polychaetes, like most marine animals, have some AsB in their tissues (e.g., AsB comprises about 60% of the arsenic in the nereidids *Hediste diversicolor* with the rest as TETRA), but some species have interesting arsenic speciation that is dominated by other less innocuous arsenic compounds. *Arenicola marina* has predominantly inorganic arsenic (70% of $\sim$50 mg kg^{-1}) and can biomethylate As(V) to DMA [179]; in contrast,

Nereis diversicolor and *Nereis virens* can biomethylate As(V) to TETRA [179,180], although in both these studies transformation via algae or bacteria could not be excluded. The speciation in *Sabella spallanzanii* is the same in the branchial crown and the body with DMA accounting for up to 85% of the total arsenic with TETRA, AsB, and AsC making up the rest. DMA also predominated when the crowns were regenerated [181] after non-axenic exposure to As(V), whereas AsB had no effect on the branchial crowns but was significantly accumulated in body tissues [182]. Other unusual arsenic compounds predominated in only a few polychaete species: AsC accounted for 60% of the arsenic present in *Perkinsiana* sp, with the remaining 40% as AsB [178]; AsB2 acccounted for 33% in *Australonuphis parateres*; and inorganic arsenic (38%) and arsenosugars (30%) were observed in *Notomastus estuarius* [183].

This wide variation in speciation in marine worms is probably species specific and is not related to external factors. It has been suggested that the high arsenic levels found in some tissues might act as a defense mechanism against predation [178].

The polychaete *Nereis diversicolor* collected from a contaminated area accumulated arsenic along with metals, and 58% of the arsenic was inorganic, compared with only 0.7% inorganic arsenic in the same worms collected from an uncontaminated area [184]. Therefore arsenic accumulation in this animal under contaminated conditions (approximately 9 times more than in control worms) does not necessarily translate into biotransformation to organoarsenicals, although much higher TETRA concentrations were measured in the contaminated worms than in the controls. When zebrafish were fed the contaminated worms, reduced reproductive output was observed, although no overall effect on population growth was noted [184].

9.3. Cnidaria: Sea Anemones, Jellyfish

The arsenic compounds found in nine species of sea anemones which contain total arsenic in the range 1.6–7.0 mg kg^{-1} (wet weight) do not include As(V), MMA, DMA, or TMAO. The main arsenicals are AsB, AsB2, AsC, and TETRA [185]. The relative amounts of these arsenicals vary markedly with the species of the anemone: for example, TETRA comprises 87% of the water soluble arsenic in *Entamacia actinostoloides*, but AsB and AsB2 were undetected. On the other hand, AsB is the main arsenical (76% of the water soluble fraction) in *Metridium senile* and AsC predominates (71%) in *Actinodendron arboretum*. This accumulation of AsC is unusual: apart from mushrooms (Section 7.3) the only other known AsC accumulator is the

Antarctic polychaete *Perkinsiana* sp. [178] (Section 9.2.2), as well as shrimp and two fish species [186] (Section 9.4.3 and 9.2.2).

AsB was the predominant water-soluble arsenical in 10 species of jellyfish and their mucus, although all jellyfish contained relatively low total arsenic concentrations (<0.7 mg kg^{-1} wet weight) [187]. The jellyfish were classified as AsC rich or poor, and only the Semaostomae order had AsC rich species with an AsC maximum of 17% of the AsB concentrations. The same species tended towards higher levels of TETRA as well, although some species of other orders had similar amounts of TETRA. Lipid soluble arsenic (not identified) constituted up to 26% of the arsenic [187] (Section 10).

9.4. Arthropoda: Crayfish, Lobsters, Crabs, Sea Lice, Shrimp

9.4.1. *Terrestrial Insects*

Few reports of arsenic in insects are available and the speciation is predominantly inorganic; like in terrestrial worms the inorganic form appears to be As(III) bound to sulfur [188,189]. Of the organoarsenicals, low or trace concentrations of AsB have been found in ants [188,190].

A recent study identified organoarsenicals in caterpillars, moths, grasshoppers, slugs, ants, spiders, mosquitoes and dragonflies from a contaminated site in Nova Scotia [188]. Predatory invertebrates had more organoarsenicals but the amount accounted for a maximum of 4% of the total arsenic. DMA was found in all invertebrates, MMA in grasshoppers and slugs, TMAO in spiders and mosquitoes, and AsB was found in slugs and spiders.

Limited research has been conducted on how invertebrates take up and biotransform arsenic [189,191,192]. Two studies showed a lack of biotransformation in invertebrate species: bark beetles ingesting an arsenic pesticide, the sodium salt of MMA, did not seem to modify the compound [193], and *Drosophila melanogaster* (fruit flies) did not have the ability to methylate inorganic arsenic, nor alter the form of DMA [191]. The moths *Mamestra configurata* Walker formed As(III) sulfur species, mentioned above, upon exposure and uptake of As(V), but no organoarsenic species were reported [189].

9.4.2. *Freshwater*

The crayfish *Procambarus Clarkii*, found in Spain, accumulates up to 8.5 mg kg^{-1} arsenic [194] with inorganic species accounting for up to 50% of the total. Methanol/water (1:1) extraction afforded one unknown (30%) and arsenosugars (22%) as major species with lower concentrations of As(III), As(V) and/or DMA, and AsB. The main species in the hepatopancreas are AsS-OH and As(III); in the tail, AsS-SO4 (80%); the "rest" contained AsS-SO3 and -PO4, and an unknown.

Williams and coworkers [195,196] studied an Australian species *Cherax destructor* known as the yabby that is gaining popularity as a food. Some of their animals came from mining impacted sites with high arsenic concentration in the sediments. They found that the total arsenic concentration in the yabbies could reach over 200 mg kg^{-1} (the Australian food standard for arsenic is 2 mg kg^{-1}) and that this accumulation was related to the arsenic concentration in the sediments rather than the water [195]. Limited speciation studies on methanol/water extracts revealed the presence of TETRA, As(III), As(V), DMA, MMA, and AsB: some arsenosugars were reported [196]. In animals from uncontaminated sites all these species are distributed fairly evenly between the hepatopancreas, the abdominal muscle, and the "rest". As the total arsenic content increases, the distribution shifts to a preponderance of inorganic arsenic and AsB, and then to almost all inorganic species. Laboratory fed animals were found to be similar with As(V) accumulating in the hepatopancreas following feeding with either As(V) or As(III).

9.4.3. *Marine*

Being the first animal from which AsB was isolated, lobster is well known to contain this compound as the major arsenical in the edible tail. The standard reference material TORT-2, lobster hepatopancreas, used to monitor quality control in total arsenic measurements, has been well characterized for arsenic species. As expected, AsB predominates, but other compounds have now been quantified in this material: inorganic arsenic, MMA, DMA, TMAO, TETRA, AsB2, AsC, and arsenosugars [197–199], as well as minor amounts of the compounds DMAA, dimethylarsinoyl propionate ($(CH_3)_2As(O)CH_2CH_2COO^-$) and DMAE [9].

AsB dominated in the crab *Callinectes sapidus*: 95% of 25 mg kg^{-1} [186,200]. AsB also dominated in the hemolymph ("blood") of Dungeness crab *Cancer magister* (97%); two arsenosugars (AsS-OH and -PO4) and DMA were also found [201]. The results were interpreted as providing evidence that ingested arsenic compounds are not fully metabolized in the gut and are partly absorbed into the hemolymph for distribution throughout the crab's body.

AsB is normally the major compound found in shrimp [6]. It is therefore surprising that AsC was reported to be the major arsenical in the shrimp

Farfantepenaeus notialis, specifically 92.9% of 16.2 mg kg^{-1} [186,200]. AsC was previously believed to be only a minor species in the marine environment [1,4]; however, it is present in substantial quantity in the leatherback turtle (Section 9.8), the Antarctic polychaete *Perkinsiana* sp (Section 9.2.2) and two fish species (Section 9.9.2). A minor (<0.1%) component of a shrimp certified reference material was identified as DMAA [9].

9.5. Gastropods

9.5.1. *Terrestrial*

Methanol/water extracts and protease digests of the freshwater snail *Stagnicola* sp. from a contaminated bay in Yellowknife (Canada) contained predominantly TETRA and inorganic arsenic, but MMA, DMA, AsS-OH, and TMAO, as well as AsB in one sample were also found in smaller proportions [109]. Snails from the family Viviparidae collected from Pender Island (BC, Canada) contain mainly AsS-OH and -PO4 in addition to lower concentrations of their thio analogues (unpublished results).

9.5.2. *Marine*

Gastropods can contain high concentrations of arsenic; for example, *Buccinun undatum* collected from Newfoundland (Canada) has more than 100 mg kg^{-1} in the foot muscle and one sample contained up to 1360 mg kg^{-1}. The major compound was AsB but there were traces of arsenosugars [202]. AsB is also the major species in the related species, *Buccinum schantaricum* but in lower concentrations in the muscle, along with TETRA (13% of the 20.5 mg kg^{-1} total arsenic) and AsC (5%). The speciation in the mid gut gland (51% of the total arsenic, 32.3 mg kg^{-1}) is similar [203].

Goessler and coworkers [204] found that 95% of the arsenic in the carnivorous gastropod *Morula marginalba* was present as AsB. This sample was obtained from a rock pool which also contained a herbivorous gastropod, *Austrocochlea constricta*, that is eaten by *M. marginalba. A. constricta* was also found to contain mainly AsB with traces of inorganic arsenic, DMA, AsC, TETRA as well as several unknowns, even though its diet was considered to be the seaweed *Hormosira banksii* (commonly known as sea grapes), containing AsS-OH. Although *A. constricta* probably does eat *H. banksii* as claimed by the authors, its diet is likely more complex since its feeding habit has been described as "moving over rocks and scraping up microalgae" [205]. Rock microalgae, analyzed more recently (2006) in a similar study, contained AsB (59%) and arsenosugars (36%) [206]. Thus the finding of AsB in *A. constricta* is probably attributable to dietary ingestion. In addition to the arsenicals previously found in *A. constricta* and *M. marginalba* [204], arsenosugars, including thioarsenosugars and **9** (in herbivores) were also identified [206]. The arsenic speciation in two other herbivorous gastropods *Bembicium nanum* and *Nerita atramentosa* was similar to that in *A. constricta* [206].

9.6. Bivalves

9.6.1. Fresh Water

In freshwater mussels *Margaritifera* sp. from Campbell River (BC, Canada) the highest concentration of arsenic was found in gills (11.8 $mg\,kg^{-1}$) and arsenosugars were the main species extracted (< 56%) from all tissues [207]. AsS-SO4 was found in some samples but not in others, and AsS-OH was present in most samples, along with DMA. In a different mussel *Anadonta* sp from Yellowknife (Canada) with 6.7 $mg\,kg^{-1}$ total arsenic, AsS-OH and AsS-PO4 predominated in the water soluble fractions (30%) and As(V) and unknowns were also present. AsB was absent in *Margaritifera* sp. and *Anadonta* sp. [207]. AsB was present at low levels, however, in recent analyses of *Margaritifera* sp. and *Anadonta* sp. from the Campbell River area, with arsenosugars AsS-OH and AsS-SO3 predominating in the identified fraction (maximum 29%) (unpublished results).

Similar results were seen in mussel samples from the Danube River, which had total arsenic concentrations in the range 3.8–12.8 $mg\,kg^{-1}$. The highest concentration was found in *Unio pictorum*. Arsenobetaine was absent, and the majority of the arsenic was unextracted (extraction efficiency 13%) [208]. The predominant extracted arsenicals were AsS-OH (0.69 $mg\,kg^{-1}$) and AsS-PO4 (0.5 $mg\,kg^{-1}$), with a smaller amount of DMA (0.09 $mg\,kg^{-1}$), and minor amounts of thioAsS-OH (0.009 $mg\,kg^{-1}$), thioAsS-phosphate (0.016 $mg\,kg^{-1}$) and As(V) (trace) (see Section 11 for more details on thioarsenosugars in shellfish).

9.6.2. Marine

The kidney of the giant clam, *Tridacna maxima*, has been the source of most of the arsenic species shown in Figure 1 [1]. It is generally believed that these are not manufactured directly by the clam but have their origin in the photosynthetic zooxanthellae that live in the mantle of the clam and lie in the blood space of the animal [209]. Their excretion products are mainly arsenosugars, which are released into the circulation system of the clam and have access to both gill and kidney tissues [209].

A related clam, *T. derasa*, was studied by McSheehy et al. [11] who were able to identify 15 arsenicals from kidney extracts, four of which were new. They found the common arsenosugars, in addition to traces of AsB and DMA. They also identified a number of species such as **5**, **6**, **14**, and **15,** which the authors suggest the clam transformed from the arsenosugars produced by the zooxanthallae via a series of oxidations and decarboxylations. Fifty percent of the arsenic was found in the form of 5-dimethylarsinoyl-2,3,4-trihydroxypentanoic acid, **14** (Figure 1).

AsB and TETRA are the main species in the clam species *Saxidomus giganteus, Schizothoerus nuttalli, Protothaca staminea*, and *Venerupis japonica* [210]; TETRA was also found in *Meretrix lusoria* [211].

AsB together with lower concentrations of TETRA and an unknown arsenical are the major water soluble species in the adductor muscles of sea scallops (*Placopectin magellanicus*) collected from a number of sites in Newfoundland (Canada) [212,213]. The arsenic speciation in the scallop gonads seems to depend on the sex and the season. AsB is found in both sexes up to 3 mg kg^{-1} but the four common arsenosugars are the major species with AsS-SO4 predominating, up to 16.5 mg kg^{-1}. It seems that the concentration of this arsenosugar is dependent on the sex of the scallop with higher concentrations in the prespawning females, up to 9.64 mg kg^{-1}. The postspawning gonads contain up to 11.4 mg kg^{-1} of the same arsenosugars with no difference in the sexes. AsB was the predominant species, as expected, in scallop kidney extract [214], in which a total of 23 arsenicals were seen, but not all were identified.

Mytilus galloprovincialis was used as an indicator species in the Adriatic Sea [215], and initially contained predominantly AsB (60–65% of arsenic), as well as AsC (20%) and TETRA (15%), with trace amounts of DMA and TMAO. A year later AsB was down to 45% with a concomitant increase in DMA (16%) and TMAO (8%). The increase of the latter was attributed to possible phytoplanktonic blooms. Interestingly, no arsenosugars were observed even though they are quite common in other *Mytilus* species and bivalves.

Unusually high levels of inorganic arsenic (up to 42% of total arsenic) have been measured in blue mussels *Mytilus edulis* L. from Norway, and when the entire dataset was examined ($n = 175$) the inorganic arsenic content was positively and highly correlated with total arsenic content [216]. A similar trend (higher percent inorganic with higher total arsenic) is suggested by limited speciation results for oysters in an earlier study [217]. In the Norwegian study, the constant and low concentration of inorganic arsenic (<8%) for total concentrations less than 3 mg kg^{-1} (wet weight), with increasing concentrations and proportions thereafter, suggests that once this body burden is reached, biotranformation of inorganic arsenic to organoarsenic may be inhibited [216]. No information was given about the sources of arsenic at the Norwegian sites, but high concentrations and proportions of inorganic arsenic were also detected in clams (*Mya arenaria*) from a location in Nova Scotia (Canada) that was highly contaminated with arsenic; organoarsenic species did not appear to increase with increasing exposure to arsenic [12].

9.7. Cephalopoda: Squid, Octopus

AsB is the predominant arsenical in the few cephalopoda studied so far. An octopus *Paractopus defleini* had more than 90% of the arsenic in its muscle as AsB [218] and the arms of 24 specimens of *Octopus vulgaris* were reported to contain almost 100% AsB, although no information about extraction efficiency was given [219]. In the latter study total arsenic concentrations reached a comparatively high 133 mg kg^{-1} dry weight.

The arsenic in the Japanese flying squid *Todarodes pacificus* [220] at less than 10 mg kg^{-1} is spread fairly evenly between the muscle, liver, reproductive organs and the gill, with AsB as the predominant water soluble species (max 6.77 mg kg^{-1} in liver) and lower amounts of DMA, TMAO, and TETRA. Lipid soluble arsenicals accounted for up to 10% of the arsenic in the liver and testes and are discussed in Section 10.

9.8. Reptilia: Frogs, Turtles

Very few reptiles have been studied and at the present results are available only for frogs (freshwater/terrestrial) and turtles (marine). Schaeffer et al. [107] reported arsenic speciation in a single frog (*Rana* sp) from the Danube River. Along with inorganic species, MMA and DMA, 23% of the arsenic in the frog was TETRA (trace amounts of TMAO, AsB, and AsC were also seen). In a recent study of amphibians (green frog *Rana* sp. and one eastern American toad *Bufo americanus*) from a contaminated area in Nova Scotia, a large proportion of TETRA was also seen: up to 14% of total arsenic (identified by XANES) in *Rana* sp (unpublished results). TETRA was found in all frog samples except for two from the uncontaminated area; TMAO was seen in several samples, and DMA and inorganic species were ubiquitous (no AsC, AsB or arsenosugars were detected, however) (unpublished results).

Arsenic compounds in the leatherback (marine) turtle *Dermochelys coriacea* were first reported in 1994 by Edmonds et al. [221], where predominantly AsB was found with up to approximately 11% of total arsenic as AsC in liver. Other species of turtles have been studied since and AsB was found in those species as well: green turtles *Chelonia mydas*, hawksbill turtles *Eretmochelys imbricate*, and loggerhead turtles *Caretta caretta* [222,223]. Other arsenicals included DMA, AsC, and TETRA in green and loggerhead turtles; in the latter species 25% of the total arsenic was AsC (compared with 55% of total arsenic as AsB; 85% of total arsenic was identified) [223]. High concentrations of TMAO were also recently found in hawksbill turtles; tissue specific speciation in the hawksbill and green turtles indicated that many of the arsenic species found in the non-digestive tissues (specifically, AsB) are likely ingested [224,225].

9.9. Fish

9.9.1. *Freshwater*

Protease digests and methanol/water extracts of fish from Yellowknife (Canada) contained AsB, arsenosugars, DMA, and unknowns [207]. Similar results were found in a later study on fish from the same location. AsB and DMA were present in all of the fish studied, with DMA predominant in many samples, and inorganic arsenic and additionally MMA found in several samples [226]. The methodology available could not be used to identify arsenosugars, TMAO, or TETRA.

AsB in some freshwater fish has been attributed to dietary uptake [227], but it is not present in all or even most fish studied to date. For example, carp reared under "natural conditions" (presumably AsB-free diet) contained inorganic arsenic, MMA and DMA, although from one location in the study AsS-PO4 predominated in the water-soluble portion (extraction efficiencies ranged from 2–29% in carp) [227]. The arsenosugars were also thought to be acquired through diet. In another study of Hungarian fish from the Danube River, AsS-PO4 was the main compound found, present in four out of five fish samples but it was not found in silver carp, which contained only TMAO. AsB was present only in trace or very low concentrations in white bream, which also had thioAsS-PO4 [107]. A large proportion of arsenic was unknown, either unidentified extracted arsenic species (as a result of the HPLC-ICPMS method used), or unextracted [107]. In another study that could not identify arsenosugars, their presence was postulated (in amounts up to 14% of total arsenic); significant proportions of TETRA, up to 35% of total arsenic in pumpkinseed *Lepomis gibbosus*, were observed [228]. In the latter study, extraction efficiencies were higher than in the other studies mentioned so far, ranging from 67 to 89%.

Cluster analysis of arsenic species (unextracted arsenic, As(III), DMA, TMAO, AsB, and an unknown cationic compound) in a limited number of freshwater fish revealed that salmonids (three species of trout), which had predominantly AsB, were in one cluster; Gadidae (burbot, one specimen), predominated by DMA, was in a second cluster; and all other groups (including catfish and three species from the Cyprinidae family), which had mostly unextracted arsenic, were in a third cluster [229].

The effect of the contamination level on the arsenic speciation of freshwater fish was studied, where fish from arsenic contaminated ponds in Thailand had substantially more DMA in their tissues than fish from uncontaminated waters. The reverse was true for inorganic arsenic. Large proportions in both were unextracted but the arsenic concentrations in contaminated fish were comparable to marine fish [230].

9.9.2. *Marine*

Most researchers report predominantly (>90%) AsB in marine fish tissues (see for example a review by Edmonds and Francesconi [6]), but the appearance and quantities of other arsenic compounds appear to be possibly dependent on the fish's position in the food chain. For example, AsS-PO4 is found in all tissues of a herbivore fish except muscle, but not in a pelagic carnivore [231]. Another herbivore contained predominantly AsS-PO4 with little AsB (maximum 15%) in tissues [232]. An earlier study showed the absence of arsenobetaine in another herbivore, the silver drummer fish, which contained predominantly TMAO [128].

A zwitterion related to arsenobetaine, trimethylarsoniopropionate (AsB2), was first isolated from *Abudefduf vaigiensis* in 2000 [233]. Although found in other animals, it is never a major constituent.

Arsenocholine was the major arsenic species found in two fish: *Haemulon* sp. at 97% of the total arsenic (26.7 mg kg^{-1}) and in *Lutjanus synagris* at 89% of the total arsenic (11.9 mg kg^{-1}) collected from Cienfuegos Bay (Cuba), in which a spill of 3.7 tons of "arsenic oxides" had occurred [186]. The AsB concentrations in all the fish samples speciated in this study were low and did not account for more than 2% of the arsenic present. Instead, the predominant compounds were AsC, as stated above, or in two fish samples with elevated arsenic concentrations (ca. 500 mg kg^{-1}), inorganic arsenic (98 and 99%). One of those fish was the same species that contained predominantly AsC (*Lutjanus synagris*) at lower total arsenic concentrations [186].

9.10. Birds

9.10.1. *Terrestrial*

Birds collected from areas both adjacent to and distant from mining operations in Yellowknife had different arsenic compounds in their tissues, depending on the bird species [234]. Whereas inorganic species and DMA predominated in migratory species like yellow-rumped warbler, American tree sparrow, and dark-eyed junco, arsenobetaine constituted up to 10% of total arsenic in gray jay tissues, and up to 36% in spruce grouse tissues. The latter two birds are non-migratory and the source of AsB is not obvious. Earthworms which can contain arsenobetaine are absent in Yellowknife, but AsB-containing mushrooms are present and cannot be discounted as a dietary source of AsB even though they do not typically form part of a spruce grouse's diet.

Chicken meat has been analyzed by several groups [24, 235–237] with consistent results of predominantly DMA and AsB. Chicken feed is often made with fish meal so it is possible that the AsB in chicken is a result of ingestion. AsB was the only detectable species in a single liver from a jungle crow *Corvus macrorhynchos* from Japan and accounted for 79% of the total arsenic (0.24 mg kg^{-1}) [223]; this terrestrial bird was also thought to obtain its AsB through diet, probably through foraging at dump sites.

Few feeding studies of birds have been carried out in recent years. When Zebra finches (*Taeniopygia guttata*) were exposed to MSMA, MMA was the predominant form in blood plasma and brain tissues, whereas DMA was the major form found in liver and kidney tissues [238,239]. When chickens were given an As_2O_3 enriched diet, arsenic species in liver extracts were predominantly DMA, with some As(III) [240]. In another study chickens were given As(V) in their drinking water, and As(III) was dominant in the auricle, DMA in meat, and AsB in fat and heart (with greater then 80% extraction, and a maximum of 160 μg kg^{-1} total arsenic). The authors stated that "AsB is formed only through microorganism activity" and thus postulated that the AsB was produced by some uncontrolled microbial activity [241].

9.10.2. *Marine*

AsB predominates in livers of two species of marine birds, black-footed albatross *Diomedea nigripes* (89% of total arsenic) and black-tailed gull *Larus crassirostris* (67% of total arsenic) [223]. Black-footed albatrosses had higher concentrations of arsenic in their livers (12 ± 11 mg kg^{-1}), on average about six times higher than gulls (2.3 ± 0.9 mg kg^{-1}). Other arsenic species extracted from albatross and gull livers included DMA, AsC, and TETRA, with 90% of total arsenic in albatross and 71% in gulls identified. Arsenic was transferred from mother black-tailed gulls to eggs as AsB (88–95%) and DMA (5–12%) but the total rate of maternal transfer of arsenic was comparatively low at 10% [242].

The albatross was an interesting case for further study because its liver concentrations were higher than most other higher trophic animals studied. Trophic transfer coefficients (ratio of body burden to stomach content concentration) for different tissues in this bird were found to be approximately 1, suggesting that although accumulation was higher than in other birds, biomagnification was not taking place [243]. This calculation was carried out for only two animals, with analysis of arsenic in the different tissues (lung, muscle, kidney, liver, pancreas, spleen, gallbladder, brain, heart, uropygical gland, gizzard, stomach, stomach content where available, intestine, intestine content, fat, feather, bone, and gonad as testis or ovary) revealing that AsB was predominant in all tissues; DMA was also present [243]. These results are similar to those for a single black-tailed gull in an earlier study, except for a relatively large proportion (21–35% of extracted arsenic) of AsC in the intestine content of the black-tailed gull [242] compared with smaller proportions (maximum 2%) in albatross tissues [243]. Low levels of TMAO in the intestine content but not stomach content of one bird (the other had an empty stomach), where total arsenic concentrations were similar, suggested to the authors that degradation of AsB in the intestine took place. An unknown compound was observed but no details about retention time or chromatographic behavior were given; it was predicted to be AsB2.

9.11. Mammals

9.11.1. Terrestrial

A breed of sheep that live on the island of North Ronaldsay, off the coast of Scotland, feed mainly on the seaweed that washes up on the shore. This food, mainly *Laminaria digitata*, is rich in arsenosugars. The arsenic content in the sheep's urine can reach 50 mg dm^{-3} [244] with the main metabolite DMA as it is for humankind, and thioarsenicals among the minor arsenicals (see Section 11) [245]. In a control study, Blackfaced sheep fed a seaweed diet showed similar compounds in their urine, and it was concluded that the metabolism of arsenic in seaweed was not unique to the North Ronaldsay sheep, even though they are adapted to a seaweed diet [246].

Inorganic arsenic and DMA are the most common arsenicals found in methanol/water extracts of tissues obtained from terrestrial mammals living near contaminated sites in Canada (unpublished data). In deer mice from Yellowknife, and meadow voles from Nova Scotia, the predominant species were As(III) and DMA, with traces of AsB detected in deer mouse livers but not in any meadow vole tissues. The AsB in deer mouse livers may have been due to dietary intake since AsB-containing mushrooms were growing at most of the mouse sampling sites in Yellowknife at the time of sampling, but no such mushrooms were observed when the meadow voles were collected.

AsB was a major and in some cases the predominant arsenical found in hares and squirrels from Yellowknife (48 and 63% of total arsenic in squirrel livers) (unpublished data). AsC (6–23% of total arsenic) was also found in hare liver but not muscle, and squirrel livers and muscle, and TMAO (7–26% of total arsenic) was found in squirrel muscle. Both hares and squirrels are known to eat mushrooms so it is possible they are also ingesting AsB (they were captured at the same time as the deer mice).

In a fox from Yellowknife, AsB and AsC were found in most tissues except for bone, nails, and teeth. These compounds were also found in stomach and intestinal contents and therefore it seems likely that the retention of these compounds followed ingestion (unpublished data).

Additional reports of arsenic speciation in terrestrial mammals collected from the natural environment are not available. However, there is a large body of literature available on controlled laboratory studies of various mammals [7] such as mice, rats, hamsters, rabbits, guinea pigs, and primates, with occasional studies of dogs and most recently horses [247]. In most of this work the primary goal was to gain information about arsenic metabolism and the mechanisms of toxic action of arsenic in humans. These publications will not be reviewed here because our primary interest is the environment not the laboratory. But for those interested in the horse study it seems that the disodium salt of MMA is sometimes used as a doping agent for race horses. The animals behave like other mammals (some primates are an exception) and metabolize MMA to DMA [247].

9.11.2. *Marine*

The predominance of AsB in marine animal tissues was found to extend to marine mammal livers (specifically, pilot whales, ringed seals, a bearded seal, and a beluga whale) more than 10 years ago [248]. However, with 25–55% of the arsenic unextracted, AsB only accounted for 31–70% of total arsenic in the livers, with smaller amounts of AsC in all livers, DMA in all but one liver, and TETRA in all seals in the 1998 study. Small amounts of an unknown compound were observed in all tissues; the chromatographic behavior of this compound matched that of a compound that was later identified in tissues of a sperm whale as AsB2 [249]. An arsenical that was thought to be AsB2 was observed in all tissues of both mother and fetus of Dall's porpoise, as well as in tissues of short-finned pilot whale, harp seal, ringed seal, loggerhead turtle, green turtle, and black-tailed gull [223,242,250–252].

Northern fur seal and ringed seals had similar speciation profiles in their livers: predominantly AsB and DMA, with some AsC (about one-tenth the concentration of AsB), TETRA, and MMA in ringed seals; extraction was $>90\%$ [253]. Similar results, except for lower extraction efficiencies ($>65\%$), were found in other marine animals, namely ringed seals in another study, in harp seals, and in short-finned pilot whales [223]. Higher hepatic arsenic concentrations ($3\times$) and AsB percentages in ringed seals from Alaska (90% AsB) and Pangnirtung (66% AsB) have been attributed to higher total arsenic concentrations, which resulted from gold mining activities in the Alaskan marine ecosystem that was sampled [248,250].

An exception to the usual pattern was noted in Dall's porpoise, which had a greater proportion of AsC and DMA in its liver (DMA was equivalent to the AsB amount) [223]. However, in a later study of a single female Dall's porpoise and her fetus, this unusual arsenic speciation was not reproduced, since AsB predominated in all tissues ($>76\%$ of total arsenic); the differences in these results have not been reconciled [251]. The arsenic compounds in the fetus generally reflected those in the mother, except that total arsenic was lower, especially in blubber (fetal arsenic blubber concentration was 13% of the maternal arsenic concentration).

Another exception was the algae-eating dugong, which has predominantly MMA and some DMA in its liver [250]. The authors drew parallels with the algae-eating sheep who metabolize arsenosugars to methylated species.

10. ARSENOLIPIDS

The existence of lipid-like fractions in marine alga had been recognized for many years (e.g., [254]) before the first full identification of such a species by Morita and Shibata in 1990 [255]. Ethanol/chloroform extraction of the brown alga *Undaria pinnatifida* followed by Sephadex chromatography led to the isolation of compound **16** (see Fig. 1), whose identity was established by two-dimensional NMR spectroscopy. Arsenosugars were also present as AsS-OH, -PO4, -SO3 [256]. Around the same time Francesconi et al. [257] isolated phosphatidylarsenocholine, **23**, from yellow-eye mullet that had been fed AsC. The compound $R=H$ was the hydrolysis product of the isolated lipid and it was also found in the animal. The authors suggested that production of the arsenolipid might be a response to the ingestion of arsenocholine and might not be a normal constituent of the animal.

Phospholipase treatment of the arsenolipid fraction from *Laminaria digitata* indicated that their structure was related to that of **16** [258]. The digestive gland of the western rock lobster *Panulirus cygnus* contains lipids based on arsenocholine and arsenosugars **23** and **16** [259].

Other lipids based on DMA have been isolated from fish oil, seal blubber and starspotted shark liver [260–262]. Recent examples of such species are shown in **17**–**22**. The six polar compounds **19** (n = 6, 7, 8, 9), **21**, and **22**, accounting for 20% of the total arsenolipids, were isolated from cod liver oil following extensive chromatography (at least nine other arsenolipid fractions were obtained). Structural assignment was aided by mass spectrometry but the double bonds in **21** and **22** are placed in positions that would be expected from the known structures of fatty acids found in the oil. The concentration of the first member of the series in the oil, **19** (n = 6), is estimated to be less than 0.02 μg As g^{-1} [263]. The authors argue that any synthetic path to these compounds which contain the equivalent of an even number of carbon atoms is unlikely to involve DMA(III) or DMA(V). The same biosynthetic conundrum is encountered in the structures of the arsenolipids **17** and **18** isolated from the oil from the capelin *Mallotus villosus*, a plankton feeder. The placing of the double bonds is again based on the known structures of fatty acids. These three compounds comprise about 70% of the total arsenic in the oil (11.7 mg kg^{-1} As) [264].

More complex DMA-based arsenolipids were found in the Japanese flying squid, *Todarodes pacificus*, a common food source in Japan [220]. These authors examined the muscle, liver, testes/ovary, and gill. The arsenic concentrations in each compartment were less than 10 mg kg^{-1} with AsB and DMA as the major contributors. The liver and testes were the main source of arsenolipids (10% of liver arsenic and 6% of testes arsenic) which were characterized, by using chemical and enzymatic hydrolysis, as phosphatidyldimethylarsinic acid, **24**, and DMA-containing sphingomyelin, **25**.

11. ORGANOARSENICALS WITH ARSENIC-SULFUR BONDS

As was noted in 1989 [1], arsenicals that have As-S or As=S moieties are to be expected in the environment. This conclusion is based on the well-known affinity of arsenic for sulfur which in turn is not based on the thermodynamic stability of the As-S bond, but on its kinetic stability [265]. Hence we easily speak of arsenic compounds binding to sulfyhdryl groups of proteins and of the facile hydrolysis of ADP-arsenate. So given the appropriate environment, all of the compounds in Figure 2 with As=O moieties might be expected to be found as their thio analogues. However, unless the appropriate environment is maintained during the analytical process, thioarsenicals will be transformed to oxy analogues and not be detected [266,267]. And even if such compounds are detected we need to consider whether they were formed by a biochemical process rather than by reaction with hydrogen sulfide. For example, reports of the production of thioarsenicals by anaerobic microflora of the mouse caecum were followed up by studies on the fate of ^{34}S-thioDMA in the same system. Labeled thioTMAO was produced without cleavage of the As-S bond. These results have been interpreted in terms of a modified Challenger pathway involving thio-DMA(III) as an intermediate [268].

In 2004 thioDMAA was found to be a significant component of the urine and wool of seaweed-eating sheep [245,269]. The compound is now known to be more toxic than DMA [45,270,271]. ThioDMA was identified as a trace component, together with other thioarsenicals, in the urine of a human volunteer who consumed 0.945 mg of AsS-OH [272]. This 2005 study, a rerun of one reported in 2002 [273], found 12 arsenic-containing metabolites that accounted for the bulk of the arsenic in the urine. Most of these were identified (in order of relative abundance): DMA (51%), thioDMAA (19%), thioDMAE (10%), DMAE (<4%), DMAA (2%), unknown, AsS-OH (traces), and thioDMA (traces). Of course, this species distribution is not to be expected in the urine of all individuals who have eaten a meal that was rich in arsenosugars. For example, DMA is seen almost immediately in the urine of some volunteers after eating Nori, a commercial seaweed product, whilst others appear to be unresponsive [274]. Individual metabolisms of arsenicals in seafood such as mussels, which are rich in arsenosugars and arsenobetaine, also show wide variations [275].

As mentioned previously in Section 1.3, DMA(III) was found to be both cytotoxic and genotoxic, much more so than As(III) and As(V), contrary to the then accepted dogma that organoarsenicals were less toxic than inorganic species [42,44]. Consequently there was considerable interest in reports establishing that DMA(III) was present in the urine of arsenic-exposed individuals (e.g., [276,277]). These first reports were usually based on the use of DMA(III) standards obtained by hydrolysis of iododimethylarsine, and identification was made by using either HPLC-ICPMS or hydride generation methods. Unfortunately some groups elected to use another method to prepare their DMA(III) standards, making the assumption that a method developed for the reduction of As(V) to As(III) [278] would work for DMA(V) to produce DMA(III). This is not a clean reaction and the main product is actually thioDMA [279], so papers based on standards prepared by the Reay and Ascher reaction should be read with caution (e.g., [280]).

Subsequently, there were claims that all reports of the finding of DMA(III) in human urine are probably in error and that the metabolites are actually thioDMA [270,279]. (The identification of either of these species is complicated by their high instability [281]). One report from Mexico [282] that is based on the use of hydride generation finds that DMA(III) is a very significant urinary metabolite in individuals living in an arsenic afflicted region. It has been suggested that some, if not all of this arsenical, is thioDMA [270].

ThioDMA was identified in the urine of Japanese men [283] and in 2007 the same research group reported that 44% of 75 women in Bangladesh who were continually exposed to arsenic-rich water excreted thioDMA in their urine [270]. The concentration of the species identified as thioDMA ranged from trace to 24 $\mu g\,dm^{-3}$ representing 0.4–5.4% of the total arsenic in the urine, which is much lower than that found for the species identified as DMA(III) in the Mexican study [282].

Ackerman et al. [284] found DMA and inorganic As in cooked rice when using trifluoroacetic acid as the extractant but enzymatic extraction revealed the presence of thioDMA. For example, instant rice contained 305 $\mu g\,kg^{-1}$ total As comprised of 29 $\mu g\,kg^{-1}$ As(V) plus As(III), 226 $\mu g\,kg^{-1}$ DMA and 40 $\mu g\,kg^{-1}$ thioDMA.

The first report of thioMMA(V) and MMA(III) in terrestrial food appeared in 2008 [158]. The species were identified in carrots that had been in storage for a number of years, since the 1980s (see Section 8 for more details). Results for one arsenic-rich carrot (total arsenic 18.7 $mg\,kg^{-1}$) are as follows ($\mu g\,kg^{-1}$): MMA(III) 2400, MMA(V) 11300, DMA 24, thioMMA 141, As(III) 65.

A standard for thioMMA was prepared from MMA and H_2S and the reaction was monitored by using IC-ICPMS. When DMA is reacted with H_2S, thioDMA is the first product to form, followed by dithioDMA together with some DMA(III). The reaction in water or methanol needs to be carefully monitored to ensure that the desired arsenical is obtained [285,286].

The anaerobic microflora from mice caecum readily convert AsS-OH to its thioanalog as a result of H_2S production. Conversion of AsS-SO4 is slower [267]; see also [287]). This conversion of arsenosugars to their thio analogs is pH-sensitive and is promoted in the range where HS^- is converted to H_2S ($pK_1 = 7$). At a 15-fold excess of sulfide at pH 4.8 the conversion to sulfide is $>80\%$. In shellfish the S:As ratio is $>200{:}1$ and therefore the finding of thioarsenosugars in such samples is expected. Chromatography conditions can influence speciation results. For example, un-neutralized extracts of butter clam contain AsS-OH (55 $\mu g\,kg^{-1}$) and thioAsS-OH (20 $\mu g\,kg^{-1}$); neutralized extracts contain no AsS-OH and more of the thio analogue (62 $\mu g\,kg^{-1}$) [266,267].

The first reports of thioarsenosugars in mollusks actually appeared in 2004 when Fricke et al. [288] found that thioAsS-PO4 is a major arsenical species in marine clams and mussels. In freshwater mussels, the total arsenic content is much the same as in marine species: 12.7 $mg\,kg^{-1}$ [208] and 8.02 $mg\,kg^{-1}$ [207]

(unpublished results); however the speciation is very different. AsB is a minor constituent, often below the detection limit; DMA and inorganic As are minor species. In mussel samples from the Danube River the four common arsenosugars are the major species accompanied by lower amounts of their thio analogues. Mussels from Quinsam River (BC, Canada) contain mainly AsS-OH and AsS-PO4, together with low amounts of their thioanalogues (unpublished results). Freshwater snails, *Stagnicola* sp from the same region contain around 7.5 $mg\,kg^{-3}$ arsenic, much of which is unextracted (2.8 mg kg^{-3}) or extracted but not detected (1.5 $mg\,kg^{-3}$). AsS-OH (1.2 $mg\,kg^{-3}$), MMA (1 $mg\,kg^{-3}$), and TETRA (1 $mg\,kg^{-3}$) are the main arsenicals together with traces of thioAsS. Snails from the family Viviparidae ("live bearing") have lower arsenic levels, around 3 $mg\,kg^{-3}$, with AsS-OH (0.35 $mg\,kg^{-3}$) and AsS-PO4 (0.3 $mg\,kg^{-3}$) as the major species along with traces of MMA, thioAsS-OH and thioAsS-SO4. The thioarsenosugar concentration increases to 0.2 $mg\,kg^{-3}$ in the unborn snails with a corresponding reduction in the oxyarsenosugar concentrations (unpublished results).

A different pair of thioarsenosugars, thioAsS-SO4 and thioAsS-SO3, are found in the gonad and muscle of the great scallop [289]. Methanol aided the extraction of these species. The concentration of thioAsS-SO4 was the greater of the two at around 0.2 $mg\,kg^{-1}$ in the muscle.

Both Meier et al. [115] and Nischwitz et al. [10] found thioarsenosugars in marine algae. The first group reported that the macro alga *Fucus vesiculosus* contains thioAsS-SO4 and thioAsS-SO3 amounting to around 10% of the total arsenic content. The same two thioarsenosugars were found in commercial kelp samples [10].

Traar and Francesconi [290] have devised an elegant synthetic route to arsenosugars that eliminates the problems associated with the polarity and water solubility of the oxyarsenosugars such as AsS-OH by replacing the oxygen with sulfur. The resulting compounds are less polar and soluble in organic solvents, allowing easier manipulation. The same principle was employed in one synthesis of thioDMA. DMA was treated with H_2S in a water/ethyl acetate mixture. The product moved into the organic phase where it was not exposed to more H_2S.

12. ARSENIC TRANSFORMATIONS

The detection of specific arsenicals in biological samples is often presented as evidence that the source organism was responsible for the production of these compounds. More realistically, the 'inventory' of organoarsenicals is usually the result of the biotransformation and/or consumption (including absorption) of arsenicals from lower down the food chain.

In the natural environment, organisms at one trophic level live in close association with other species from lower levels that are capable of biomodifying arsenic compounds. For most animals, microbial associates, especially within the digestive system, are likely to carry out the biotransformation. Epiphytes are probably important – these can be bacteria, fungal, animals, including zooxanthalla, and alga – in providing arsenicals to aquatic plants, macroalgae, terrestrial plant roots (and possibly shoots), and higher fungi (i.e., those that form fruiting bodies from myceliar structures in the soil). Some lower trophic level organisms (e.g., cyanobacteria) inhabit both terrestrial and marine environments and can form simple methylated arsenicals and arsenosugars; we would not be surprised to learn that such organisms can also produce arsenobetaine.

Mechanistically, the Challenger pathway (Figure 2) provides logical initial steps for the formation of all of the dimethylated arsenicals shown in Figure 1; however, this does not mean that all the subsequent steps take place in one organism. As noted earlier (Section 1.3), the methylation of inorganic arsenic was long thought to be a detoxification process, but new information regarding the toxicity of, especially, MMA(III) and DMA(III), which are putative intermediates in the Challenger pathway, has dispelled this notion. It does appear, however, that this pathway is operative to some degree in many organisms, so these toxic species are probably not normally found "free" in living cells; however, they have been detected in both fresh and salt water (Section 5.2).

The formation of TETRA, which is reasonably widespread in the environment, can be accounted for by the full Challenger process, but this would involve "free" trimethylarsine as an intermediate, something that is difficult to contemplate in a given organism. It is possible that TETRA arises from the degradation of AsB, but the route is not at all obvious, although TETRA is produced in AsB-containing food on cooking [291].

The fact that SAM can provide a sugar-containing group, in addition to a methyl group, provides a route to the formation of arsenosugars (Figure 3). Once compound **3** is formed, reasonable sequences of biochemical pathways are available to account for many of the compounds listed in Figure 1, such as the arsenolipids, **8**, **13**, and AsS-PO4 [6]. However there is a dearth of evidence to show that a given organism synthesizes arsenosugars from inorganic arsenic. Most likely they start from readily available DMA and its reactive reduction product DMA(III). Notably, most photosynthetic organisms contain arsenosugars and SAM is important in the photosynthetic process.

Several pathways have been proposed for the production of arsenobetaine [6]. These include formation from dimethylated arsenosugars either by conversion to DMAA (Figure 3) or via DMAE and AsC. A related route might involve trimethylated sugars (such as **8** and **9**) which could be converted to AsC and then to AsB, but the low occurrence of these sugars in the environment makes this route unlikely and certainly not dominant. The presence of arsenosugars and AsB in organisms from deep sea vents [292], and AsB but no arsenosugars in mangrove swamps [293] has been used as an argument against the arsenosugar precursor pathway [293] but 'ocean snow' (including copepods) provides nutrients and organic matter to these locations and could easily be a source of many arsenicals, including arsenosugars and arsenobetaine. Lastly, an alternative route involving DMA(III) and glyoxylate (Figure 3) offers a conceptually more direct but multistep route to AsB [6] via simple methylated compounds widespread in the environment. The production of simple methylated arsenicals up to TMAO by pasteurized compost and the finding of similar compounds as well as AsB in the fruiting body of mushrooms provides some evidence for this route as no arsenosugars were found in either treatment (Section 7.3).

The use of radiotracers is one of the best ways of establishing biosynthetic pathways yet little has been done along these lines with arsenicals. One early study involved exposing *Mytilus californianus* to [^{3}H]-MMA in a static seawater system. The label became distributed over the whole animal, even the byssal threads, with most in the vicera, gills, foot and muscle. Methanol extracted 75% of the activity and the solution contained labeled [^{3}H]-MMA, [^{3}H]-AsB and two labeled unknowns (possibly arsenosugars). The authors conclude that AsB is either accumulated from water and/or food ([^{3}H]-AsB was found in the water, even in the absence of mussels), or is synthesized from arsenicals other than MMA within the mussel itself [34]. Similar experiments with *Mytilus edulis* led to similar conclusions [294]. More work of this kind is needed.

AsB was regarded as being more prevalent in the marine environment than the terrestrial but it is now being found in more and more samples from freshwater and terrestrial ecosystems as the range of sampling is increased. Primary productivity in the ocean is mainly dependent on upwelling of nitrogen, whereas in freshwater environments it depends on the availability of phosphorus and during freshwater blooms, this phosphate may compete with arsenate uptake. There is a low, but consistent, arsenic and phosphorus supply in ocean waters. The concentration of arsenic in freshwater is generally much lower, although it can increase locally in response to the surrounding geology and/or anthropogenic input. These differences may result in a relatively greater amount of arsenic uptake by marine phytoplankton where the Challenger pathway provides a plausible pathway to arsenosugars and possibly eventually to AsB (Figure 3). In freshwater systems we generally detect less arsenosugars and AsB but probably this is the result of a generally lower arsenic intake rather than the absence of methylation pathways. However, we should point out the normal response of an organism to an above normal exposure to inorganic arsenic is to accumulate the arsenic without methylation, presumably because the Challenger pathway becomes saturated.

It would be interesting to examine both phyto- and zoo-plankton in a freshwater system with a high dissolved arsenic concentration.

An important chemical difference between freshwater and marine environments is salinity, which results in an organism's need to osmoregulate (to maintain the osmotic potential of cells or tissues in hypertonic media, e.g., saline environments). Edmonds and Francesconi proposed some time ago [3] that AsB can act as an osmoregulator. We now have some supporting evidence. AsB concentrations were significantly negatively correlated with glycinebetaine concentrations in six species of marine animals (two seal species, two seabirds species, and two turtle species) suggesting that AsB can replace glycinebetaine (the nitrogen analogue of AsB) [253]. Thus, selective retention of AsB may account for its presence in many marine organisms. In the terrestrial environment arsenobetaine may play a similar role. High concentrations have been found in some, but not all, mushrooms. In one instance the AsB was located in the cap and outer stalk, suggesting that the AsB may accompany other osmolytes to maintain turogor pressure [28]. In earthworms, arsenobetaine is absent in the body wall but localized in the chloragogenous tissue [171] which may be involved in osmoregulation [295]. The easy loss of arsenosugars from macroalgae when exposed to different salinities may indicate a similar role for these arsenicals (Section 5.3).

In conclusion, it seems that arsenic transformation in the marine environment is a consequence of the uptake of arsenate via the phosphate transport mechanism. The normal cell response is reduction and elimination of the arsenic as arsenite. However, some of the arsenic in the cells is methylated in a random process initiating the Challenger pathway and subsequent transformations (Figure 2 and 3). In the terrestrial environment, organoarsenicals are produced in a similar fashion, but the bulk of the arsenic is retained in an inorganic form that is not easily extracted.

ACKNOWLEDGMENT

We are grateful to the Natural Sciences and Engineering Research Council of Canada for some financial support. Special mention must be made of Elizabeh Varty, who produced the figures.

ABBREVIATIONS

For the structural formulas of the arsenic species see Figures 1 and 2.

ADP	adenosine 5′-diphosphate
AsB2	trimethylarsoniopropionate

AsC	arsenocholine
CCA	chromated copper arsenate
DMA	dimethylarsinic acid
DMAA	dimethylarsinoylacetic acid
DMAE	dimethylarsinoylethanol
EE	extraction efficiency
ESI-ITMS	electrospray ionization ion trap mass spectrometry
ESI-MS	electrospray ionization mass spectrometry
GC-MS	gas chromatography mass spectrometry
GS	glutathione
HG	hydride generation
HPLC	high performance liquid chromatography
ICPMS	inductively coupled plasma mass spectrometry
MMA	monomethylarsonic acid
MSMA	monosodium methylarsonate
NMR	nuclear magnetic resonance
SAM	*S*-adenosylmethionine
SPME	solid phase microextraction
TETRA	tetramethylarsonium ion
TMAO	trimethylarsine oxide
XANES	X-ray absorption near edge structure
XAS	X-ray absorption spectroscopy

REFERENCES

1. W. R. Cullen and K. J. Reimer, *Chem. Rev.*, 1989, **89**, 713–764.
2. K. A. Francesconi and D. Kuehnelt, in *Environmental Chemistry of Arsenic*, Ed. W. T. Frankenberger, Marcel Dekker, New York, 2002, pp. 51–94.
3. J. S. Edmonds and K. A. Francesconi, *Mar. Pollut. Bull.*, 1993, **26**, 665–674.
4. K. A. Francesconi and J. S. Edmonds, in *Advances in Inorganic Chemistry*, Academic Press, San Diego, 1997, pp. 147–189.
5. K. A. Francesconi and J. S. Edmonds, in *Oceanography and Marine Biology An Annual Review: 31*, Ed. H. Barnes, A. D. Ansell and R. N. Gibson, University College London Press, London, 1993, pp. 111–151.
6. J. S. Edmonds and K. A. Francesconi, in *Organometallic Compounds in the Environment*, Ed. P. J. Craig, John Wiley and Sons Ltd, Chichester UK, 2003, pp. 195–222.
7. D. Kuehnelt and W. Goessler, in *Organometallic Compounds in the Environment*, Ed. P. J. Craig, John Wiley and Sons, Ltd,, Chichester, UK, 2003, pp. 223–275.
8. W. R. Cullen, in *Is Arsenic an Aphrodisiac? The Sociochemistry of an Element*, Royal Society of Chemistry, Cambridge, UK., 2008.
9. J. J. Sloth, E. H. Larsen and K. Julshamn, *Rapid Commun. Mass Spectrom.*, 2005, **19**, 227–235.

10. V. Nischwitz, K. Kanaki and S. A. Pergantis, *J. Anal. At. Spectrom.*, 2006, **21**, 33–40.
11. S. McSheehy, J. Szpunar, R. Lobinski, V. Haldys, J. Tortajada and J. S. Edmonds, *Anal. Chem.*, 2002, **74**, 2370–2378.
12. I. Koch, K. McPherson, P. Smith, L. Easton, K. G. Doe and K. J. Reimer, *Mar. Pollut. Bull.*, 2007, **54**, 586–594.
13. E. Schmeisser, W. Goessler, N. Kienzl and K. A. Francesconi, *Anal. Chem.*, 2004, **76**, 418–423.
14. R. Regmi, B. F. Milne and J. Feldmann, *Anal. Bioanal. Chem.*, 2007, **388**, 775–782.
15. J. Szakova, P. Tlustos, W. Goessler, D. Pavlikova and J. Balik, *Appl. Organomet. Chem.*, 2005, **19**, 308–314.
16. J. T. Van Elteren, Z. Slejkovec, M. Kahn and W. Goessler, *Anal. Chim. Acta*, 2007, **585**, 24–31.
17. K. A. Mir, A. Rutter, I. Koch, P. Smith, K. J. Reimer and J. S. Poland, *Talanta*, 2007, **72**, 1507–1518.
18. S. Foster, W. Maher, F. Krikowa and S. Apte, *Talanta*, 2007, **71**, 537–549.
19. S. Foster, D. Thomson and W. Maher, *Mar. Chem.*, 2008, **108**, 172–183.
20. D. Thomson, W. Maher and S. Foster, *Appl. Organomet. Chem.*, 2007, **21**, 396–411.
21. P. G. Smith, I. Koch and K. J. Reimer, *Sci. Total Environ.*, 2008, **390**, 188–197.
22. K. Bluemlein, A. Raab and J. Feldmann, *Anal. Bioanal. Chem.*, 2009, **393**, 357–366.
23. J. H. Huang and G. Ilgen, *Int. J. Environ. Anal. Chem.*, 2006, **86**, 347–358.
24. I. Pizarro, M. Gomez, C. Camara and M. A. Palacios, *Anal. Chim. Acta*, 2003, **495**, 85–98.
25. I. Koch, A. Duso, C. Haug, C. Miskelly, M. Sommerville, P. Smith and K. J. Reimer, *Environ. Forensics*, 2005, **6**, 335–344.
26. I. J. Pickering, R. C. Prince, M. J. George, R. D. Smith, G. N. George and D. E. Salt, *Plant Physiol.*, 2000, **122**, 1171–1177.
27. C. J. Langdon, A. A. Meharg, J. Feldmann, T. Balgar, J. Charnock, M. Farquhar, T. G. Piearce, K. T. Semple and J. Cotter-Howells, *J. Environ. Monit.*, 2002, **4**, 603–608.
28. P. G. Smith, I. Koch and K. J. Reimer, *Environ. Sci. Technol.*, 2007, **41**, 6947–6954.
29. P. G. Smith, I. Koch and K. J. Reimer, *Sci. Total Environ.*, 2008, **390**, 198–204.
30. K. Bluemlein, A. Raab, A. A. Meharg, J. M. Charnock and J. Feldmann, *Anal. Bioanal. Chem.*, 2008, **390**, 1739–1751.
31. K. A. Francesconi, J. S. Edmonds and M. Morita, in *Arsenic in the Environment, Part 1, Cycling and Characterization*, Ed. J. O. Nriagu, John Wiley, New York, 1994, pp. 189–219.
32. W. R. Cullen, L. G. Harrison, H. Li and G. Hewitt, *Appl. Organomet. Chem.*, 1994, **8**, 313–324.
33. H. Hasegawa, Y. Sohrin, K. Seki, M. Sato, K. Norisuye, K. Naito and M. Matsui, *Chemosphere*, 2001, **43**, 265–272.
34. W. R. Cullen and S. A. Pergantis, *Appl. Organomet. Chem.*, 1993, **7**, 329–334.

35. I. Mancini, G. Guella, M. Frostin, E. Hnawia, D. Laurent, C. Debitus and F. Pietra, *Chem. Eur. J.*, 2006, **12**, 8989–8994.
36. K. J. Irgolic, D. Spall, B. K. Puri, D. Ilger and R. A. Zingaro, *Appl. Organomet. Chem.*, 1991, **5**, 117–124.
37. B. K. Puri and K. J. Irgolic, *Environ. Geochem. Health*, 1989, **11**, 95–99.
38. Z. Slejkovec and T. A. Kanduc, *Environ. Sci. Technol.*, 2005, **39**, 3450–3454.
39. R. H. Fish, W. Walker and R. S. Tannous, *Energy Fuels*, 1987, **1**, 243–247.
40. R. H. Fish, R. S. Tannous, W. Walker, C. S. Weiss and F. E. Brinckman, *J. Chem. Soc. Chem. Commun.*, 1983, 490–492.
41. S. P. Cramer, M. Siskin, L. D. Brown and G. N. George, *Energy Fuels*, 1988, **2**, 175–180.
42. M. J. Mass, A. Tennant, B. C. Roop, W. R. Cullen, M. Styblo, D. J. Thomas and A. D. Kligerman, *Chem. Res. Toxicol.*, 2001, **14**, 355–361.
43. J. S. Petrick, B. Jagadish, E. A. Mash and H. V. Aposhian, *Chem. Res. Toxicol.*, 2001, **14**, 651–656.
44. T. Schwerdtle, I. Walter, I. Mackiw and A. Hartwig, *Carcinogenesis*, 2003, **24**, 967–974.
45. T. Ochi, K. Kita, T. Suzuki, A. Rumpler, W. Goessler and K. A. Francesconi, *Toxicol. Appl. Pharmacol.*, 2008, **228**, 59–67.
46. W. R. Cullen and R. Bentley, *J. Environ. Monit.*, 2005, **7**, 11–15.
47. J. Matschullat, *Sci. Total Environ.*, 2000, **249**, 297–312.
48. D. A. Bright, M. Dodd and K. J. Reimer, *Sci. Total Environ.*, 1996, **180**, 165–182.
49. H. Hasegawa, Y. Sohrin, M. Matsul, M. Hojo and M. Kawashima, *Anal. Chem.*, 1994, **66**, 3247–3252.
50. H. Hasegawa, *Appl. Organomet. Chem.*, 1996, **10**, 733–740.
51. D. A. Bright, S. Brock, W. R. Cullen, G. M. Hewitt, J. Jafaar and K. J. Reimer, *Appl. Organomet. Chem.*, 1994, **8**, 415–422.
52. A. S. Howard and S. D. W. Comber, *Appl. Organomet. Chem.*, 1989, **3**, 510–514.
53. H. Hasegawa, M. A. Rahman, T. Matsuda, T. Kitahara, T. Maki and K. Ueda, *Sci. Total Environ.*, 2009, **407**, 1418–1425.
54. S. Khokiattiwong, W. Goessler, S. N. Pedersen, R. Cox and K. A. Francesconi, *Appl. Organomet. Chem.*, 2001, **15**, 481–489.
55. M. J. Ellwood and W. A. Maher, *Anal. Chim. Acta*, 2003, **477**, 279–291.
56. W. R. Cullen, B. C. McBride and J. Reglinski, *J. Inorg. Biochem.*, 1984, **21**, 45–60.
57. M. Takeuchi, A. Terada, K. Nanba, Y. Kanai, M. Owaki, T. Yoshida, T. Kuroiwa, H. Nirei and T. Komai, *Appl. Organomet. Chem.*, 2005, **19**, 945–951.
58. D. C. Chilvers and P. J. Peterson, in *Lead, Mercury, Cadmium and Arsenic in the Environment, Scope 31*, Ed. T. C. Hutchinson and K. M. Meema, John Wiley and Sons, Chichester, UK, 1987, pp. 279–301.
59. W. T. Frankenberger, *Soil Biol. Biochem.*, 1998, **30**.
60. M. Pantsar-Kallio and A. Korpela, *Anal. Chim. Acta*, 2000, **410**, 65–70.
61. R. Pongratz, *Sci. Total Environ.*, 1998, **224**, 133–141.
62. H. Mukai, Y. Ambe, T. Muku, K. Takeshita and T. Fukuma, *Nature*, 1986, **324**, 239–241.

63. F. Challenger, *Adv. Enzymol. Relat. Areas Mol. Biol.*, 1951, **12**, 429–491.
64. V. Puntoni, *Arsenioschizomiceti. Ann. Ig.*, 1917, **27**, 293–303.
65. F. Challenger, C. Higgenbottom and L. Ellis, *J. Chem. Soc.*, 1933, **1933**, 95–101.
66. F. Challenger, *Chem. Rev.*, 1945, **36**, 315–361.
67. R. Bentley and T. G. Chasteen, *Microbiol. Molecular Biol. Rev.*, 2002, **66**, 250.
68. W. R. Cullen, H. Li, G. Hewitt, K. J. Reimer and N. Zalunardo, *Appl. Organomet. Chem.*, 1994, **8**, 303–311.
69. B. C. McBride and R. S. Wolfe, *Biochemistry*, 1971, **10**, 4312–4317.
70. P. T. S. Wong, Y. K. Chau, L. Luxon and G. A. Bengert, in *Trace Substances in Environmental Health–XI*, Ed. D. D. Hemphill, University of Missouri, Columbia, 1977, pp. 100–106.
71. K. Michalke, E. B. Wickenheiser, M. Mehring, A. V. Hirner and R. Hensel, *Appl. Environ. Microbiol.*, 2000, **66**, 2791–2796.
72. C. F. Harrington, E. I. Brima and R. O. Jenkins, *Chem. Speciation Bioavailability*, 2008, **20**, 173–180.
73. A. W. Ritchie, J. S. Edmonds, W. Goessler and R. O. Jenkins, *FEMS Microbiol. Lett.*, 2004, **235**, 95–99.
74. R. O. Jenkins, A. W. Ritchie, J. S. Edmonds, W. Goessler, N. Molenat, D. Kuehnelt, C. F. Harrington and P. G. Sutton, *Arch. Microbiol.*, 2003, **180**, 142–150.
75. V. Devesa, A. Loos, M. A. Suner, D. Velez, A. Feria, A. Martinez, R. Montoro and Y. Sanz, *J. Agric. Food Chem.*, 2005, **53**, 10297–10305.
76. J. S. Edmonds, K. A. Francesconi and J. A. Hansen, *Experientia*, 1982, **38**, 643–644.
77. P. Pengprecha, M. Wilson, A. Raab and J. Feldmann, *Appl. Organomet. Chem.*, 2005, **19**, 819–826.
78. A. V. Hirner, J. Feldmann, R. Goguel, S. Rapsomanikis, R. Fischer and M. O. Andreae, *Appl. Organomet. Chem.*, 1994, **8**, 65–69.
79. J. Feldmann and A. V. Hirner, *Internat. J. Environ. Anal. Chem.*, 1995, **60**, 339–359.
80. S. Maillefer, C. R. Lehr and W. R. Cullen, *Appl. Organomet. Chem.*, 2003, **17**, 154–160.
81. K. Michalke and R. Hensel, in *Organic Metal and Metalloid Species in the Environment*, Ed. A. V. Hirner and H. Emons, Springer, Berlin, Germany, 2004, pp. 137–150.
82. R. A. Diaz–Bone, B. Menzel, A. Barrenstein and A. V. Hirner, in *Organic Metal and Metalloid Species in the Environment*, Ed. A. V. Hirner and H. Emons, Springer, Berlin, Germany, 2004, pp. 97–110.
83. E. A. Woolson and P. C. Kearney, *Environ. Sci. Technol.*, 1973, **7**, 47–50.
84. E. A. Woolson, *Weed Sci.*, 1977, **25**, 412–416.
85. C. N. Cheng and D. D. Focht, *Appl. Environ. Microbiol.*, 1979, **38**, 494.
86. M. Shariatpanahi, A. C. Anderson, A. A. Abdelghani and A. J. Englande, in *Biodeterioration*, Ed. T. A. Oxley and S. Barry, John Wiley and Sons, Chichester, UK, 1983, pp. 268–277.
87. R. Turpeinen, M. Pantsar-Kallio and T. Kairesalo, *Sci. Total Environ.*, 2002, **285**, 133–145.

88. J. Meyer, A. Schmidt, K. Michalke and R. Hensel, *Syst. Appl. Microbiol.*, 2007, **30**, 229–238.
89. S. M. A. Islam, K. Fukushi, K. Yamamoto and G. C. Saha, *Arch. Environ. Contam. Toxicol.*, 2007, **52**, 332–338.
90. B. Planer-Friedrich, C. Lehr, J. Matschullat, B. J. Merkel, D. K. Nordstrom and M. W. Sandstrom, *Geochim. Cosmochim. Acta*, 2006, **70**, 2480–2491.
91. J. Qin, C. R. Lehr, C. Yuan, X. C. Le, T. R. McDermott and B. P. Rosen, *Proc. Natl. Acad. Sci. USA*, 2009, **106**, 5213–5217.
92. C. R. Lehr, E. Polishchuk, M. C. Delisle, C. Franz and W. R. Cullen, *Hum. Exptl. Toxicol.*, 2003, **22**, 325–334.
93. T. Maki, N. Takeda, H. Hasegawa and K. Ueda, *Appl. Organomet. Chem.*, 2006, **20**, 538–544.
94. K. Hanaoka, S. Tagawa and T. Kaise, *Appl. Organomet. Chem.*, 1991, **5**, 435–438.
95. K. Hanaoka, N. Araki, S. Tagawa and T. Kaise, *Appl. Organomet. Chem.*, 1994, **8**, 201–206.
96. K. Hanaoka, Y. Dote, K. Yosida, T. Kaise, T. Kuroiwa and S. Maeda, *Appl. Organomet. Chem.*, 1996, **10**, 683–688.
97. T. Maki, H. Hasegawa, H. Watarai and K. Ueda, *Anal. Sci.*, 2004, **20**, 61–68.
98. C. G. Rosal, G. M. Momplaisir and E. M. Heithmar, *Electrophoresis*, 2005, **26**, 1606–1614.
99. J. R. Garbarino, A. J. Bednar, D. W. Rutherford, R. S. Beyer and R. L. Wershaw, *Environ. Sci. Technol.*, 2003, **37**, 1509–1514.
100. J. F. Stolz, E. Perera, B. Kilonzo, B. Kail, B. Crable, E. Fisher, M. Ranganathan, L. Wormer and P. Basu, *Environ. Sci. Technol.*, 2007, **41**, 818–823.
101. E. Fisher, A. M. Dawson, G. Polshyna, J. Lisak, B. Crable, E. Perera, M. Ranganathan, M. Thangavelu, P. Basu and J. F. Stolz, *Ann. NY Acad. Sci.*, 2008, **1125**, 230–241.
102. J. Miot, G. Morin, F. Skouri-Panet, C. Facrard, E. Aubry, J. Briand, Y. Wang, G. Ona-Nguema, F. Guyot and G. F. brown, *Environ. Sci. Technol.*, 2008, **42**, 5342–5347.
103. L. A. Murray, A. Raab, I. L. Marr and J. Feldmann, *Appl. Organomet. Chem.*, 2003, **17**, 669–674.
104. S. Maeda, A. Ohki, K. Kusadome, T. Kuroiwa, I. Yoshifuku and K. Naka, *Appl. Organomet. Chem.*, 1992, **6**, 213–219.
105. T. Kuroiwa, A. Ohki, K. Naka and S. Maeda, *Appl. Organomet. Chem.*, 1994, **8**, 325–333.
106. F. L. Hellweger, *Appl. Organomet. Chem.*, 2005, **19**, 727–735.
107. R. Schaeffer, K. A. Francesconi, N. Kienzl, C. Soeroes, P. Fodor, L. Varadi, R. Raml, W. Goessler and D. Kuehnelt, *Talanta*, 2006, **69**, 856–865.
108. I. Koch, J. Feldmann, L. X. Wang, P. Andrewes, K. J. Reimer and W. R. Cullen, *Sci. Total Environ.*, 1999, **236**, 101–117.
109. I. Koch, Ph.D. Thesis, University of British Columbia, Vancouver, British Columbia, Canada, 1998.
110. V. W. M. Lai, W. R. Cullen, C. F. Harrington and K. J. Reimer, *Appl. Organomet. Chem.*, 1997, **11**, 797–803.

111. V. M. Dembitsky and T. Rezanka, *Plant Sci.*, 2003, **165**, 1177–1192.
112. R. Tukai, W. A. Maher, I. J. McNaught, M. J. Ellwood and M. Coleman, *Mar. Freshwater Res.*, 2002, **53**, 971–980.
113. V. Nischwitz and S. A. Pergantis, *J. Agric. Food Chem.*, 2006, **54**, 6507–6519.
114. V. W. M. Lai, W. R. Cullen, C. F. Harrington and K. J. Reimer, *Appl. Organomet. Chem.*, 1998, **12**, 243–251.
115. J. Meier, N. Kienzl, W. Goessler and K. A. Francesconi, *Environ. Chem*, 2005, **2**, 304–307.
116. S. C. R. Granchinho, E. Polishchuk, W. R. Cullen and K. J. Reimer, *Appl. Organomet. Chem.*, 2001, **15**, 553–560.
117. K. Bellman, B.Sc. Thesis, University of British Columbia, Vancouver, British Columbia, Canada, 2002.
118. S. C. R. Granchinho, C. M. Franz, E. Polishchuk, W. R. Cullen and K. J. Reimer, *Appl. Organomet. Chem.*, 2002, **16**, 721–726.
119. A. Geiszinger, W. Goessler, S. N. Pedersen and K. A. Francesconi, *Environ. Toxicol. Chem.*, 2001, **20**, 2255–2262.
120. J. S. Edmonds and K. A. Francesconi, *Experientia*, 1987, **43**, 553.
121. R. G. Wuilloud, J. C. Altamirano, P. N. Smichowski and D. T. Heitkemper, *J. Anal. At. Spectrom.*, 2006, **21**, 1214–1223.
122. S. G. Salgado, M. A. Q. Nieto and M. M. B. Simon, *Talanta*, 2008, **75**, 897–903.
123. U. Kohlmeyer, J. Kuballa and E. Jantzen, *Rapid Commun. Mass Spectrom.*, 2002, **16**, 965–974.
124. V. Nischwitz and S. A. Pergantis, *Analyst*, 2005, **130**, 1348–1350.
125. Z. Slejkovec, E. Kapolna, I. Ipolyi and J. T. Van Elteren, *Chemosphere*, 2006, **63**, 1098–1105.
126. M. Grotti, F. Soggia, C. Lagomarsino, W. Goessler and K. A. Francesconi, *Environ. Chem.*, 2008, **5**, 171–175.
127. Y. Shibata, M. Sekiguchi, A. Otsuki and M. Morita, *Appl. Organomet. Chem.*, 1996, **10**, 713–719.
128. J. S. Edmonds, Y. Shibata, K. A. Francesconi, R. J. Rippingale and M. Morita, *Appl. Organomet. Chem.*, 1997, **11**, 281–287.
129. H. R. Smith and E. J. Cameron, *Ind. Eng. Chem. (Anal.)*, 1933, **5**, 400–401.
130. F. Challenger and C. Higginbottom, *J. Chem. Soc.*, 1935, 1757–1778.
131. W. Merrill and D. W. French, *Proc. Minnesota Acad. Sci.*, 1964, **31**, 105–106.
132. R. Zussman, E. Vicher and I. lyon, *J. Bacteriol.*, 1961, **81**, 157.
133. D. P. Cox and M. Alexander, *Bull. Environ. Contam. Toxicol.*, 1973, **9**, 84–87.
134. D. P. Cox and M. Alexander, *Appl. Microbiol.*, 1973, **25**, 408–413.
135. K. D. Huysmans and W. T. Frankenberger, *Sci. Total Environ.*, 1991, **105**, 13–28.
136. W. R. Cullen and J. Nelson, *Appl. Organomet. Chem.*, 1992, **6**, 179–183.
137. W. R. Cullen, H. Li, S. A. Pergantis, G. K. Eigendorf and A. A. Mosi, *Appl. Organomet. Chem.*, 1995, **9**, 507–515.
138. P. Andrewes, K. T. Kitchin and K. Wallace, *Chem. Res. Toxicol.*, 2003, **16**, 994–1003.
139. M. Hiroki and Y. Yoshiwara, *Soil Sci. Plant Nutr.*, 1993, **39**, 237–243.

140. V. U. Y. Ultra, S. Tanaka, K. Sakurai and K. Iwasaki, *Soil Sci. Plant Nutr.*, 2007, **53**, 499–508.
141. V. U. Ultra, S. Tanaka, K. Sakurai and K. Iwasaki, *Plant Soil*, 2007, **290**, 29–41.
142. A. Raab, H. Schat, A. A. Meharg and J. Feldmann, *New Phytol.*, 2005, **168**, 551–558.
143. Z. Slejkovec, A. R. Byrne, T. Stijve, W. Goessler and K. J. Irgolic, *Appl. Organomet. Chem.*, 1997, **11**, 673–682.
144. Z. Slejkovec, A. R. Byrne, W. Gossler, D. Kuehnelt, K. J. Irgolic and F. Pohleven, *Acta Chim. Slov.*, 1996, **43**, 269–283.
145. C. Soeroes, N. Kienzl, I. Ipolyi, M. Dernovics, P. Fodor and D. Kuehnelt, *Food Control*, 2005, **16**, 459–464.
146. I. Koch, L. X. Wang, K. J. Reimer and W. R. Cullen, *Appl. Organomet. Chem.*, 2000, **14**, 245–252.
147. D. Kuehnelt, J. Lintschinger and W. Goessler, *Appl. Organomet. Chem.*, 2000, **14**, 411–420.
148. T. Mrak, Z. Slejkovec and Z. Jeran, *Talanta*, 2006, **69**, 251–258.
149. A. Machado, Z. Slejkovec, J. T. Van Elteren, M. C. Freitas and M. S. Baptista, *J. Atmos. Chem.*, 2006, **53**, 237–249.
150. M. M. Farinha, Z. Slejkovec, J. T. Van Elteren, H. T. Wolterbeek and M. C. Freitas, *J. Atmos. Chem.*, 2004, **49**, 343–353.
151. T. Mrak, Z. Slejkovec, Z. Jeran, R. Jacimovic and D. Kastelec, *Environ. Pollut.*, 2008, **151**, 300–307.
152. I. Koch, L. X. Wang, C. A. Ollson, W. R. Cullen and K. J. Reimer, *Environ. Sci. Technol.*, 2000, **34**, 22–26.
153. J. H. Huang and E. Matzner, *Environ. Sci. Technol.*, 2007, **41**, 1564–1569.
154. D. Thomson, W. Maher and S. Foster, *Appl. Organomet. Chem.*, 2007, **21**, 381–395.
155. J. Zheng, H. Hintelmann, B. Dimock and M. S. Dzurko, *Anal. Bioanal. Chem.*, 2003, **377**, 14–24.
156. M. J. Ruiz-Chancho, J. F. Lopez-Sanchez, E. Schmeisser, W. Goessler, K. A. Francesconi and R. Rubio, *Chemosphere*, 2008, **71**, 1522–1530.
157. R. Zhao, M. X. Zhao, H. Wang, Y. Taneike and X. R. Zhang, *Sci. Total Environ.*, 2006, **371**, 293–303.
158. S. K. V. Yathavakilla, M. Fricke, P. A. Creed, D. T. Heitkemper, N. V. Shockey, C. Schwegel, J. A. Caruso and J. T. Creed, *Anal. Chem.*, 2008, **80**, 775–782.
159. Y. J. Zavala, R. Gerads, H. Gurleyuk and J. M. Duxbury, *Environ. Sci. Technol.*, 2008, **42**, 3861–3866.
160. N. P. Vela, D. T. Heitkemper and K. R. Stewart, *Analyst*, 2001, **126**, 1011–1017.
161. P. N. Williams, A. H. Price, A. Raab, S. A. Hossain, J. Feldmann and A. A. Meharg, *Environ. Sci. Technol.*, 2005, **39**, 5531–5540.
162. N. M. Smith, R. Lee, D. T. Heitkemper, K. D. Cafferky, A. Haque and A. K. Henderson, *Sci. Total Environ.*, 2006, **370**, 294–301.
163. L. Thrall, K. Christen, B. Booth, R. Renner and P. D. Thacker, *Environ. Sci. Technol.*, 2006, **40**, 2072–2078.
164. P. N. Williams, A. Raab, J. Feldmann and A. A. Meharg, *Environ. Sci. Technol.*, 2007, **41**, 2178–2183.

165. A. Raab, K. Ferreira, A. A. Meharg and J. Feldmann, *J. Exp. Bot.*, 2007, **58**, 1333–1338.
166. W. R. Cullen, D. Hettipathirana and J. Reglinski, *Appl. Organomet. Chem.*, 1989, **3**, 515–521.
167. T. Kunito, R. Kubota, J. Fujihara, T. Agusa and S. Tanabe, in *Reviews of Environmental Contamination and Toxicology 195*, Ed. D. M. Whitacre, Springer, New York, 2008, pp. 31–69.
168. K. Shiomi, M. Aoyama, H. Yamanaka and T. Kikuchi, *Comp. Biochem. Physiol., C*, 1988, **90**, 361–365.
169. Y. Yamaoka, M. L. Carmona, J. M. Oclarit, K. Z. Jin and Y. Shibata, *Appl. Organomet. Chem.*, 2001, **15**, 261–265.
170. Y. Yamaoka, M. L. Carmona, J. M. Oclarit, K. Jin and Y. Shibata, *Appl. Organomet. Chem.*, 2006, **20**, 545–548.
171. C. J. Langdon, C. Winters, S. R. Sturzenbaum, A. J. Morgan, J. M. Charnock, A. A. Meharg, T. G. Piearce, P. H. Lee and K. T. Semple, *Environ. Sci. Technol.*, 2005, **39**, 2042–2048.
172. A. Geiszinger, W. Goessler, D. Kuehnelt, K. Francesconi and W. Kosmus, *Environ. Sci. Technol.*, 1998, **32**, 2238–2243.
173. M. J. Watts, M. Button, T. S. Brewer, G. R. T. Jenkin and C. F. Harrington, *J. Environ. Monit.*, 2008, **10**, 753–759.
174. C. J. Langdon, T. G. Piearce, J. Feldmann, K. T. Semple and A. A. Meharg, *Environ. Toxicol. Chem.*, 2003, **22**, 1302–1308.
175. A. E. Geiszinger, W. Goessler and W. Kosmus, *Appl. Organomet. Chem.*, 2002, **16**, 473–476.
176. K. W. Lin, S. Behl, A. Furst, P. Chien and R. F. Toia, *Toxicol. in vitro*, 1998, **12**, 197–199.
177. J. Waring and W. Maher, *Appl. Organomet. Chem.*, 2005, **19**, 917–929.
178. D. Fattorini, A. Notti, M. N. Halt, M. C. Gambi and F. Regoli, *Mar. Ecol. – Evolut. Perspect.*, 2005, **26**, 255–264.
179. A. E. Geiszinger, W. Goessler and K. A. Francesconi, *Mar. Environ. Res.*, 2002, **53**, 37–50.
180. A. E. Geiszinger, W. Goessler and K. A. Francesconi, *Environ. Sci. Technol.*, 2002, **36**, 2905–2910.
181. D. Fattorini and F. Regoli, *Environ. Toxicol. Chem.*, 2004, **23**, 1881–1887.
182. A. Notti, D. Fattorini, E. M. Razzetti and F. Regoli, *Environ. Toxicol. Chem.*, 2007, **26**, 1186–1191.
183. J. Waring, W. Maher, S. Foster and F. Krikowa, *Environ. Chem.*, 2005, **2**, 108–118.
184. D. Boyle, K. V. Brix, H. Amlund, A. K. Lundebye, C. Hogstrand and N. R. Bury, *Environ. Sci. Technol.*, 2008, **42**, 5354–5360.
185. T. D. Ninh, Y. Nagashima and K. Shiomi, *Chemosphere*, 2008, **70**, 1168–1174.
186. D. Fattorini, C. M. Alonso–Hernandez, M. Diaz-Asencio, A. Munoz-Caravaca, F. G. Pannacciulli, M. Tangherlini and F. Regoli, *Mar. Environ. Res.*, 2004, **58**, 845–850.
187. K. Hanaoka, H. Ohno, N. Wada, S. Ueno, W. Goessler, D. Kuehnelt, C. Schlagenhaufen, T. Kaise and K. J. Irgolic, *Chemosphere*, 2001, **44**, 743–749.

188. M. Moriarty, I. Koch, R. A. Gordon and K. J. Reimer, *Environ. Sci. Technol.*, 2009, **43**, 4818–4823.
189. R. Andrahennadi and I. J. Pickering, *Environ. Chem.*, 2008, **5**, 413–419.
190. D. Kuehnelt, W. Goessler, C. Schlagenhaufen and K. J. Irgolic, *Appl. Organomet. Chem.*, 1997, **11**, 859–867.
191. M. Rizki, E. Kossatz, A. Velazquez, A. Creus, M. Farina, S. Fortaner, E. Sabbioni and R. Marcos, *Environ. Mol. Mutagen.*, 2006, **47**, 162–168.
192. S. P. Hopkin, in *Ecophysiology of Metals in Terrestrial Invertebrates*, Elsevier Applied Science Publishers, London, 1989, pp. 1–366.
193. C. A. Morrissey, C. A. Albert, P. L. Dods, W. R. Cullen, V. W. M. Lai and J. E. Elliott, *Environ. Sci. Technol.*, 2007, **41**, 1494–1500.
194. V. Devesa, M. A. Suner, V. W. M. Lai, S. C. R. Granchinho, D. Velez, W. R. Cullen, J. M. Martinez and R. Montoro, *Appl. Organomet. Chem.*, 2002, **16**, 692–700.
195. G. Williams, J. A. West and E. T. Snow, *Environ. Toxicol. Chem.*, 2008, **27**, 1332–1342.
196. G. Williams, J. M. West, I. Koch, K. J. Reimer and E. T. Snow, *Sci. Total Environ.*, 2009, **407**, 2650–2658.
197. S. Hirata, H. Toshimitsu and M. Aihara, *Anal. Sci.*, 2006, **22**, 39–43.
198. R. Wahlen, S. McSheehy, C. Scriver and Z. Mester, *J. Anal. At. Spectrom.*, 2004, **19**, 876–882.
199. R. Schaeffer, P. Fodor and C. Soeroes, *Rapid Commun. Mass Spectrom.*, 2006, **20**, 2979–2989.
200. D. Fattorini, A. Notti and F. Regoli, *Chem. Ecol.*, 2006, **22**, 405–414.
201. U. Norum, V. W. M. Lai, S. A. Pergantis and W. R. Cullen, *J. Environ. Monit.*, 2005, **7**, 122–126.
202. V. W. M. Lai, A. S. Beach, W. R. Cullen and K. J. Reimer, *Appl. Organomet. Chem.*, 2002, **16**, 458–462.
203. T. D. Ninh, Y. Nagashima and K. Shiomi, *Food Addit. Contam.*, 2006, **23**, 1299–1307.
204. W. Goessler, W. Maher, K. J. Irgolic, D. Kuehnelt, C. Schlagenhaufen and T. Kaise, *Fresenius J. Anal. Chem.*, 1997, **359**, 434.
205. A. J. Underwood and M. G. Chapman, in *Seashores: A Beachcomber's Guide*,Reprint edn, New South Wales University Press, Sydney Australia, 1993, p. 53.
206. S. Foster, W. Maher, E. Schmeisser, A. Taylor, F. Krikowa and S. Apte, *Environ. Chem.*, 2006, **3**, 304–315.
207. I. Koch, K. J. Reimer, A. Beach, W. R. Cullen, A. Gosden and V. W. M. Lai, in *Arsenic Exposure and Health Effects IV*, Ed. W. R. Chappell, C. O. Abernathy and R. L. Calderon, Elsevier Science, Oxford, UK, 2001, pp. 115–123.
208. C. Soeroes, W. Goessler, K. A. Francesconi, E. Schmeisser, R. Raml, N. Kienzl, M. Kahn, P. Fodor and D. Kuehnelt, *J. Environ. Monit.*, 2005, **7**, 688–692.
209. A. A. Benson, *Z. Naturforsch*, 1990, **45c**, 793–796.
210. W. R. Cullen and M. Dodd, *Appl. Organometal. Chem.*, 1989, **3**, 79–88.
211. K. Shiomi, M. Orii, H. Yamanaka and T. Kikuchi, *Bull. Japanese Soc. Sci. Fish.*, 1987, **53**, 103–108.

212. V. W. M. Lai, W. R. Cullen and S. Ray, *Mar. Chem.*, 1999, **66**, 81–89.
213. V. W. M. Lai, W. R. Cullen and S. Ray, *Appl. Organomet. Chem.*, 2001, **15**, 533–538.
214. J. J. Sloth, E. H. Larsen and K. Julshamn, *J. Anal. At. Spectrom.*, 2003, **18**, 452–459.
215. D. Fattorini, A. Notti, R. Di Mento, A. M. Cicero, M. Gabellini, A. Russo and F. Regoli, *Chemosphere*, 2008, **72**, 1524–1533.
216. J. J. Sloth and K. Julshamn, *J. Agric. Food Chem.*, 2008, **56**, 1269–1273.
217. N. J. Valette-Silver, G. F. Riedel, E. A. Crecelius, H. Windom, R. G. Smith and S. S. Dolvin, *Mar. Environ. Res.*, 1999, **48**, 311–333.
218. K. Shiomi, A. Shinagawa, H. Yamanaka and T. Kikuchi, *Bull. Japanese Soc. Sci. Fish.*, 1983, **49**, 79–83.
219. S. Seixas, P. Bustamante and G. J. Pierce, *Chemosphere*, 2005, **59**, 1113–1124.
220. T. D. Ninh, Y. Nagashima and K. Shiomi, *J. Agric. Food Chem.*, 2007, **55**, 3196–3202.
221. J. S. Edmonds, Y. Shibata, R. I. T. Prince, K. A. Francesconi and M. Morita, *J Mar. Biol. Assoc. UK*, 1994, **74**, 463–466.
222. K. Saeki, H. Sakakibara, H. Sakai, T. Kunito and S. Tanabe, *Biometals*, 2000, **13**, 241–250.
223. R. Kubota, T. Kunito and S. Tanabe, *Environ. Toxicol. Chem.*, 2003, **22**, 1200–1207.
224. T. Agusa, K. Takagi, H. Iwata and S. Tanabe, *Mar. Pollut. Bull.*, 2008, **57**, 782–789.
225. T. Agusa, K. Takagi, R. Kubota, Y. Anan, H. Iwata and S. Tanabe, *Environ. Pollut.*, 2008, **153**, 127–136.
226. S. de Rosemond, Q. L. Xie and K. Liber, *Environ. Monit. Assess.*, 2008, **147**, 199–210.
227. C. Soeroes, W. Goessler, K. A. Francesconi, N. Kienzl, R. Schaeffer, P. Fodor and D. Kuehnelt, *J. Agric. Food Chem.*, 2005, **53**, 9238–9243.
228. J. Zheng and H. Hintelmann, *J. Anal. At. Spectrom.*, 2004, **19**, 191–195.
229. Z. Slejkovec, Z. Bajc and D. Z. Doganoc, *Talanta*, 2004, **62**, 931–936.
230. P. Jankong, C. Chalhoub, N. Kienzl, W. Goessler, K. A. Francesconi and P. Visoottiviseth, *Environ. Chem.*, 2007, **4**, 11–17.
231. J. Kirby and W. Maher, *Appl. Organomet. Chem.*, 2002, **16**, 108–115.
232. S. Foster, W. Maher and F. Krikowa, *Environ. Chem.*, 2008, **5**, 176–183.
233. K. A. Francesconi, S. Khokiattiwong, W. Goessler, S. N. Pedersen and M. Pavkov, *Chem. Comm.*, 2000, 1083–1084.
234. I. Koch, J. V. Mace and K. J. Reimer, *Environ. Toxicol. Chem.*, 2005, **24**, 1468–1474.
235. D. Sanchez-Rodas, J. L. Gomez-Ariza and V. Oliveira, *Anal. Bioanal. Chem.*, 2006, **385**, 1172–1177.
236. A. Polatajko and J. Szpunar, *J. AOAC Int.*, 2004, **87**, 233–237.
237. E. Sanz, R. Munoz-Olivas and C. Camara, *J. Chromatogr., A*, 2005, **1097**, 1–8.
238. C. A. Albert, T. D. Williams, C. A. Morrissey, V. W. M. Lai, W. R. Cullen and J. E. Elliott, *Environ. Toxicol. Chem.*, 2008, **27**, 605–611.

239. C. Albert, T. D. Williams, C. A. Morrissey, V. W. M. Lai, W. R. Cullen and J. E. Elliott, *J. Toxicol. Environ. Health Part A*, 2008, **71**, 353–360.
240. I. Falnoga, V. Stilbilj, M. Tusek-Znidaric, Z. Slejkovec, D. Mazej, R. Jacimovic and J. Scancar, *Biol. Trace Elem. Res.*, 2000, **78**, 241–254.
241. I. Pizarro, M. M. Gomez, P. Fodor, M. A. Palacios and C. Camara, *Biol. Trace Elem. Res.*, 2004, **99**, 129–143.
242. R. Kubota, T. Kunito, S. Tanabe, H. Ogi and Y. Shibata, *Appl. Organomet. Chem.*, 2002, **16**, 463–468.
243. J. Fujihara, T. Kunito, R. Kubota, H. Tanaka and S. Tanabe, *Mar. Pollut. Bull.*, 2004, **48**, 1153–1160.
244. H. R. Hansen, A. Raab, K. A. Francesconi and J. Feldmann, *Environ. Sci. Technol.*, 2003, **37**, 845–851.
245. H. R. Hansen, R. Pickford, J. Thomas-Oates, M. Jaspars and J. Feldmann, *Angew. Chem. Int. Ed.*, 2004, **43**, 337–340.
246. S. J. Martin, C. Newcombe, A. Raab and J. Feldmann, *Environ. Chem.*, 2005, **2**, 190–197.
247. R. A. Assis, I. L. Kuchler, N. Miekeley and M. B. Tozzi, *Anal. Bioanal. Chem.*, 2008, **390**, 2107–2113.
248. W. Goessler, A. Rudorfer, E. A. Mackey, P. R. Becker and K. J. Irgolic, *Appl. Organomet. Chem.*, 1998, **12**, 491–501.
249. A. Geiszinger, S. Khokiattiwong, W. Goessler and K. A. Francesconi, *J. Mar. Biol. Assoc. UK*, 2002, **82**, 165–168.
250. R. Kubota, T. Kunito and S. Tanabe, *Mar. Pollut. Bull.*, 2002, **45**, 218–223.
251. R. Kubota, T. Kunito, J. Fujihara, S. Tanabe, J. Yang and N. Miyazaki, *Mar. Pollut. Bull.*, 2005, **51**, 845–849.
252. K. Ebisuda, T. Kunito, R. Kubota and S. Tanabe, *Appl. Organomet. Chem.*, 2002, **16**, 451–457.
253. J. Fujihara, T. Kunito, R. Kubota and S. Tanabe, *Comp. Biochem. Physiol., C*, 2003, **136**, 287–296.
254. G. Lund, in *Report on Technological Research Concerning the Norwegian Fish Industry*, Directorate of Fisheries, Bergen, Norway, 1972.
255. M. Morita and Y. Shibata, *Appl. Organomet. Chem.*, 1990, **4**, 181–190.
256. Y. Shibata, M. Morita and K. Fuwa, *Adv. Biophys.*, 1992, **28**, 31–80.
257. K. A. Francesconi, R. V. Stick and J. S. Edmonds, *Experientia*, 1990, **46**, 454–456.
258. S. Devalla and J. Feldmann, *Appl. Organomet. Chem.*, 2003, **17**, 906–912.
259. J. S. Edmonds, Y. Shibata, K. A. Francesconi, J. Yoshinaga and J. Morita, *Sci. Total Environ.*, 1992, **122**, 321–335.
260. U. Kohlmeyer, S. Jakubik, J. Kuballa and E. Jantzen, *Microchim. Acta*, 2005, **151**, 249–255.
261. K. Ebisuda, T. Kunito, J. Fujihara, R. Kubota, Y. Shibata and S. Tanabe, *Talanta*, 2003, **61**, 779–787.
262. K. Hanaoka, W. Goessler, K. Yoshida, Y. Fujitaka, T. Kaise and K. J. Irgolic, *Appl. Organomet. Chem.*, 1999, **13**, 765–770.
263. A. Rumpler, J. S. Edmonds, M. Katsu, K. B. Jensen, W. Goessler, G. Raber, H. Gunnlaugsdottir and K. A. Francesconi, *Angew. Chem., Int. Ed.*, 2008, **47**, 2665–2667.

264. M. S. Taleshi, K. B. Jensen, G. Raber, J. S. Edmonds, H. Gunnlaugsdottir and K. A. Francesconi, *Chem. Comm.*, 2008, 4706–4707.
265. C. D. Baer, J. O. Edwards and P. H. Rieger, *Inorg. Chem.*, 1981, **20**, 905–907.
266. S. D. Conklin, P. A. Creed and J. T. Creed, *J. Anal. At. Spectrom.*, 2006, **21**, 869–875.
267. S. D. Conklin, M. W. Fricke, P. A. Creed and J. T. Creed, *J. Anal. At. Spectrom.*, 2008, **23**, 711–716.
268. K. M. Kubachka, M. C. Kohan, K. Herbin-Davis, J. T. Creed and D. J. Thomas, *Toxicol. Appl. Pharmacol.*, 2009, in press.
269. A. Raab, D. R. Genney, A. A. Meharg and J. Feldmann, *Appl. Organomet. Chem.*, 2003, **17**, 684–692.
270. R. Raml, A. Rumpler, W. Goessler, M. Vahter, L. Li, T. Ochi and K. A. Francesconi, *Toxicol. Appl. Pharmacol.*, 2007, **222**, 374–380.
271. K. Kuroda, K. Yoshida, M. Yoshimura, Y. Endo, H. Wanibuchi, S. Fukushima and G. Endo, *Tox. Appl. Pharmacol.*, 2004, **198**, 345–353.
272. R. Raml, W. Goessler, P. Traar, T. Ochi and K. A. Francesconi, *Chem. Res. Toxicol.*, 2005, **18**, 1444–1450.
273. K. A. Francesconi, R. Tanggaard, C. J. McKenzie and W. Goessler, *Clin. Chem.*, 2002, **48**, 92–101.
274. X. C. Le, W. R. Cullen and K. J. Reimer, *Clin. Chem.*, 1994, **40**, 617–625.
275. V. W. M. Lai, Y. Sun, E. Ting and W. R. Cullen, *Toxicol. Appl. Pharmacol.*, 2004, **198**, 297–306.
276. X. C. Le, X. Lu, M. Ma, W. R. Cullen, H. V. Aposhian and B. Zheng, *Anal. Chem.*, 2000, **72**, 5172–5177.
277. L. M. DelRazo, G. G. GarciaVargas, H. Vargas, A. Albores, M. E. Gonsebatt, R. Montero, P. Ostrosky Wegman, M. Kelsh and M. E. Cebrian, *Arch. Toxicol.*, 1997, **71**, 211–217.
278. P. F. Reay and C. J. Ascher, *Anal. Biochem.*, 1977, **78**, 557–560.
279. H. R. Hansen, A. Raab, M. Jaspars, B. F. Milne and J. Feldmann, *Chem. Res. Toxicol.*, 2004, **17**, 1086–1091.
280. B. K. Mandal, Y. Ogra and K. T. Suzuki, *Chem. Res. Toxicol.*, 2001, **14**, 371–378.
281. Z. Gong, X. Lu, W. R. Cullen and X. C. Le, *J. Anal. At. Spectrom.*, 2001, **16**, 1409–1413.
282. O. L. Valenzuela, V. H. Borja-Aburto, G. G. Garcia-Vargas, E. A. Garcia-Montalvo, E. S. Calderon-Aranda and L. M. D. Razo, *Environ. Health Perspect.*, 2005, **113**, 250–254.
283. R. Raml, W. Goessler and K. A. Francesconi, *J. Chromatogr., A*, 2006, **1128**, 164–170.
284. A. H. Ackerman, P. A. Creed, A. N. Parks, M. W. Fricke, C. A. Schwegel, J. T. Creed, D. T. Heitkemper and N. P. Vela, *Environ. Sci. Technol.*, 2005, **39**, 5241–5246.
285. M. W. Fricke, M. Zeller, H. Sun, V. W. M. Lai, J. A. Shoemaker, M. R. Witkowski and J. T. Creed, *Chem. Res. Toxicol.*, 2005, **18**, 1821–1829.
286. M. Fricke, M. Zeller, W. Cullen, M. Witkowski and J. Creed, *Anal. Chim. Acta*, 2007, **583**, 78–83.

287. H. R. Hansen, M. Jaspars and J. Feldmann, *Analyst*, 2004, **129**, 1058–1064.
288. M. W. Fricke, P. A. Gould, A. N. Parks, L. J. A. Shoemaker, C. A. Schwegel and J. T. Creed, *J. Anal. Spectrom.*, 2004, **19**, 1–8.
289. M. Kahn, R. Raml, E. Schmeisser, B. Vallant, K. A. Francesconi and W. Goessler, *Environ. Chem.*, 2005, **2**, 171–176.
290. P. Traar and K. A. Francesconi, *Tetrahedron Lett.*, 2006, **47**, 5293–5295.
291. K. Hanaoka, W. Goessler, H. Ohno, K. J. Irgolic and T. Kaise, *Appl. Organomet. Chem.*, 2001, **15**, 61–66.
292. E. H. Larsen, C. R. Quetel, R. Munoz, A. Fiala-Medioni and O. F. X. Donard, *Mar. Chem.*, 1997, **57**, 341.
293. K. Francesconi, *Environ. Chem,*. 2009, in press.
294. W. R. Cullen and J. C. Nelson, *Appl. Organomet. Chem.*, 1993, **7**, 319–327.
295. E. Fischer and L. Molnar, *Soil Biol. Biochem.*, 1992, **24**, 1723–1727.

Met. Ions Life Sci. **2010**, *7*, 231–265

7

Organoarsenicals. Uptake, Metabolism, and Toxicity*

Elke Dopp,[a] ***Andrew D. Kligerman,***[b] ***and Roland A. Diaz-Bone***[c]

[a]Institute of Hygiene and Occupational Medicine, University Hospital Essen, Hufelandstrasse 55, D-45122 Essen, Germany
<elke.dopp@uni-due.de>

[b]National Health and Environmental Effects Research Laboratory, Office of Research and Development, US Environmental Protection Agency, Research Triangle Park, NC, 27709, USA
<kligerman.andrew@epamail.epa.gov>

[c]Institute of Environmental Analytical Chemistry, University of Duisburg-Essen, Universitätsstrasse 3-5, D-45141 Essen, Germany
<roland.diaz@uni-due.de>

*This article has been reviewed by the National Health and Environmental Effects Research Laboratory, US Environmental Protection Agency, and approved for publication. Approval does not signify that the contents necessarily reflect the views and policies of the Agency nor does mention of trade names or commercial productions does constitute endorsement or recommendation for use.

Metal Ions in Life Sciences, Volume 7 Edited by Astrid Sigel, Helmut Sigel, and Roland K. O. Sigel

Published by the Royal Society of Chemistry, www.rsc.org DOI: 10.1039/9781849730822-00231

ABSTRACT: Arsenic is categorized by the WHO as the most significant environmental contaminant of drinking water due to the prevalence of geogenic contamination of groundwaters. Arsenic and the compounds which it forms are considered to be carcinogenic. The mechanism of toxicity and in particular of carcinogenicity of arsenic is still not well understood. The complexity originates from the fact that arsenic can form a rich variety of species, which show a wide variability in their toxicological behavior. The process of biomethylation was for many years regarded as a detoxification process; however, more recent research has indicated that the reverse is in fact the case. In this book chapter we give a summary of the current state of knowledge on the toxicities and toxicological mechanisms of organoarsenic species in order to evaluate the role and significance of these regarding their adverse effects on human health.

KEYWORDS: Carcinogenicity · DNA methylation · metabolism · organoarsenicals · toxicity · uptake

1. INTRODUCTION

In spite of huge research efforts in the investigation of arsenic-induced malignancies over more than a century, the mechanism of toxicity and in particular of carcinogenicity of arsenic is still not well understood. The complexity originates from the fact that arsenic can form a rich variety of species, which show a wide variability in their toxicological behavior. As arsenic undergoes rapid metabolism in the human body, the differentiation of the effects of the various species is difficult.

Historically, methylation of As has long been considered a detoxification process. Acute toxicity of iAs^{III} is orders of magnitude higher in comparison to pentavalent methylated species, which are mainly excreted via urine.

Thus, the assumption that iAs^{III} is the main actor in genotoxicity was common until the end of the 1990's.

The situation has changed fundamentally with the discovery of the high toxicity of trivalent methylated species (MMA^{III} and DMA^{III}), which are intermediates of the methylation process [1] and have been detected in small quantities in human urine. In the last few years it has been shown that these species are more cyto- and genotoxic (e.g., [2–7]) and more potent enzyme inhibitors (e.g., (8–10]) than their pentavalent counterparts and the inorganic arsenic species. In addition to the oxoforms of methylated arsenic species, methylated thioforms of arsenic were detected in human urine, which show toxicity and damaging effects at similar concentrations to trivalent methylated species [11].

In this chapter we give a summary of the current state of knowledge on the toxicities and toxicological mechanisms of organoarsenic species in order to evaluate the role and significance of these regarding their adverse effects on human health.

1.1. Arsenic Species of Interest

Arsenic is ubiquitous in the biosphere and occurs naturally in both organic and inorganic forms. While arsenic can be found to a small extent in the elemental form, the most important inorganic arsenic compounds are arsenic trioxide, sodium arsenite, arsenic trichloride (i.e., trivalent forms), and arsenic pentoxide, arsenic acid and arsenates, such as, lead and calcium arsenates (i.e., pentavalent forms).

The most important forms of organic arsenic compounds are methylated species in the oxidation states of +III and +V, which are intermediates in the process of biomethylation. Arsenobetaine (AsBet) and arsenocholine (AsCol) are the most predominant organoarsenicals in marine animals. Due to the advancement of analytical methodology, the number of arsenic containing sugars and phospholipids discovered in the environment is steadily growing [12].

Although arsenic compounds (Table 1) were commonly used in the past as drugs, their main uses today are as pesticides, veterinary drugs and in industrial applications, such as the manufacture of integrated circuits and the production of alloys [13].

2. SYSTEMIC TOXICITY AND CARCINOGENICITY OF ARSENIC

Arsenic causes a wide range of very different effects in the human body leading to a multitude of different systemic effects. Most strikingly, the

Table 1. Arsenic species of interest.

Low toxic species	Molecular formula	Abbreviation
Arsenate	AsO_4^{3-}	iAs^{V}
Monomethylarsonic acid	$(CH_3)AsO(OH)_2$	MMA^{V}
Dimethylarsinic acid	$(CH_3)_2AsO(OH)$	DMA^{V}
Trimethylarsine oxide	$(CH_3)_3AsO$	$TMAO^{V}$
Arsenobetaine	$(CH_3)_3As^{+}CH_2COO^{-}$	AsBet
Arsenocholine	$(CH_3)_3As^{+}CH_2CH_2OH$	AsCol
Arsenosugars	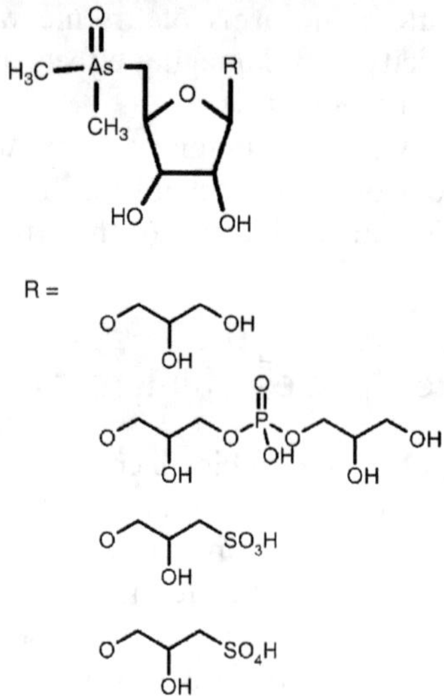	AsSug

Highly toxic species	Molecular formula	Abbreviation
Arsenite	AsO_3^{3-}	iAs^{III}
Monomethylarsonous acid	$(CH_3)As(OH)_2$	MMA^{III}
Dimethylarsinous acid	$(CH_3)_2As(OH)$	DMA^{III}
Dimethylmonothioarsinic acid	$(CH_3)_2AsS(OH)$	$DMMTA^{V}$
Dimethyldithioarsinic acid	$(CH_3)_2AsS(SH)$	$DMDTA^{V}$
Monomethylarsine	$(CH_3)AsH_2$	MMAH
Dimethylarsine	$(CH_3)_2AsH$	DMAH
Trimethylarsine	$(CH_3)_3As$	TMA

effects of arsenic from long-term exposure through drinking water are very different from acute poisoning [14]. Immediate symptoms of acute poisoning typically include vomiting, esophageal and abdominal pain, and bloody "rice water" diarrhea. Long-term exposure to arsenic in drinking water is causally related to increased risks of cancer in the skin, lungs, urinary bladder, and kidney. Arsenic is considered to be genotoxic in humans on the basis of both clastogenicity in exposed individuals and *in vitro* findings [13]. Clear exposure-response relationships have been shown between arsenic exposure and the risk of cancer [13,14]. Case-control studies indicate that a long latent stage between exposure and cancer diagnosis exists [15–17]. Because large numbers of arsenic-contaminated tube-wells have been installed in the last decades, a major increase of arsenic-related diseases is to be expected in the coming years [18].

In addition to carcinogenic effects, exposure to arsenic has been associated with several different vascular effects in both large and small vessels. Strong evidence has been gathered for a role for arsenic in inducing hypertension and cardiovascular disease. The best studied endemic peripheral vascular disease (PVD) is blackfoot disease (BFD), which is characterized by numbness in one or both feet followed by ulceration, black discoloration, and dry gangrene [13]. While BFD has only been documented in Taiwan, in studies from several other countries, other forms of PVD have been shown to be caused by arsenic.

In comparison to carcinogenic and vascular effects, the causality is less certain in the relationship between arsenic and diabetes or arsenic and reproductive effects [13]. Although there is good evidence that acute arsenic poisoning causes neurological effects, especially in the peripheral nervous system, there is little evidence of neurological effects from long-term lower-level environmental or occupational arsenic exposure [13].

For investigation of the carcinogenic activity of arsenic compounds, suitable animal models are needed. Cohen et al. have reviewed the carcinogenic activity of methylated arsenicals in rodents and humans [19]. The authors concluded that good animal models have not yet been found. They summarized that DMA^V is a urinary bladder carcinogen only in rats and only when administered in the diet or drinking water at high doses. The trivalent arsenicals that are cytotoxic and genotoxic *in vitro* are formed to only a small extent in an organism exposed to MMA^V or DMA^V because of poor cellular uptake and limited metabolism of the ingested compounds. Furthermore, the authors suggest a non-linear dose-response relationship for the biological processes involved in the carcinogenicity of arsenicals.

In a review by Wanibuchi et al., it is discussed that DMA^V has a profound multi-organ tumor-promoting activity in different rodent species with different administration protocols and is a complete carcinogen in the rat urinary bladder, although the doses required to produce effects are relatively high [20]. The authors conclude from their own studies that promoting activity requires chronic exposure. While hyperplasia of the uroepithelium was induced by MMA^V, MMA^V alone did not result in bladder tumor formation, indicating that arsenic carcinogenesis is species specific ($DMA^V \gg MMA^V$), at least for urinary bladder tumors. In four different genotypes of mice, DMA^V showed strong cancer-promoting characteristics. Strain-species differences in the carcinogenicity profile of DMA^V could correlate with differences in metabolic pathways of arsenic compounds in different animal species and could potentially explain the differences in the susceptibility to DMA^V between rats and mice. The authors summarize that the pentavalent forms of MMA^V and DMA^V are less reactive with tissue constituents, are therefore less toxic, and are more readily excreted in the urine than inorganic arsenic, especially the trivalent form iAs^{III}. The latter is highly reactive with tissue components, due to its strong affinity for sulfhydryl groups.

3. UPTAKE AND METABOLISM OF ARSENIC SPECIES

In addition to gastrointestinal, dermal or pulmonary uptake, exposure to organic arsenic species originates from methylation of inorganic arsenic inside the human body. Thus, the exposure and uptake of both organic and inorganic arsenic will be briefly described here.

3.1. Human Exposure to Organic and Inorganic Arsenic Species

Arsenic is present in the environment at an average concentration of 2 mg/kg. In nature, arsenic-bearing minerals undergo oxidation and release arsenic to water. Due to the uneven distribution of arsenic in minerals, worldwide concentrations of arsenic in groundwater vary by several orders of magnitude. Whereas the concentrations of arsenic in unpolluted surface water and groundwater as well as open sea water are typically in the range of 1–10 μg/L, elevated concentrations in groundwater (up to >1 mg/L) of geochemical origins have been found in Taiwan, West Bengal, India, most districts of Bangladesh, Chile, northern Mexico, several areas of Argentina, parts of the Peoples Republic of China (Xinjiang and Inner Mongolia) and the United States of America (California, Utah, Nevada, Washington and Alaska) [13].

The daily intake of total arsenic from food and beverages is generally between 20 and 300 μg/day; pulmonary exposure has been estimated to contribute up to approximately 10 μg/day in smokers and about 1 μg/day in non-smokers [13]. While in some geogenic contaminated areas arsenic in drinking water constitutes the principal contributor to the daily arsenic intake, food is generally considered the principal contributor to the daily intake of total arsenic [13]. For European countries and the United States dietary intake of arsenic has been investigated in detail [21–23]. Highest arsenic concentrations in food are usually detected in seafood, but the main arsenic species in seafood, AsBet and AsCol, are relatively non-toxic. Comparing the arsenic speciation in different foodstuffs, the relative proportion of inorganic arsenic is highly variable. While meat, poultry, dairy products, and cereals contain mainly inorganic arsenic, organic species predominate in fruits, vegetables, and seafood. For a North American diet, approximately 25% of the daily intake of dietary arsenic is estimated to be inorganic [24]. In contrast, rice and other grains, which are the principal contributors to dietary arsenic intake for non-seafood diets, contain high levels of inorganic arsenic including trivalent arsenic [25]. High arsenic levels in rice and rice products from paddy rice fields irrigated with arsenic-contaminated water can significantly contribute to arsenic exposure even in areas with arsenic-contaminated drinking water [26–29]. The majority of arsenic in groundwater is iAs^{III} or iAs^{V}, but also methylated species have been observed in some groundwaters [30]. Cooking of food can significantly alter the levels as well as the speciation of arsenic in food and should therefore be considered in risk assessment [31–34].

Contamination by ingestion of soils is an important exposure route for environmental contaminants and, in particular, is a problem for children [35,36]. Therefore, it is an important pathway in assessing public health risks associated with exposure to arsenic-contaminated soils [37]. Furthermore, burning of arsenic-rich coals, which occurs in some parts of China, is a severe health hazard affecting approximately 300,000 people in China alone [38,39].

3.2. Uptake and Biotransformation in the Gastrointestinal Tract

Both pentavalent and trivalent arsenic compounds can be rapidly and extensively absorbed in the gastrointestinal tract when administered in soluble form. In controlled ingestion studies in humans, between 45% and 75% of the ingested dose of trivalent forms of arsenic were excreted in the urine within a few days [13]. In comparison to inorganic species, ingested organoarsenicals such as MMA^{V}, DMA^{V} and arsenobetaine are much less extensively metabolized in the human body and more rapidly eliminated in urine than inorganic arsenic in both laboratory animals and humans [13]. After oral administration of radiolabelled arsenobetaine to rabbits, mice, and rats, 75% (rabbits) and 98% (mice and rats) was excreted in the urine unchanged within three days [40]. Organic arsenic species in fish are also rapidly absorbed; less than 5% was found to be eliminated in feces [41].

The bioavailability of arsenic from soils was significantly lower (0.6%–68%) when tested in various animal models [13]. In addition to the solubility of the arsenical compound itself, the matrix in which it is ingested (food, water, soil) as well as the presence of other food constituents and nutrients in the gastrointestinal tract can influence the bioavailability of ingested arsenic [13]. Gonzalez et al. demonstrated that uptake of pentavalent arsenic is carried out by a saturable transport process and that addition of phosphate markedly decreased arsenic absorption, most likely because iAs^{V} and phosphate can share the same transport mechanism [42].

Risk assessment of ingested arsenic might consider not only the bioavailability and toxicity of the initially ingested arsenic species, but also the changes in bioavailability and speciation during digestion in the human intestine. In order to estimate arsenic bioaccessibility and the deriving of human health risk from the ingestion of arsenic-contaminated foodstuff, soils and mine tailings, several *in vitro* gastrointestinal models were developed simulating the chemical and enzymatic solubilization in the stomach and small intestine [32,33,37,43–47]. Lowering the gastric pH was found to significantly increase the bioaccessible arsenic fraction [43].

Surprisingly, little attention has yet been paid to the role of the intestinal microbiocenosis. Herbel et al. demonstrated that arsenic-reducing prokaryotes (DARPs) in slurried hamster feces are able to reduce arsenate and may thereby promote the intestinal resorption of arsenite [48]. Laird et al. [165] investigated the effect of colon microorganisms on the bioaccessibility of arsenic from mine tailings using a microbial model system of the gastrointestinal tract and found a significant increase in bioaccessibility during the colon passage [10]. Rat and mouse cecal microorganisms can transform up to 50% of inorganic arsenic to methylated species within 21 hours [49,50].

Kuroda et al. showed that *Escherichia coli* strains isolated from rat cecal contents after long-term oral administration of $DMAs^{V}$ are able to metabolize $DMAs^{V}$ to TMAO as well as sulfur-containing arsenic species [51], which were later identified as methylated thio species [52]. These metabolites were shown to be highly cyto- and genotoxic [53]. As these metabolites were also found in the urine of rats after oral, but not intraperitoneal administration of $DMAs^{V}$, the authors concluded that this process also occurs *in vivo* [54].

Recently, the formation of volatile arsenic species by human colon microorganisms was studied by Diaz-Bone and Van de Wiele [55]. In addition to TMA and the highly toxic arsine, hitherto undescribed volatile arsenic/sulfur species were identified [56]. This process is of particular importance due to the ability of volatile metal(loid) species to pass cell membranes and hence be distributed through the entire body.

The degradation of ingested organic arsenic species by intestinal microorganisms has not been studied to any great extent. Recently, the capability of intestinal microorganisms to metabolize AsBet to di- and trimethyl arsenate as well as dimethylarsinoylacetate in the human intestine was shown, but only under aerobic conditions [11].

3.3. Cellular Uptake and Extrusion

One of the key aspects to explain the toxicity of arsenic species is their ability to pass through cellular membranes. Different studies have shown large differences depending not only upon the arsenic compound investigated, but also on the cell type and the concentration levels used. In mammalian systems, iAs^{III} is taken up into cells through aquaporin isozyme 7 or 9 (AQP7/9), a member of the aquaglyceroporin family [57–59]. In the case of iAs^{V}, however, phosphate transporters are thought to act to incorporate arsenate into cells [60].

For inorganic arsenic, the transport processes and the relevant carriers have been well characterized. Liu et al. suggested that mammalian aquaglyceroporins (membrane transport proteins) may be a major route of iAs^{III} uptake into mammalian cells because the passive permeation of iAs^{III} is energetically unfavorable [57]. Rosen showed that mammalian aquaglyceroporins catalyze uptake of trivalent metalloids [61]. He also stated that cytosolic iAs^{III} is detoxified by removal from the cytosol.

Tatum and Hood investigated the iAs^{III} uptake in rat hepatocytes (primary culture) and in three established rat cell lines [62]. The authors found a concentration-dependent arsenic uptake. Variability in cellular uptake was observed with a maximum uptake after an exposure period of from 4 h to 8 h. The intracellular iAs^{III} concentrations were similar in all cell types [62]. Other authors also propose that higher/faster uptake of iAs^{III} may be responsible for its increased cytogenetic and genotoxic potency compared to iAs^{V}. In recent studies by Hirano et al., the differences in cytotoxicity and uptake rate of iAs^{III} and iAs^{V} were investigated *in vitro* [63]. iAs^{III} was more cytotoxic than iAs^{V}, and the trivalent form was taken up by the endothelial cells 6 to 7 times faster than the pentavalent form. The authors suggested that the difference in cellular uptake of arsenic is not due to the ionic charge of arsenic but due to some transport mechanisms in the plasma membrane that allow a faster uptake of iAs^{III} compared to iAs^{V} [63].

In addition to the methylation process itself (see below), the formation of glutathione complexes has important implications for the efflux of arsenic. Arsenite triglutathione [$As(SG)_3$] and $MMA^{III}(SG)_2$, but not $DMA^{III}(SG)$, are transported out by multidrug-resistance proteins (MRPs) [64,65]. A proposed pathway of transporters for uptake and efflux of arsenites and enzymes responsible for arsenic excretion into extracellular space in

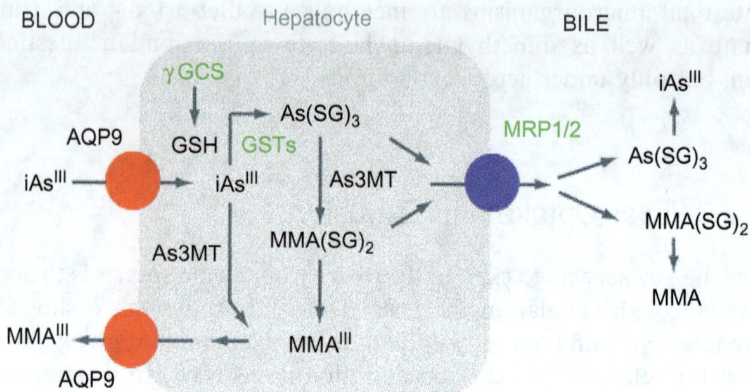

Figure 1. Proposed pathways of transporters for uptake and efflux of arsenites and enzymes responsible for arsenic excretion into extracellular space in hepatocytes. iAs^{III}, inorganic arsenite; MMA^{III}, monomethylarsonous acid; $As(SG)_3$, arsenite triglutathione; $MMA(SG)_2$, monomethylarsonic diglutathione; As3MT, arsenic methyltransferase; γGCS, γ-glutamylcysteine synthase; GSTs, glutathione *S*-transferases; GSH, glutathione; AQP9, aquaglyceroporin 9. Proteins (*green*) are regulated by Nrf2. Adapted from [164] with permission from the *Annual Review of Pharmacology and Toxicology*, copyright (2007).

hepatocytes is shown in Figure 1 and was recently published by Kumagai and Sumi [164].

Similar to inorganic arsenic, the uptake of organic arsenic compounds is also highly dependent upon the cell type. By comparing the uptake capabilities of fibroblasts (CHO-9) and hepatic cells, Dopp et al. [66] demonstrated that organic and inorganic arsenicals are taken up to a higher degree by the non-methylating fibroblasts compared to the methylating hepatoma cells. The authors observed an increased resistance to intracellular accumulation of arsenic in the hepatic cells when compared to CHO-9 cells, which was either due to an increased resistance at the uptake level or to an enhanced efflux rate [66]. DMA^{III} proved to be the most membrane-permeable arsenic species in all studies (up to 16% uptake from the external medium), probably because of its neutral charge which allows it to diffuse easily into cells. In contrast, the pentavalent methylated arsenic species are negatively charged at physiological pH and were poorly taken up by all tested cell lines (0% to max. 2%) [66].

Dopp et al. have shown that the highest arsenic uptake was detectable at relatively low concentrations [iAs^{III}: 500 nM, iAs^{V}:1 μM], and this percentage decreases with increasing arsenic concentrations in the external medium [67]. A defense mechanism seems to exist: the extrusion of iAs^{III} from cells and the prevention of uptake at higher concentrations. Wang and Rossman concluded from their results on iAs^{III} -treated Chinese hamster cells (V79) that mammalian cells contain an iAs^{III} pump, the activity of which may be modulated by prior exposure to iAs^{III} [68]. In another study of Wang et al., the authors demonstrated that an energy-dependent arsenic efflux pump exists in mammalian cells [69]. Also, the authors showed that iAs^{V} is intracellularly reduced to iAs^{III}.

In experiments from Dopp et al. [67] the cellular uptake of different arsenic species was compared. With regard to the methylated arsenic species, the pentavalent ones were less membrane-permeable than the trivalent forms. After incubation of CHO cells for 1 h with MMA^{V}, DMA^{V}, and TMAO, respectively, less than 0.1% of substrate was detected intracellularly. The authors suggested that the trivalent arsenic compounds are more membrane-permeable in comparison to the other arsenic species. The order of cellular uptake for the arsenic compounds in trivalent state was: $DMA^{III} > MMA^{III} > iAs^{III}$ and for the arsenic compounds in the pentavalent state: $iAs^{V} > MMA^{V} > DMA^{V} > TMAO^{V}$.

3.4. Biotransformation of Arsenic by Mammalian Cells

The metabolism of arsenic in mammalian cells is of central importance for understanding its toxicological mode of action (MOA). Three different processes with high toxicological importance occur in human cells: first the reduction of pentavalent to trivalent arsenic species, second the methylation, and third the replacement of hydroxyl by thiol groups (thiolation). Both the metabolic pathways and the role of arsenic metabolism for arsenic toxicity are currently the subject of intensive debate. Following uptake, inorganic arsenic can undergo biotransformation to mono- (MMA^{III}, MMA^{V}) and dimethylated metabolites (DMA^{III}, DMA^{V}). Trimethylarsine oxide (TMAO) is the final metabolite of inorganic arsenicals in some animal species such as rats and hamsters and has been found in trace amounts in human urine after consumption of oxo-arsenosugar [70,71]. In addition to these methylated oxoforms, the formation of thiolated methylarsenicals has recently been demonstrated in rat liver and red blood cells [72,73]. The formation of methylated thiospecies has been postulated by exchange of oxygen by sulfur subsequent to methylation.

The central site for arsenic methylation in the human body is the liver. Methylation of inorganic arsenic facilitates the excretion of arsenic from the body, as the end-products MMA^{V} and DMA^{V} are readily excreted in urine. The mammalian enzyme responsible for the transfer of the methyl group from the methyl donor *S*-adenosyl-methionine (SAM) to arsenic has been identified and was initially named Cyt19, later arsenic methyltransferase (As3MT). By using RNA silencing of As3MT expression in human hepatic cells, Drobna et al. were able to demonstrate that this protein is the major enzyme in this pathway, although their data hint at a contribution from other processes [74]. As3MT was first isolated from rat liver cytosol [75] and more recently from mouse neuroblastoma cell lines [76]. Furthermore, As3MT has been cloned and expressed using *E. coli* [77]. The variability of the gene sequence of human As3MT has been intensively studied, and inter-individual variances in this protein have been proposed to be responsible for differences in the sensitivity to arsenic exposure [78].

While the methyl transfer system is well established, the pathways of biomethylation are currently under debate. Two pathways have been proposed, which are both illustrated in Figure 2. The long-accepted

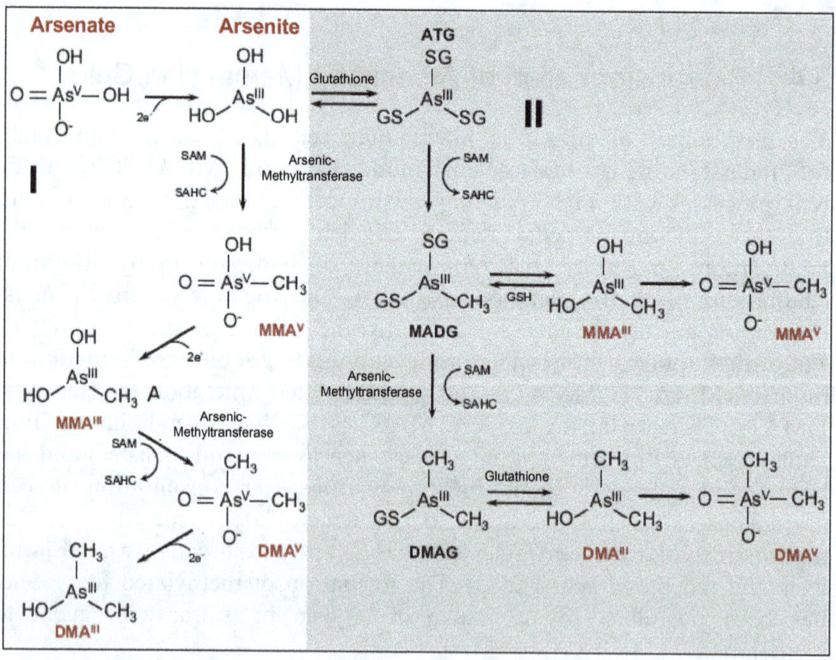

Figure 2. Biotransformation of inorganic arsenic in humans. Discussed are two alternative pathways (I, II). Main metabolites of arsenic found in human urine are marked with red. ATG, arsenite triglutathione; MADG, monomethylarsonic diglutathione; DMAG, dimethylarsinous glutathione; SAHC, *S*-adenosyl homocysteine; SAM, *S*-adenosyl methionine. Adapted from [168] with permission from *Nachrichten aus der Chemie*, copyright (2009).

pathway of arsenic biotransformation consists of a series of reductions of pentavalent to trivalent arsenic species and subsequent oxidative methylation with the sulfur atom from SAM as redox partner (Figure 2, I) [79,168].

Arsenate reductases, such as the omega isoform of GSH *S*-transferase (GSTomega) [80–82] and purine nucleoside phosphorylase (PNP) [83,84], can catalyze the reduction of arsenate species, including organic arsenicals to arsenite. Because trivalent species are more toxic than arsenates, variation in the enzyme activity of GSTomega isoform 1, which is identical to monomethylarsonate (MMA^V) reductase, could influence arsenic toxicity, as suggested by Aposhian and his associates [85a]. However, in a later study by this group, it was suggested that each step of the biotransformation of inorganic arsenic has an alternative enzyme to biotransform the arsenic substrate [85b]. Also, reduction of arsenic can occur via sulfhydryl groups from moieties such as GSH [166].

Recently, a new and much cited metabolic pathway for arsenic biotransformation was proposed, in which trivalent arsenic species bound to glutathione are methylated without being oxidized (Figure 2, II) [86]. Hayakawa et al. suggested this mechanism as they found arsenic glutathione complexes to be the preferable substrate for methylation [86]. They postulated the nucleophilic attack by the sulfur of arsenic-bound glutathione towards the cationic sulfur in SAM, but the postulated product *S*-adenosyl-glutathionyl-homocysteine has not been verified yet. In contrast, a simple explanation, which has not been considered by the authors, is that the arsenic-glutathione complex can also serve as a substrate for oxidative methylation similar to the Challenger mechanism. In a recent review Thomas and coworkers showed that glutathione is not essential but can be replaced by other reducing systems yielding much higher conversion rates [87]. Thus, Thomas et al. proposed that GSH has an indirect role in the methylation of arsenic, possibly by reduction of cysteine residues in As3MT.

In urine predominantly pentavalent methylated metabolites (mainly DMA^V) are excreted, and a proportion of the inorganic arsenicals is excreted without further metabolization. Trivalent (+3) methylated metabolites are detected in urine to a much lesser extent than the +5 species and the inorganic arsenicals [88,89]. Dimethyldithioarsinic acid ($DMDTA^V$) and monomethylmonothioarsonic acid ($MMMTA^V$) were found to be common in the urine of arsenic-exposed humans and animals [11,90]. Studies in humans suggest the existence of a wide difference in the activity of methyltransferases, and the existence of polymorphism has been hypothesized. Factors such as dose, age, gender, and smoking contribute only minimally to the large interindividual variation in arsenic methylation observed in humans [13].

4. MODES OF ACTION OF ORGANOARSENICALS

4.1. Introduction

How arsenicals cause genetic changes, toxicity, and ultimately cancer is an extremely complex and intensively researched field; however, there is no consensus yet on what are the most important factors in these processes as they relate to arsenicals. Describing a MOA is an attempt to identify key events in the carcinogenic process that will enable one to have an understanding on how cancer is induced by a particular agent. One of the difficulties in investigating the MOA of arsenicals, and in particular organoarsenicals, is that arsenicals induce a plethora of responses in cells. Arsenic is a potent inducer of multiple types of DNA damage including chromosome breakage, aneuploidy, and single and double DNA strand breaks. It is a weak or poor inducer of sister chromatid exchanges (SCEs) and point mutations. Arsenicals inhibit DNA repair, influence methylation patterns, induce oxidative stress, bind to proteins, but they do not directly cause DNA adducts. Some arsenicals are highly toxic causing cell death, cell turnover, and cell cycle delay. Others interfere with cell signalling pathways. Arsenic can act as a tumor promoter. Thus, the MOA of arsenicals may involve several key events. Several authors have suggested that the methylated arsenic species do not even share a common mechanism for the induction of DNA damage [91–94].

For cancer to occur, genetic change is necessary. The next section will concentrate on how organoarsenicals affect genotoxicity and DNA repair. Although the authors of this chapter believe that these are the more important key events in the induction of cancer by arsenicals, we realize that other investigators may have equally valid beliefs supporting other key events and MOAs. Thus, short summaries of other, maybe equally important, key events will be briefly addressed in later sections of this chapter.

4.2. Genotoxicity

Genotoxicity, by which we mean here the ability of a chemical to interact with the genetic material or interfere with processes that control the faithful replication, transmission, or translation of the genetic material has been extensively investigated with regard to inorganic arsenicals over the course of several decades. Inorganic arsenicals were generally found to be genotoxic, capable of causing chromosome breakage, micronucleus induction, and DNA strand breakage as well as inhibiting DNA repair. The inorganic arsenicals will not be reviewed here, but only mentioned when necessary for comparison with their methylated forms. What follows is a review of the genotoxicity of the organoarsenicals including the oxo-arsenicals, marine arsenicals, and the thioarsenicals.

4.2.1. Tri- and Pentavalent Methylated Oxoarsenicals

As early as 1929, a study by Dustin and Piton [95] showed that both DMA^V and MMA^V acted as a mitotic poison (i.e., blocking the completion of mitosis) after injection into mice. This was confirmed by King and Ludford [96] in mouse fibroblasts and further validated for DMA^V by Endo et al. [97] and Eguchi et al. [98] using V79 cells. They also reported that trimethylarsine oxide inhibited mitoses at a threefold higher concentration than DMA^V. In 1989, Yamanaka et al. [99] administered DMA^V by gavage at 1500 mg/kg and found DNA single strand breaks in the lung and other organs 12 hours later. By trapping volatile metabolites in the breath of mice and through *in vitro* studies they apparently determined that the causative DNA strand breaking agent was dimethylarsine, a metabolite of DMA^V. This was one of the first clues that the trivalent methylated arsenicals were actually potent DNA damaging agents. (However, there is some question to the source of the arsenic activity; this will be addressed in Section 4.2.4). Later studies by Sordo et al. [100] showed that iAs^{III}, MMA^V, and DMA^V induced little or no DNA damage as measured by the single cell gel electrophoresis (SCGE) assay in unstimulated leukocytes, but in stimulated lymphocytes, DMA^V showed a modest response that was greater than that of both iAs^{III} and MMA^V.

In the mid-1990's studies were published that showed organic arsenicals might induce several types of chromosome damage aside from acting to disrupt mitoses. This was mentioned in an abstract by Endo et al. [101] who stated (without giving data) that DMA^V could induce SCEs. Oya-Ohta et al. [102] showed that DMA^V, MMA^V, and $TMAO^V$ could all induce chromosome breakage in human fibroblasts at relatively high concentrations; however, they were all less potent than iAs^{III} and iAs^V. Moore at al. [103] tested several arsenicals in the L5178Y/$TK^{+/-}$ mouse lymphoma assay and determined that iAs^V and iAs^{III} were active at low micromolar concentrations, while MMA^V and DMA^V were only active at millimolar concentrations. They concluded from the size of the mutant colonies that the majority of the mutations were caused by chromosome breakage and not point mutations. In a later somewhat parallel study *in vivo*, Noda et al. [104] used Muta™ mouse to determine if DMA^V and arsenic trioxide could induce point mutations and/or induce micronuclei in peripheral blood recticulocytes. The authors concluded that neither compound caused a statistically significant increase in point mutations in the lung, kidney, bladder, or bone marrow; and only iAs^{III} caused an increase in micronuclei. Rasmussen and Menzel [105] using a lymphoblastoid cell line found that DMA^V and iAs^V were inactive in inducing SCEs and that iAs^{III} was a weak SCE-inducer.

Though Cullen et al. [106] had shown that MMA^{III} was more toxic towards the yeast, *Candida humicola*, than iAs^{III}, it was not until trivalent methylated arsenicals were found in human urine [107,108] and Styblo et al. [2,109] and Petrick et al. [110] showed that trivalent methylated arsenicals were indeed more toxic than their pentavalent arsenical counterparts in mammalian cells in culture that research on the toxicology of these compounds burgeoned.

Styblo et al. suggested that exposures to methylated trivalent arsenicals are associated with a variety of adverse effects that have a profound impact on cell viability and proliferation [111]. The known effects include inhibition of several key enzymes, damage to DNA structure, and activation of AP-1-dependent gene transcription.

Using the SCGE assay in human lymphocytes and the ΦX174 RFI DNA nicking assay, Mass et al. [112] reported that MMA^{III} and DMA^{III} were orders of magnitude more potent than iAs^{III} and iAs^{V} and that DMA^{V} and MMA^{V} were essentially inactive. This was followed by a study of Nesnow et al. [113] implicating reactive oxygen species as the causative agent in inducing DNA damage by MMA^{III} and DMA^{III} in the ΦDNA nicking assay. Schwerdtle et al. came to a similar conclusion using the alkaline unwinding technique [91]. They concluded that iAs^{III}, MMA^{III}, and DMA^{III} induced high levels of oxidative DNA damage in cultured human cells as measured by DNA strand breakage and FPG-sensitive sites. At approximately two orders of magnitude higher concentrations, the authors found that the pentavalent methylated forms induced low levels of strand breakage but pronounced increases in FPG-sensitive sites. They concluded that lesions are generated *in vitro* not by the arsenicals themselves, but rather by reactive species formed inside the cell.

In an extensive *in vitro* study of the genotoxicity of three trivalent and three pentavalent arsenicals, Kligerman et al. [114] evaluated SCE induction, chromosome breakage, DNA damage as measured by the SCG assay, and mutagenicity using *Salmonella*, the prophage induction assay (DMA^{III} and MMA^{III}, only) and the L5178Y/$TK^{+/-}$ mouse lymphoma assay (DMA^{III} and MMA^{III}, only). iAs^{III}, iAs^{V}, MMA^{III}, MMA^{V}, and DMA^{V} were at best very weak SCE-inducers in human lymphocytes. DMA^{III} was the most potent SCE inducer of the six compounds tested but still only induced about 1 SCE/μM. All six arsenicals were clastogenic, with DMA^{III} and MMA^{III} the most potent, followed by iAs^{III}. The methylated pentavalent forms were much less potent by several orders of magnitude. None of the arsenicals induced mutations in TA98, TA100, or TA104 in the presence or absence of metabolic activation (e.g., S9). Both trivalent methylated arsenicals did not induce significant prophage induction but were highly mutagenic in the mouse lymphoma assay, inducing primarily small colony mutants indicative of chromosome breakage events. The authors concluded that the trivalent methylated arsenicals were the most potent forms of the six arsenicals tested and that the genotoxicity signature was suggestive of chemicals that act through the generation of reactive oxygen species (ROS).

These results were verified and extended upon by Dopp et al. [67]. They found that DMA^V, MMA^V, and TMAO did not induce SCEs in CHO cells; MMA^{III} and DMA^{III} were much more potent SCE inducers than iAs^{III} and iAs^V. A similar pattern was seen with the induction of chromosome aberrations. The cytochalasin B block micronucleus assay was also used to investigate the genotoxicity of the aforementioned seven arsenicals. iAs^{III} and iAs^V caused a small but statistically non-significant increase in micronuclei, but DMA^{III} and MMA^{III} were potent micronuclei inducers at low micromolar concentrations. MMA^V, DMA^V, and TMAO failed to induce micronuclei at concentrations up to 5 mM. Similarly, Colognato et al. [115] examined the effects of several arsenicals in the cytochalasin B block micronucleus test and found that MMA^{III} was about 250 times more potent than MMA^V; DMA^V and TMAO were essentially inactive. They also concluded that MMA^{III} showed clear aneugenic effects using fluorescent centromere analysis.

Aneuploidy, the loss or gain of one or more chromosomes with respect to the normal chromosome complement, is a prominent characteristic of most tumors. In addition, the gain of whole chromosome sets can occur leading to polyploidy. Whether these are a cause of tumors or part of the process in the progression of a mutated cell to a neoplasia is still not settled. In fact, it is still a subject of debate on whether or not aneuploidy should be considered a genotoxic event.

However, many arsenicals are spindle poisons, as some of the first researchers on the toxicity of arsenicals have shown, leading to the induction of polyploidy, aneuploidy, and cell cycle arrest. Kligerman et al. [116] reviewed much of the literature in this area [97,100,117–120], while also reporting on the arsenicals' mitotic poison potential as well as their effects on tubulin polymerization. Pentavalent arsenicals were found to be relatively weak inducers of mitotic arrest, except at high concentrations (>5 mM) and were not effective in inhibiting tubulin polymerization. Methylated trivalent arsenicals were found to have potent colchicine-like effects (mitotic arrest) and to be highly effective in inhibiting tubulin polymerization at low concentrations.

4.2.2. *Methylated Thioarsenicals*

Over the last several years, investigations have discovered a new class of arsenicals in the urine of sheep [121] and humans [90,122,123]. These were termed thioarsenicals, and two of these, dimethylmonothioarsinic acid ($DMMTA^V$) and dimethyldithioarsinic acid ($DMDTA^V$) were studied by Ochi et al. [124] for their genotoxic potential. $DMMTA^V$, but not $DMDTA^V$, was a potent clastogen *in vitro* producing predominantly chromatid breaks and exchanges. In addition, $DMMTA^V$ induced cell cycle arrest and apparent aneuploidy. These results were consistent with the study from Naranmandura et al. [125], who compared the effects of $DMMTA^V$ with iAs^{III}, iAs^V, DMA^{III} and DMA^V. They found that $DMMTA^V$ is one of the most toxic arsenic metabolites, increasing the level of reactive oxygen species and inducing cell cycle perturbation.

4.2.3. *Marine Organic Arsenicals*

There are several organic arsenic compounds that have been found in marine organisms; however, only a limited number of genotoxicity studies have been conducted on these chemicals. In general, they have been inactive when tested. Cannon et al. [126] found that AsBet was non-mutagenic with and without S9 in four different strains of *Salmonella* in the Ames assay. Kaise et al. [127] looked at the clastogenic and SCE-inducing potential of a marine AsSug, 1-(2′,3′-dihydroxypropyl)-5-deoxyribosyldimethylarsine oxide, and AsBet in fibroblasts cells as well as iAs^V, iAs^{III}, MMA^V, and DMA^V. None of the compounds induced SCEs, and the AsSug and AsBet were very weak clastogens (when gaps were included); weaker than both MMA^V and DMA^V, which were themselves only weak inducers of chromosome breakage.

To date the only other study on the genotoxicity of AsSug was by Andrewes et al. [128]. They examined the pentavalent form investigated by Kaise et al. [127] and compared it to its trivalent form using the DNA nicking assay and the preincubation assay with *Salmonella* strain TA104. The trivalent form was found to nick DNA and be approximately as active as DMA^{III}, but the pentavalent form was inactive. Both failed to induce mutations in *Samonella.* Guillamet et al. [129] found that AsBet was marginally genotoxic at best, up to a concentration of 10 mM in the single cell gel assay. Soriano et al. [130] replicated the results of Moore et al. [103] with MMA^V and DMA^V, and extended them to show that AsBet failed to induce point mutations in the mouse lymphoma assay at concentrations up to 10 mM. In general, the studies reported to date seem to indicate that these mainly marine organic arsenicals are either inactive or very weakly active in genotoxicity assays. The main concern is if the pentavalent forms are reduced *in vivo* to potentially more active trivalent forms. Whether this can happen to any appreciable extent is unknown at present.

4.2.4. *Volatile Arsenic Species*

The genotoxicity of volatile arsines has been the subject of several studies. Yamanaka et al. [99] explained the induction of DNA single strand breaks in the lung and other organs after oral administration of 1500 mg/kg DMA^V by the formation of DMAH. Identification of DMAH was based on trapping volatile metabolites in the breath of mice in 5% H_2O_2 and subsequent analysis by thin layer chromatography, which showed an oxidized analyte co-eluting to DMA^V. In addition to the analytical ambiguity of this identification protocol, due to the oral administration, it is likely that the volatile compound was formed by intestinal bacteria.

Even though the origin and nature of the volatile metabolite cannot unambiguously be determined, subsequent studies revealed that DMAH induced DNA damage by formation of peroxyl radicals [131]. Furthermore, Kato et al. showed that TMA induced micronuclei in the bone marrow of mice after intraperitoneal injections of 8.5 and 14.7 mg/kg [132]. These findings were confirmed by Andrewes et al. [133] who investigated the DNA-damaging potential of MMAH, DMAH, and TMA using supercoiled DNA. They concluded that the latter two arsines are about 100 times more potent than DMA^{III}. Thus, while the formation of volatile arsines by human cells, as yet, has not been proven, the high genotoxicity of volatile species has to be considered if generated by intestinal bacteria.

4.3. Inhibition of DNA Repair

In addition to direct damage of DNA, the inhibition of the DNA repair mechanisms is an important pathway that can lead to the fixation of genetic damage leading to cell death, mutation, and tumor formation. Several investigations have shown that inorganic arsenic, in particular arsenite can inhibit DNA repair.

Schwerdtle et al. treated A549 human lung cells with +-anti BPDE to produce DNA adducts and either performed no further treatment or treated the cells with arsenite, MMA^{III}, MMA^V, DMA^{III}, or DMA^V to study these arsenicals' effects on DNA repair [134]. MMA^{III} caused a significant increase in BP-DNA adducts; DMA^{III} and MMA^V and DMA^V did not cause an increase in BP-DNA adducts. MMA^{III}, DMA^{III}, and MMA^V and DMA^V all inhibited DNA repair, but the trivalent methylated arsenicals did so at a 100-fold lower concentration (2.5 μM versus 250 μM). The investigators also studied zinc release from a synthesized XPAzf DNA repair protein as a measure of an arsenical's potential interference with DNA repair. Both MMA^{III} and DMA^{III} caused a concentration-related increase in zinc release from a synthesized XPAzf protein; while the pentavalent methylated forms were essentially inactive up to 10 mM. Inorganic arsenic had an intermediate effect. Reactions of arsenicals with thiols could be responsible for inactivating zinc finger motifs on repair proteins, but the authors believe that further investigations are needed to see if this takes place in whole cells at low concentrations. Additional studies were conducted to determine what effects, if any, arsenicals had on formamidopyrimidine glycosylase (Fpg) activity. Fpg is involved in the recognition of several oxidative bases. Oxidatively-damaged PM2 DNA was used as a substrate, and DNA strand breakage was used as a measure of the Fpg activity. Arsenite and the pentavalent methylated forms were inactive up to 10 mM in inhibiting Fpg. However, MMA^{III} and DMA^{III} produced substantial inhibition at relatively low concentrations of 1 mM and 100 μM, respectively. Overall, these results strongly indicate that methylated trivalent arsenicals are potent inhibitors of DNA repair proteins, but the authors conclude that cellular uptake and arsenic speciation may affect results.

In a follow up paper from this group and collaborators, Piatek et al. [135] using a cellular system with a synthetic polypeptide, showed that MMA^{III} binds much more readily to the XPAzf synthetic polypeptide than arsenite, forming monomethyl and dimethyl derivatives and causing the oxidation of unprotected thiols to intramolecular dithiols. The affinity of MMA^{III} for thiol groups on the XPAzf is 30 times higher than that for arsenite, which, if this occurred *in vivo* would inhibit DNA repair possibly leading to carcinogenesis.

Because poly(ADP-ribose) polymerase-1 (PARP-1) is involved in base excision repair (and probably nucleotide excision repair), binds to DNA strand breaks via two zinc finger motifs, and because methylated trivalent arsenicals were previously found to release zinc from DNA repair protein XPA, it was logical to investigate the effects of several arsenicals on poly (ADP-ribosyl)ation in cultured human cells [136]. HeLaS3 cells were exposed to 100 μM hydrogen peroxide for 5 min to induce poly(ADP-ribosyl)ation, which occurs shortly after DNA strand breakage. MMA^{III} and DMA^{III} decreased poly(ADP-ribosyl)ation in a concentration-dependent manner starting at concentrations as low as 1 nM. The pentavalent methylated arsenicals had no effect on poly(ADP-ribosyl)ation at 500 μM and 250 μM, respectively. These were low, non-cytotoxic concentrations, 10 times lower than that needed for arsenite to produce an equivalent effect. Neither pentavalent (100 μM) nor trivalent arsenicals (0.1 μM) had an effect on gene expression of PARP-1 after an 18 h exposure, and MMA^{III} and DMA^{III} at 10 μM inhibited isolated recombinant PARP-1.

Shen et al. [137] used a similar approach to that used by Schwerdtle et al. [134] to try to determine how arsenicals affect DNA repair. Normal human fibroblasts were treated with anti-BPDE, and the effect of arsenicals was monitored by measuring the removal of BPDE-DNA adducts. Trivalent arsenic compounds, DMA^{III} and MMA^{III}, as wells as iAs^{III} to a lesser extent, inhibit BPDE-DNA adduct repair at low concentrations. At 1 μM there was a 45% and 37% reduction in adduct removal for MMA^{III} and DMA^{III}, respectively. Repair inhibition was observable within 4 h of arsenical treatment. In contrast there was no significant reduction with 2.5 μM iAs^{III}. They also examined expression levels of several common genes involved in DNA repair. Expression levels of p48, XPC, p62, and XPA were not affected by MMA^{III}. However, methylated arsenicals inhibited p53 accumulation, which is needed for efficient global nucleotide excision repair. MMA^{III} inhibited phosphorylation of p53 at serine-15, which led to reduced p53 stability. The p53 null cell line failed to show repair inhibition by MMA^{III}. p21 expression was also reduced, probably due to the effect of MMA^{III} on p53. Thus, they concluded that the effects of arsenicals on NER are due to suppression of p53.

In total, all of these studies indicate that arsenicals can inhibit DNA repair processes. And again, the trivalent methylated forms were much more potent than the inorganic or pentavalent methylated arsenicals when tested in similar systems.

An overview of the principal arsenic-induced cellular responses is given in Figure 3 and described shortly also in the following sections. Most investigations were carried out with inorganic arsenic.

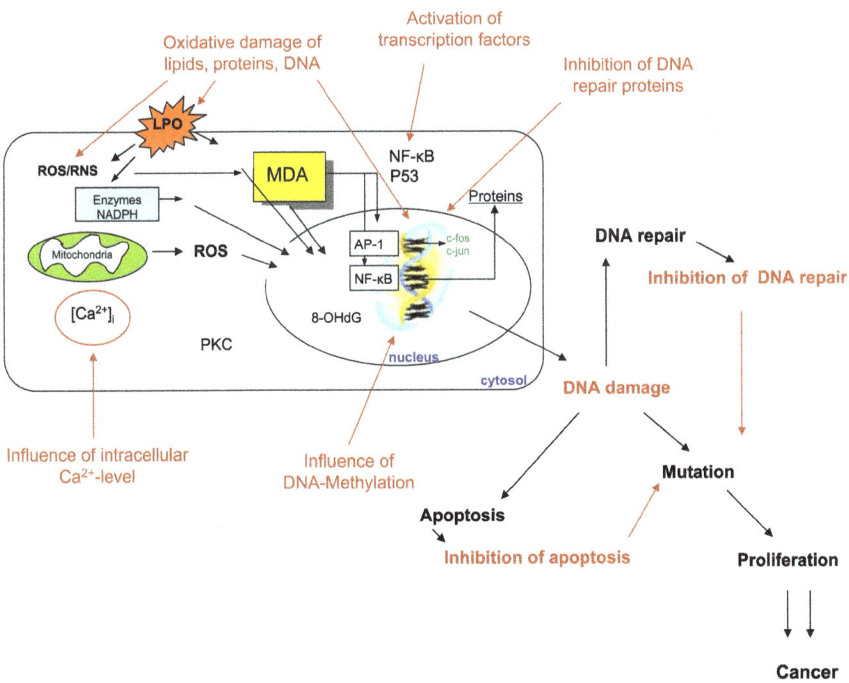

Figure 3. Overview about possible cellular effects caused by arsenic compounds. LPO, lipid peroxidation; MDA, malondialdehyde; $[Ca^{2+}]_i$, intracellular calcium level; PKC, protein kinase C; 8-OHdG, 8-hydroxy-2′-deoxyguanosine; AP-1, activator protein 1 (transcription factor).

4.4. DNA Methylation

Exposure to arsenic can induce both DNA hypomethylation and hypermethylation. DNA methylation changes are typically observed in cancer, in which global methylation is reduced, but some gene-specific promoter methylation is increased [138]. Long-term low-dose arsenic exposure induces global loss of DNA methylation in cultured rat liver cells [139]. Investigations about DNA methylation caused by organoarsenicals were not found in the literature. Arsenic-induced global DNA hypomethylation is associated with the depletion of SAM pool and suppression of DNA methyltransferases DNMT1 and DNMT3A [139,140].

Specific hypomethylation of the estrogen receptor-α (ER-α) gene promoter is seen in arsenic-exposed mouse livers and may result in aberrant ER-α expression and aberrant estrogen signaling [141], which is potentially involved in arsenic hepatocarcinogenesis. Liver steatosis (fatty liver, a preneoplastic change associated with methyl deficiency) is also a frequent observation following chronic arsenic exposure and associated with methyl insufficiency and DNA methylation loss in cells or animals [140,141]. Arsenic-induced alterations in DNA methylation could enhance genomic instability, such as chromosomal instability in mammalian cells [142]. Of note is that individual gene hypermethylation can occur concomitantly with global DNA hypomethylation. In this regard, the loss of p16 expression is observed in arsenic-transformed liver cells, which could be due to both the hypermethylation of the p16 gene and the homozygous deletion of p16 [143]. Both inorganic arsenite and arsenate produced hypermethylation of the p53 gene in human lung adenocarcinoma A549 cells [144]. Thus, altered DNA methylation status could affect genetic stability and gene expression, and could be a key factor in arsenic carcinogenesis.

4.5. Apoptotic Tolerance

Arsenic-intoxicated cells can be eliminated through apoptosis if the damage is severe enough. However, during chronic arsenic exposure, adaptation to the effects of arsenic occurs, including apoptosis, and this frequently results in a generalized tolerance. Apoptotic resistance is a common phenomenon in cells malignantly transformed by arsenic, including rat liver epithelial cells [145]. Tolerance to apoptosis may be an important factor for arsenic carcinogenesis because it may allow the damaged cells that otherwise would be eliminated to survive and to transmit genetic or epigenetic lesions (see Figure 3). Apoptotic tolerance is often associated with increased cell proliferation, as evidenced by proliferative changes *in vivo* frequently seen with chronic arsenic exposure [141]. Arsenic often induces overexpression of cell proliferation-related genes, such as *cyclin D1* and proliferating cell nuclear antigen (PCNA), as seen in arsenic-treated mouse liver cells [141,146].

Ochi et al. studied the induction of apoptosis caused by methylated arsenic species *in vitro* [147]. The authors showed that DMA^V induced apoptosis in cultured human HL-60 cells at concentrations of 1–5 mM after an incubation period of 18 h. *In vivo* administration of DMA^V, however, resulted in cytotoxicity with necrosis, followed by regenerative hyperplasia of the bladder epithelium [148].

4.6. Further Possible Effects

Regulation of intercellular and intracellular signaling is fundamental for survival and death in biologic organisms; the systems that control ion movements across cell membranes are essential for cell survival. A deregulation of channels or pumps can cause events that lead to cell death. Apoptosis can be caused by loss of Ca^{2+} homeostatic control but can also be positively or negatively controlled by changes in Ca^{2+} distribution within intracellular compartments. It was shown that even non-disruptive changes in Ca^{2+} signaling could have adverse effects, including alterations in cell proliferation and differentiation, as well as in the modulation of apoptosis [149].

Florea et al. assessed inorganic iAs^{III} and iAs^V, as well as MMA^V, DMA^V, and $TMAO^V$ for early disturbances in calcium homeostasis in HeLa S3 cells within the first few seconds after application [150]. A drop in the fluorescence signal of the dye was recorded by confocal laser scanning microscopy. The drop was transient for iAs^{III}, iAs^V and MMA^V, and the signal returned rapidly to the initial level within 20 sec. The authors concluded that the calcium signals might occur as active efflux from the cell to the exterior (energy consuming) or as deregulation of other ion transports. A mechanism via membrane receptor activation or membrane damage was suggested.

$[Ca^{2+}]_n$ is involved in the regulation of many events also in the nucleus, including gene expression, DNA replication, DNA repair, chromatin fragmentation in apoptosis, and modulation of an intranuclear contractile system. The importance of a precise cellular Ca^{2+} level regulation for an optimal DNA repair process was demonstrated already by Gafter et al. [151]. Bugreev and Mazin showed that the human Rad51 protein, which plays a key role in homologous recombination and DNA repair, is dependent upon the intracellular calcium level [152].

From several studies it is known that arsenic can enhance the mutagenicity of other carcinogens [142]. iAs^{III} enhances the mutagenicity and/or clastogenicity of UV, *N*-methyl-*N*-nitrosourea, diepoxybutane, X-rays, and methylmethane sulfonate in mammalian cells [153]. Arsenic inhibits the repair of DNA adducts caused by benzo[*a*]pyrene in rats [154]. Because of its inhibitory effects on DNA repair, arsenic acts as a very efficient cocarcinogen.

The influence of arsenic on signaling pathways was also studied in the literature. Aberrant estrogen receptor signaling pathways were observed in liver carcinogenesis induced by arsenic [155]. Intense expression of ER-α is observed in liver tumors and tumor-surrounding normal tissues after gestational arsenic exposure in mice [156]. The most important evidence for a promoting effect of arsenic in aberrant estrogen signaling related to cancer development *in utero* came from a study of Waalkes et al. [156]. The combined treatment of mice with arsenic and diethylstilbestrol, a synthetic estrogen, synergistically increased liver tumor in male offspring, and increased liver tumor incidence in females [156].

5. ARSENIC CARCINOGENESIS AND OXIDATIVE STRESS

Arsenicals are known to produce oxidative stress as a mechanism of hepatotoxicity and carcinogenicity [157,167]. Hepatic lipid peroxidation and glutathione depletion are observed in chronic arsenic-treated animals [158]. A number of oxidative stress-related genes, such as those of heme oxygenase-1 and metallothionein, are often increased following acute, high-dose arsenic exposure [159]. However, expressions of these stress-related genes were not increased during low-dose, chronic exposures [160]. Various adaptive mechanisms that reduce acute arsenic toxicity are often induced to protect against arsenic-induced oxidative stress [161]. One of these adaptive mechanisms is the induction of hepatic glutathione *S*-transferase, which in turn plays a key role in ameliorating arsenic-induced oxidative damage and helping transport arsenic out of the liver cell [159]. Increases in hepatic DNA 8-hydroxydeoxyguanosine levels, a biomarker for oxidative DNA damage, have been associated with hepatocarcinogenesis induced by methylated arsenicals [20,162].

Oxidative damage induced by iAs^{III} as well as the methylated arsenic species can also occur *via* indirect mechanisms. Both the inhibition of important detoxifying enzymes [93] and the depletion of cellular glutathione levels have been proposed. MMA^{III} and DMA^{III} are potent inhibitors of glutathione reductase suggesting that the effect is due to the interaction of trivalent arsenic with critical thiol groups, thus altering the cellular redox status. The weak or insignificant SCE induction by these compounds, in contrast to their potent clastogenicity and cytotoxicity, is indicative of agents that act through an ROS mechanism.

DMA^V-induced lung-specific DNA damage in mice can be attributed to free radicals, particularly peroxyl, superoxide or hydroxyl radicals, arising from the reaction of DMA^V with molecular oxygen *in vivo* [163]. Depletion in cellular glutathione may be correlated with oxidative stress mediated by reactive oxygen/nitrogen species. The reaction and interaction of these reactive species with target molecules lead to oxidative stress, lipid peroxidation, DNA damage, and activation of signaling cascades associated with tumor promotion and/or progression [82]. Antioxidants can inhibit, reduce, or scavenge the production of reactive oxygen and nitrogen species induced by arsenic. These cannot only decrease direct cellular damage such as lipid peroxidation, enzyme inactivation and DNA oxidation caused by arsenic, but they can also ameliorate cell injuries or death by redox signaling pathways activated by arsenic exposure [82].

Arsenic-induced oxidative stress can cause DNA damage/chromosome breakage and cell death followed by regenerative cell proliferation. This could cause cell initiation and progression leading to cancer. This genetic damage could be enhanced due to the effects of arsenicals on DNA repair. Figure 4 shows a scheme on how this may occur. Trivalent organoarsenicals induce reactive oxygen species that can induce single-strand DNA breaks either directly or through the inhibition of DNA repair enzymes. These breaks would normally be repaired quite rapidly without error. However, if there is scant time for DNA repair, either because the cells are rapidly proliferating (proliferative regeneration) or the cells are damaged during S-phase of the cell cycle, or because DNA repair is inhibited by arsenic itself, the single-strand breaks can be converted into double-strand breaks during S-phase leading to chromatid-type chromosomal aberrations. Though not shown to keep the schematic relatively simple, chromatid-type exchanges can lead to derived translocations in the subsequent cell division. In addition, double-strand breaks could be induced before DNA synthesis through the action of endonucleases or during the process of repair of closely spaced single-strand breaks. These could cause the formation of chromosome-type chromosome aberrations such as translocations. Chromosomal events such as translocations are a prominent characteristic of many tumors.

Thus, organoarsenicals, through their action of inducing reactive oxygen species can produce cytotoxicity and accompanying regenerative proliferation. Through their ability to also induce DNA damage and at the same time inhibit DNA repair, they can lead to the fixation of mutations necessary for cancer induction, and through their action on the spindle apparatus can produce aneuploidy and cellular changes leading to progression and cellular instability eventually producing neoplasia.

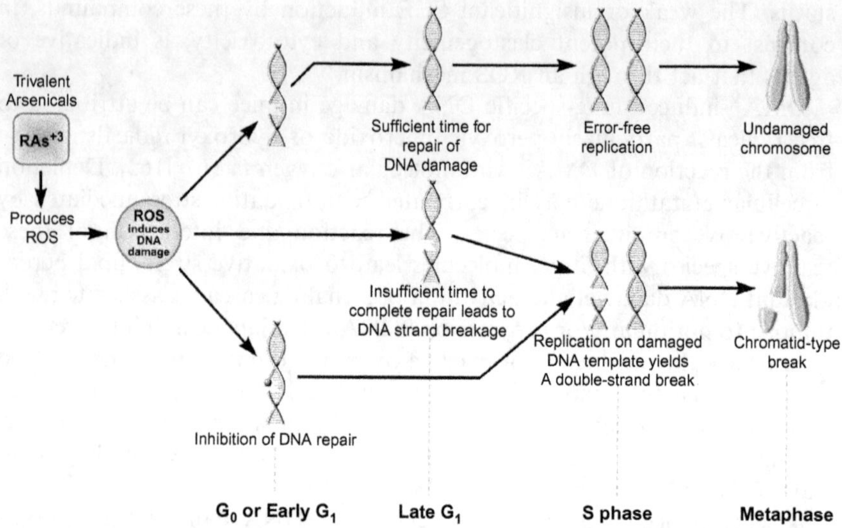

Figure 4. Hypothesis of how active trivalent organic arsenicals (RAs^{+3}) may induce chromosome damage. RAs^{+3} produces reactive oxygen species (ROS) that directly induce DNA single strand breaks or damaged bases that lead to DNA repair-induced strand breakage. If there is sufficient time for completion of DNA repair (G_0 or early G_1 treatment), then cells proceed to metaphase without visible chromosome damage. If RAs^{+3} treatment occurs in late G_1 or S phase of the cell cycle or if DNA repair is inhibited, DNA containing single strand breaks or base damage are replicated leading to DNA double strand breaks and chromatid-type aberrations visible at metaphase.

ABBREVIATIONS

8-OHdG	8-hydroxy-2′-deoxyguanosine
γGCS	γ-glutamylcysteine synthase
AP-1	activator protein 1
AQP7/9	aquaporin isozyme 7 or 9
As3MT	arsenic (+3 oxidation state) methyltransferase
AsBet	arsenobetaine
AsCol	arsenocholine
AsLip	arsenolipids
$As(SG)_3$	= ATG
AsSug	arsenosugars
ATG	arsenite triglutathione
BFD	blackfoot disease
BP	benzo[*a*]pyrene

BPDE	benzo[*a*]pyrene diolepoxide
$[Ca^{2+}]_i$	intracellular calcium level
CHO cells	Chinese hamster ovary cells
DARP	arsenic-reducing prokaryotes
DMAH	dimethylarsine
DMA^{III}	dimethylarsinous acid
DMA^{V}	dimethylarsinic acid
DMAG	dimethylarsinous glutathione (= $DMA^{III}(SG)$)
$DMA^{III}(SG)$	= DMAG
$DMDTA^{V}$	dimethyldithioarsinic acid
$DMMTA^{V}$	dimethylmonothioarsinic acid
DNMT	DNA methyltransferase
ER-α	estrogen receptor-α
Fpg	formamidopyrimidine glycosylase
GSTomega	omega isoform of glutathione *S*-transferase
iAs^{III}	inorganic arsenite
iAs^{V}	inorganic arsenate
LPO	lipid peroxidation
MADG	monomethylarsonic diglutathione (= $MMA(SG)_2$)
MDA	malondialdehyde
$MMA(SG)_2$	= MADG
MMAH	monomethylarsine
MMA^{III}	monomethylarsonous acid
MMA^{V}	monomethylarsonic acid
$MMMTA^{V}$	monomethylmonothioarsonic acid
MOA	mode of action
MRP	multidrug-resistance proteins
NADPH	nicotinamide adenine dinucleotide phosphate
NER	nucleotide excision repair
NF-κB	nuclear factor κ-light-chain-enhancer of activated B cells
PARP-1	poly(ADP-ribose) polymerase-1
PcNA	proliferating cell nuclear antigen
PKC	protein kinase C
PNS	purine nucleoside phosphorylase
PVD	peripheral vascular disease
RAs^{+3}	trivalent organic arsenical
SAHC	*S*-adenosyl homocysteine
SAM	*S*-adenosyl methionine
SCE	sister chromatid exchange
SCGE	single cell gel electrophoresis
SG/GS/GSH	glutathione
TMA	trimethylarsine

$TMAO^V$	trimethylarsine oxide
XPA	*Xeroderma pigmentosum* group A complementing protein
XPAzf	XPA zinc finger

REFERENCES

1. H. V. Aposhian, *Annu. Rev. Pharmacol. Toxicol.*, 1997, **37**, 397–419.
2. M. Styblo, L. M. Del Razo, L. Vega, D. R. Germolec, E. L. LeCluyse, G. A. Hamilton, W. Reed, C. Wang, W. R. Cullen and D. J. Thomas, *Arch. Toxicol.*, 2000, **74**, 289–299.
3. J. S. Petrick, B. Jagadish, E. A. Mash and H. V. Aposhian, *Chem. Res. Toxicol.*, 2001, **14**, 651–656.
4. M. J. Mass, A. Tennant, B. C. Roop, W. R. Cullen, M. Styblo, D. J. Thomas and A. D. Kligerman, *Chem. Res. Toxicol.*, 2001, **14**, 355–361.
5. A. D. Kligerman, C. L. Doerr, A. H. Tennant, K. Harrington-Brock, J. W. Allen, E. Winkfield, P. Poorman-Allen, B. Kundu, K. Funasaka, B. C. Roop, M. J. Mass and D. M. DeMarini, *Environ. Mol. Mutagen.*, 2003, **42**, 192–205.
6. H. V. Aposhian, R. A. Zakharyan, M. D. Avram, M. J. Kopplin and M. L. Wollenberg, *Toxicol. Appl. Pharmacol.*, 2003, **193**, 1–8.
7. E. Dopp, L. M. Hartmann, A. M. Florea, U. von Recklinghausen, R. Pieper, B. Shokouhi, A. W. Rettenmeier, A. V. Hirner and G. Obe, *Toxicol. Appl. Pharmacol.*, 2004, **201**, 156–165.
8. M. Styblo, S. V. Serves, W. R. Cullen and D. J. Thomas, *Chem. Res. Toxicol.*, 1997, **10**, 27–33.
9. M. Schuliga, S. Chouchane and E. T. Snow, *Toxicol. Sci.*, 2002, **70**, 183–192.
10. K. N. Chang, T. C. Lee, M. F. Tam, Y. C. Chen, L. W. Lee, S. Y. Lee, P. J. Lin and R. N. Huang, *Biochem. J.*, 2003, **371**, 495–503.
11. H. Naranmandura, N. Suzuki, K. Iwata, S. Hirano and K. T. Suzuki, *Chem. Res. Toxicol.*, 2007, **20**, 616–624.
12. V. M. Dembitsky and D. O. Levitsky, *Prog. Lipid Res.*, 2004, **43**, 403–448.
13. A. Gomez-Caminero, P. Howe, M. Hughes, E. Kenyon, D. R. Lewis, M. Moore, J. Ng, A. Aitio and G. Becking, *Environmental Health Criteria. Arsenic and Arsenic Compounds,*World Health Organization, 2001.
14. C. Abernathy and A. Morgan, in *United Nations Synthesis Report on Arsenic in Drinking Water*, UN Organization, 2001.
15. M. N. Bates, O. A. Rey, M. L. Biggs, C. Hopenhayn, L. E. Moore, D. Kalman, C. Steinmaus and A. H. Smith, *Am. J. Epidemiol.*, 2004, **159**, 381–389.
16. H. Y. Chiou, S. T. Chiou, Y. H. Hsu, Y. L. Chou, C. H. Tseng, M. L. Wei and C. J. Chen, *Am. J. Epidemiol.*, 2001, **153**, 411–418.
17. C. Steinmaus, Y. Yuan, M. N. Bates and A. H. Smith, *Am. J. Epidemiol.*, 2003, **158**, 1193–1201.

18. A. H. Smith, E. O. Lingas and M. Rahman, *Bull. World Health Organization*, 2000, **78**, 1093–1103.
19. S. M. Cohen, L. L. Arnold, M. Eldan, A. S. Lewis and B. D. Beck, *Crit. Rev. Toxicol.*, 2006, **36**, 99–133.
20. H. Wanibuchi, E. I. Salim, A. Kinoshita, J. Shen, M. Wei, K. Morimura, K. Yoshida, K. Kuroda, G. Endo and S. Fukushima, *Toxicol. Appl. Pharmacol.*, 2004, **198**, 366–376.
21. H. Robberecht, R. Van Cauwenbergh, D. Bosscher, R. Cornelis and H. Deelstra, *Eur. Food Res. Technol.*, 2002, **214**, 27–32.
22. R. A. Schoof, L. J. Yost, J. Eickhoff, E. A. Crecelius, D. W. Cragin, D. M. Meacher and D. B. Menzel, *Food Chem. Toxicol.*, 1999, **37**, 839–846.
23. S. S. H. Tao and P. M. Bolger, *Food Addit. Contam.*, 1999, **16**, 465–472.
24. L. J. Yost, R. A. Schoof and R. Aucoin, *Hum. Ecol. Risk Assess.*, 1998, **4**, 137–152.
25. P. N. Williams, A. H. Price, A. Raab, S. A. Hossain, J. Feldmann and A. A. Meharg, *Environ. Sci. Technol.*, 2005, **39**, 5531–5540.
26. A. A. Meharg, G. X. Sun, P. N. Williams, E. Adomako, C. Deacon, Y. G. Zhu, J. Feldmann and A. Raab, *Environ. Pollution*, 2008, **152**, 746–749.
27. G. X. Sun, P. N. Williams, A. M. Carey, Y. G. Zhu, C. Deacon, A. Raab, J. Feldmann, R. M. Islam and A. A. Meharg, *Environ. Sci. Technol.*, 2008, **42**, 7542–7546.
28. A. A. Meharg, *Trends Plant Sci.*, 2004, **9**, 415–417.
29. A. J. Signes-Pastor, K. Mitra, S. Sarkhel, M. Hobbes, F. Burlo, W. T. De Groot and A. A. Carbonell-Barrachina, *J. Agric. Food Chem.*, 2008, **56**, 9469–9474.
30. A. Shraim, N. C. Sekaran, C. D. Anuradha and S. Hirano, *Appl. Organomet. Chem.*, 2002, **16**, 202–209.
31. J. M. Laparra, D. Velez, R. Barbera, R. Farre and R. Montoro, *J. Agric. Food Chem.*, 2005, **53**, 8829–8833.
32. J. M. Laparra, D. Velez, R. Montoro, R. Barbera and R. Farre, *J. Agric. Food Chem.*, 2003, **51**, 6080–6085.
33. J. M. Laparra, D. Velez, R. Montoro, R. Barbera and R. Farre, *Appl. Organomet. Chem.*, 2004, **18**, 662–669.
34. V. Devesa, D. Velez and R. Montoro, *Food Chem. Toxicol.*, 2008, **46**, 1–8.
35. E. J. Calabrese, E. J. Stanek, R. C. James and S. M. Roberts, *Environ. Health Persp.*, 1997, **105**, 1354–1358.
36. J. H. Vanwijnen, P. Clausing and B. Brunekreef, *Environ. Res.*, 1990, **51**, 147–162.
37. R. R. Rodriguez, N. T. Basta, S. W. Casteel, F. P. Armstrong and D. C. Ward, *J. Environ. Quality*, 2003, **32**, 876–884.
38. J. Liu, B. S. Zheng, H. V. Aposhian, Y. S. Zhou, M. L. Chen, A. H. Zhang and M. P. Waalkes, *Environ. Health Persp.*, 2002, **110**, 119–122.
39. A. Shraim, X. Cui, S. Li, J. C. Ng, H. P. Wang, Y. L. Jin, Y. C. Liu, L. Guo, D. S. Li, S. Q. Wang, R. Z. Zhang and S. Hirano, *Toxicol. Lett.*, 2003, **137**, 35–48.
40. M. Vahter, E. Marafante and L. Dencker, *Sci. Tot. Environ.*, 1983, **30**, 197–211.

41. M. Vahter, in *Biological Effect of Arsenic*, Ed. B. Fowler, Elsevier Science Publ., Amsterdam, 1983.
42. M. J. Gonzalez, M. V. Aguilar and M. C. M. Para, *Vet. Human Toxicol.*, 1995, **37**, 131–136.
43. K. C. Makris, S. Quazi, R. Nagar, D. Sarkar, R. Datta and V. L. Sylvia, *Environ. Sci. Technol.*, 2008, **42**, 6278–6284.
44. P. Pouschat and G. J. Zagury, *Environ. Sci. Technol.*, 2006, **40**, 4317–4323.
45. R. R. Rodriguez and N. T. Basta, *Environ. Sci.Technol.*, 1999, **33**, 642–649.
46. M. V. Ruby, A. Davis, R. Schoof, S. Eberle and C. M. Sellstone, *Environ. Sci. Technol.*, 1996, **30**, 422–430.
47. J. K. Yang, M. O. Barnett, P. M. Jardine and S. C. Brooks, *Soil Sediment Contamination*, 2003, **12**, 165–179.
48. M. J. Herbel, J. S. Blum, S. E. Hoeft, S. M. Cohen, L. L. Arnold, J. Lisak, J. F. Stolz and R. S. Oremland, *Fems Microbiol. Ecol.*, 2002, **41**, 59–67.
49. L. L. Hall, S. E. George, M. J. Kohan, M. Styblo and D. J. Thomas, *Toxicol. Appl. Pharmacol.*, 1997, **147**, 101–109.
50. I. R. Rowland and M. J. Davies, *J. Appl. Toxicol.*, 1981, **1**, 273–283.
51. K. Kuroda, K. Yoshida, A. Yasukawa, H. Wanibuchi, S. Fukushima and G. Endo, *Appl. Organomet. Chem.*, 2001, **15**, 548–552.
52. K. T. Suzuki, B. K. Mandal, A. Katagiri, Y. Sakuma, A. Kawakami, Y. Ogra, K. Yamaguchi, Y. Sei, K. Yamanaka, K. Anzai, M. Ohmichi, H. Takayama and N. Aimi, *Chem. Res. Toxicol.*, 2004, **17**, 914–921.
53. K. Kuroda, K. Yoshida, M. Yoshimura, Y. Endo, H. Wanibuchi, S. Fukushima and G. Endo, *Toxicol. Appl. Pharmacol.*, 2004, **198**, 345–353.
54. K. Yoshida, K. Kuroda, Y. Inoue, H. Chen, Y. Date, H. Wanibuchi and S. Fukushima, *Appl. Organomet. Chem.*, 2001, **15**, 539–547.
55. R. A. Diaz-Bone and T. R. Van de Wiele, *Environ. Sci. Technol.*, in press.
56. R. A. Diaz-Bone, M. Hollmann, O. Wuerfel and D. Pieper, *J. Anal. At. Spectrom.*, DOI 10.1039/B822968F.
57. Z. J. Liu, J. Shen, J. M. Carbrey, R. Mukhopadhyay, P. Agre and B. P. Rosen, *Proc. Natl. Acad. Sci. USA*, 2002, **99**, 6053–6058.
58. H. Bhattacharjee, J. Carbrey, B. P. Rosen and R. Mukhopadhyay, *Biochem. Biophys. Res. Commun.*, 2004, **322**, 836–841.
59. Z. J. Liu, J. M. Carbrey, P. Agre and B. P. Rosen, *Biochem. Biophys. Res. Commun.*, 2004, **316**, 1178–1185.
60. I. Csanaky and Z. Gregus, *Toxicol. Sci.*, 2001, **63**, 29–36.
61. B. P. Rosen, *FEBS Lett.*, 2002, **529**, 86–92.
62. F. M. Tatum and R. D. Hood, *Toxicol. Sci.*, 1999, **52**, 20–25.
63. S. Hirano, X. Cui, S. Li, S. Kanno, Y. Kobayashi, T. Hayakawa and A. Shraim, *Arch. Toxicol.*, 2003, **77**, 305–312.
64. S. V. Kala, M. W. Neely, G. Kala, C. I. Prater, D. W. Atwood, J. S. Rice and M. W. Lieberman, *J. Biol. Chem.*, 2000, **275**, 33404–33408.
65. E. M. Leslie, A. Haimeur and M. P. Waalkes, *J. Biol. Chem.*, 2004, **279**, 32700–32708.

66. E. Dopp, L. M. Hartmann, U. von Recklinghausen, A. M. Florea, S. Rabieh, U. Zimmermann, B. Shokouhi, S. Yadav, A. V. Hirner and A. W. Rettenmeier, *Toxicol. Sci.*, 2005, **87**, 46–56.
67. E. Dopp, L. M. Hartmann, A. M. Florea, U. von Recklinghausen, R. Pieper, B. Shokouhi, A. W. Rettenmeier, A. V. Hirner and G. Obe, *Toxicol. Appl. Pharmacol.*, 2004, **201**, 156–165.
68. Z. L. Wang and T. G. Rossman, *Toxicol. Appl. Pharmacol.*, 1993, **118**, 80–86.
69. (a) Z. L. Wang, S. Dey, B. P. Rosen and T. G. Rossman, *Toxicol. Appl. Pharmacol.*, 1996, **137**, 112–119; (b) Y. F. Lin, A. R. Walmsley and B. P. Rosen, *Proc. Natl. Acad. Sci. USA*, 2006, **103**, 15617–15622.
70. K. A. Francesconi, R. Tanggaard, C. J. McKenzie and W. Goessler, *Clin. Chem.*, 2002, **48**, 92–101.
71. R. Raml, W. Goessler, P. Traar, T. Ochi and K. A. Francesconi, *Chem. Res. Toxicol.*, 2005, **18**, 1444–1450.
72. H. Naranmandura and K. T. Suzuki, *Toxicol. Appl. Pharmacol.*, 2008, **227**, 390–399.
73. K. T. Suzuki, K. Iwata, H. Naranmandura and N. Suzuki, *Toxicol. Appl. Pharmacol.*, 2007, **218**, 166–173.
74. Z. Drobna, W. B. Xing, D. J. Thomas and M. Styblo, *Chem. Res. Toxicol.*, 2006, **19**, 894–898.
75. S. Lin, Q. Shi, F. B. Nix, M. Styblo, M. A. Beck, K. M. Herbin-Davis, L. L. Hall, J. B. Simeonsson and D. J. Thomas, *J. Biol. Chem.*, 2002, **277**, 10795–10803.
76. J. P. P. John, J. E. Oh, A. Pollak and G. Lubec, *Amino Acids*, 2008, **35**, 355–358.
77. D. J. Thomas, S. B. Waters and M. Styblo, *Toxicol. Appl. Pharmacol.*, 2004, **198**, 319–326.
78. A. Hernandez and R. Marcos, *Pharmacogenomics*, 2008, **9**, 1113–1132.
79. F. Challenger, *Chem. Revi.*, 1945, **36**, 315–361.
80. R. A. Zakharyan and H. V. Aposhian, *Chem. Res. Toxicol.*, 1999, **12**, 1278–1283.
81. R. A. Zakharyan, A. Sampayo-Reyes, S. M. Healy, G. Tsaprailis, P. G. Board, D. C. Liebler and H. V. Aposhian, *Chem. Res. Toxicol.*, 2001, **14**, 1051–1057.
82. R. A. Zakharyan, G. Tsaprailis, U. K. Chowdhury, A. Hernandez and H. V. Aposhian, *Chem. Res. Toxicol.*, 2005, **18**, 1287–1295.
83. T. R. Radabaugh, A. Sampayo-Reyes, R. A. Zakharyan and H. V. Aposhian, *Chem. Res. Toxicol.*, 2002, **15**, 692–698.
84. Z. Gregus and B. Nemeti, *Toxicol. Sci.*, 2002, **70**, 13–19.
85. (a) L. L. Marnell, G. G. Garcia-Vargas, U. K. Chowdhury, R. A. Zakharyan, B. Walsh, M. D. Avram, M. J. Kopplin, M. E. Cebrian, E. K. Silbergeld and H. V. Aposhian, *Chem. Res. Toxicol.*, 2003, **16**, 1507–1513; (b) U. K. Chowdhury, R. A. Zakharyan, A. Hernandez, M. D. Avram, M. J. Kopplin and H. V. Aposhian, *Toxicol. Appl. Pharmacol.*, 2006, **216**, 446–457.
86. T. Hayakawa, Y. Kobayashi, X. Cui and S. Hirano, *Arch. Toxicol.*, 2005, **79**, 183–191.

87. D. J. Thomas, J. X. Li, S. B. Waters, W. B. Xing, B. M. Adair, Z. Drobna, V. Devesa and M. Styblo, *Exp. Biol. Med.*, 2007, **232**, 3–13.
88. H. V. Aposhian and M. M. Aposhian, *Chem. Res. Toxicol.*, 2006, **19**, 1–15.
89. H. V. Aposhian, E. S. Gurzau, X. C. Le, A. Gurzau, S. M. Healy, X. F. Lu, M. S. Ma, L. Yip, R. A. Zakharyan, R. M. Maiorino, R. C. Dart, M. G. Tircus, D. Gonzalez-Ramirez, D. L. Morgan, D. Avram and M. M. Aposhian, *Chem. Res. Toxicol.*, 2000, **13**, 693–697.
90. R. Raml, A. Rumpler, W. Goessler, M. Vahter, L. Li, T. Ochi and K. A. Francesconi, *Toxicol. Appl. Pharmacol.*, 2007, **222**, 374–380.
91. T. Schwerdtle, I. Walter, I. Mackiw and A. Hartwig, *Carcinogenesis*, 2003, **24**, 967–974.
92. A. Basu, J. Mahata, S. Gupta and A. K. Giri, *Mutat. Res.-Rev. Mutat. Res.*, 2001, **488**, 171–194.
93. D. J. Thomas, M. Styblo and S. Lin, *Toxicol. Appl. Pharmacol.*, 2001, **176**, 127–144.
94. M. F. Hughes, *Toxicol. Lett.*, 2002, **133**, 1–16.
95. M. Dustin and M. Piton, *Bull. Acad. Roy. Med. Belg.*, 1929, **9**, 26–37.
96. H. King and R. J. Ludford, *Bull. Environ. Contam. Toxicol.*, 1950, **8**, 2086–2088.
97. G. Endo, K. Kuroda, A. Okamoto and S. Horiguchi, *Bull. Environ. Contam. Toxicol.*, 1992, **48**, 131–137.
98. N. Eguchi, K. Kuroda and G. Endo, *Arch. Environ. Contam. Toxicol.*, 1997, **32**, 141–145.
99. K. Yamanaka, A. Hasegawa, R. Sawamura and S. Okada, *Biochem. Biophys. Res. Commun.*, 1989, **165**, 43–50.
100. M. Sordo, L. A. Herrera, P. Ostrosky-Wegman and E. Rojas, *Teratog. Carcinog. Mutagen.*, 2001, **21**, 249–260.
101. G. Endo, K. Kuroda, I. Kiyota and S. Horiguchi, *Mutat. Res.*, 1988, **203**, 370–371.
102. Y. Oya-Ohta, T. Kaise and T. Ochi, *Mutat. Res.*, 1996, **357**, 123–129.
103. L. E. Moore, A. H. Smith, C. Hopenhayn-Rich, M. L. Biggs, D. A. Kalman and M. T. Smith, *Cancer Epidemiol. Biomarkers Prev.*, 1997, **6**, 31–36.
104. Y. Noda, T. Suzuki, A. Kohara, A. Hasegawa, T. Yotsuyanagi, M. Hayashi, T. Sofuni, K. Yamanaka and S. Okada, *Mutat. Res.*, 2002, **513**, 205–212.
105. R. E. Rasmussen and D. B. Menzel, *Mutat. Res.*, 1997, **386**, 299–306.
106. W. R. Cullen, B. C. McBride, H. Manji, A. W. Pickett and J. Reglinski, *Appl. Oranomet. Chem.*, 1989, **3**, 71–78.
107. H. V. Aposhian, B. Zheng, M. M. Aposhian, X. C. Le, M. E. Cebrian, W. Cullen, R. A. Zakharyan, M. Ma, R. C. Dart, Z. Cheng, P. Andrewes, L. Yip, G. F. O'Malley, R. M. Maiorino, W. Van Voorhies, S. M. Healy and A. Titcomb, *Toxicol. Appl. Pharmacol.*, 2000, **165**, 74–83.
108. X. C. Le, M. Ma, W. R. Cullen, H. V. Aposhian, X. Lu and B. Zheng, *Environ. Health Perspect.*, 2000, **108**, 1015–1018.
109. M. Styblo, L. M. Del Razo, E. L. LeCluyse, G. A. Hamilton, C. Wang, W. R. Cullen and D. J. Thomas, *Chem. Res. Toxicol.*, 1999, **12**, 560–565.
110. J. S. Petrick, F. Ayala-Fierro, W. R. Cullen, D. E. Carter and H. V. Aposhian, *Toxicol. Appl. Pharmacol.*, 2000, **163**, 203–207.

111. M. Styblo, Z. Drobna, I. Jaspers, S. Lin and D. J. Thomas, *Environ. Health Persp.*, 2002, **110**, 767–771.
112. M. J. Mass, A. Tennant, B. C. Roop, W. R. Cullen, M. Styblo, D. J. Thomas and A. D. Kligerman, *Chem. Res. Toxicol.*, 2001, **14**, 355–361.
113. S. Nesnow, B. C. Roop, G. Lambert, M. Kadiiska, R. P. Mason, W. R. Cullen and M. J. Mass, *Chem. Res. Toxicol.*, 2002, **15**, 1627–1634.
114. A. D. Kligerman, C. L. Doerr, A. H. Tennant, K. Harrington-Brock, J. W. Allen, E. Winkfield, P. Poorman-Allen, B. Kundu, K. Funasaka, B. C. Roop, M. J. Mass and D. M. DeMarini, *Environ. Mol. Mutagen.*, 2003, **42**, 192–205.
115. R. Colognato, F. Coppede, J. Ponti, E. Sabbioni and L. Migliore, *Mutagenesis*, 2007, **22**, 255–261.
116. A. D. Kligerman, C. L. Doerr and A. H. Tennant, *Mol. Cell. Biochem.*, 2005, **279**, 113–121.
117. T. Ochi, S. Meguro, M. Namikoshi, Y. Oya-Ohta and T. Kaise, *Appl. Organomet. Chem.*, 2002, **16**, 432–436.
118. T. Ochi, F. Nakajima and N. Fukumori, *Arch. Toxicol.*, 1998, **72**, 566–573.
119. T. Ochi, F. Nakajima and M. Nasui, *Toxicology*, 1999, **136**, 79–88.
120. E. Kashiwada, K. Kuroda and G. Endo, *Mutat. Res.*, 1998, **413**, 33–38.
121. H. R. Hansen, R. Pickford, J. Thomas-Oates, M. Jaspars and J. Feldmann, *Angew. Chem. Int. Ed. Engl.*, 2004, **43**, 337–340.
122. R. Raml, W. Goessler and K. A. Francesconi, *J. Chromatogr. A*, 2006, **1128**, 164–170.
123. R. Raml, W. Goessler, P. Traar, T. Ochi and K. A. Francesconi, *Chem. Res. Toxicol.*, 2005, **18**, 1444–1450.
124. T. Ochi, K. Kita, T. Suzuki, A. Rumpler, W. Goessler and K. A. Francesconi, *Toxicol. Appl. Pharmacol.*, 2008, **228**, 59–67.
125. H. Naranmandura, K. Ibata and K. T. Suzuki, *Chem. Res. Toxicol.*, 2007, **20**, 1120–1125.
126. J. R. Cannon, J. B. Saunders and R. F. Toia, *Sci. Total Environ.*, 1983, **31**, 181–185.
127. T. Kaise, Y. Oya-Ohta, T. Ochi, T. Okubo, K. Hanaoka, K. J. Irgolic, T. Sakurai and C. Matsubara, *J. Food Hyg. Soc. Japan.*, 1996, **37**, 135–141.
128. P. Andrewes, D. M. Demarini, K. Funasaka, K. Wallace, V. W. Lai, H. Sun, W. R. Cullen and K. T. Kitchin, *Environ. Sci. Technol.*, 2004, **38**, 4140–4148.
129. E. Guillamet, A. Creus, J. Ponti, E. Sabbioni, S. Fortaner and R. Marcos, *Mutagenesis*, 2004, **19**, 129–135.
130. C. Soriano, A. Creus and R. Marcos, *Mutat. Res.*, 2007, **634**, 40–50.
131. K. Yamanaka, M. Hoshino, M. Okamoto, R. Sawamura, A. Hasegawa and S. Okada, *Biochem. Biophys. Res. Commun*, 1990, **168**, 58–64.
132. K. Kato, K. Yamanaka, A. Hasegawa and S. Okada, *Mutat. Res.-Genetic Toxicol. Environ. Mutagen.*, 2003, **539**, 55–63.
133. P. Andrewes, K. T. Kitchin and K. Wallace, *Chem. Res. Toxicol.*, 2003, **16**, 994–1003.
134. T. Schwerdtle, I. Walter and A. Hartwig, *DNA Repair (Amst)*, 2003, **2**, 1449–1463.

135. K. Piatek, T. Schwerdtle, A. Hartwig and W. Bal, *Chem. Res. Toxicol.*, 2008, **21**, 600–606.
136. I. Walter, T. Schwerdtle, C. Thuy, J. L. Parsons, G. L. Dianov and A. Hartwig, *DNA Repair (Amst)*, 2007, **6**, 61–70.
137. S. Shen, J. Lee, M. Weinfeld and X. C. Le, *Mol. Carcinog.*, 2008, **47**, 508–518.
138. S. B. Baylin and J. G. Herman, *Trends in Genetics*, 2000, **16**, 168–174.
139. C. Q. Zhao, M. R. Young, B. A. Diwan, T. P. Coogan and M. P. Waalkes, *Proc. Natl. Acad. Sci. USA*, 1997, **94**, 10907–10912.
140. J. F. Reichard, M. Schnekenburger and A. Puga, *Biochem. Biophys. Res. Commun.*, 2007, **352**, 188–192.
141. H. Chen, S. F. Li, J. Liu, B. A. Diwan, J. C. Barrett and M. P. Waalkes, *Carcinogenesis*, 2004, **25**, 1779–1786.
142. T. G. Rossman, *Mutat. Res.*, 2003, **533**, 37–65.
143. J. Liu, L. Benbrahim-Tallaa, X. Qian, L. M. Yu, Y. X. Xie, J. Boos, W. Qu and M. P. Waalkes, *Toxicol. Appl. Pharmacol.*, 2006, **216**, 407–415.
144. M. J. Mass and L. J. Wang, *Mutat. Res.-Rev. Mutat. Res.*, 1997, **386**, 263–277.
145. W. Qu, C. D. Bortner, T. Sakurai, M. J. Hobson and M. P. Waalkes, *Carcinogenesis*, 2002, **23**, 151–159.
146. M. P. Waalkes, J. M. Ward and B. A. Diwan, *Carcinogenesis*, 2004, **25**, 133–141.
147. T. Ochi, F. Nakajima, T. Sakurai, T. Kaise and Y. OyaOhta, *Arch. Toxicol.*, 1996, **70**, 815–821.
148. S. M. Cohen, L. L. Arnold, E. Uzvolgyi, M. Cano, M. S. John, S. Yamamoto, X. F. Lu and X. C. Le, *Chem. Res. Toxicol.*, 2002, **15**, 1150–1157.
149. S. Orrenius, B. Zhivotovsky and P. Nicotera, *Nature Rev. Mol. Cell Biol.*, 2003, **4**, 552–565.
150. A. M. Florea, E. N. Yamoah and E. Dopp, *Environ. Health Persp.*, 2005, **113**, 659–664.
151. U. Gafter, T. Malachi, Y. Ori and H. Breitbart, *J. Lab. Clin. Med.*, 1997, **130**, 33–41.
152. D. V. Bugreev and A. V. Mazin, *Proc. Natl. Acad. Sci. USA*, 2004, **101**, 9988–9993.
153. J. Liu and M. P. Waalkes, *Toxicol. Sci.*, 2008, **105**, 24–32.
154. H. P. Tran, A. S. Prakash, R. Barnard, B. Chiswell and J. C. Ng, *Toxicol. Lett.*, 2002, **133**, 59–67.
155. R. B. Dickson and G. M. Stancel, *J. Natl. Cancer Inst. Monogr.*, 2000, **27**, 135–145.
156. M. P. Waalkes, J. Liu, J. M. Ward and B. A. Diwan, *Toxicol. Appl. Pharmacol.*, 2006, **215**, 295–305.
157. International Agency for Research on Cancer, *Some Drinking Water Disinfectants and Contaminants, including Arsenic*, IARC Monographs, Vol. 84, IARC Press, Lyon, 2004.
158. D. N. G. Mazumder, *Toxicol. Appl. Pharmacol.*, 2005, **206**, 169–175.
159. J. Liu, M. B. Kadiiska, Y. Liu, T. Lu, W. Qu and M. P. Waalkes, *Toxicol. Sci.*, 2001, **61**, 314–320.

160. J. Liu, Y. X. Xie, D. M. K. Ducharme, J. Shen, B. A. Diwan, B. A. Merrick, S. F. Grissom, C. J. Tucker, R. S. Paules, R. Tennant and M. P. Waalkes, *Environ. Health Persp.*, 2006, **114**, 404–411.
161. Y. Xie, K. Trouba, J. Liu, M. Waalkes and D. Germolec, *Environ. Health Persp.*, 2004, **112**, 1255–1263.
162. A. Kinoshita, H. Wanibuchi, M. Wei, T. Yunokl and S. Fukushima, *Toxicol. Appl. Pharmacol.*, 2007, **221**, 295–305.
163. A. Kinoshita, H. Wanibuchi, K. Morimura, M. Wei, D. Nakae, T. Arai, O. Minowa, T. Noda, S. Nishimura and S. Fukushima, *Cancer Sci.*, 2007, **98**, 803–814.
164. Y. Kumagai and D. Sumi, *Ann. Rev. Pharmacol. Toxicol.*, 2007, **47**, 243–262.
165. B. D. Laird, T. R. Van de Wiele, M. C. Corriveau, H. E. Jamieson, M. B. Parsons, W. Verstraete and S. D. Siciliano, *Environ. Sci. Technol.*, 2007, **41**, 5542–5547.
166. W. R. Cullen, B. C. McBride and J. Reglinski, *J. Inorg. Biochem.*, 1984, **21**, 179–194.
167. M. F. Hughes and K. T. Kitchin, *Arsenic, Oxidative Stress and Carcinogenesis*, in *Oxidative Stress, Disease and Cancer*, Ed. K. K. Singh, Imperial College Press, London, 2006, pp. 825–850.
168. T. Schwerdtle and A. Hartwig, *Trendbericht Lebensmittelchemie 2008*, in *Nachrichten aus der Chemie*, **57**, 312–316 (2009).

Met. Ions Life Sci. **2010**, *7*, 267–301

8

Alkyl Derivatives of Antimony in the Environment

Montserrat Filella

Institute F.-A. Forel, University of Geneva, Route de Suisse 10, CH-1290 Versoix, Switzerland

< montserrat.filella@unige.ch >

Metal Ions in Life Sciences, Volume 7 Edited by Astrid Sigel, Helmut Sigel, and Roland K. O. Sigel

Published by the Royal Society of Chemistry, www.rsc.org DOI: 10.1039/9781849730822-00267

ABSTRACT: The presence of methylated antimony species has been reported in surface waters, sediments, soils, and biota, mainly detected using hydride generation techniques. Compared to other elements, relatively few studies have been published. Monomethyl-, dimethyl-, and trimethylantimony species have been found, always at very low concentrations. It is important to point out that (i) it has been proved that the identity of some of the published species might be uncertain due to possible artefacts during the analytical process; (ii) existing analytical methods do not reveal the oxidation state of the antimony in the detected species. Volatile methylated species have also been detected in landfill and sewage fermentation gases. Laboratory culture experiments have indicated that biomethylation can result from bacterial, yeast, and fungal activity, in both aerobic and anaerobic conditions. Antimony is methylated much less rapidly and less extensively than arsenic and it has been suggested that antimony biomethylation could be a fortuitous rather than a detoxification process.

KEYWORDS: antimony · biomethylation · dimethylantimony · monomethylantimony · speciation · trimethylantimony

1. INTRODUCTION

Antimony is a naturally occurring element of current industrial significance, especially through its role in fire retardants. It belongs to group 15 of the periodic table. Antimony can exist in a variety of oxidation states (–III, 0, III, V). However, in environmental and biological media it is mainly found in oxidation states III and V. It has no known biological role and has largely been overlooked as an element of environmental concern. General aspects of antimony behavior in the environment, its solution chemistry, and the role of biota have been thoroughly reviewed [1–3]. In addition, a critical overview of the current state of the research of antimony has very recently been published [4].

Until the mid 1990's, there was little evidence for the existence of organoantimony species in environmental media. Initial studies were fuelled by the experience gained by studying arsenic and an interest in finding antimony analogues of organoarsenic compounds in the environment. In the 90s the suggestion that there might be a link between sudden infant death syndrome (SIDS) and volatile toxic hydrides of group 15 elements in cot mattress foam [5,6] triggered a strong interest in methylated antimony compounds. But despite this, there are still far fewer studies on organoantimony species in the environment compared to those on arsenic and other elements of environmental concern. The field is characterized by the limited number of research groups active in it.

Organometallic species may be found in the natural environment either because they have been formed there or because they have been introduced as a result of human use. In the case of antimony, although some applications of alkyl compounds have been described, no important uses are known to exist. It is therefore safe to assume that organoantimony species detected in environmental systems have been formed within those systems, most probably by biomethylation.

A wide variety of compounds containing the Sb-C bond is known and there is a vast body of literature of interest to synthetic and mechanistic organometallic chemists. However, only methylated antimony compounds are of relevance in the environment and they will be the only ones discussed here. In this chapter, the terms monomethylantimony (MMA), dimethylantimony (DMA), and trimetylantimony (TMA) will be used to refer to any antimony compound containing one, two or three methyl groups, respectively. However, these names imply nothing about the oxidation state of antimony in the compound or the number and type of inorganic substituents.

The data available has been presented in tabular form rather than in running text. An effort has been made to collate the relevant information in a consistent format, which is easy to read and compare. General issues such as the main gaps in knowledge and methodological problems are discussed in the text. Given that Chapter 2 of this book is devoted to analytical aspects, no analytical section has been included. However, analytical methods are detailed in the tables and analytical aspects are discussed in the corresponding sections where relevant.

2. PHYSICAL AND CHEMICAL CHARACTERISTICS OF METHYLANTIMONY COMPOUNDS

Good knowledge of the characteristics and, in particular, of the stability and reactivity of methylantimony compounds is a prerequisite for anyone interested in studying antimony biomethylation in environmental systems, but a detailed review of the literature on the synthesis, reactivity and physical and chemical properties of these compounds largely exceeds the scope of this chapter. Nonetheless, a brief overview of the main characteristics of methylated antimony compounds similar to the species that might exist in natural systems, or that have been used to study them, can be found in Table 1 [7–39]. Further information can easily be found in a number of publications ([40–42] and Gmelin database). Unfortunately, many aspects, particularly those regarding speciation and behavior in solution and in diluted conditions, remain insufficiently studied.

Organoantimony compounds can be broadly divided into Sb(III) and Sb(V) compounds. The former may contain from one to four organic groups, while the latter contain from one to six. In general, Sb(V)

Table 1. Main properties of methylantimony compounds.

Compound, CAS number	Formula	Synthesis references	State	Melting point (°C)
Pentavalent				
Methylstibonic acid[a] 78887-52-2	$CH_3SbO(OH)_2$	[7]	white X-ray amorphous solid [7]	
Dimethylstibinic acid[a] 35952-95-5	$(CH_3)_2SbO(OH)$	[8–10]	colorless solid [10]	does not melt [10]
Dimethylantimony trichloride 7289-79-4	$(CH_3)_2SbCl_3$	[8,11]	white crystalline solid [8]	105–110° with gas production [8] decomposition: 106–110° [11]
Dimethylantimony tribromide 149442-29-5	$(CH_3)_2SbBr_3$	[8,13][a]	yellowish-white crystalline solid [8]	
Trimethylantimony oxide 19727-40-3	$(CH_3)_3SbO$	[14–16]	hygroscopic crystalline solid	95° [17]
Trimethylantimony dihydroxide 19727-41-1	$(CH_3)_3Sb(OH)_2$	[14,18,19]	slightly hygroscopic colorless crystalline solid [18]	98–100° incongruent melting [16]
Trimethylantimony dichloride 13059-67-1	$(CH_3)_3SbCl_2$	[18,22–24] CA[c]	colorless crystalline solid [18]	[d]
Trimethylantimony dibromide 5835-64-3	$(CH_3)_3SbBr_2$	[23,26] CA[c]		[d]
Trivalent				
Monomethylstibine 23362-09-6	CH_3SbH_2	[30–32]	colorless liquid [30]	
Monomethylstibine dichloride 42496-23-1	CH_3SbCl_2	[8,11]	oil [8], transparent, highly refractive liquid [11]	
Monomethylstibine dibromide 54533-06-9	CH_3SbBr_2	[8]	greyish-white needles [8]	42° [8]
Dimethylstibine 23362-10-9	$(CH_3)_2SbH$	[30,31]	colorless liquid [30]	
Dimethylstibine chloride 18380-68-2	$(CH_3)_2SbCl$	[8]	colorless oil [8]	
Dimethylstibine bromide 53234-94-9	$(CH_3)_2SbBr$	[8]	yellow oil, solidifies slowly [8]	40/89° [8]
Trimethylstibine 594-10-5	$(CH_3)_3Sb$	[33–35] CA[c]		−87.6° [36], −62.0° [37]
[f]	$(CH_3)_3Sb^+CH_2COO^-$	[39]	white crystalline solid [39]	

[a]Stibonic and stibinic acids are very weak acids and IUPAC classifies them as oxide hydroxides rather than as acids and names them accordingly.
[b]The compound prepared is $(CH_3)PR^+Me_2SbBr_2^-$ with $R = C_6H_5$ or n-$CH_3(CH_2)_3$.
[c]CA = commercially available.
[d]Although some melting points have been published, according to [23] they are not reliable because these substances lose methyl halide upon heating.
[e]The author titrates $(CH_3)_3SbBr_2$ but makes the hypothesis that this compound hydrolyzes to $(CH_3)_3SbO$ to which the pK corresponds.
[f]Antimony analogue of arsenobetaine.

Boiling point (°C)	Stability	Water solubility	Solution
		soluble only when freshly synthesized [7]	
	high thermal stability [10]		
	unstable at room T [8]	soluble	monomeric [12]
	very unstable at room T [8]		
		soluble	
	stable [18]	soluble	$pK = 9.14$ [20] $Me_3Sb(OH)^+$, main species in aqueous solution [21]
	stable at room T, decomposes only at 150–200 °C [25]	soluble	extensive hydrolysis [20,26] $Me_3Sb(OH)^+$, main species in aqueous solution [21,27]
	stable at room T, decomposes at 50 °C [28]	soluble	extensive hydrolysis [20,26,29] $pK = 5.64^e$ (20 °C) [29]
41° [30]	stable at −78 °C, decomposes slowly above [30]		
115–120° (60 Torr) [8]	decomposes in water [8]		
	not inflammable, not oxidized in air; decomposes in water [8]		
60.7° [30]	stable at −78 °C, decomposes slowly above [30]		
155–160° [8]	oxidizable; spontaneously inflammable at 40 °C [8]		
	extremely oxidizable in air; spontaneously inflammable at 50 °C [8]		
79.4° [34], 80.6° [37]	readily oxidized, spontaneously inflammable [8], may explode [38]		

compounds are solids while Sb(III) compounds are rather unstable, readily oxidizable, volatile liquids.

Monomethyl Sb(V) compounds have proved to be very difficult to synthesize and remain largely unstudied. For instance, the synthesis and isolation of methylstibonic acid (MSA), the only alkylstibonic acid known with certainty, was not reported until 1990 [7], while dimethylstibinic acid (DMSA) had already been synthesized in 1926 [8]. Previous attempts to synthesize MSA had either failed or been inconclusive. Monomethyl Sb(V) standards have not been used in environment-related studies except by the authors who detected for the first time the presence of organoantimony species in an environmental compartment [43]. The purity of this MSA standard has been the subject of some controversy ever since (Section 3.1).

Trimethyl Sb(V) compounds are more soluble than monomethyl and dimethyl compounds, which seem to readily polymerize in solution. Trimethyl dihalides, the best known Sb(V) methylated compounds, are extensively hydrolyzed and the resulting compounds, probably trimethylantimony oxide or dihydroxide, act as weak bases. Trimethyl dihalides are readily reduced to the corresponding stibines. For this reason, trimethylantimony dichloride (TMC) has been extensively used to generate stibines in analytical methods (Section 3).

Trialkylstibines are powerful reducing agents; they are all readily oxidized and the lower members are spontaneously inflammable in air. Although fast oxidation of trimethylstibine (TMS) has been proposed [44,45], its oxidation at low concentrations is probably much slower, as confirmed by the fact that it is possible to find TMS in landfill gas samples collected some days earlier [46]. According to Craig and coworkers [47], the oxidation of TMS in air, at environmentally relevant concentrations, produces a complex series of products (trimethylstibine oxide and a range of cyclic and linear oligomers), but does not lead to any significant antimony-carbon bond cleavage, as had been suggested by Parris and Brinckman [45].

3. OCCURRENCE IN THE ENVIRONMENT

3.1. Waters

The first organoantimony compounds to be detected in the environment were found in natural waters over 25 years ago (Table 2) [43,48–56]. Stibine, MMS and DMS were detected in natural waters using AAS after derivatization of the samples with borohydride by Andreae and coworkers [43,48,50], who claimed that the waters contained MSA and DMSA on the basis of the derivatization response of these two standard compounds. However, it is now known that (i) the experimental acidic conditions used are likely to produce artefacts, namely methyl group redistribution during the hydride generation (HG) process [19,57]; (ii) the reference compounds used contained impurities and doubt has been cast on the identity itself of one of the compounds (MSA) [19]. More important, even in the absence of these problems, the HG method does not make it possible to establish either the antimony oxidation state or the inorganic or organic counterparts in the methyl species. Therefore, there is no doubt that methylantimony species were present in the samples analyzed by Andreae and coworkers [43,48,50], but their identity is open to discussion. Similar considerations apply to the results obtained by Bertine and Lee in applying the same approach to the seawater and sediment porewaters of Saanich Inlet [49] and by Cutter [51] in the Black Sea. In a later study, Cutter and coworkers [53] acknowledged that the technique used was incapable of identifying the species exactly and reported that MMA rather than MSA was present. In this study, relatively constant concentrations were found over a transect of 11,000 km in the Atlantic Ocean, implying either uniform production or long subsurface-water residence time to allow mixing. In a more recent study in the North Pacific Ocean, Cutter and Cutter [56] measured one profile where MMA displayed conservative behavior throughout the entire water column. According to the authors, this behavior, observable thanks to the correction of a previously unknown nitrite/nitrate interference and never reported before, "radically change[s] the known biogeochemical cycle of antimony". However, reporting vertical profiles of antimony methylated species was not really new; they had already been measured in the past [43,48–50].

Ellwood and Maher [54] found MMA, DMA, and TMA along three surface transects in the Chatham Rise region east of New Zealand. The flow injection HG conditions used did not fully prevent TMA demethylation but the extent of the problem was measured using trimethylantimony bromide and dimethylantimony chloride standards and was found not to be severe (86% TMA recovered). This study was the first to report the presence of TMA species in marine samples. These authors postulated that the batch HG conditions used in previous studies, where demethylation had not been tested, might have degraded any TMA present. This might well have been the case but it should also be noted that in all previous studies MMA and DMA standards had been used, while TMA had not.

DMA and TMA were the species found in mine effluent runoff (Yellowknife, BC, Canada) [52]. It should be noted that no methylated antimony species were detected in any other water sample in this system, even when high concentrations of antimony were present. In this study, HG was performed without the addition of acid or buffers to minimize the abovementioned artefact problem. The identity of the methylated species was

Table 2. Reported methylantimony species in natural waters.

System	Detected Sb species	Concentration/ nmol Sb L^{-1}	Sampling and conservation
US and German rivers	MSA, DMSA	MSA: ND-0.019 DMSA: ND	Filtration not mentioned
Ochlockonee Bay estuary		MSA: 0.007–0.103 DMSA: ND-0.012	Storage dark, room T, 4 d
Gulf of Mexico, Apalachee Bay		MSA: 0.044, 0.070 DMSA: 0.026, ND	
Saanich Inlet, Canada water column sediment pore waters	MSA	 0.02–0.03 up to 4.9 in the methane zone	Not mentioned
Baltic Sea (5 profiles)	MSA	Profile 1: 0.006–0.082 Profile 2: 0.008–0.066 Profile 3: 0.013–0.034 Profile 4: <0.005–0.09 Profile 5: <0.005–0.07	Not mentioned
Black Sea (profiles 0–2200 m depth)	MSA	ND-0.06	0.4 μm fitration Acidification to pH <2 (HCl)
Mine effluent runoff (standing water), Yellowknife, BC, Canada	DMA TMA	0.335 ± 0.007 (n = 2) 0.13 ± 0.05 (n = 2)	Not mentioned
Western Atlantic Ocean (a 11,000 km surface transect and 6 profiles)	MMA	Transect: 0.13 ± 0.07	0.4 μm fitration Acidification to pH 1.6 (HCl), analysis on board
Chatham Rise, New Zealand (3 surface transects)	MMA DMA TMA	0.06–0.07 0.015–0.025 0.005–0.015	0.2 or 0.4 μm fitration Acidification to pH <2, storage 4 °C
North Pacific Ocean (a 15,000 km surface transect and 9 profiles)	MMA	Profile: 0.037 ± 0.006	0.2 or 0.4 μm fitration Refrigeration until analysis on board

[a]These authors reported values of methylated species for a few natural water samples when developing an analytical method [55]. These values have not been included in this table.

Analytical method	Comments	Ref.
HG-CT-GC-AAS pH HG: 30 mM HCl Standards: MSA, DMSA	Present throughout the water column Methylated Sb = 10% total Sb Probable source: biological, and in particular, algal activity	[43,48]
HG-AAS pH HG: not given Relationships in [48] applied	Present throughout the water column Below 145 cm MSA becomes the second Sb species in pore waters	[49]
HG-CT-GC-AAS pH HG: probably as in [43] Standards: MSA, DMSA	Present throughout the water column Methylated Sb = 10% total Sb Probable source: bacterial production; no methylated compounds detected in algae	[50]
HG-CT-GC-AAS pH HG: as in [48] Standards used?	Detection only in the upper 65 m	[51]
HG-CT-GC-AAS pH HG: no acid added Semiquantitative calibration: inter-element based, internal liquid standard Species confirmation by HG-GC-MS (stibines formed by HG of TMC)	Methylated Sb found only in one of the 6 water samples analyzed	[52]
HG-CT-GC/PID pH HG: as in [48] Standards used?	Only detected in surface waters; relatively constant in transect Methylated Sb = 10% total Sb	[53]
HG-CT-ICP-MS pH HG: 0.06 M HCl Standards: TMB, TMC Demethylation checked	Methylated Sb = 8% total Sb No methylated species below 25 m Probable source: phytoplankton, bacteria or fungi	[54][a]
HG-CT-GC/PID pH HG: 0.5 M HCl Standards used? Sulfanilamide added to remove a nitrite/nitrate interference	MMA behaves conservatively throughtout the water column (in one profile)	[56]

confirmed by HG-GC-MS using a mixture of stibine species (MMS, DMS, TMS) formed by HG from a TMC standard. This method takes advantage of the above-mentioned enhanced demethylation of TMS when HG is performed at acidic pH values. It has been used extensively, particularly in laboratory incubation studies.

3.2. Soils and Sediments

Very few studies report the presence of methylantimony compounds in soils and sediments (Table 3) [58–65]. The bulk of these have been carried out in heavily polluted systems. MMA, DMA, and TMA species were detected in studies where HG was applied, while IC-based studies found traces of a substance that had the same retention time as trimethylantimony oxide. Methylantimony species were extracted from the samples using different extractants when determined by IC-ICP-MS or FI-HG-ICP-MS and were directly volatilized from the soils and sediments in the other studies, either by direct derivatization of samples with borohydride in acidic solution [58,60] or by derivatization according to a pH-gradient [64,65]. This method optimizes simultaneous volatilization conditions of different elements in one run and minimizes artefacts [62]. Demethylation was not tested for in any of the HG studies, even though acidic pH conditions are known to favor it [57]. Results should therefore be considered with some caution because the formation of artefact species cannot be completely excluded. Moreover, reported values are only semi-quantitative because quantification was performed by using interelement calibration.

3.3. Biota

Results from the few studies where methylantimony species have been detected in biota are shown in Table 4 [52,66–70]. The analytical methods used in all of the studies except one were based on HG. The specimens examined always came from systems which had been heavily impacted by mining.

When measuring speciation in plants, organometallic species need to be extracted beforehand. The choice of the ideal extractant, i.e., the one that gives high yields while preserving speciation, remains a critical issue in this type of measurements. Three studies opted for acetic acid extraction [66–68], while a water-methanol mixture [52] and citric acid [69,70] were used in two others. Acetic acid extracts from pondweed contained TMA on its own in one lake, or along with DMA and MMA species in a second one [67], while the same type of extracts from plants sampled close to an old antimony mine contained DMA species only [68]. The authors of both studies rigorously checked that no molecular rearrangement occurred during the HG process. DMA was also the only species detected by HG-GC-ICP-MS in a moss from a zone affected by gold mining activities [52]. In a more recent study, a different analytical method was applied, IC-UV-HG-AFS, and only the presence of TMA was reported [69,70] but in concentrations much higher than any methylated species in previous works. Unfortunately, the diversity of extraction procedures applied and plants studied, as well as the low number of existing studies, precludes any possibility of extracting general conclusions about antimony biomethylation in plants.

Methylantimony species were found for the first-and so far only-time in an animal, the snail *Stagnicola* sp. from Yellowknife, Canada [52].

For years, it has generally been accepted that, as established by Bailly and coworkers [71], inorganic antimony is not methylated *in vivo* in rats and in human beings. However, Krachler and Emons [72] reported the detection of TMA by HPLC-HG-ICP-MS in urine samples from persons occupationally exposed to antimony. The presence of trace amounts of MMA, DMA, and TMA in human urine was also reported in a study on the presence of metalloid species after fish consumption [73] but the values found were extremely low (less than 10 ng Sb L^{-1}) compared with inorganic antimony (up to 2000 ng Sb L^{-1}) or even methylated arsenic species (up to 1940 ng As L^{-1} for only 240 ng As L^{-1} as inorganic arsenic). The presence of methylated antimony in human urine needs further investigations to be confirmed.

3.4. Gases from Landfills and Water Treatment Plants

The presence of antimony oxide deposits in biogas burners indicates the formation of volatile antimony species in fermentation gases from landfills and water treatment plants [74]. Direct evidence for volatile antimony species in such systems was obtained in a series of studies in Germany (Table 5) [75–79] where TMS was detected in landfill and sewage gas by using LTGC coupled with ICP-MS detection. Confirmation of the identity of the species, initially identified by measuring their retention times, was obtained by using GC-MS to analyze sewage gas from Canadian sites [78] and comparing sample mass spectra with the ones of TMS generated by HG of TMC.

Condensed water samples, obtained from the outlet of the landfill gas collection pipeline, were found to contain MMA and DMA, and possibly TMA and triethylantimony species, by using HG-GC-ICP-MS [75]. Methylated antimony species were reported in the standing water on a landfill site by applying the same technique [57]. The presence of a methylantimony species in liquid phases from fouling and sewage sludges was

Table 3. Reported methylantimony species in soils and sediments.

System	Detected Sb species	Concentration/ μg Sb kg^{-1} dry weight
40 river sediment samples of different locations, Germany	MMA DMA TMA	0.2–9.8 0.1–1.2 0.1–0.9
Strongly polluted by industrial waste soils, Bitterfeld, Germany	Traces of a substance that has the same rt as trimethylantimony oxide	
13 contaminated soils (shredder, domestic waste, gas station, industrial site, coal mining/processing), Germany	MMA DMA TMA	0.070–0.430 0.006–0.350 0.010–0.560
Strongly polluted by industrial waste soils, Bitterfeld, Germany	Traces of a substance that behaves like trimethylantimony oxide	
Urban soils (arable, gardening, abandoned industrial, flood plain), Ruhr basin, Germany	MMA DMA TMA	<DL-56 <DL-7.6 <DL-0.28 (DL = 0.007)
Six sediments	MMA, DMA, TMA	
>2000		2.92, 0.33, 0.02
2000–630		2.62, 0.35, 0.01
630–180		1.53, 0.14, 0.01
180–63		4.86, 0.13, 0.01
63–20		6.72, 0.24, 0.01
<20		12.0, 0.40, 0.06

Analytical method	Comments	Reference
HG-LTGC-ICP-MS pH HG: 2 Identification: bp-rt correlation Semiquantitative calibration	Possible presence of a triethylantimony	[58]
Methanol:water and acetic acid extractions IC-ICP-MS		[59]
HG-LTGC-ICP-MS pH HG: 2 Identification: bp-rt correlation Semiquantitative calibration	All < DL in shredder; MMA detected in 11 samples, DMA in 10 and TMA in 5	[60]
Water extraction FI-HG-ICP-AES with fluoride as a modifier		[61]
Sieving (2 mm) HG-PT-GC-ICP-MS pH HG: pH gradient [62] Species confirmation: HG-GC-EI-MS/ICP-MS [63] Semiquantitative calibration	Highest concentrations in agricultural and garden soils	[64]
Sieving (2 mm) + cryomilling HG-PT-GC-ICP-MS pH HG: pH gradient [62] Semiquantitative calibration	Only mean values quoted here Concentrations increase when particle size decreases	[65]

Table 4. Reported methylantimony species in biota.

System	Detected Sb species	Concentration/ µg Sb kg^{-1} dry weight	Extraction
Marine algae from San Diego Bay, CA, US: *Ulva* sp., *Enteromorpha* sp., *Sargassum* sp.	No methylantimony detected	–	Acetic acid
Pondweed (*Potamogetan pectinatus*) from two Canadian lakes:		Not reported	Acetic acid
Kam Lake	TMA		
Keg Lake	MMA, DMA, TMA		
Biota close to an old Sb mine, Louisa, Scotland, UK:	DMA		Acetic acid
Plant (liverwort)		181 (RSD: 26, n = 4)	
Moss		101 (RSD: 15, n = 4)	
Biota from Yellowknife, Canada:			Methanol: water (1:1)
Drepanocladus sp. (moss)	DMA		
June		46	
August		44	
August (standing water location)		170 ± 10 (n = 2)	
Stagnicola sp. (snail)	DMA	5	
	TMA	24	
Biota close to an old Sb mine, Pyrenees, Catalonia, Spain:	TMA		Citric acid
Hydnum cupressiforme (moss)		2870 ± 320 (n = 3)	
Dryopteris filix-max (fern) (2 samples)		<DL, 890 ± 50 (n = 3)	
Stellaria halostea		300 ± 50 (n = 3)	
Chaenorhinum asarina (figwort)		2270 ± 140 (n = 3)	

Analytical method	Comments	Reference
HG-AAS Method as in [49] but no standard apparently used		[66]
HG-CT-GC-MS HG pH: 1 mol L^{-1} HCl added Molecular rearrangements checked (standards: $(CH_3)_3Sb(OH)_2$, TMC)	DMA is the main species in Keg Lake	[67]
HG-CT-AAS HG pH: 2.4 (HCl) Molecular rearrangements checked (standards: TMB, TMC)	Proportion of organoantimony: 0.3–0.5% Moss not affected by Sb mining: 0 DMA	[68]
HG-CT-GC-AAS pH HG: no acid added Semiquantitative calibration: inter-element based, internal liquid standard Species confirmation by HG-GC-MS (stibines formed by HG of TMC)	7 other plant species and 3 species of lichen were tested and no methylated Sb found, neither in *Minulus* sp. from Meager Creek (hydrothermal zone), Canada	[52]
IC-UV-HG-AFS Standard additions of TMC		[69,70]

Table 5. Reported methylantimony species in gases from landfills, sewage treatment plants and hydrothermal systems.

System	Detected Sb species	Concentration/ $\mu g\,Sb\,m^{-3}$
Landfill gas (domestic waste deposit, Aβlar, Hessen, Germany)	TMS	23.9–71.6 (n = 8)
Landfill gas (two municipal waste deposits, Germany)	Volatile Sb compounds	0.040–2.4 (n = 8)
Sewage gas at 56 °C and 35 °C (municipal sewage treatment plant, Germany)	TMS	0.618–14.72
Landfill gas from municipal waste deposits and gas from a mesophilic sewage sludge digester (Vancouver, Canada)	TMS	Landfill: 0.00408–0.0171 Digester: similar to [77]
Geothermal springs (Meager Creek, BC, Canada)	TMS	Not reported

Analytical method	Comments	Reference
Sampling: cryogenic trapping (–80 °C) LTGC-ICP-MS Identification: bp-rt correlation Semiquantitative calibration: inter-element based, internal liquid standard	Concentrations are for total volatile Sb	[75]
Sampling: cryogenic trapping (–80 °C) Desorption into the Ar plasma of the ICP-MS Semiquantitative calibration: same approach as in [75]	Concentrations are for total volatile Sb	[76]
Sampling: cryogenic trapping (–80 °C) LTGC-ICP-MS Identification: comparison with rt of Sb standards Semiquantitative calibration: same approach as in [75]	Concentrations are for total volatile Sb	[77]
Sampling: Tedlar bags CT-LTGC-ICP-MS Identification: matching rt, isotopic fingerprints with Sb standard (TMS formed by HG of TMC) Confirmation: CGC-EI-MS-MS (same standard) Calibration: not described		[78]
Sampling: Tedlar bags LTGC-ICP-MS	TMS detected above and within algal mats	[79]

detected by CE-ICP-MS [80]; a small peak in the electropherogram had the same retention time as a standard of TMC.

3.5. Hydrothermal Systems

Since hydrothermal systems are well-known for being rich in bacteria and metals, they are particularly interesting to explore for the presence of methylated species. MMA, DMA, and TMA species were detected by HG-GC-ICP-MS in geothermal waters from various New Zealand locations [79]. However, the reliability of these results is subject to the limitations concerning the possibility of demethylation described earlier. Traces of TMA were measured by HG-GC-AAS in one hot spring in Meager Creek, BC, Canada [52] but were not analyzed in any of the other six water samples.

4. MICROBIAL TRANSFORMATIONS OF ANTIMONY COMPOUNDS

4.1. Laboratory Experiments

A wide variety of organisms have been shown to be capable of antimony methylation. These are: a few aerobic filamentous fungi (*Scopulariopsis brevicaulis* and *Phaeolus schweinitzii*), some strictly anaerobic prokaryotes (anaerobic bacteria: *Clostridium collagenovorans*, *Desulfovibrio vulgaris*, and methanogenic archaea: *Methanobacterium formicicum, Methanobacterium thermoautrophicum, Methanosarcina barkeri*), one strictly aerobic bacterium (*Flavobacterium* sp.), and one aerobic yeast (*Cryptococcus humicolus*). Undefined mixed cultures of bacteria growing under anaerobic conditions have also shown antimony methylation activity. Thus, both aerobic and anaerobic organisms, including aerobic prokaryotes, seem to be capable of methylating antimony. Published results are summarized in Table 6 [28,81–105].

Antimony(III) compounds have been used as substrates in most of the published studies. The most commonly used of these is potassium antimony tartrate (PAT). Antimony(III) trioxide (ATO) has been used occasionally, but always in addition to PAT. The preference for PAT is most probably due to the higher solubility of this compound. ATO has sometimes been added as a saturated suspension, which makes the calculation of available Sb(III) uncertain. Potassium hexahydroxyantimonate (PHA) has been used as an Sb(V) source, except in a couple of cases where APO was added. No attention seems to have been paid to the consequences of the choice of the initial Sb(III) compound. It is well known that, although it is true that

Sb(III) is more soluble when added as tartrate, it remains largely complexed by this ligand in solution. In consequence, the speciation of Sb(III) in such solutions is radically different from the speciation of 'pure' Sb(III) solutions, even in the case of equal total Sb(III) concentrations. The implications of this fact on the bioavailability of Sb(III) (i.e., lower 'free' Sb(III) concentrations but higher concentrations of a complex of unknown bioavailability) have been systematically ignored in all studies. On the other hand, although it is well known that the ligands present in the culture media can deeply change the speciation and bioavailability of any element, this fact has never been taken into account in any of the published studies and no attempt has been made to estimate 'true' antimony speciation in the culture media. Finally, it should be mentioned that the redox status of antimony in the cultures, in the absence of microorganisms, usually has not been checked. However, it is extremely probable that antimony, initially present in a culture as Sb(III), oxidizes after several days or weeks in aerobic conditions, which often comprise continuous aeration of the culture media.

When Sb(III) and Sb(V) substrates are compared, Sb(III) seems to be preferentially methylated, at least by some organisms. For instance, Sb(V) has been reported either not to be methylated at all [85] or less efficiently than Sb(III) [87,88] by *Scopulariopsis brevicaulis*. *Phaeolus schweintzii* also was less efficient at biomethylating Sb(V) [97]. However, Sb(V) was biomethylated by *Cryptococcus humicolus* [100,101] and by soil and sewage sludge bacteria [28,102].

Production of both volatile and involatile methylated antimony compounds has been reported. Initial studies, which focused largely on *Scopulariopsis brevicaulis*, showed the formation of only one volatile species, TMS. This compound was also found to be formed by undefined mixed cultures of bacteria growing under anaerobic conditions [28,84,86] and to be the transformation product of trimethylantimony dibromide (TMB) by the aerobic bacteria *Pseudomonas fluorescens* [28]. Formation of stibine in culture headspace gases has been reported together with MMS, DMS, and TMS for *Methanobacterium formicicum* [96], and with DMS and TMS for *Cryptococcus humicolus* [100] and for anaerobic cultures of alluvial soil samples [103]. Involatile MMA, DMA, and TMA species (one, two or all three of them) have been detected in various proportions in the culture media of various microorganisms, always in very low concentrations and within the above mentioned analytical limitations. Clearly, more results are needed before the existing information can be assembled to give a more general overview.

4.2. Biomethylation Mechanism

It is generally accepted that arsenic biomethylation follows the pathway proposed by Challenger and Ellis [81]. This mechanism involves a series of

Table 6. Reported methylantimony species in laboratory cultures.

Organism	Culture details	Initial Sb compound	Detected volatile Sb species[a]
Scopulariopsis brevicaulis		PAT	None
Scopulariopsis brevicaulis, Penicillium notatum	Aerobic	$KSbO_3$, phenylstibonic acid Na salt	Sb possibly detected in air over the cultures
Thalassiosira nana (marine diatom)		$^{125}SbCl_3$	NM
Pseudomonas fluorescens K27	Anaerobic, 30 °C, 24 h	PAT, TMC, PHA	No volatile Sb
Soils: sewage plant, backyard of auto repair shop (Huntsville, TX, US); As-contaminated (Dübendorf, Switzerland)	Anaerobic, 30 °C, 2 weeks	PAT, PHA	TMS
7 aerobes isolated from cot matresses and 4 human oral facultative anaerobes[b]	Aerobic: plate and flask cultures, 28 °C and 37 °C	PAT, ATO	No methylated compounds formed
Mixed cultures of anaerobes in cot mattresses and pond sediments	Anaerobic: deep cultures, 28 °C and 37 °C		TMS
Scopulariopsis brevicaulis	Aerobic Small scale flask and large-scale bioreactor experiments	PAT, ATO, TMC, PHA, phenylstibonic acid	Irreproducible formation at ultratrace levels
UK soils: garden topsoil, black sediment pond, tannery polluted soil, auto-garage soil, petrochemical-contaminated soil	Anaerobic, 3 culture media, 25 °C or 30 °C, dark, 5-8 weeks	PAT	TMS
Scopulariopsis brevicaulis	A- Aerobic, 25 °C, 8 d B-Biphasic: aerobic (6 d), anaerobic (3d)	PAT, ATO, APO	TMS

Detected involatile Sb species[a]	Analytical technique	Comments	Reference
NM			[81]
NM	Marsh and Gutzeit tests	Positive results with *P. notatum* only	[82]
Stibnolipid	Radioautographs of paper chromatograms of methanol cell suspensions and comparison with As	Sb is bound to three methyl groups and one O in the stibnolipid	[83]
NM	GC-fluorine-induced chemiluminescence detector; calibration: TMS standard GC-MS	TMS in 24 of 48 soils amended *P. fluorescens* produced TMS from TMC but did not methylate PAT, PHA	[28]
NM	Adsorption on $HgCl_2$ soaked glass fiber papers Thermal desorption + MS	TMS in 3 cultures from one pond (PAT); total number of pond cultures: 78	[84]
DMA, TMA	Volatile: GC-ICP-MS Non-volatile: SPE + HG-GC-AAS, HG-GC-ICP-MS	DMA, TMA: low yields No methylation of Sb(V) compounds	[85]
NM	PT (cryogenic) GC-AAS, GC-MS Standard: HG TMC[c]	TMS in 12 cultures out of a total of 104 TMS not detected in garden and auto-garage top-soil	[86]
NM	(A) PT (nitric acid) + ICP-MS (B) PT (Tenax) + GC-ET-AAS, GC-MS (standard: HG TMC[c])	Rapid oxidation of TMS in aerobic conditions Methylation of Sb(V) but "less readily"	[87]

Table 6. (*Continued*).

Organism	Culture details	Initial Sb compound	Detected volatile Sb species[a]
Scopulariopsis brevicaulis 8 cot mattress isolates[d]	A- Liquid aerobic, 25 °C, 8 d B- Liquid biphasic: aerobic (6 d), anaerobic (3 d) C- Solid in air, 25 °C, 18d or CO_2 33 °C, 18 d	PAT, ATO, PHA, APO	TMS
Scopulariopsis brevicaulis *Phaeolus schweinitzii* (wood decay fungus)	Plate cultures, 7 d	PAT, ATO	No Sb volatilization reliably detected
Scopulariopsis brevicaulis	Aerobic, 26 °C, dark	PAT, TMC	TMS
Scopulariopsis brevicaulis	Aerobic, 26 °C, 1 month	PAT + $^{13}CD_3$-L-methionine	NM
Scopulariopsis brevicaulis	Aerobic, 28 °C, 5 or 8 d	PAT	TMS
Scopulariopsis brevicaulis	Aerobic, 26 °C, 1 month	PAT, ATO, PHA + Na_3AsO_3, Na_3AsO_4	NM
Scopulariopsis brevicaulis	Aerobic, 26 °C, 1 month	PAT or $NaAsO_3$ + $^{13}CD_3$-L-, $^{13}CD_3$-D-methionine	NM
Inoculum of porcine feces (1 mL of 10% suspension) Different monoseptic cultures[e]	Anaerobic cultures of PVC foam mattresses with human urine, 33 °C (feces) or 28 °C (cultures), 4 weeks	PVC Sb-containing leachate PAT (only feces)	No volatile Sb

Detected involatile Sb species[a]	Analytical technique	Comments	Reference
NM	(A, C) PT (nitric acid) + ICP-MS (A, B) PT (Tenax) + GC-MS (standard: HG TMC[c])	Highest production in solid media Reduced production in CO_2 Other organisms do not produce TMS Methylation of Sb(V) but "less readily"	[88]
NM	Adsorption on $AgNO_3$ filter papers HG-AAS		[89]
NM	PT (cryogenic) GC-ICP-MS Standard: HG TMC[c]	High amounts of substrate required Sb yields much lower than of As (no As added)	[90]
DMA, TMA	SPE HG-CGC-MS Standard: HG TMC[c]	DMA and TMA contained $^{13}CD_3$	[91]
NM	PT (Tenax) GC-ET-AAS, GC-MS Standard: HG TMC[c]	TMS in headspace of 75% cultures (5 d); 25% (8 d)	[92]
TMA	SPE HG-GC-AAS Standard: HG TMC[c]	Sb(III), but not Sb(V), inhibits As methylation; As(III) enhances PAT methylation	[93]
DMA, TMA	SPE HG-GC-AAS, HG-GC-MS	Similar $^{13}CD_3$ incorporation from methionine to As and Sb	[94]
MMA, DMA, TMA	Volatile: PT (Tenax) or syringe + GC-MS (standard: HG TMC[c]) Involatile: HG-GC-AAS, GC-MS	Involatile results correspond to the incubation of foam, no organisms added	[95]

Table 6. (*Continued*).

Organism	Culture details	Initial Sb compound	Detected volatile Sb species[a]
Sewage sludge, municipal wastewater treatment plant, Germany	Anaerobic, 37 °C, dark, 1 week	$SbCl_3$	TMS
3 methanogenic archaea, 2 sulfate-reducing bacteria, a peptolytic bacterium[f]	Anaerobic, 5 d, 2 d or overnight		SbH_3, MMS, DMS, TMS
Phaeolus schweinitzii (wood rotting fungus)	Aerobic, 26 °C, 40 d	PAT, ATO, PHA	NM
Flavobacterium sp.	Aerobic, 25 °C, 14 d	PAT + Na_3AsO_3	NM
Corynebacterium xerosis			
Proteus vulgaris			
Escherichia coli			
Flavobacterium sp.			
Pseudomonas fluorescens			
Soil enriched cultures (*Clostridia* growth promotion)	Anaerobic, 3 culture media, dark, 28 °C, 4–6 weeks	PAT	TMS in cooked meat media only
Clostridia[g]	Anaerobic, dark, 28 °C, 28 d		No volatiles
Cryptococcus humicolus	Biphasic: aerobic (6 d), anaerobic (18 d)	PAT	TMS
		PHA	SbH_3, DMS, TMS
Cryptococcus humicolus	Aerobic, 28 °C, 28 d	PAT	NM
		ATO	
		PHA	

Detected involatile Sb species[a]	Analytical technique	Comments	Reference
NM	PT GC-ICP-MS Identification: bp-rt correlation Semiquantitative calibration	No methylation by *D. gigas* All organisms produced only TMS except *M. formicicum*	[96]
TMA, DMA	SPE HG-GC-AAS (standard: HG TMC[c]) HG-GC-MS	More efficient than *S. brevicaulis* More TMA than DMA Inefficient methylation of Sb(V) compounds	[97]
MMA, DMA, TMA	SPE HG-AAS Standard: HG TMC[c]	Methylation only by *Flavobacterium* sp. Sb $<20\,mg\,L^{-1}$: only MMA, DMA; at $30\,mg\,L^{-1}$, TMA predominant As(III) enhanced Sb(III) methylation	[98]
NM MMA, DMA, TMA	Volatiles: PT (Tenax) + GC-MS Involatiles: SPE + HG-GC-AAS Standard: HG TMC[c]	MMA, DMA transient species, TMA final one	[99]
NM	SPME + GC-MS PT (cryogenic) + GC-AAS Standard: HG TMC[c]		[100]
MMA, DMA, TMA MMA, DMA, TMA DMA, TMA	HG-GC-AAS Standard: HG TMC[c]	At $<50\,mg\,Sb\,L^{-1}$, TMA predominant; at $>100\,mg\,Sb\,L^{-1}$, DMA As up to 100-fold more efficient methylation As influences Sb methylation	[101]

Table 6. (*Continued*).

Organism	Culture details	Initial Sb compound	Detected volatile Sb species[a]
Sewage sludge	Anaerobic, 37 °C, 14 d	Isotopically enriched ^{123}Sb(V)	TMS
	Methanogenic archaea and SRB stimulation, 7 and 21 d	PAT	TMS
Alluvial soil samples, near River Ruhr, Germany	Anaerobic, 37 °C, dark, 3 months	$SbCl_3$	SbH_3, DMS, TMS
Isolated strain ASI-1	Anaerobic, 37 °C, dark, 3 d		TMS
Clostridium glycolicum			No volatile Sb
Sediment pore water from a maturation pond in a wastewater facility, Bochum, Germany	Sediment and fauna incubation experiment; aerobic, 20–25 °C, dark, 76 d	PHA	NM
Feces from 14 human volunteers before and after ingesting 215 mg Bi	Anaerobic, 37 °C, dark, up to 4 weeks	–	SbH_3, MMS, TMS

[a]NM, not measured.

[b]Aerobes from cot mattresses: *Scopulariopsis brevicaulis, Bacillus amyloliquifaciens, B. subtilis, B. firmus, B. pumulus, B. megaterium, B. licheniformis*. Oral facultative anaerobes: *Actinomyces odontolyticus, Lactobacillus casei, Porphyromonas gingivalis*.

[c]HG TMC = generation of a mixture of stibine species (MMS, DMS, TMS) by HG of a TMC standard (see Section 3.1).

[d]Cot mattress isolates: *Penicillium* spp, *Aspergillus niger, A. fumigatus, Alternaria* sp., *Bacillus licheniformis, B. subtilis, B. megaterium*.

[e]Monoseptic cultures: *Clostridium sporogenes, Escherichia coli, Enterobacter aerogenes, Salmonella gallinarum, Serratia marcescens, Proteus vulgaris*.

[f]Methanogenic archaea: *Methanobacterium formicicum, Methanosarcina barkeri, Methanobacterium thermoautrophicum*; peptolytic bacterium: *Clostridium collagenovorans*; sulfate-reducing bacteria: *Desulfovibrio vulgaris, D. gigas*.

[g]*Clostridium acetobutylicum, C. butyricum, C. cochlearium, C. sporogenes*, two isolates from enrichment culture.

Detected involatile Sb species[a]	Analytical technique	Comments	Reference
MMA, DMA, TMA MMA, DMA, TMA	Volatiles: Tedlar bags + GC-ICP-MS (standard: TMS) Involatiles: HG-GC-ICP-MS	64% of TMS originates from the spiked $^{123}Sb(V)$ Involatiles measured in filtrate and in sludge; only 1/10 in the filtrate High production of MMA Stepwise methylation confirmed by ^{123}Sb MMA, DMA, TMA contents Methanogenic archaea probably involved	[102]
NM	PT GC-ICP-MS Identification: bp – rt correlation		[103]
MMA, DMA, TMA	HG-PT-GC-ICP-MS	DMA predominant Eutrophication and acidification favor methylation	[104]
NM	PT-GC-ICP-MS Identification: no details, only reference [96] given		[105]

reductive methylation and oxidation steps, with trimethylarsine as the final product; monomethyl, dimethyl, and trimethyl species of As(III) and As(V) occur as intermediates. Because of the chemical similarities between arsenic and antimony, the hypothesis that antimony biomethylation follows the same biomethylation pathway as arsenic has been explored by various authors. One of the lines of investigation pursued has been the search for the expected intermediates. As discussed in the previous sections, MMA, DMA, and TMA have indeed been found in environmental compartments and in laboratory cultures, although, as mentioned, some of these species may have been formed as a result of TMS demethylation in the HG process. The origin of the DMA species detected by some authors has been the subject of some controversy at the end of the 90s, with Craig and coworkers long supporting the hypothesis that, in the absence of analytical artefacts, DMA was formed from TMS oxidation [92]. Later, Cullen and coworkers performed experiments that, in their opinion, proved that DMA species are not readily formed by TMS oxidation [94] and that, therefore, they are intermediates in the pathway to TMS. In a more recent study, the inoculation of sewage sludge with isotopically labelled Sb(V) showed that, at least in the system investigated, antimony methylation was occurring in steps from MMA to DMA and TMA [102], in line with Challenger's hypothesis. The fact that methionine, which is a precursor for *S*-adenosylmethionine (Challenger's methyl donor), has been identified as a methyl donor for antimony biomethylation in *Scopulariopsis brevicaulis* [90,94] further substantiates this hypothesis. On the other hand, glutathione and methylcobalamin have been suggested to play a role in the abiotic methylation process of Sb(V) in digested sewage sludge from a wastewater treatment plant [106].

Some other aspects that need to be considered in relation to antimony biomethylation, and that have so far received scant attention, are: (i) Sb(III) and Sb(V) uptake transport mechanisms by organisms, (ii) intracellular antimony oxidation and reduction processes, and (iii) the removal of antimony species from cells. These aspects are, in general, incompletely known and have mainly been studied in relation either to the development of bacterial tolerance mechanisms or to the use of antimony in the treatment of leishmaniasis (caused by a protozoan of the genus *Leishmania*) [3] but not in relation to antimony interactions with the organisms involved in biomethylation.

Extremely low yields of methylated antimony species in laboratory incubation experiments have led several authors to suggest that antimony biomethylation is a fortuitous process [85,98] rather than a detoxification mechanism. Moreover, it has been observed that the presence of small quantities of As(III) can stimulate the biomethylation of antimony [93], and that arsenic, and preferentially, cells pre-incubated with As(III), not only enhances the methylation of antimony but also alters the speciation of the methylantimony biotransformation products [101]. Both observations support the hypothesis that antimony methylation could be a fortuitous process, catalyzed at least in part by enzymes responsible for arsenic methylation.

5. ECOTOXICITY

The potential for metalloid organic compounds to adversely affect ecosystems and human health is well documented for many elements [107]. However, no ecotoxicological studies exist for antimony and even published toxicity studies are few and far apart. Those that exist all point to a very low toxicity of methylantimony compounds. As early as 1939, Seifter performed experiments to determine the acute toxicity of TMS to animals and concluded that "trimethylstibine possesses no great or pronounced acute toxicity to animals" [38]. The fungal toxicity of some diphenyl-, triphenyl-, and trimethylantimony compounds has been determined; only diphenylantimony compounds had EC_{50} values less than $30\,mg\,Sb\,L^{-1}$ [108].

Recently, stibine and TMS have been found to be genotoxic [109]. However, the minimum concentration in solution required to cause DNA damage was $200\,\mu mol\,L^{-1}$. This concentration is many orders of magnitude greater than the typical trace quantities of TMS found in fermentation gases (Table 6). Curiously, TMS is nearly as genotoxic as trimethylarsine, while arsine is not genotoxic at all, but stibine is. TMC is poorly membrane-permeable and does not induce cyto- and genotoxic effects under normal exposure conditions [110]. From the scarce existing (eco)toxicological information, and considering how low the concentrations of methylated antimony species detected in the environment are, it seems unlikely that they could be of any great concern.

6. CONCLUDING REMARKS

Methylated antimony species have been detected in various environmental compartments at very low levels of concentration. The number of published studies (seawaters: 7 [43,48–51,53,54,56], freshwaters: 2 [43,52], soils: 4 [59–61,64], sediments: 2 [58,65], biota: [52,66–70]) is fairly low as compared to other elements. Monomethyl, dimethyl, and trimethyl species have been reported to exist in the various systems, but these results, along with those from laboratory incubations, have always been haunted by the possibility of artefacts during analysis, in particular when HG techniques – by far the most commonly applied – are used. Alternative methods, such as

HPLC-based methodologies, have not yet been used much and, curiously, where they have been applied – always using a TMA standard – the only species detected has been TMA. When MSA and DMSA standards were used in seawater studies, only MMA and DMA were found. The fact that the results obtained are so dependent on the techniques and standards used merit some investigation. More data, obtained in a larger variety of environmental systems, and as free as possible from analytical uncertainties, are needed in order to ascertain the importance of methylated compounds in the biogeochemical cycle of antimony.

Not much is known about the properties and reactivity of alkylantimony species, and even less in conditions close to environmental ones. As is clear from the short overview in Section 2, data on physical and chemical properties of these compounds are fragmentary and old. Moreover, 'pure chemists' are used to working either with pure compounds or at concentration levels in solution which are much higher than the low concentrations found in natural systems, while in fact reactivity may be strongly dependent on concentration. As mentioned above, this point has been already discussed concerning reactivity in the gas phase in relation to TMS oxidation, but the same considerations apply to aqueous solutions. Additionally, nothing is known about the binding of methylated antimony by natural ligands, whether those with low molecular mass or colloidal ones (e.g., natural organic matter, clays, iron oxyhydroxides, etc). Further work is undoubtedly needed on all these fundamental issues in order to gain a better understanding of the role that methylantimony species may play in the various ecosystems and to reconcile puzzling facts such as the constant concentrations of methylantimony species found in surface oceanic waters and the low yields of antimony biomethylation obtained in laboratory studies performed in conditions that should, in principle, favor that process (i.e., high substrate concentrations, chosen microorganisms, etc.).

ABBREVIATIONS

AAS	atomic absorption spectrometry
AES	atomic emission spectrometry
AFS	atomic fluorescence spectrometry
APO	antimony pentoxide, Sb_2O_5
ATO	antimony trioxide, Sb_2O_3
bp	boiling point
CAS	Chemical Abstract Services
CE	capillary electrophoresis
CGC	capillary gas chromatography

CT	cold trap
DL	detection limit
DMA	dimethylantimony species
DMS	dimethylstibine, $(CH_3)_2SbH$
DMSA	dimethylstibinic acid, $(CH_3)_2SbO(OH)$
EC_{50}	effective concentration, 50%
EI	electron ionization
ESI	positive ion electrospray
ET	electrothermal
FI	flow injection
GC	gas chromatography
HG	hydride generation
HPLC	high performance liquid chromatography
IC	ion chromatography
ICP	inductively coupled plasma
LT	low temperature
MMA	monomethylantimony species
MMS	monomethylstibine, CH_3SbH_2
MS	mass spectrometry
MSA	methylstibonic acid, $CH_3SbO(OH)_2$
ND	not detected
NM	not measured
PAT	potassium antimony tartrate, $KSbOC_4H_4O_6 \cdot \frac{1}{2}H_2O$
PHA	potassium hexahydroxyantimonate, $K[Sb(OH)_6]$
PID	photoionization detection
PT	purge and trap
RSD	relative standard deviation
rt	retention time
SIDS	sudden infant death syndrome
SPE	solid-phase extraction
SPME	solid phase microextraction
SRB	sulfate reducing bacteria
STB	stibine, SbH_3
TMA	trimethylantimony species
TMB	trimethylantimony dibromide, $(CH_3)_3SbBr_2$
TMC	trimethylantimony dichloride, $(CH_3)_3SbCl_2$
TMS	trimethylstibine, $(CH_3)_3Sb$

REFERENCES

1. M. Filella, N. Belzile and Y.-W. Chen, *Earth Sci. Rev.*, 2002, **57**, 125–176.
2. M. Filella, N. Belzile and Y.-W. Chen, *Earth Sci. Rev.*, 2002, **59**, 265–285.

3. M. Filella, N. Belzile and M.-C. Lett, *Earth Sci. Rev.*, 2007, **80**, 195–217.
4. M. Filella, P. A. Williams and N. Belzile, *Environ. Chem.*, 2009, **6**, 95–105.
5. B. A. Richardson, *Lancet*, 1990, **335**, 670.
6. B. A. Richardson, *J. Forensic Soc.*, 1994, **34**, 199–204.
7. M. Wieber and J. Walz, *Z. Naturforsch.*, 1990, **45b**, 1615–1617.
8. G. T. Morgan and G. R. Davies, *Proc. R. Soc., Ser. A*, 1926, **110**, 523–534.
9. F. A. Paneth and H. Loleit, *J. Chem. Soc.*, 1935, 366–371.
10. H. A. Meinema and J. G. Noltes, *J. Organometal. Chem.*, 1972, **36**, 313–322.
11. O. J. Scherer, P. Hornig and M. Schmidt, *J. Organometal. Chem.*, 1966, **6**, 259–264.
12. K. Dehnicke and H. G. Nadler, *Chem. Ber.*, 1976, **109**, 3034–3038.
13. G. E. Graves and J. R. van Wazer, *J. Inorg. Nucl. Chem.*, 1977, **39**, 1101.
14. G. G. Long, G. O. Doak and L. D. Freedman, *J. Am. Chem. Soc.*, 1964, **86**, 209–213.
15. M. Shindo and R. Okawara, *J. Organometal. Chem.*, 1966, **5**, 537–544.
16. G. Lang, K. W. Klinkhammer, Ch. Recker and A. Schmidt, *Z. anorg. allg. Chem.*, 1998, **624**, 689–693.
17. B. A. Arbuzov, *Izvestiya Akademii Nauk SSSR, Seriya Khimicheskaya*, 1987, 1380–1383.
18. G. T. Morgan and V. E. Yarsley, *Proc. R. Soc., Ser. A*, 1926, **110**, 534–537.
19. M. Dodd, S. L. Grundy, K. J. Reimer and W. R. Cullen, *Appl. Organomet. Chem.*, 1992, **6**, 207–211.
20. T. M. Lowry and J. H. Simons, *Ber. dtsch. chem. Ges.*, 1930, **63**, 1595–1602.
21. J. Zheng, A. Takeda and N. Furuta, *J. Anal. At. Spectrom.*, 2001, **16**, 62–67.
22. H. Landolt, *J. Prakt. Chem.*, 1861, **84**, 328–354.
23. G. O. Doak, G. G. Long and M. E. Key, *Inorg. Synt.*, 1967, **9**, 92–97.
24. H. Althaus, H. J. Breunig and E. Lork, *Organometallics*, 2001, **20**, 586–589.
25. J. W. Dale, H. J. Emeléus, R. N. Haszeldine and J. H. Moss, *J. Chem. Soc.*, 1957, 3708–3713.
26. A. Hantzsch and H. Hibbert, *Ber. dtsch. chem. Ges.*, 1907, **40**, 1508–1519.
27. J. Lintschinger, O. Schramel and A. Kettrup, *Fresenius J. Anal. Chem.*, 1998, **361**, 96–102.
28. H. Gürleyük, V. Van Fleet-Stalder and T. G. Chasteen, *Appl. Organomet. Chem.*, 1997, **11**, 471–483.
29. P. Nylén, *Z. anorg. allgm. Chem.*, 1941, **246**, 227–330.
30. A. B. Burg and L. R. Grant, *J. Am. Chem. Soc.*, 1959, **81**, 1–5.
31. N. A. D. Carey, H. A. Meinema and J. G. Noltes, *Inorg. Chem.*, 1968, **7**, 2643–2645.
32. A. L. Rheingold, P. Choudhury and M. F. El-Shazly, *Synth. React. Inorg. Met.-Org. Chem.*, 1978, **8**, 453–465.
33. H. Landolt, *Justus Liebigs Ann. Chem.*, 1851, **78**, 91–96.
34. H. Hibbert, *Ber. dtsch. chem. Ges.*, 1906, **39**, 160–192.
35. L. H. Long and J. F. Sackman, *Trans. Faraday Soc.*, 1955, **51**, 1062–1064.
36. L. H. Long and J. F. Sackman, *Research Correspondence*, 1955, **8**, S23–S24.
37. C. H. Bamford, D. L. Levi and D. M. Newitt, *J. Chem. Soc.*, 1946, 468–471.
38. J. Seifter, *J. Pharmacol.*, 1939, **66**, 366–377.

39. G. Balàzs, L. Balàzs, H. J. Breunig and E. Lork, *Appl. Organomet. Chem.*, 2002, **16**, 155–159.
40. G. O. Doak and L. D. Freedman, *Organometallic Compounds of Arsenic, Antimony and Bismuth*, Wiley Interscience, New York, 1970.
41. L. D. Freedman, G. O. Doak, G. G. Long, T. Mahmood and C. B. Lindhal, in *Kirk-Othmer Encyclopedia of Chemical Technology*, 5th edn., Vol. 3, J. Wiley & Sons, Chichester UK, 2004, pp. 56–87.
42. *Gmelin Handbook of Inorganic and Organometallic Chemistry*, 8th ed., Springer, Berlin and Heidelberg, 1996.
43. M. O. Andreae, J.-F. Asmodé, P. Foster and L. Van't dack, *Anal. Chem.*, 1981, **53**, 1766–1771.
44. C. H. Bamford and D. M. Newitt, *J. Chem. Soc.*, 1946, 695–701.
45. G. E. Parris and F. E. Brinckman, *Environ. Sci. Technol.*, 1976, **10**, 1128–1134.
46. P. Andrewes and W. R. Cullen, in *Organometallic Compounds in the Environment*, 2nd edn., Ed. P. J. Craig, J. Wiley & Sons, Chichester, UK, 2003, pp. 277–303.
47. P. J. Craig, S. A. Foster, R. O. Jenkins, G. Lawson, D. Miller and N. Ostah, *Appl. Organomet. Chem.*, 2001, **15**, 527–532.
48. M. O. Andreae, in *Trace Metals in Sea Water*, Ed. C. S. Wong, E. Boyle, K. W. Bruland, J. D. Burton and E. D. Goldberg, NATO Adv. Res. Inst., Plenum Press, New York, 1983, pp. 1–19.
49. K. K. Bertine and D. S. Lee, in *Trace Metals in Sea Water*, Ed. C. S. Wong, E. Boyle, K. W. Bruland, J. D. Burton and E. D. Goldberg, NATO Adv. Res. Inst., Plenum Press, New York, 1983, pp. 21–38.
50. M. O. Andreae and P. N. Froelich, *Tellus*, 1984, **36B**, 101–117.
51. G. A. Cutter, *Deep-Sea Res.*, 1991, **38**, S825–S843.
52. I. Koch, L. Wang, J. Feldmann, P. Andrewes, K. J. Reimer and W. R. Cullen, *Intern. J. Environ. Anal. Chem.*, 2000, **77**, 111–131.
53. G. A. Cutter, L. S. Cutter, A. M. Featherstone and S. E. Lohrenz, *Deep-Sea Res. II*, 2001, **48**, 2895–2915.
54. M. J. Ellwood and W. A. Maher, *Deep-Sea Res. I*, 2002, **49**, 1971–1981.
55. M. J. Ellwood and W. A. Maher, *J. Anal. At. Spectrom.*, 2002, **17**, 197–203.
56. G. A. Cutter and L. S. Cutter, *Geochem. Geophys. Geosyst.*, 2006, **7**, 1–12; doi 10.1029/2005GC001159.
57. I. Koch, J. Feldmann, J. Lintschinger, S. V. Serves, W. R. Cullen and K. J. Reimer, *Appl. Organomet. Chem.*, 1998, **12**, 129–136.
58. E. M. Krupp, R. Grümping, U. R. R. Furchtbar and A. V. Hirner, *Fresenius J. Anal. Chem.*, 1996, **354**, 546–549.
59. N. Ulrich, *Anal. Chim. Acta*, 1998, **359**, 245–253.
60. A. V. Hirner, U. M. Grüter and J. Kresimon, *Fresenius J. Anal. Chem.*, 2000, **368**, 263–267.
61. N. Ulrich, *Anal. Chim. Acta*, 2000, **417**, 201–209.
62. R. A. Diaz-Bone and M. Hitzke, *J. Anal. At. Spectrom.*, 2008, **23**, 861–870.
63. J. Kösters, J. Hippler, R. A. Diaz-Bone and A. V. Hirner, *J. Anal. At. Spectrom.*, 2005, **20**, 996–999.

64. L. Duester, R. A. Diaz-Bone, J. Kösters and A. V. Hirner, *J. Environ. Monit.*, 2005, **7**, 1186–1193.
65. L. Duester, L. M. Hartmann, L. Luemers and A. V. Hirner, *Appl. Organometal. Chem.*, 2007, **21**, 441–446.
66. R. Kantin, *Limnol. Oceanogr.*, 1983, **28**, 165–168.
67. M. Dodd, S. A. Pergantis, W. R. Cullen, H. Li, G. K. Eigendorf and K. J. Reimer, *Analyst*, 1996, **121**, 223–228.
68. P. J. Craig, S. N. Foster, R. O. Jenkins and D. Miller, *Analyst*, 1999, **124**, 1243–1248.
69. R. Miravet, E. Bonilla, J. F. López-Sánchez and R. Rubio, *J. Environ. Monit.*, 2005, **7**, 1207–1213.
70. R. Miravet, E. Bonilla, J. F. López-Sánchez and R. Rubio, *Appl. Organomet. Chem.*, 2006, **20**, 12–19.
71. R. Bailly, R. Lauwerys, J. P. Buchet, P. Mahieu and J. Konings, *British J. Ind. Med.*, 1991, **48**, 93–97.
72. M. Krachler and H. Emons, *J. Anal. At. Spectrom.*, 2001, **16**, 20–25.
73. J. Kresimon, U. M. Grüter and A. V. Hirner, *Fresenius J. Anal. Chem.*, 2001, **371**, 586–590.
74. D. Glindemann, P. Morgenstern, R. Wennrich, U. Stottmeister and A. Bergmann, *Environ. Sci. Pollut. Res.*, 1996, **3**, 75–77.
75. J. Feldmann, R. Grümping and A. V. Hirner, *Fresenius J. Anal. Chem.*, 1994, **350**, 228–234.
76. A. V. Hirner, J. Feldmann, R. Goguel, S. Rapsomanikis, R. Fischer and M. O. Andreae, *Appl. Organomet. Chem.*, 1994, **8**, 65–69.
77. J. Feldmann and A. V. Hirner, *Int. J. Environ. Anal. Chem.*, 1995, **60**, 339–359.
78. J. Feldmann, I. Koch and W. R. Cullen, *Analyst*, 1998, **123**, 815–820.
79. A. V. Hirner, J. Feldmann, E. Krupp, R. Grümping, R. Goguel and W. R. Cullen, *Org. Geochem.*, 1998, **29**, 1765–1778.
80. B. Michalke and P. Schramel, *J. Chromat. A*, 1999, **834**, 341–348.
81. F. Challenger and L. Ellis, *J. Chem. Soc.*, 1935, 396–400.
82. D. Barnard, *Studies on the metabolism of certain Aspergilli and Penicillia*, PhD Thesis, University of Leeds, UK, 1947.
83. A. A. Benson, in *The Biological Alkylation of Heavy Metals*, Ed. P. J. Craig and F. Glocking, Special Publication 66, Royal Society of Chemistry, Cambridge, 1988, pp. 135–137.
84. P. N. Gates, H. A. Harrop, J. B. Pridham and B. Smethurst, *Sci. Total Environ.*, 1997, **205**, 215–221.
85. P. Andrewes, W. R. Cullen, J. Feldmann, I. Koch, E. Polishchuk and K. J. Reimer, *Appl. Organomet. Chem.*, 1998, **12**, 827–842.
86. R. O. Jenkins, P. J. Craig, D. P. Miller, L. C. A. M. Stoop, N. Ostah and T.-A. Morris, *Appl. Organomet. Chem.*, 1998, **12**, 449–455.
87. R. O. Jenkins, P. J. Craig, W. Goessler and K. J. Irgolic, *Human Experimental Toxicol.*, 1998, **17**, 231–238.
88. R. O. Jenkins, P. J. Craig, W. Goessler, D. Miller, N. Ostah and K. J. Irgolic, *Environ. Sci. Technol.*, 1998, **32**, 882–885.

89. R. B. Pearce, M. E. Callow and L. E. Macaskie, *FEMS Microbiol. Lett.*, 1998, **158**, 261–265.
90. P. Andrewes, W. R. Cullen and E. Polishchuk, *Appl. Organomet. Chem.*, 1999, **13**, 659–664.
91. P. Andrewes, W. R. Cullen, J. Feldmann, I. Koch and E. Polishchuk, *Appl. Organomet. Chem.*, 1999, **13**, 681–687.
92. P. J. Craig, R. O. Jenkins, R. Dewick and D. P. Miller, *Sci. Tot. Environ.*, 1999, **229**, 83–88.
93. P. Andrewes, W. R. Cullen and E. Polishchuk, *Environ. Sci. Technol.*, 2000, **34**, 2249–2253.
94. P. Andrewes, W. R. Cullen and E. Polishchuk, *Chemosphere*, 2000, **41**, 1717–1725.
95. R. O. Jenkins, T.-A. Morris, P. J. Craig, W. Goessler, N. Ostah and K. M. Wills, *Human Exper. Toxicol.*, 2000, **19**, 693–702.
96. K. Michalke, E. B. Wickenheiser, M. Mehring, A. V. Hirner and R. Hensel, *Appl. Environ. Microbiol.*, 2000, **66**, 2791–2796.
97. P. Andrewes, W. R. Cullen, E. Polishchuk and K. J. Reimer, *Appl. Organomet. Chem.*, 2001, **15**, 473–480.
98. R. O. Jenkins, S. N. Forster and P. J. Craig, *Arch. Microbiol.*, 2002, **178**, 274–278.
99. L. M. Smith, P. J. Craig and R. O. Jenkins, *Chemosphere*, 2002, **47**, 401–407.
100. L. M. Smith, W. A. Maher, P. J. Craig and R. O. Jenkins, *Appl. Organomet. Chem.*, 2002, **16**, 287–293.
101. L. M. Hartmann, P. J. Craig and R. O. Jenkins, *Arch. Microbiol.*, 2003, **180**, 347–352.
102. S. Wehmeier and J. Feldmann, *J. Environ. Monit.*, 2005, **7**, 1194–1199.
103. J. Meyer, A. Schmidt, K. Michalke and R. Hensel, *System. Appl. Microbiol.*, 2007, **30**, 229–238.
104. L. Duester, J. P. M. Vink and A. V. Hirner, *Environ. Sci. Technol.*, 2008, **42**, 5866–5871.
105. K. Michalke, A. Schmidt, B. Huber, J. Meyer, M. Sulkowski, A. V. Hirner, J. Boertz, F. Mosel, P. Dammann, G. Hilken, H. J. Hedrich, M. Dorsch, A. W. Rettenmeier and R. Hensel, *Appl. Environ. Microbiol.*, 2008, **74**, 3069–3075.
106. S. Wehmeier, A. Raab and J. Feldmann, *Appl. Organomet. Chem.*, 2004, **18**, 631–639.
107. E. Dopp, L. M. Hartmann, A.-M. Florea, A. W. Rettenmeier and A. V. Hirner, *Crit. Rev. Toxicol.*, 2004, **34**, 301–333.
108. R. E. Burrell, C. T. Corke and R. G. Goel, *J. Agric. Food Chem.*, 1983, **31**, 85–88.
109. P. Andrewes, K. T. Kitchin and K. Wallace, *Toxicol. Appl. Pharmacol.*, 2004, **194**, 41–48.
110. E. Dopp, L. M. Hartmann, A.-M. Florea, U. von Recklinghausen, S. Rabieh, B. Shokouhi, A. V. Hirner and A. W. Rettenmeier, *Toxicol. in Vitro*, 2006, **20**, 1060–1065.

Met. Ions Life Sci. **2010**, *7*, 303–317

9

Alkyl Derivatives of Bismuth in Environmental and Biological Media

Montserrat Filella

Institute F.-A. Forel, University of Geneva, Route de Suisse 10, CH-1290 Versoix, Switzerland

< montserrat.filella@unige.ch >

ABSTRACT: Knowledge about methylated species of bismuth in environmental and biological media is very limited. The presence of volatile trimethylbismuthine has been unequivocally detected in landfill and sewage fermentation gases but the trace concentrations of methylated bismuth species reported in a few polluted soils and sediments probably require further confirmation. In contrast to arsenic and antimony, no

Metal Ions in Life Sciences, Volume 7 Edited by Astrid Sigel, Helmut Sigel, and Roland K. O. Sigel

Published by the Royal Society of Chemistry, www.rsc.org DOI: 10.1039/9781849730822-00303

methylated bismuth species have ever been found in surface waters and biota. Volatile monomethyl-, dimethyl- and trimethylbismuthine have been produced by some anaerobic bacteria and methanogenic archaea in laboratory culture experiments. Bismuth methylation differs significantly from the one of arsenic and antimony because no Bi(V) compound is known to be formed in biological and environmental media. Moreover, alkylbismuth compounds are rather instable due to the easy cleavage of the weak Bi-C bond.

KEYWORDS: bismuth · biomethylation · trimethylbismuth · trimethylbismuthine

1. INTRODUCTION

Bismuth is a naturally occurring element. It is the heaviest stable element in the periodic table. It belongs to group 15 together with nitrogen, phosphorus, arsenic, and antimony. Bismuth can exist in a variety of oxidation states (−III, 0, III, V) but is mainly found in oxidation state III in environmental and biological samples. Bismuth(V) is a powerful oxidant in aqueous solution. Little information exists on the transformation and transport of bismuth in the different environmental compartments. Even information on total bismuth content in the various media is scarce and often contradictory. Bismuth has no known biological function and appears to be relatively benign for humans. However, it is toxic to prokaryotes and bismuth compounds have been used since the Middle Ages to treat ailments resulting from bacterial infections. It is still widely used to treat gastric and duodenal ulcers. Although the mechanism of action has not been completely elucidated, the effectiveness of bismuth has been partly attributed to its bactericidal action against *Helicobacter pylori.*

According to the classical review of Gilman and Yale [1], the synthesis of triethylbismuthine in 1850 by Löwig and Schweizer [2] inaugurated the study of the chemistry of organobismuth compounds. However, the spontaneous inflammability of these trialkyl derivatives limited investigations in the field until Michaelis and Polis prepared triphenylbismuthine in 1887 [3]. This aromatic compound was stable in air. From 1913 to 1934, the research by Challenger and his coworkers made an important contribution to the field of organobismuth compounds (see [4]). These studies preceded the work on biomethylation that are considered to be Challenger's main scientific legacy.

Though outside the scope of this chapter, there is a vast organometallic bismuth chemistry of interest to synthetic and mechanistic organometallic chemists, but it is of little significance in an environmental or biological context. It is well-known that organometallic species of some elements (e.g., lead, tin) are found in the natural environment derived directly from human use, but this does not seem to be the case for any alkyl or aryl derivative of bismuth. As is the case for most elements, only methyl-containing species have been found in natural systems and this review will focus on them.

Methylbismuth species had not been detected and quantified in environmental media until relatively recently (mid-90's) and only in a few studies carried out by the same research group (see below and in Tables 2 and 3 in Sections 4 and 5, respectively) or, in the only case when not, by using the same approach. In spite of the limited information that exists, a section on bismuth methylation is found in all recent reviews on biomethylation (e.g., [5–7]) and even a significant part of a chapter in a book [8] has been devoted to it, undoubtedly amplifying the impact of the few experimental observations carried out to date.

2. PHYSICAL AND CHEMICAL CHARACTERISTICS OF METHYLBISMUTH COMPOUNDS

Bismuth differs from arsenic and antimony in the lower stability of the pentavalent oxidation state relative to the trivalent one. There are no known monomethyl and dimethyl compounds of bismuth(V). Although the crystal structure of trimethylbismuth dichloride has been characterized by low-temperature X-ray diffraction analysis [9], this compound is thermally unstable and decomposes rapidly at room temperature.

Trialkylbismuth compounds are highly refractive, colorless or pale yellow, oily liquids. The methyl and ethyl compounds have an unpleasant odor [1]. The enthalpy of formation of trimethylbismuthine (TMB) is largely endothermic because of the very weak Bi-C bond, the weakest of the main group metals [10]. The reactivity of TMB and other alkyl bismuth compounds is largely characterized by the weakness of this bond. Lower members of the trialkylbismuth compounds, such as TMB, are spontaneously inflammable in air, confirming the ease of oxidative cleavage of the Bi-C bond by molecular oxygen. Because of their inflammability in air it is recommended that these compounds be isolated under an inert atmosphere. It is important to mention however, that, at low concentrations, such as the ones found in environmental and biological systems, the oxidation of TMB might be significantly slower, as is the case for other elements [11]. This would explain the relatively high recovery of TMB sampled in Tedlar bags after 8 h of storage [12]. However, in this study recoveries were lower than for methylated species of other elements and they were better in samples from anaerobic systems such as sewage sludge digester gases, indicating that oxidative breakdown remains an important depletion process for TMB.

Monomethylbismuthine (MMB), $Bi(CH_3)H_2$, and dimethylbismuthine (DMB), $Bi(CH_3)_2H$, are liquids which are stable at –60° but not stable at room temperature and decompose giving BiH_3 and TMB [13].

Not much is known about methylated bismuth halides. The crystal structure of CH_3BiCl_2 has been studied recently by Althaus and coworkers [14]. These authors also synthesized CH_3BiBr_2. Both compounds had already been prepared by Marquardt in 1887 [15]. The dichloro compound is a yellow solid (melting point: 242 °C [15], 246–249 °C [14]), air-stable both in solution and in the solid state; the dibromo compound, also a yellow solid (melting point: 214 °C [15], 195–197 °C [14]), decomposes in solution but is air-stable as a solid. CH_3BiI_2 crystallizes as dark red needles and appears also to be relatively air stable [16] (melting point: 225°C [15]). The dimethyl halides, also synthesized by Marquardt in 1887 [15], have been less studied. All these compounds might be useful to study the behavior of methylated bismuth compounds in the environment.

Published normal boiling points of TMB, extrapolated from vapor pressure measurements, are shown in Table 1 [17–21]. Long and Sackman [21] reported the melting point of TMB as –107.7 °C; this value is about 22°C lower than the melting point of –85.8 °C reported by Bamford and coworkers [20]. No reason for this discrepancy has been given.

The C-Bi bonds have a very low degree of polarity. This gives compounds that have a very small dipolar moment and will not be very soluble in water [1,15]. However, Sollmann and Seifter [22] reported that a freshly made saturated and filtered solution of TMB in water contained 0.5162 mg of Bi per mL (0.0024 molar solution) which seems quite high for an insoluble substance.

Table 1. Published trimethylbismuthine normal boiling point values and related information.

Vapor pressure-temperature relationship (p/torr) and (T/K)	Boiling point (°C)[a]	Latent heat of vaporization/ $kcal\,mol^{-1}$	Reference
$\log p = -A/T + B$ $A = 1815$, $B = 7.659$ Measured: −10 °C to 84 °C	107.1	8.308	[20]
$\log p = -A/T + B$ $A = 1816$, $B = 7.6280$ Measured: –25 °C to 15 °C	109.3	8.31	[21]
$\log p = -A/T - B \log T + C$ $A = 2225.7$, $B = 2.749$, $C = 15.8011$ Measured: –58 °C to 107 °C	108.8	8.3768	[13][b]

[a]Other published boiling point values are: 108 °C [17], 110 °C [18], 102–106 °C [19].

[b]This author also estimated boiling points by extrapolation of vapor pressure measurements (in parentheses the range of T measurements in °C) for the following substances: MMB, 72.0 °C (–87 to –15) and DMB, 103.0 °C (–67 to –23).

With the exception of a few Lewis acid-base reactions, there are virtually no trialkylbismuth compound reactions which do not involve cleavage of the carbon-bismuth bond. However, according to Doak and Freedman [23], in general they are not affected by water or aqueous bases but are hydrolyzed by inorganic and organic acids.

3. DETECTION AND QUANTIFICATION

The analytical technique used to study methylated species of bismuth in environmental and laboratory gas samples has been gas chromatography (GC) coupled with detection by inductively coupled plasma mass spectrometry (ICP-MS). The identification of the metal species is based on the combination of the temperature-based chromatographic separation with the element-specific detection (ICP-MS). The species associated with the peaks on the m/z 209 trace of the ICP-MS have usually been identified by calculating theoretical boiling points (bp) from to the measured retention times (rt) by using pre-established bp-rt correlations and the theoretical bp for the methylated bismuth species. The identity of TMB has sometimes been confirmed by matching the retention time of a TMB standard or by GC-MS. Quantification is a problem in this type of samples because of the difficulty of working with gaseous standards at low concentrations and the unavailability of reference standards. A method for semiquantification where an aqueous sample is used as a calibrant has been applied instead [24]. An internal standard, usually ^{103}Rh, is aspirated during the analysis in this approach.

In the few studies, where waters, soils, and sediments have been analyzed, the same measuring technique was applied to the gases generated by direct hydrogenation of the samples with $NaBH_4$. However, this method is well-known for generating analytical artefacts by demethylation (see Chapter 8 in this book). It is likely that the same problem occurs in bismuth: the headspace of a TMB standard dissolved in diethyl ether gave only one peak by GC-ICP-MS but four peaks after hydride generation of the same solution [25]. Demethylation would not be surprising considering that Bi-C bonds are easily cleaved by acid and that acidic conditions are often used in the hydride generation process.

4. OCCURRENCE IN ENVIRONMENTAL AND BIOLOGICAL MEDIA

TMB has been detected in landfill and sewage sludge fermentation gases. Published values are shown in Table 2. No data exists for natural waters and

Table 2. Reported methylbismuthine concentrations in gases from landfills and sewage treatment plants.[a]

System	TMB/μg m^{-3}	Sampling
Landfill gas (domestic waste deposit, Aβlar, Hessen, Germany)	0.312–0.892 (n = 8)	Cryogenic trapping (–80 °C)
Landfill gas (two municipal waste deposits, Germany)	0.0002–0.0065[b] (n = 8)	Cryogenic trapping (–80 °C)
Sewage gas at 56 °C and at 35 °C (municipal sewage treatment plant, Germany)	0.016–1.056[c]	Cryogenic trapping (–80 °C)
Landfill gas from municipal waste deposits and gas from a mesophilic sewage sludge digester (Vancouver, Canada)	Landfill: 0.013–0.030 Digester: "at least 3 orders of magnitude higher than in landfill gas"	Tedlar bags
Sewage gas A 1997[d]	5.00 ± 1.29 (n = 5)	Cryogenic trapping (–78 °C to –80 °C) except for H and M (Tedlar bags)
Sewage gas A 1998	5.53 ± 1.59 (n = 6)	
Sewage gas B	1.67 ± 0.16 (n = 3)	
Sewage gas C	24.2 ± 1.58 (n = 5)	
Sewage gas D	6.24 ± 1.37 (n = 3)	
Sewage gas E	4.29 ± 0.65 (n = 5)	
Sewage gas F	0.003 – 0.016 (n = 5)	
Sewage gas H	1–5	
Landfill gas J 1998	0.168	
Landfill gas M	0.01–0.03 (n = 6)	
Gas wells, landfill N	0.01–0.404 (n = 9)	
Soil gas 100 m from landfill N	0–0.034 (n = 6)	
Landfill gas, Vancouver site, Canada	Detected	Tedlar bags
Compost heap	Not detected	
Experimental compost mixtures	0.00002–0.0001	Tedlar bags

[a]Only values from peer-reviewed publications are considered.

[b]In subsequent publications by the same authors, these values are quoted as being TMB but no species is ever mentioned in this article.

[c]Values for landfill gases shown in a table of this article already published in [26].

[d]Locations: sewage treatment plants A to F in North Rhine-Westfalia, Germany; H and M in Vancouver, Canada; landfill J is in the Palatinate, Germany (also studied in [26]) and N in North Rhine-Westfalia, Germany.

[e]A reference is given but is probably wrong.

Analytical method	Comments	Ref.
LTGC-ICP-MS Identification: bp-rt correlation Semiquantitative calibration: interelement-based, internal liquid standard	One peak on m/z 209 Concentrations are for total volatile Bi	[26]
Desorption into the Ar plasma of the ICP-MS Semiquantitative calibration: same approach as in [26]	Concentrations are for total volatile Bi	[27]
LTGC-ICP-MS Identification: bp-rt correlation Semiquantitative calibration: same approach as in [26]	One peak on m/z 209 Concentrations are for total volatile Bi	[28]
CT-LTGC-ICP-MS Identification: bp-rt correlation Confirmation: CGC-EI-MS-MS (in digester gas only) Calibration: not described	One peak on m/z 209	[29]
GC-ICP-MS or PT-ICP-MS depending on sample Confirmation: GC-EI-MS Semiquantitative calibration, not described[e]	One peak on m/z 209	[25]
GC-MS and GC-ICP-MS Identification: rt	TMB masked by volatile organic compounds in GC-MS One peak on m/z 209 in GC-ICP-MS	[30]
CT-LTGC-ICP-MS Identification: bp-rt correlation Semiquantitative calibration: same approach as in [26]		[44]

biota. The presence of non-volatile methylbismuth species in polluted sediments [25,31] (monomethyl) and soils [32,33] (trimethyl in two soils and monomethyl, dimethyl and trimethyl in a third one) has been detected. However, these results should be considered with caution because the concentrations measured were always very low, a semi-quantitative method was used for calibration, and analytical artefacts are possible with the approach taken (Section 3). Negative results have been reported for condensed waters of pipelines in municipal landfills [25,26].

There is not enough experimental data to explain the absence of methylated bismuth species in environmental media except in fermentation gases. Numerous reasons can be cited and it is important to realise that some of them are independent of any biomethylation process but are directly related to the properties of the element, e.g., very low concentration levels of bismuth in the environment, low solubility of alkylbismuth compounds in water, chemical instability of these compounds, etc.

5. MICROBIAL TRANSFORMATIONS OF BISMUTH COMPOUNDS

5.1. Laboratory Experiments

Results from laboratory fermentation experiments are shown in Table 3. Pure cultures of some methanogenic archaea (*Methanobacterium formicicum, Methanobrevibacter smithii*) and anaerobic bacteria (*Clostridium collagenovorans, Desulfovibrio piger, Eubacterium eligens, Lactobacillus acidophilus*) have been shown to be capable of biomethylating bismuth. Undefined bacteria growing under anaerobic conditions from contaminated river sediments mixed with uncontaminated pond sludge, sewage sludge and soils have also shown bismuth methylation activity. Compared with methanoarchaea, anaerobic bacterial strains produced a more restricted spectrum of volatilized derivatives and the production rates of volatile bismuth derivatives were lower. Recently, human feces and isolated gut segments of mice were shown to be capable of producing TMB when incubated anaerobically, thus suggesting that human gut microbiota might catalyze this transformation in the human body [38].

It is important to point out that, even though it is well known that the bioavailability of any element is a function of its speciation and not of the total concentration present, none of the laboratory studies took into account the actual speciation of bismuth in the culture media. For this reason, when interpreting these results, unfortunately it is impossible to go much further than describing whether or not methyl bismuth species are produced in the headspace of the various cultures. In fact, all culture media contained a high number of substances (e.g., at least 32 were added in [35]), many of which are potential complexants of bismuth (e.g., cysteine). Furthermore, in some cases, bismuth complexants were even added in the bismuth spike itself (e.g., EDTA [35,37]). Therefore, the actual concentrations of 'free' bismuth or of any other potentially bioavailable species formed in the culture media were completely unknown.

5.2. Biomethylation Mechanism

One of the most frequently cited biomethylation mechanisms, the biomethylation of arsenic [39] involves reductions of pentavalent to trivalent arsenic and oxidative methylations in alternating order. As mentioned above, bismuth differs from arsenic in that the stability of the pentavalent oxidation state is much lower relative to the trivalent state and methylated Bi(V) compounds are not formed. As such, biomethylation of bismuth thorough the Challenger mechanism does not seem likely. Biomethylation of bismuth probably involves non-oxidative methyl transfer, where methylcobalamin could be the methyl source. A few published results support this hypothesis: (i) treatment of cell extracts of *Methanobacterium formicicum* with *S*-adenosylmethionine failed to yield any TMB but treatment of those extracts with methylcobalamin did form this compound [35]; (ii) *in vitro* treatment of bismuth nitrate with methylcobalamin also yielded TMB [35]. However, not only biogenic methyl sources exist and can be used in biomethylation: for instance, *Methanosarcina barkeri*, isolated from sewage sludge samples, has been shown to produce TMB in solutions containing low-molecular-weight silicones [40].

6. TOXICITY

In 1939 Sollmann and Seifter published [22] a lengthy account of the toxicology of TMB based on experiments with invertebrates (paramecia, earthworms, Daphnia), excised or exposed organs (motor nerve, skeletal muscle, motor nerve endings, sensory nerves, frog's heart), cold blooded vertebrates (goldfish, intact frogs), warm-blooded animals (humans, dogs, cats, rats, pigeons, rabbits). They described a long list of effects depending on the dose and the organism or organ considered.

Triphenylbismuth has shown a slight degree of cytotoxicity on human embryonic lung fibroplast tissue cells [41] and on rat thymocytes [42] but these results cannot be extrapolated to TMB because it has very different

Table 3. Reported methylbismuth species in laboratory cultures.

Organism/system	Culture details	Initial Bi compound	Detected Bi species
Contaminated river sediments mixed with uncontaminated pond sludge (1:1), Germany[a]	Anaerobic, 30 °C, 2 weeks	–	TMB
Sewage sludge, municipal wastewater treatment plant, Germany	Anaerobic, 37 °C, dark, 1 week	$Bi(NO_3)_3$ (20, 100 μM)	TMB
Pure cultures: *Methanobacterium formicicum* *Clostridium collagenovorans*	Anaerobic, 37 °C, dark, 1 week Exponential growth phase cultures		
Methanobacterium formicicum	Anaerobic, 37 °C, dark, 40 d Early exponential growth phase cultures	$Bi(NO_3)_3$ (0.01–20 μM) Bismofalk, (1 μM) Noemin (1 μM)	TMB (BH_3, MMB, DMB)
Alluvial soil samples, near river Ruhr, Germany	Anaerobic, 37 °C, dark, 3 months	$Bi(NO_3)_3$ (10 μM)	MMB, DMB, TMB
Isolated strain ASI-1	Anaerobic, 37 °C, dark, 3 d		Not detected
Clostridium glycolicum	Exponential growth phase cultures		Not detected
Methanobrevibacter smithii	Anaerobic, 37 °C, dark, up to 14 d	$Bi(NO_3)_3$ (1 μM)	MMB, DMB, TMB
Desulfovibrio piger	Early exponential growth phase cultures		TMB
Eubacterium eligens			TMB
Lactobacillus acidophilus			TMB
Feces from 14 human volunteers before and after ingestion of CBS tablets (215 mg Bi)	Anaerobic, 37 °C, dark, up to 4 weeks	–	BiH_3, MMB, DMB, TMB
Colon segments of mice (*Mus musculus*) fed for 7 d with standard or Bi-containing diet	Anaerobic, 37 °C, dark, up to 3 weeks		

[a]Klein Dalzig, Weisse-Elster, Saale, creek near Bitterfeld, Cu mine waste deposit.

[b]*Methanosarcina barkeri*, *Methanobacterium thermoautotrophicum*, *Desulfovibrio vulgaris*, and *D. gigas*.

[c]*Bacillus alcalophilus*, *Bacteroides coprocola*, *Bacteroides thetaiotaomicron*, *Bacteroides vulgatus*, *Bifidobacterium bifidum*, *Butyrivibrio crossotus*, *Clostridium aceticum*, *Clostridium leptum*, *Collinsella intestinalis*, *Eubacterium biforme*, and *Ruminococcus hansenii*.

Analytical method	Comments	Ref.
PT-GC-ICP-MS See entry for this reference in Table 2	No correlation between TMB production and total Bi sediment contents or Bi volatilized by hydride generation	[25]
PT-GC-ICP-MS Identification bp-rt correlation and comparison with rt of a TMB standard Semiquantitative calibration [24]	No production by *C. collagenovorans* at 100 μM No production was observed for other microorganisms[b]	[34]
PT-GC-ICP-MS Identification MMB, DMB, TMB: bp-rt correlation; TMB confirmed with a TMB standard Semiquantitative calibration [24]	BH_3, MMB, DMB only detected in late exponential growth phase and for low Bi concentrations Maximum conversion: 2.6% in 1 μM solutions	[35]
PT-GC-ICP-MS Identification: bp-rt correlation	Low concentrations found TMB produced by ASI-1 only in the presence of As or Sb	[36]
PT-GC-ICP-MS Identification by parallel ICP-MS and EI-MS	No production was observed for other microorganisms[c] Se conversion rates were generally higher	[37]
PT-GC-ICP-MS Identification, quantification: no details, only reference given [34]	No general correlation between feces Bi content and production rate of Bi derivatives Colon segments from germfree mice did not produce TMB	[38]

physical and chemical characteristics [1]. Very recently, the cellular uptake of monomethylbismuth (inorganic counterion not mentioned) by three different human cells (hepatocytes, lymphocytes, and erythrocytes) and its cytotoxic and genotoxic effects were studied [43]. The uptake of monomethylbismuth was appreciably higher in erythrocytes than in lymphocytes (17%) and practically non-existent in hepatocytes. Cytotoxic effects were detectable in erythrocytes at concentrations higher than 4 $\mu mol\,L^{-1}$ but only at more than 130 and 430 $\mu mol\,L^{-1}$ in hepatocytes and lymphocytes, respectively (24 h exposure). Significantly, increases of chromosomal aberrations and sister chromatoid exchanges were observed in lymphocytes when exposed at 250 $\mu mol\,L^{-1}$ monomethylbismuth for 1 h. Bismuth citrate and bismuth glutathione did not show any of these effects. These results show that, as expected, this methylated bismuth species is more membrane-permeable than the other compounds studied. It is, however, unclear whether these high concentrations of monomethylbismuth may exist in natural conditions.

7. CONCLUDING REMARKS

Published data do not support the widespread presence of methylated bismuth species in environmental and biological systems. However, the detection of methylated species in landfill and sewage gases and in anaerobic cultures suggests that bismuth biomethylation, even if not widespread, takes place in particular media where the formation and/or the stability of the methylated species formed is favored. In order to identify such systems and to better understand the mechanisms behind bismuth biomethylation, further research in some areas, partially beyond the strict biomethylation field, is needed, namely in: (i) speciation of bismuth in environmental and biological media, (ii) stability and speciation of methylbismuth species in diluted solutions, (iii) bismuth uptake by biota, (iv) bismuth toxicity against prokaryotes.

As mentioned in the introduction, bismuth is an element that is relatively non-toxic to humans but toxic to some prokaryotes. For this reason, bismuth compounds have been used for a long time to treat bacterial infections. Nowadays, colloidal bismuth subcitrate (CBS) is successfully used in the treatment of both gastric and duodenal ulcer disease. Its effectiveness has been attributed, at least partially, to its bactericidal action against *Helicobacter pylori* and a lot of research has been devoted to the understanding of the toxicity mechanism [45–47]. Current and future research in this field might help to understand some aspects of bismuth biomethylation.

ABBREVIATIONS

bp	boiling point
CBS	colloidal bismuth subcitrate
CGC	capillary gas chromatography
CT	cold trap
DL	detection limit
DMB	dimethylbismuthine, $(CH_3)_2BiH$
EI	electron ionization
GC	gas chromatography
ICP	inductively coupled plasma
LT	low temperature
MMB	monomethylbismuthine, CH_3BiH_2
MS	mass spectrometry
PT	purge and trap
rt	retention time
TMB	trimethylbismuthine, $(CH_3)_3Bi$

REFERENCES

1. H. Gilman and H. L. Yale, *Chem. Rev.*, 1942, **30**, 281–320.
2. C. Löwig and E. Schweizer, *Justus Liebigs Ann. Chem.*, 1850, **75**, 315–355.
3. A. Michaelis and A. Polis, *Ber.*, 1887, **20**, 54–57.
4. T. G. Chasteen and R. Bentley, *Appl. Organomet. Chem.*, 2003, **17**, 201–211.
5. R. Bentley and T. G. Chasteen, *Microb. Mol. Biol. Rev.*, 2002, **66**, 250–271.
6. J. S. Thayer, *Appl. Organomet. Chem.*, 2002, **16**, 677–691.
7. E. Dopp, L. M. Hartmann, A.-M. Florea, A. W. Rettenmeier and A. V. Hirner, *Crit. Rev. Toxicol.*, 2004, **34**, 301–333.
8. J. Feldmann, in *Organometallic Compounds in the Environment*, 2nd edn., Ed. P. J. Craig, J. Wiley & Sons, Chichester, UK, 2003, pp. 353–389.
9. S. Wallenhauer and K. Seppelt, *Angew. Chem. Int. Ed. Engl.*, 1994, **33**, 976–978.
10. K. Dill and E. L. McGown, in *The Chemistry of Organic Arsenic, Antimony and Bismuth Compounds*, Ed. S. Patai, Wiley, New York, 1994, pp. 695–713.
11. P. J. Craig, G. Eng and R. O. Jenkins, in *Organometallic Compounds in the Environment*, 2nd edn., Ed. P. J. Craig, J. Wiley & Sons, Chichester, UK, 2003, pp. 1–55.
12. K. Haas and J. Feldmann, *Anal. Chem.*, 2000, **72**, 4205–4211.
13. E. Amberger, *Ber. dtsch. chem. Ges.*, 1961, **94**, 1447–1452.
14. H. Althaus, H. J. Breunig and E. Lork, *Organometallics*, 2001, **20**, 586–589.
15. A. Marquardt, *Ber.*, 1887, **20**, 1516–1523.
16. S. Wang, D. B. Mitzi, G. A. Landrum, H. Genin and R. Hoffmann, *J. Am. Chem. Soc.*, 1997, **119**, 724–732.

17. A. L. Allred and A. L. Hensley, *J. Inorg. Nucl. Chem.*, 1961, **17**, 43–54.
18. T. N. Bell, B. J. Pullman and B. O. West, *Aust. J. Chem.*, 1963, **16**, 636–646.
19. O. J. Scherer, P. Hornig and M. Schmidt, *J. Organomet. Chem.*, 1966, **6**, 259–264.
20. C. H. Bamford, D. L. Levi and D. M. Newitt, *J. Chem. Soc.*, 1946, 468–471.
21. L. H. Long and J. F. Sackman, *Research Correspondence*, 1955, **8**, S23–S24.
22. T. Sollmann and J. Seifter, *J. Pharmacol. Exp. Ther.*, 1939, **67**, 17–49.
23. G. O. Doak and L. D. Freedman, *Organometallic Compounds of Arsenic, Antimony, and Bismuth*, Wiley-Interscience, New York, 1970, pp. 419–461.
24. J. Feldmann, *J. Anal. At. Spectrom.*, 1997, **12**, 1069–1076.
25. J. Feldmann, E. M. Krupp, D. Glindemann, A. V. Hirner and W. R. Cullen, *Appl. Organomet. Chem.*, 1999, **13**, 739–748.
26. J. Feldmann, R. Grümping and A. V. Hirner, *Fresenius J. Anal. Chem.*, 1994, **350**, 228–234.
27. A. V. Hirner, J. Feldmann, R. Goguel, S. Rapsomanikis, R. Fischer and M. O. Andreae, *Appl. Organomet. Chem.*, 1994, **8**, 65–69.
28. J. Feldmann and A. V. Hirner, *Int. J. Environ. Anal. Chem.*, 1995, **60**, 339–359.
29. J. Feldmann, I. Koch and W. R. Cullen, *Analyst*, 1998, **123**, 815–820.
30. S. Maillefer, C. R. Lehr and W. R. Cullen, *Appl. Organomet. Chem.*, 2003, **17**, 154–160.
31. E. M. Krupp, R. Grümping, U. R. R. Furchtbar and A. V. Hirner, *Fresenius J. Anal. Chem.*, 1996, **354**, 546–549.
32. A. V. Hirner, U. M. Grüter and J. Kresimon, *Fresenius J. Anal. Chem.*, 2000, **368**, 263–267.
33. U. M. Grüter, J. Kresimon and A. V. Hirner, *Fresenius J. Anal. Chem.*, 2000, **368**, 67–72.
34. K. Michalke, E. B. Wickenheiser, M. Mehring, A. V. Hirner and R. Hensel, *Appl. Environ. Microbiol.*, 2000, **66**, 2791–2796.
35. K. Michalke, J. Meyer, A. V. Hirner and R. Hensel, *Appl. Organomet. Chem.*, 2002, **16**, 221–227.
36. J. Meyer, A. Schmidt, K. Michalke and R. Hensel, *System. Appl. Microbiol.*, 2007, **30**, 229–238.
37. J. Meyer, K. Michalke, T. Kouril and R. Hensel, *System. Appl. Microbiol.*, 2008, **31**, 81–87.
38. K. Michalke, A. Schmidt, B. Huber, J. Meyer, M. Sulkowski, A. V. Hirner, J. Boertz, F. Mosel, P. Dammann, G. Hilken, H. J. Hedrich, M. Dorsch, A. W. Rettenmeier and R. Hensel, *Appl. Environ. Microbiol.*, 2008, **74**, 3069–3075.
39. F. Challenger, *Chem. Rev.*, 1945, **36**, 315–361.
40. K. Michalke, J. Meyer and R. Hensel, *Appl. Environ. Microbiol.*, 2006, **72**, 6819–6821.
41. H. R. Rawls, M. V. Marshall, H. L. Cardenas, H. R. Bhagat and I. Cabasso, *Dent. Mater.*, 1992, **8**, 54–59.
42. T. Arata, Y. Oyama, K. Tabaru, M. Satoh, H. Hayashi, S. Ishida and Y. Okano, *Environ. Toxicol.*, 2002, **17**, 472–477.
43. U. von Recklinghausen, L. M. Hartmann, S. Rabieh, J. Hippler, A. V. Hirner, A. W. Rettenmeier and E. Dopp, *Chem. Res. Toxicol.*, 2008, **21**, 1219–1228.

44. P. Pinel-Raffaitin, C. Pécheyran and D. Amouroux, *Atmospheric Environment*, 2008, **42**, 7786–7794.
45. M. V. Bland, S. Ismail, J. A. Heinemann and J. I. Keenan, *Antimocrob. Agents Chemother.*, 2004, **48**, 1983–1988.
46. R. Ge, X. Sun, Q. Gu, R. M. Watt, J. A. Tanner, B. C. Y. Wong, H. H. Xia, J. D. Huang, Q. -Y. He and H. Sun, *J. Biol. Inorg. Chem.*, 2007, **12**, 831–842.
47. S. Cun, H. Li, R. Ge, M. C. M. Lin and H. Sun, *J. Biol. Chem.*, 2008, **283**, 15142–15151.

Met. Ions Life Sci. **2010**, *7*, 319–364

10

Formation, Occurrence, Significance, and Analysis of Organoselenium and Organotellurium Compounds in the Environment

Dirk Wallschläger [a] ***and Jörg Feldmann*** [b]

[a]Environmental & Resource Sciences Program and Department of Chemistry, Trent University, 1600 West Bank Dr., Peterborough, ON K9J 7B8, Canada
<dwallsch@trentu.ca>

[b]Trace Element Speciation Laboratory (TESLA), College of Physical Science, University of Aberdeen, Meston Walk, Aberdeen, Scotland, AB24 3UE, UK
<j.feldmann@abdn.ac.uk>

Metal Ions in Life Sciences, Volume 7 Edited by Astrid Sigel, Helmut Sigel, and Roland K. O. Sigel

Published by the Royal Society of Chemistry, www.rsc.org DOI: 10.1039/9781849730822-00319

ABSTRACT: Among all environmentally-relevant trace elements, selenium has one of the most diverse organic chemistries. It is also one of the few trace elements that may biomagnify in food chains under certain conditions. Yet, the exact chemical forms of selenium involved in the uptake into organisms and transfer to higher trophic levels, as well as the biochemical mechanisms that lead to their subsequent metabolism in organisms, are still not well understood. This is in part due to the analytical challenges associated with measuring the myriad of discrete Se species occurring in organisms. While there are generalized concepts of selenium metabolism, there is a lack of conclusive analytical evidence supporting the existence of many postulated intermediates. Likewise, there is a disconnect between the major selenium species encountered in abiotic compartments (waters, soils, and sediment), and those found in organisms, which renders the qualitative and quantitative description of the bioaccumulation process uncertain. Here, we summarize the knowledge on important selenium and tellurium species in all environmental compartments, and identify gaps and uncertainties in the existing body of knowledge, with emphasis on problems associated with past and current analytical methodology.

KEYWORDS: amino acids · bioaccumulation · natural organic matter · proteins · speciation analysis · volatilization

1. INTRODUCTION

Selenium and tellurium occur in the environment as trace elements. They are both classical metalloids in the group 16 of the periodic table of the elements. Although the metallic character in the group increases with elemental mass, the general chemistry of both elements exhibits some resemblance to the chemistry of the non-metal sulfur. All three elements occur mainly in the oxidation states –II, 0, +IV and +VI. While in the oxidation states +IV and + VI, they form mainly oxo-acids or their corresponding anions, in their reduced oxidation states (–II, 0), they can form either metal salts and complexes or bind to organic moieties. The oxo-acids of selenium are selenous acid/selenite [oxidation state Se(IV): $H_2SeO_3/HSeO_3^-/SeO_3^{2-}$] and selenic acid/selenate [oxidation state Se(VI): $H_2SeO_4/HSeO_4^-/SeO_4^{2-}$]. For tellurium, the oxo-acids tellurous acid/tellurite [oxidation state Te(IV): $H_2TeO_3/HTeO_3^-/TeO_3^{2-}$] and telluric acid/tellurate exist, but the latter has the general structure $Te(OH)_6$ [oxidation state Te(VI): $H_6TeO_6/H_5TeO_6^-/H_4TeO_6^{2-}$], which differs from its sulfur and selenium analogs [1].

Since Te is less electronegative than C, H, and S, the oxidation state of tellurium in organo-Te compounds is always + II, unless a compound has a Te-Te bond, in which case the oxidation state becomes + I. By contrast, the assignment of a formal Se oxidation state in organo-Se compounds becomes more ambiguous, because Se has a very similar electronegativity to those of S and C [1]. Consequently, the formal Se oxidation states in the two simplest and most common organo-Se species, CH_3-Se-CH_3 and CH_3–Se-Se-CH_3, could be assigned any value between –II and + II. Therefore, we will not refer to organo-Se species by oxidation state in this chapter, and it should be understood that when others have discussed organic Se compounds as Se(0) or Se(–II) species, we have substituted those expressions with the term "organo-Se species". The abbreviations and structures of identified organoselenium compounds are listed in Table 1.

The selenium- and tellurium-carbon bonds get weaker when the oxidation state of the chalcogen increases, due to the larger gap of orbital energies or the polarity of the bond. Hence, this chapter will focus mainly on reduced organo-Se and -Te species, since these are the most stable under environmental conditions and show a large natural variety, particularly for selenium. Accordingly, no organotellurium compound with higher oxidation state than + II has been identified in the environment so far, and there are only a few examples of naturally occurring organoselenium compounds, e.g., methylseleninic acid (MeSe(IV)) and selenocysteic acid (Se(IV)Cys), in which selenium has an oxidation state $> +$II, which distinguishes the chemistry of selenium and tellurium significantly from that of sulfur.

2. ORGANOSELENIUM SPECIES

It is generally assumed that organic Se species exist in ambient waters, soils, and sediments, and that they play a key role in the bioaccumulation of Se. However, there are two distinctly different classes of chemical compounds that are described as "organoselenium compounds" in the literature: discrete molecules (i.e., such to which one unique chemical structure can be assigned)

Table 1. Structures of selenium and organoselenium compounds.

Name	Abbreviation	Structure
Selenium	Se^0	Se^0
Selenide	Se^{2-}	Se^{2-}
Selenate (selenic acid)	Se(VI)	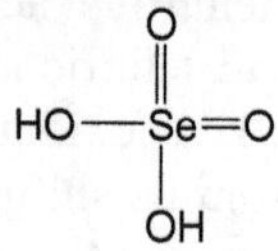
Selenite (selenous acid)	Se(IV)	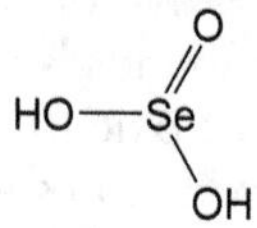
Selenocyanide	$SeCN^-$	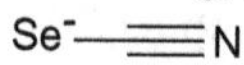
Methylselenol	MeSeH	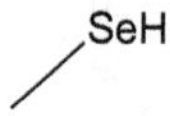
Methylseleninic acid	MeSe(IV)	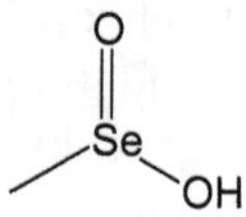
Methylselenenic acid	MeSe(II)	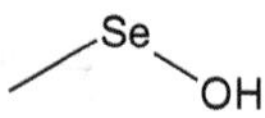
Dimethylselenide	DMSe	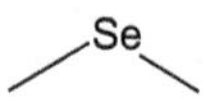
Dimethyldiselenide	DMDSe	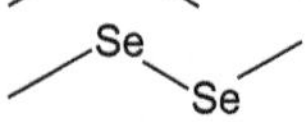
Dimethylselenenyl sulfide	DMSeS	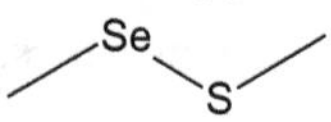
Dimethyselenenyl disulfide	DMSeDS	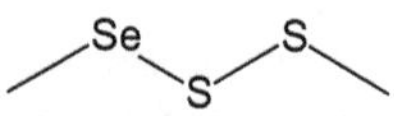
Methylethylselenide	EMSe	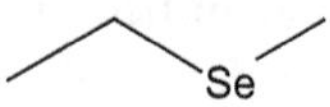
Diethylselenide	DESe	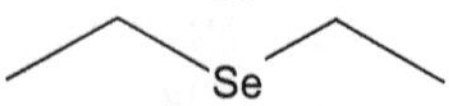
Methylallylselenide	MeAllSe	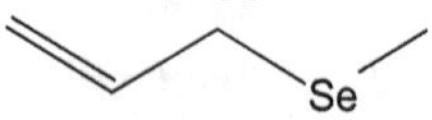
Bis(methylthio)selenide	MeSSeSMe	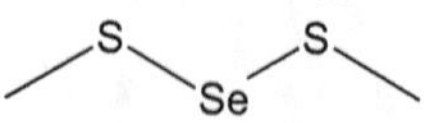
Methylthio-allylthioselenide	MeSSeSAll	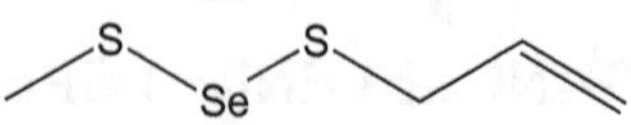
Trimethylselenonium	$TMSe^+$	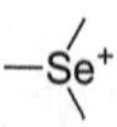
Dimethylselenonium propionate	DMSeP	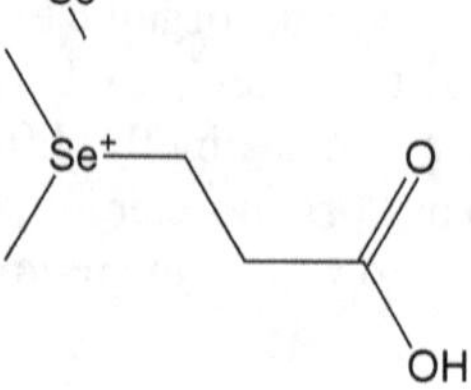

Table 1. (*Continued*).

Name	Abbreviation	Structure
Seleno(IV)cysteic acid	Se(IV)Cys	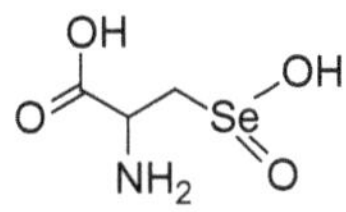
Se-cysteine	SeCys	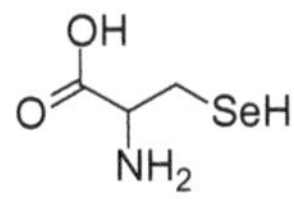
Se-methyl-seleno-cysteine	SeMeSeCys	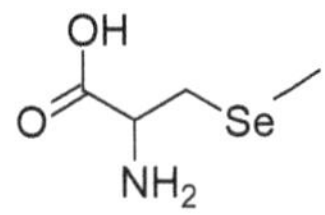
Se-allyl-seleno-cysteine	SeAllSeCys	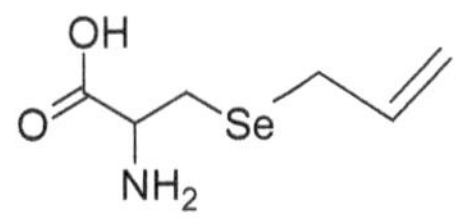
Se-methyl-seleno-cysteine-seleniumoxide	SeMeSeCysSe(IV)	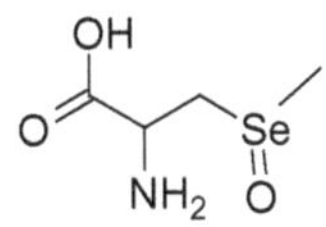
Seleno-methionine	SeMet	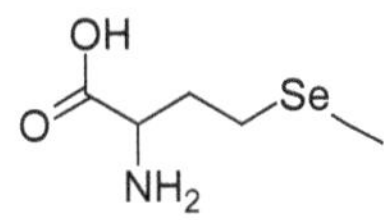
Se-methyl-seleno-methionine (dimethyl-(3-amino-3-carboxy-1-propyl) selenonium)	SeMeSeMet	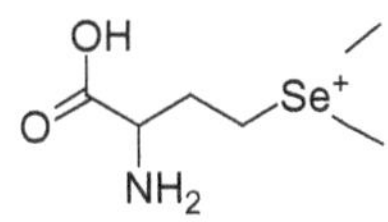
Seleno-homocysteine	SeHcys	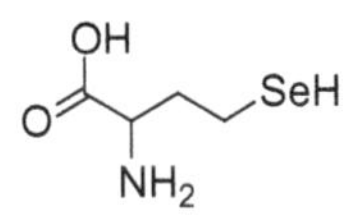
Seleno-cystine	$(SeCys)_2$	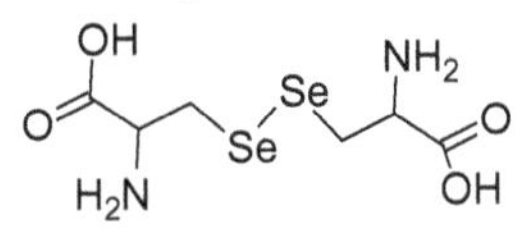
Cysteine-selenocysteine	CysSSeCys	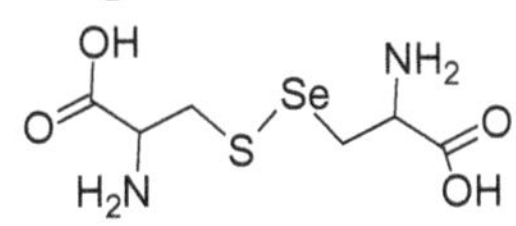
Seleno-homocystine	$(SeHcys)_2$	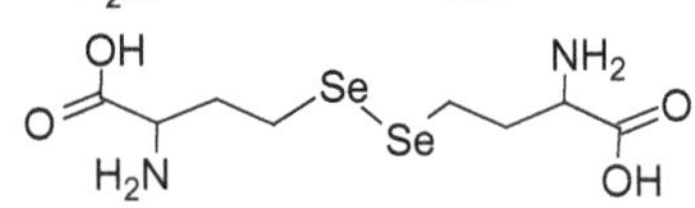

Table 1. (*Continued*).

Name	Abbreviation	Structure
Se-oxo-selenomethionine	Se(IV)Met	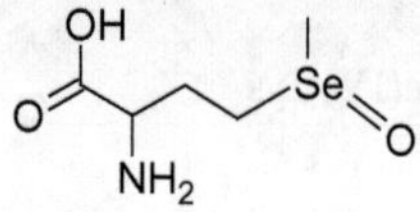
S-methyl-seleno-cysteine	SMeSeCys	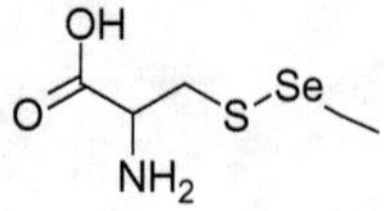
Selenocystamine	SeCyst	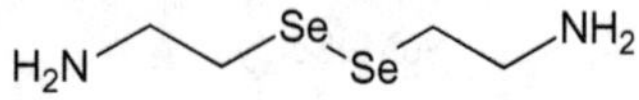
3-Butenyl isoselenocyanate	BuNCSe	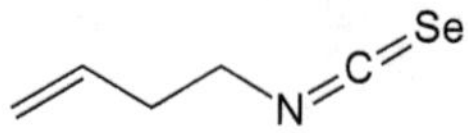
Selenourea	SeU	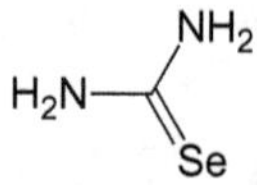
Selenobetaine	SeBet	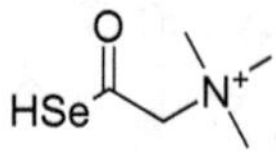
Se-cystathionine	SeCT	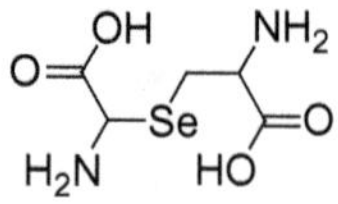
γGlutamyl-seleno-cystathionine	γGluSeCT	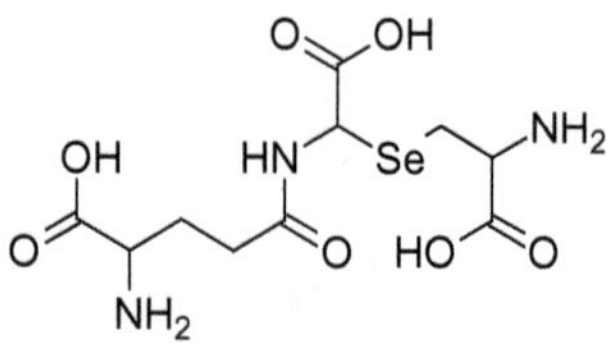
γGlutamyl-seleno-methyl-selenocysteine	γGluSeMeSeCys	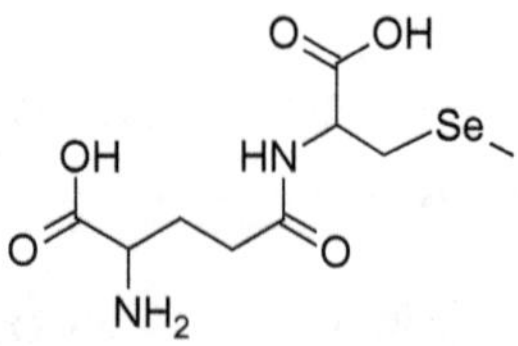
γGlutamyl-seleno-methionine	γGluSeMet	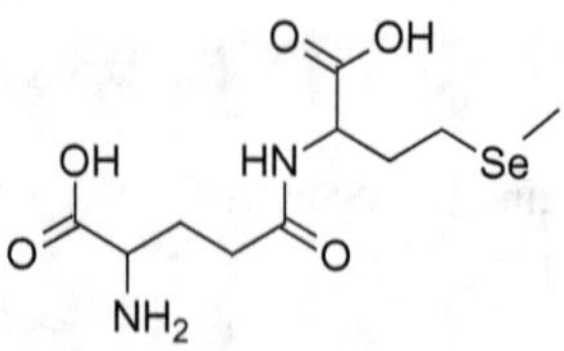

Table 1. (*Continued*).

Name	Abbreviation	Structure
Se-adenosyl-selenohomocysteine	SeAdoSeHcys	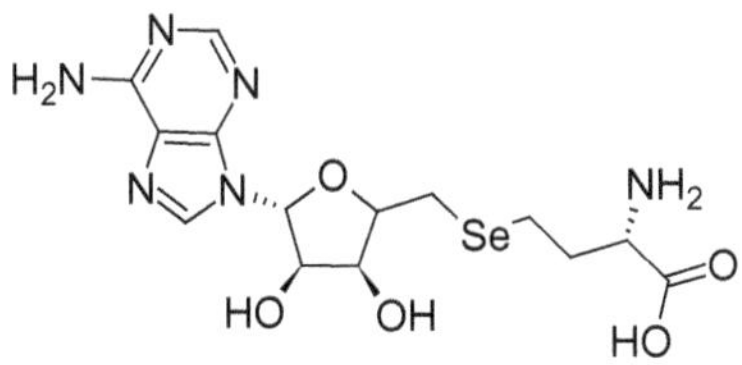
Se-adenosyl-methyl-selenomethionine	SeAdoMeSeMet	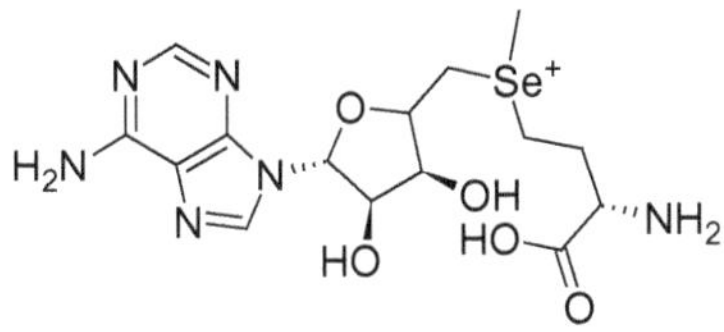
Cysteinyl-Se-glutathione	CysSeSG	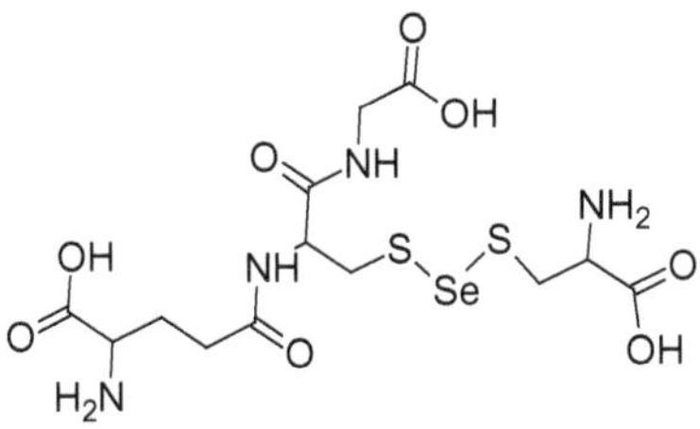
Serine-seleno-cysteinyl-glutathione	SerSeCysSG	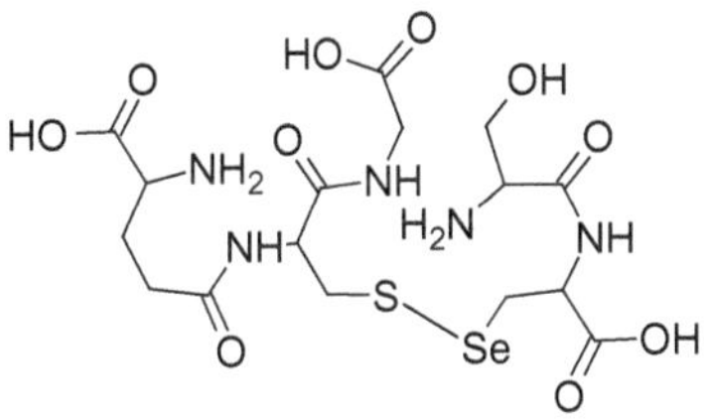
Seleno-phytochelatin 2	$SePC_2$	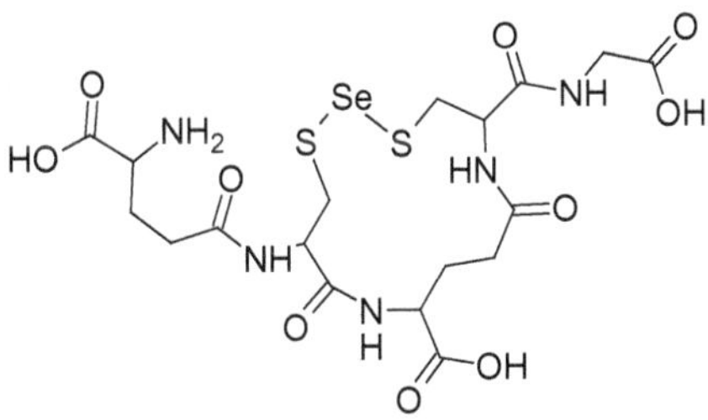
Glutathione-selenol	GSSeH	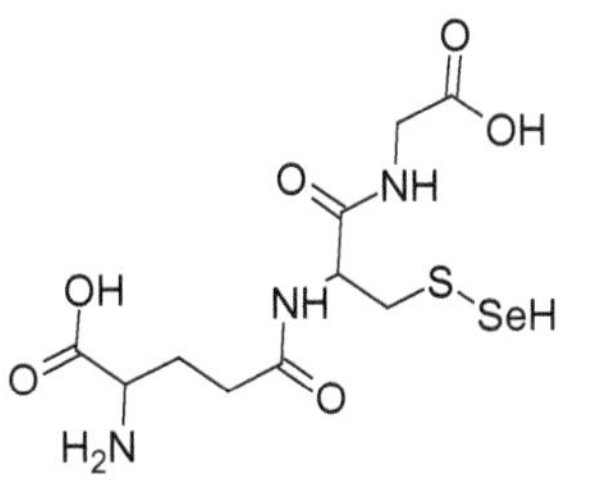

Table 1. (*Continued*).

Name	Abbreviation	Structure
Di-glutathione-selenide	GSSeSG	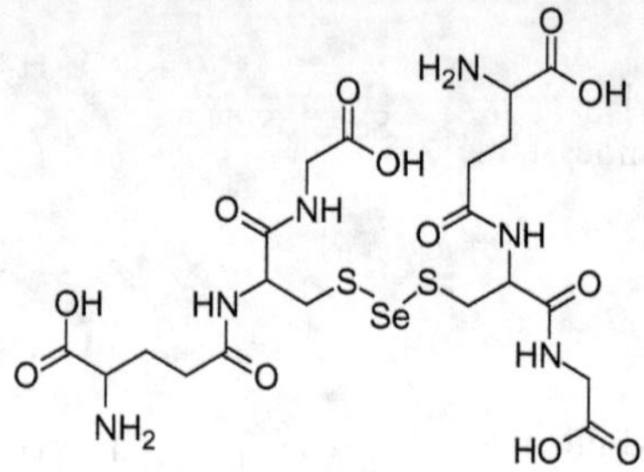
Methyl-selenide-glutathione	MeSeSG	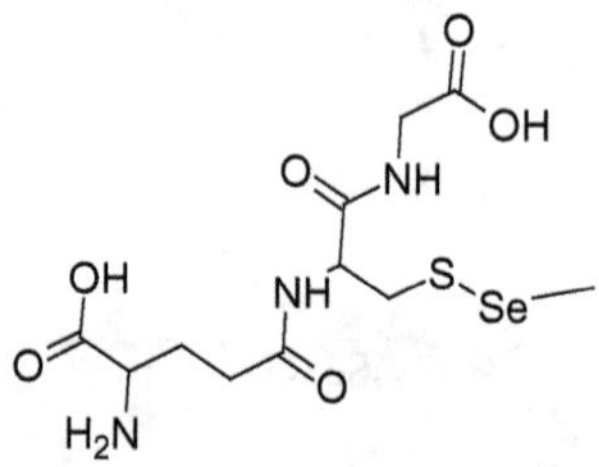
Glutathione-seleno-*N*-acetylgalactosamine	GSSeGalNAc	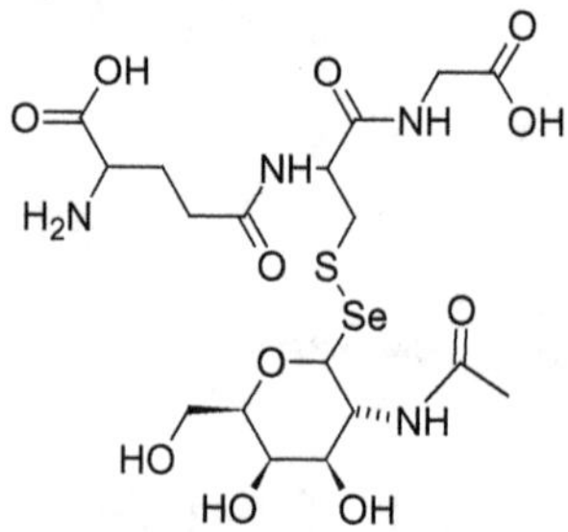
Se-methyl-seleno-*N*-acetylgalactosamine (selenosugar 1)	MeSeGalNAc	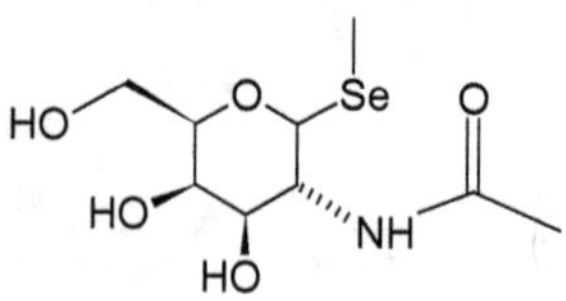
Se-methyl-seleno-*N*-acetylglucosamine (selenosugar 2)	MeSeGluNAc	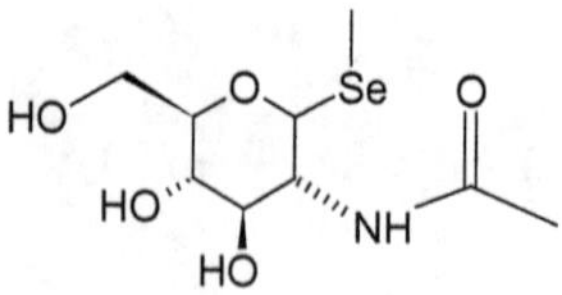
Se-methyl-seleno-galactosamine (selenosugar 3)	$MeSeGalNH_2$	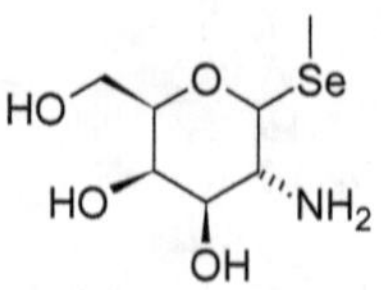

Table 1. (*Continued*).

Name	Abbreviation	Structure
Selenosinigrin		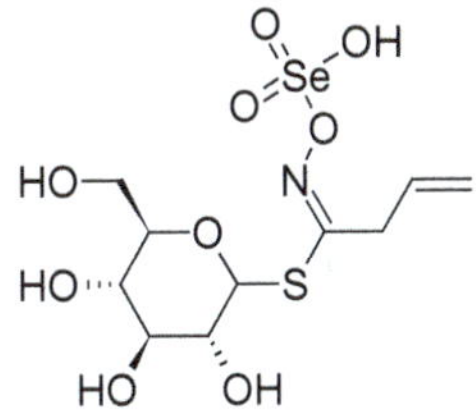
4-Selenouridine		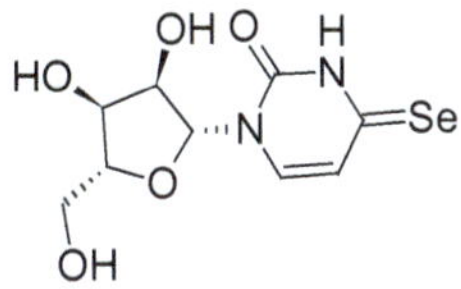
Selenobiotin		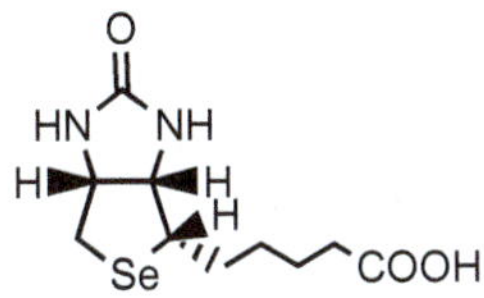
Seleno-bis(*S*-glutathionyl) arsinium ion	$GS_2As\text{-}Se^-$	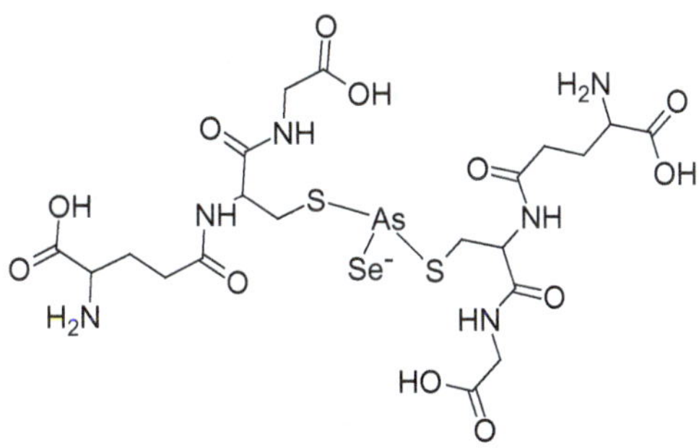
Seleno-proteins (SeCys replaces Cys in proteins)		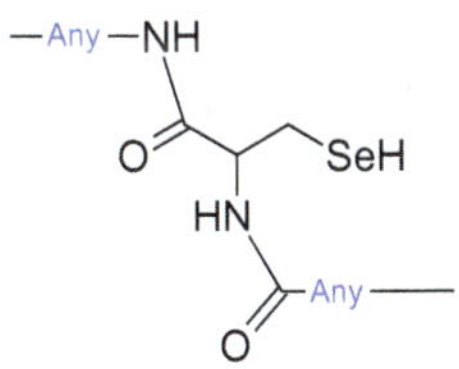
Selenium containing proteins (SeMet replaces Met in proteins)		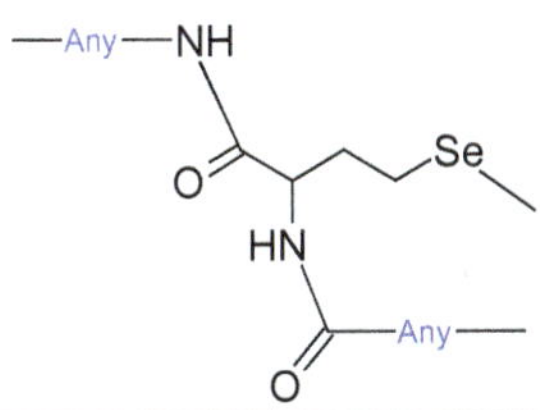

in which Se is bound to at least one carbon atom (which makes them "true" organometalloid compounds), and natural organic matter (NOM) including Se in its structure ("NOM-Se"). While each NOM-Se molecule has a discrete structure, it will generally be different from that of any other NOM-Se molecule, and therefore it is a futile effort to assign specific chemical structures to this group of Se species (although, of course, generalized structural features and molecular weight distributions can be used to characterize them). Since NOM-Se species represent the biological breakdown products of discrete organo-Se species originally present in tissues, they will generally retain their original association with at least one carbon atom (and thus be "true" organo-Se compounds).

Additionally, it is also possible that NOM molecules originally not containing Se will bind Se via their functional groups. In the resulting compounds, Se would generally be bound to either O, N or S (which constitute the vast majority of NOM functional groups), and consequently, these molecules would not be "true" organo-Se species. Although textbook geochemical knowledge assumes that inorganic Se species do not bind to common NOM functional groups, because both are typically negatively charged at ambient pH, there is some evidence that Se binds to dissolved NOM molecules [2], so this sub-type of "organic" Se species cannot be entirely ignored in environmental studies. Since these two classes of organo-Se species, i.e., discrete organo-Se species and Se-NOM (regardless of whether Se was originally incorporated into the NOM structure, or binds to it at a later point in time) are very different from one another, they require equally different analytical methods for their determination, so they will be discussed separately in the following.

2.1. Methods for the Analysis of Organic Selenium Species

2.1.1. Analysis of Discrete Organoselenium Species

The analysis of discrete organo-Se species requires at least the combination of a chromatographic separation with a Se-specific detector, so that each species can be identified by its unique retention time in the chromatogram, and its identity as a Se species can be verified by the fact that it yields a detector response. For small molecular weight Se species, gas chromatography (GC) or liquid chromatography (LC) are the most suitable separation methods, and inductively-coupled plasma-mass spectrometry (ICP-MS) is rapidly becoming the most popular Se-specific detector. As for the analysis of Se species in tissues, co-elution of a Se species found in an environmental sample with a standard is considered insufficient for proof of identity, and should be confirmed independently, either by a second chromatographic separation employing a different separation principle, or by obtaining a molecular mass spectrum of the Se species in the environmental sample [3].

When molecular mass spectrometry is not available or not sensitive enough to confirm the identity of a Se species, the fact that a substance eluting from a chromatographic separation is indeed a Se species should be confirmed via a second Se isotope (or more) when ICP-MS is used for detection, or by using another different detection principle, e.g., atomic emission spectrometry (AES), atomic fluorescence spectrometry (AFS) or atomic absorption spectrometry (AAS), where possible. When quantification of the encountered Se species is desired, then the two independent analyses, using either two different separations or two different detection modes, should agree within the margin of analytical error. While these criteria represent ideal conditions and can often not be realized in studies, they will be applied in the following to separate questionable observations reported in previous studies on the determination of discrete Se species in environmental waters, soils, and sediments from those that are verified beyond reasonable doubt.

2.1.2. *Direct Analysis of Natural Organic Matter: Selenium in Waters, Soils, and Sediments*

The analysis of NOM-Se is a challenging task when one wants to establish an actual chemical association between Se and an NOM molecule, rather than just establishing co-occurrence in an operationally-defined sample fraction (see next section) or simple statistical correlations. Since separation of individual NOM molecules from one another is an almost impossible task, at the very least, one needs to employ a direct speciation analysis method for this purpose which separates different NOM size fractions from one another and from other sample constituents, and then determine both organic carbon (OC) and Se in this fraction. The preferable way of doing this is by using a chromatographic (or similar) separation coupled on-line to both an organic carbon analyzer and an ICP-MS, and observing co-eluting signals for OC and Se.

Suitable separation methods include field flow fractionation (FFF) and gel chromatography, which is known by several synonymous names, including size exclusion chromatography (SEC), gel filtration (GF) and gel permeation chromatography (GPC). Other Se-selective detection methods could be substituted for ICP-MS, provided they do not require Se to be present in any specific chemical form. Other non-chromatographic NOM fractionation techniques, such as ultrafiltration (UF) could also be used. Strictly speaking, though, even these approaches would not prove conclusively the chemical link between Se and NOM (no matter whether Se is bound to the NOM functional groups or incorporated into the bulk NOM molecule) because they still rely on the co-occurrence of OC and Se in a given (chromatographic) sample fraction. It is consequently conceivable that Se bound to some other sample constituent (e.g., a colloidal mineral particle) co-elutes with a certain NOM size fraction without being chemically associated with any NOM molecule. Nonetheless, this approach would yield much higher certainty about NOM-Se association than any other of the mentioned approaches.

2.1.3. Operationally-Defined Determination of "Organic" Selenium in Waters

The vast majority of the previous studies that have suggested the presence of an "organic" Se fraction in ambient waters used selective sequential hydride generation (SSHG), generally with AAS detection, as the method of analysis. This approach is based on the fundamental assumption that selenite ($HSeO_3^-$) is the only Se species that forms a volatile product (in that case: hydrogen selenide H_2Se) upon reaction with borohydride (BH_4^-) under acidic conditions. It furthermore assumes that Se in ambient waters is present either as selenite (Se(IV)), selenate ((Se(VI)) or reduced Se species. The operationally-defined separation of these three Se species is then accomplished by three separate analyses: direct determination of selenite, determination of selenate after pre-reduction with boiling concentrated HCl, and determination of "reduced Se" after oxidation. Although these three analyses could theoretically be performed successively on only one sample aliquot, they are often performed in parallel on three separate sample aliquots, yielding measurements of selenite, total inorganic Se ("TISe" = selenite + selenate) and total Se ("TSe"); selenate and "reduced Se" are then calculated by difference (TISe – selenite or TSe – TISe, respectively).

It is important to point out that in the original method [4] the term "dissolved organic selenide" is used instead of "reduced Se"; although it was not shown that specific individual species that fit the general description appear only in the "reduced Se" fraction and not in the "selenite" or TISe fractions. While Se in organo-Se species is present in reduced oxidation states, there are also reduced inorganic Se species that could (partially) appear in this operationally-defined fraction, as has been shown for selenocyanate ($SeCN^-$) [5].

Unfortunately, many authors, e.g., Fio and Fujii [6], have used the term "organic Se" synonymous with "reduced Se" when SSHG was used as the analytical method in their studies, so that this fraction is now generally believed to represent organic Se species, even though the method, by virtue of its operationally-defined nature, provides no positive structural information about any Se species detected in this fraction. Considering (for illustrative purposes) the case of an ambient water containing a significant fraction of colloidal elemental Se (oxidation state 0), one would expect this Se species to be determined in the reduced Se fraction (although the behavior of Se^0 during the different sample pre-treatment steps and the hydride generation procedure has, to our knowledge, not been studied), which would lead to a fundamental misinterpretation of the obtained Se speciation pattern.

It is also important to realize that no commonly employed quality control (QC) measure would be able to identify this problem. To make matters worse, some studies have shown that the recovery of selenate in the TISe analysis can be incomplete (around 80%) [7]. If "reduced" Se is determined by difference (as usual), then this would lead to an overestimation of "reduced" (or "organic" Se). For these reasons, we believe that "organic" Se fractions reported in studies using the SSHG approach without further analytical evidence should be evaluated very critically, and certainly not be interpreted as discrete Se species. However, in defence of the results obtained in previous studies using the SSHG, it has to be conceded that just as much as it is unproven that the "reduced" Se fraction actually contains discrete organo-Se species, it is equally unproven that there are any significant fractions of reduced inorganic Se species present in ambient waters, and that these end up in and constitute the majority of Se detected in the "reduced" Se fraction. To circumvent the problems associated with the indirect determination of "organic" Se fractions by difference, a variant of the SSHG approach has been described recently [7] in which organic Se species are determined in the second analytical step after UV-assisted decomposition to selenite, before selenate is determined in the third step.

Conversely, the SSHG procedure may potentially also hide the presence of actual organic Se species in ambient waters. There is evidence [7] that some organic Se species partially break down to Se(IV) during the TISe pre-treatment step (involving boiling with HCl), which would make them appear as "Se(VI)" in the procedure.

Furthermore, considering simple methylated Se species as an example, inherently volatile compounds like dimethylselenide (CH_3-Se-CH_3, DMSe) would presumably be measured in the selenite fraction because they would be purged from solution during the HG reaction. Likewise, the frequently discussed Se(IV) species CH_3-SeO_2^- could possibly form the volatile hydride CH_3-Se-H during the HG reaction (again, we are not aware of a study that has tested the HG behavior of this species), and also be volatilized in the "selenite" analysis. These hypothetical problems could easily be prevented by using a GC separation between the HG step and the detector, as was suggested in the original method by Cutter [8] and is commonly done for arsenic speciation analysis. However, this is often not done for Se speciation analyses in ambient waters, so it is conceivable that discrete organic Se species might remain undetected because they appear in the wrong fraction of the SSHG procedure.

2.1.4. *Operationally-Defined Determination of "Organic" Selenium in Soils and Sediments*

Se speciation in soils and sediments is generally assessed in two different ways: direct spectroscopic analysis using various X-ray absorption spectroscopy (XAS) methods, and sequential extraction procedures (SEPs). Due to the fact that XAS methods have only recently become available and sensitive enough to study Se speciation at environmentally-relevant levels and require the use of a synchrotron facility, most of the existing body of knowledge was generated using various SEPs. In an SEP, it is attempted to successively solubilize the individual major constituents of a soil or sediment (e.g., organic matter or various types of minerals) by using a sequence of increasingly aggressive leaching solutions, and thereby releasing the fractions of trace elements associated with these constituents. In each step, it is intended to leach one solid phase (and its associated trace elements) completely and selectively without attacking or changing the other remaining solid phases (and their associated trace elements).

Discrete organo-Se species are generally not assessed by SEPs because they would have to be associated with a specific solid phase and would have to remain intact during this particular extraction step, so that they could then be determined by an LC-based speciation analysis method. Normally, only the total concentration of a trace element is determined in each extract; therefore, typically no information is generated about the individual Se species leached in each step of a SEP. Instead, the determination of "organic" Se in soils and sediments by SEP generally aims at NOM-Se, despite the fact that the binding of Se species to NOM is sometimes questioned. This is due to the fact that many studies on Se speciation in soils or sediments have adopted a generic SEP approach developed for cationic trace metals [9], which obviously have a very different environmental chemistry than Se. In these SEPs, NOM is solubilized by one of two general approaches: oxidative destruction in acidic medium, or alkaline leaching. Both approaches are associated with some fundamental problems, and can therefore lead to erroneous results.

Oxidation of NOM has the advantage that it can mobilize Se associated with either of the three principal NOM size fractions (fulvic acids, humic acids, and humins) because all of them are converted to CO_2 (ideally) under these conditions. The fundamental disadvantage of this approach is that it can release Se species associated with other phases, e.g., oxidation of Se^0 or acidic dissolution of sulfide or carbonate minerals. Therefore, this approach can only work if all other Se species or Se-containing solid phases that can dissolve under acidic oxidizing conditions have been removed in the preceding SEP steps. By comparison, alkaline NOM leaching (either with NaOH or Na-pyrophosphate solutions) of intact NOM molecules does not create the problems associated with acidic pH and oxidizing conditions, but is unsuitable for Se associated with humins (the largest molecular weight fraction of NOM) [10], which are insoluble in water over the entire pH range. If this shortcoming is accepted, then NOM-Se only needs to be distinguished from other easily-leachable Se species, such as adsorbed selenite and selenate, which can be accomplished using LC-based speciation analysis methods for the determination of discrete Se species in these extracts [11].

If any Se associated with humins is to be analyzed as well, the humin fraction may be extracted with organic solvents [12] in the next step, but care must be taken not to extract other Se species soluble in organic solvents simultaneously (e.g., certain Se^0 allotropes) [13]. The generic SEP for trace elements [9] does not account for any of these complications, so Se speciation patterns obtained using this approach [14,15] can be misleading and may not reflect the actual Se speciation in the studied soil or sediment. However, some Se-specific SEPs have been developed [16,17] and provide more accurate information on "organic" Se fractions in soils and sediments. By nature, SEPs also provide some information on the mobility of different Se fractions (including "organic" Se) in soils and sediments, which can very carefully be put in qualitative relation to bioavailability.

XAS techniques eliminate most fundamental problems associated with SEPs because no extraction steps are involved, since Se speciation is measured directly in the solid sample. However, XAS methods suffer from two other fundamental shortcomings: the lack of sensitivity (compared to extraction-based methods using atomic spectrometry measurements) and the critical dependence of the results on the number and quality of available standard Se species. While the first is gradually overcome by instrumental improvements, the second is method-inherent. XAS spectra are interpreted by comparison to standard compounds, and the Se speciation in the sample of interest is expressed as a linear combination of the available standards. Therefore, if we do not know a priori which Se species are present in soils or sediments, the choice and availability of standards may limit how accurately the actual Se speciation can be described with them.

Of the two most commonly employed XAS methods, X-ray absorption near-edge spectroscopy (XANES) distinguishes only between Se species based on their average oxidation states, and is consequently not able to differentiate between specific similar Se compounds. The XANES spectra of selenomethionine (SeMet), selenocysteine (SeCys), selenocystine $(SeCys)_2$, and sulfo-selenocystine (CysSSeCys) show very small differences (around 0.1 eV) between the peak positions for SeMet versus SeCys, and for $(SeCys)_2$ versus CysSSeCys [18]. While these small differences are theoretically suitable for distinguishing between the two compounds in each pair of organo-Se species, the absolute energy accuracy in XANES measurements is typically also on the order of 0.2 eV (even with energy calibration relative to a standard substance) [19]. Additionally, the absorption signals for Se in these spectra are quite broad (around 2.5 eV), so it is probably not practically possible to distinguish between these pairs of Se species, especially not when they are present in mixtures.

Furthermore, XANES does not provide structural information, so it is impossible to distinguish between SeMet/SeCys and any other Se species that contains the same structural feature, i.e., a Y-Se-C unit, where Y is either an H atom or another C atom. Likewise, it is impossible to distinguish between $(SeCys)_2$/CysSSeCys and any other Se species that contains a Y-Se-S(e)-Y structural unit. This shortcoming of XANES is important to keep in mind when interpreting the spectra recorded for natural samples, because the Se fractions that match the XANES spectra of SeMet or $(SeCys)_2$ are often inappropriately equated to those exact species. This overinterpretation may have significant implications, since SeMet is often discussed as a key species involved in Se bioaccumulation, but "its" XANES signal could equally stem from a completely different Se species, e.g., DMSe [20]. By analogy, dimethyldiselenide (CH_3-Se-Se-CH_3, DMDSe) could be "mistaken" for $(SeCys)_2$, despite their obvious chemical differences. Extended X-ray absorption fine structure spectroscopy (EXAFS) could resolve some of these ambiguities, but it requires much higher Se concentrations than XANES, which yields interpretable spectra in solids containing 1–10 mg kg^{-1} (dw) total Se [21].

Despite these ambiguities, XANES can distinguish between organic selenides (or selenols) and organic diselenides (or sulfoselenides), and also differentiate both from the commonly studied inorganic Se species Se^0, selenite, and selenate. We were unable to find a XANES study that directly compares the spectra of organic and inorganic selenides, but FeSe and $FeSe_2$ show XANES absorption peaks very close to those of Se^0 [22] and should thus be distinguishable from the organic Se compounds discussed above. However, the same study shows the absorption peak for ZnSe significantly (2 eV) higher than that of FeSe, which would bring it right into the range where the organic selenides have their edge positions. This may be a potential problem when trying to study soils or sediments in which both inorganic and organic reduced Se species can occur, so future studies are warranted to check if these classes of Se species can be distinguished by XANES. In systems where only one or the other type of reduced Se species occurs, e.g., tissues [21] or mineral adsorption studies [22], this problem is avoided, and yields informative results.

The complementary method of EXAFS yields information on the coordination of Se atoms (number and chemical identity of neighboring atoms), and is therefore capable of differentiating between more similar Se species, but this method requires higher Se concentrations in the sample, and is currently not universally applicable to the measurement of Se speciation in soils and sediments yet. Even with EAXFS, though, it is impossible to distinguish between molecules that have functional differences more than three bonds away from the Se atom. This apparent shortcoming of XAS methods (both XANES and EXAFS) is however also advantageous because it helps to integrate individual Se species in a sample into a small number of more generalized groups with distinct Se-containing "functional groups", which may be very helpful especially in the case of NOM-Se species (where the bulk of the molecule may be of little consequence for the behavior of Se).

Contrary to SEPs, no information is obtained about the molecular size or mobility of Se species, and so a combination of SEP and XAS methods is useful for characterizing Se speciation in soils and sediments [10]. Specifically, XAS can be used to identify and eliminate certain typical problems associated with SEPs, including changes in speciation caused by preceding extraction steps and re-adsorption of extracted Se fractions on other solid phases.

2.2. Occurrence of Organoselenium Species in Abiotic Compartments

2.2.1. *Air*

Although several additional volatile organo-Se species can be produced by biotic and abiotic processes (as discussed in the following sections), only DMSe and DMDSe have been detected unequivocally in ambient air samples [23,24]. The atmospheric chemistry of organo-Se species is not studied very well, but a significant build-up of organoselenium compounds in the atmosphere is not expected, since the atmospheric lifetime of those volatile organoselenium species in the presence of common atmospheric oxidants like O_3, $OH^\bullet$ and $NO_3^\bullet$ is only between 5 min and 6 h [25]. It has been suggested that methylated oxidized selenium species, e.g., dimethylselenonium oxide, might be generated as intermediates during the atmospheric oxidation of DMSe and DMDSe to selenite and selenate [26], but no such degradation product has ever been identified in the ambient atmosphere.

In a laboratory study, Rael and Frankenberger [27] studied the reactions of CH_3-Se-CH_3 with the common atmospheric oxidants O_3, $OH^\bullet$ and $NO_3^\bullet$. Ozone transformed CH_3-Se-CH_3 almost quantitatively into CH_3-Se(O)-CH_3, while the reactions with the two radicals led to significant (40–60%)

demethylation and the formation of ionic methylated products. These products were speculated to be $[CH_3\text{-}Se(OH)_2]^+$ and $[(CH_3)_2SeOH]^+$ (as their nitrate salts), which could formally be derived from the reactions of CH_3-Se(O)OH and CH_3-Se(O)-CH_3 with HNO_3. The results indicate the possibility of finding ionic methylated Se species in wet precipitation, for which some preliminary analytical evidence exists [5].

2.2.2. *Water*

Total selenium concentrations in ambient waters are quite low compared to many other trace elements (generally below 0.1 μL^{-1}). The background Se concentration in seawater is around 0.05 $\mu g\,L^{-1}$, and fresh waters appear to have similar background Se concentrations, unless they are impacted by geological or anthropogenic Se sources, such as process waters from oil refineries, mining operations or coal-fired power plants. The main dissolved selenium species in impacted ambient waters ($>1\,\mu g\,L^{-1}$) are typically selenite and selenate. At concentrations approaching the background, significant proportions of "organic" Se have been reported using the operationally-defined hydride generation-based speciation analysis methods [4,7].

Open ocean seawater (in the Atlantic Ocean) was reported to contain around 40 $ng\,L^{-1}$ total Se near the surface, most of which was present as "organic Se" [28]. A large part ($81 \pm 63\%$) of this "organic Se" was tentatively identified as selenoamino acids using a procedure that employs acidic hydrolysis of water-soluble peptides and adsorption of the liberated amino acids on a Cu^{2+}-charged chelating resin [4]. Waters from five lakes were analyzed by both the original [4] and the modified [7] hydride generation-based speciation analysis approach and showed significant fractions of "organic Se" with both methods. At total Se concentrations of $338 \pm 137\,ng\,L^{-1}$, the average "organic Se" fractions were $66 \pm 9\,ng\,L^{-1}$ with the modified and $73 \pm 10\,ng\,L^{-1}$ with the original method; there was a small average positive bias of $6.5 \pm 6.2\,ng\,L^{-1}$ more "organic Se" found with the original method [7]. The authors also noted that their standard "organic Se" compounds (Se-methionine, Se-methyl-selenocysteine, Se-cystine and Se-urea) converted substantially to TISe in the original speciation analysis procedure, but that the "NOM-Se" in the lake waters did not yield the corresponding expected positive bias, from which they concluded that the "NOM-Se" in the lake waters was probably comprised of different organic Se species [7].

So far, the only discrete organo-Se species detected in marine and fresh waters are the volatile species DMSe, DMDSe, and DMSeS [29,30] which are produced by biotic reactions. The identity of these species was confirmed by GC-MS [31], and the mass spectral evidence provided positively distinguished DMSeS from dimethylselenone, which had previously been observed evolving from soils and sewage sludge [32], despite the fact that both species have the same nominal molecular mass. There is also evidence that methylselenol exists in seawater [33], but this identification was only based on co-elution in GC-AFS, and not confirmed by molecular mass spectrometry. Recent unpublished results have also provided GC-MS evidence for the existence of dimethylselenenyl disulfide (DMSeDS), along with the other volatile dimethylated Se species, in a selenium-polluted estuary in New South Wales [34].

The concentration of these volatile Se species in waters is typically only around 0.1% of the total dissolved Se concentration [23,35], but this may still have significant consequences for the environmental cycling of selenium, because those selenium species can volatilize from water bodies such as hot springs [36], from saline lakes [37] or constructed wetlands [38]. To illustrate this point, it was estimated that the annual Se volatilization from the Great Salt Lake (UT) is 1,455 kg, which accounts for about 93% of the annual load [35], albeit only for about 0.01% of the lake's total waterborne Se inventory. Likewise, a constructed treatment wetland was able to remove $>80\%$ of the total Se in the discharge from an oil refinery, and it was estimated that 10–30% of the removed Se was volatilized in the wetland [38].

In a survey of the surface waters in three large European estuaries, it was found that the concentrations of volatile dimethylated Se species decreased in the order DMSe $\gg$ DMSeS $>$ DMDSe, and because the volatility of the species also decreases in the same order, DMSe is by far the major species contributing to Se volatilization from the estuaries [23]. Although, once again, the absolute concentrations of the volatile Se species were only a small fraction of the total dissolved Se concentrations, all three estuaries showed significant Se volatilization fluxes, often much larger than the Se transport by the rivers into the estuaries. Globally it has been estimated that the formation of these volatile organoselenium compounds accounts for 45–80% of natural selenium flux into the atmosphere [39,40].

While it is well known that aquatic organisms, e.g., algae [31,41], can generate these volatile organo-Se species in the environment, some aspects of their formation mechanisms remain speculative (cf. Section 2.3). Amouroux et al. [42] studied the potential environmental precursors for the formation of the volatile organo-Se species in laboratory experiments using synthetic sea water containing humic substances and algal exudates. They found that when selenite or selenate were used as the source of Se, no Se methylation and no Se volatilization were observed in the dark or under (artificial) sunlight. By contrast, when seleno amino acids (SeMet or $(SeCys)_2$) were used as precursors, formation of volatile methylated Se species was observed [42]. This suggests that there might be an important mechanistic link between the observed environmental Se volatilization process and the "organic Se" fraction (presumably consisting of water-soluble Se-bearing proteins) measured in ambient waters [28].

There is laboratory evidence suggesting that other classes of "NOM-Se" species, aside from Se associated with water-soluble proteins, might exist in the environment. Ferri and Sangiorgio [43] conducted a voltammetric investigation of selenite binding to polysaccharides, and found large complexation constants between log $\beta = 7.2$ and 9.6, indicating that such complexes might be stable under environmental conditions. Kamei-Ishikawa et al. [2] studied the binding of selenite to a synthetic commercial humic acid (HA). Although this HA was mostly insoluble (<0.7%), some of the Se remaining in solution associated with the dissolved HA fractions (67–464 mg/L) and UF experiments suggested that these Se-HA associates have a nominal molecular weight (NMW) > 10,000 (50–60% of all Se remaining in solution), 5,000–10,000 (30–60%) or 3,000–5,000 (10–50%). It was not reported how much Se passed through the smallest UF membrane (3,000 NMW cut-off), so it is not possible to calculate from these experiments what concentrations of soluble Se-NOM were produced in these experiments. Although synthetic HA materials are generally not thought to be a close analog to natural HA, this indicates that selenite may associate with natural HA as well, and provides a potential explanation for some of the "organic Se" fraction encountered in natural waters.

Bruggeman et al. [44] studied the interaction of selenite and selenate with humic substances (HS) in aqueous sediment extracts, and found that selenate did not undergo any transformation reactions over a period of three months. By contrast, selenite was lost from solution within one month; most of it (87 and 96%, respectively, for two different study sites) transformed into insoluble Se species, which could be precipitated by centrifugation (indicating a particle diameter >25 nm, according to the authors), over seven months, but some of it (up to 30 or 55%, respectively, for the two study sites) was intermittently (between one and three months) transformed into soluble Se species (<25 nm) that did not elute from an anion-exchange chromatography (AEC) separation. GPC studies showed a co-elution of Se and dissolved organic matter (DOM) in these extracts, and UF studies showed that >70% of the original selenite was transformed to species >30,000 NMW (for one study site) or >300,000 NMW (for the other). These results strongly suggest selenite association with large molecular weight (MW) NOM molecules.

Although biotic processes are clearly important for the formation of organo-Se species in the environment, it has recently also been shown in laboratory experiments that DMSe and diethylselenide (DESe) can be formed from inorganic Se species by UV-irradiation in the presence of formic, acetic, propionic or malonic acids [45]. This pathway should be tested for its environmental relevance to complete our understanding of the formation and fate of organo-Se species in ambient waters.

2.2.3. *Sediments and Soils*

Most studies on the speciation of selenium in soils and sediments have focussed on its inorganic forms. It is generally found that the oxidation state of Se depends strongly on the redox conditions, with the lower oxidation states Se(0) and Se(–II) (= elemental selenium (Se^0) and selenide (Se^{2-})) predominating in anaerobic and acidic conditions, while the higher oxidation states Se(IV) and Se(VI) are favored under alkaline and aerobic conditions [15,46]. While some advances have been made recently regarding the determination of exact inorganic binding forms in soils and sediments by XAS techniques [10,22], there is little knowledge on the molecular nature of "organic Se" in the same matrices beyond the fact that organic Se is present in reduced oxidation states resembling organic mono- and diselenides.

It is well established that selenium is often strongly correlated with organic matter in soils and sediments, which is frequently interpreted as indicating the presence of organoselenium compounds, specifically "NOM-Se". For example, in the Kesterson pond (CA, USA) the organic C in the soil material shows a good linear correlation with the sum of the selenium species ($R^2 = 0.96$ at $P = 0.05$) [47]. In many cases, association between NOM and Se in soils or sediments has been inferred from co-extraction during the "organic" step of sequential extraction procedures, but often no provisions were taken to distinguish between elemental Se and NOM-Se in this step, so the obtained results cannot determine conclusively if Se was indeed associated with NOM in the soil or sediment. In fact, Ponce de Léon et al. [11] showed by SEC-ICP-MS that in a wetland sediment extract (made with 1 mmol L^{-1} pyrophosphate at pH 9), Se and humic substances were not associated, but it is of course possible that they dissociated during the extraction procedure.

A systematic study [17] that compared the results obtained with different SEPs found that the "organic Se" fraction extracted from sediments by oxidation (here with NaOCl solution) overestimated in many cases the actual amount of NOM-Se (extracted with NaOH solution) because it solubilized a significant amount of elemental Se^0. However, it was also found that a procedure for extracting the elemental Se^0 with Na_2SO_3 solution [48] solubilized some of the NOM-Se present in the sediments, and was therefore unsuitable for removing Se(0) prior to an oxidative extraction of NOM-Se. We conclude, therefore, that existing information on the "organic Se" fraction in soils and sediments is quantitatively inaccurate because studies have either overestimated NOM-Se by employing only an oxidation procedure to estimate it (and included some reduced inorganic Se species) [15] or underestimated NOM-Se by trying to extract Se^0 before solubilizing the true NOM-Se fraction (and inadvertently extracted some NOM-Se in this step) [49]. Since it has recently been suggested that elemental Se^0 can be more selectively extracted with CS_2 [13], it should be tested in future studies if this can be combined with subsequent extraction steps for NOM-Se to obtain more accurate Se speciation results for soils and sediments using SEPs.

Despite these apparent quantitative inaccuracies regarding the determination of "organic Se" in soils and sediments, it is unquestioned that Se may often be associated with NOM in such matrices. In fact, a recent study [10] combining SEP and XAS showed that a large fraction (53–93%) of the total Se in river sediments was not extractable with the used SEP (specifically, neither with NaOH nor with Na_2SO_3), and concluded based on the parallel XAS results that this "nonextractable Se" was likely bound to refractory organic matter ("humin"). In support of this, Kamei-Ishikawa et al. [2] showed in a laboratory study that selenite adsorbed to a synthetic commercial HA (which remained undissolved in the conducted experiments), following a Freundlich isotherm with $K_F = 372$ and $\alpha = 0.82$, which indicates strong binding and at least two different binding sites. No analytical evidence for the binding mechanism was provided. As analytical capabilities improve, we feel that it is important to revise our current geochemical concepts regarding the mechanisms and quantitative importance of Se binding to NOM in soils and sediments.

One important aspect of Se-NOM association in soils and sediments is its dynamic nature with respect to geochemical master variables like redox potential and pH. For example, it has been shown repeatedly [46,50] that reduced Se species (presumably including significant fractions of NOM-Se) in soils and sediments convert to Se oxyanions when the matrix becomes oxidized. It is suspected that the organoselenium compounds encountered under reducing conditions stem from selenium-containing biomolecules in organisms [51], and that the decay of those organisms under anaerobic conditions will lock up the selenium in the resulting NOM, but that oxidation leads to degradation of the organic matter and/or weakens the Se-NOM association.

Since Se speciation is often studied in industrially-impacted ecosystems, it is possible that in certain situations, organic Se in soils or sediments may stem directly from the original natural resource processed, and not be formed *in situ*. Examples of such scenarios include the mining of chalk, shale and bentonite [50] or coal. Sequential extraction data suggest that only minor amounts of selenium were associated with the (organic) kerogen fraction in bentonite, but 42% and 35% of the total selenium, respectively, in chalk and shale [50]. The information on Se speciation in coal is very rudimentary, largely due to the fact that Se concentrations in coal are typically quite low (<10 mg/kg), which makes it difficult to obtain good XAS spectra. Older XANES data indicate that some Se in coal may be present in oxidation states <0, but it was not possible to distinguish between organic and inorganic Se forms in those oxidation states [52]. In a more recent XANES study, the majority of selenium in coal appeared to be in oxidation states< +IV, but here no distinction between elemental Se and more reduced species could be accomplished [53]. Additionally, SEP data show that over 50% of Se in coal are not soluble in nitric acid [54], which indicates association with refractory organic matter.

As for waters, little is known about discrete Se species in soils and sediments. Again, most analytical evidence to date focuses on the volatile dimethylated Se species, due to their importance for Se volatilization. The production of volatile species in soils amended with SeMet has been demonstrated by GC-MS, but so far not in non-spiked soil [55]. The volatile species generated from soil were DMSe, DMDSe, and DMSeS [56]. GC-MS analysis of a Se compound found volatilizing from soils and sewage sludge [32] indicated a molecular formula of $C_2H_6SeO_2$, but the authors were unable to distinguish analytically between two potential structures, CH_3-SeO_2-CH_3 and CH_3-Se(O)-OCH_3. It has been shown that the selenium volatilization rate from contaminated soils increased by more than tenfold (from 25 mg Se m^{-2} d^{-1}) when the soils were amended in the field with organic carbon substrate (methionine or casein) [57], indicating the importance of microbes for the volatilization process.

Decomposing Se-bearing organic matter is encountered in all soils and sediments, but the same biogeochemical processes can also be encountered in much more "concentrated" form in organic waste disposal processes, which are characterized by higher organic matter concentration, temperature and biological activity than in ambient soils and sediments, and may sometimes (e.g., in mixed landfills) also contain unusual other chemicals, with which the Se species can react. In a recent study, duck manure compost was analyzed for volatile selenium compounds [58]. The compost gas contained between <0.001 and 2 μg m^{-3} of volatile selenium species, and besides the common methylated Se species DMSe and DMDSe, the ethylated Se species DESe and methylethylselenide (EMSe) were also positively identified by GC-MS. EMSe made up more than 20% of all volatile species in some samples, and four additional selenium species were only tentatively identified by using element-specific detection and retention time boiling point correlations. By comparison, landfill gas from a municipal waste deposit facility contained DMSe as the only volatile selenium compound, and it was present in much lower concentration range than in the compost gas (<0.005 μg m^{-3}) [59,60]. Finally, in the anaerobic sludge bioreactor of a sewage treatment plant, selenate is biomethylated to DMSe or DMDSe [61], but this does not lead to the desired immobilization of selenium under anaerobic conditions because the methylated species remain mobile and do not form insoluble selenides with metals. This demonstrates that volatile methylated Se species are not only important for the mobility of selenium at the interfaces of air with water or soil, but also at the interfaces between anaerobic and aerobic environments.

Contrary to statements made in the literature, we were unable to find any unambiguous evidence of the existence of other (non-volatile) discrete organo-Se species in soils or sediments. Studies in which SeMet was identified in soils or sediments by GC-MS relied on derivatization techniques, and it was not conclusively demonstrated that the measured derivates could not have been produced from another original Se species. Martens and Suarez [62] reported that Se amino acids spiked to aerobic soils are unstable, and degrade within weeks. To determine SeMet (and other non-volatile discrete organo-Se species) in soils and sediments, it is necessary to use HPLC separation without derivatization, but this has not been successful to date. For example, Ponce de Léon et al. [11] found that in wetland sediment extracts (made with either 0.1 mol/L KH_2PO_4/K_2HPO_4 buffer at pH 7, 1 mol/L HNO_3, 1 mol/L HCl or 5% TMAH), a peak occurred in AEC-ICP-MS chromatograms that matched the retention time of SeMet, which was close to the dead volume. However, analysis of the same extracts by ion pairing chromatography (IPC)-ICP-MS proved that this peak was not SeMet, demonstrating the importance of confirming the identity of Se species by two independent chromatographic separations, particularly when they elute in or close to the dead volume.

2.3. Occurrence of Organoselenium Species in Biota

Most of the efforts related to the identification and quantification of organo-Se species in the environment have been devoted to biota because of selenium's propensity to bioaccumulate and cause ecotoxicological effects in higher organisms, such as water-using birds and predatory fish. Selenium bioaccumulates in aquatic food chains (i.e., Se concentrations in aquatic organisms are many orders of magnitude higher than in the surrounding water), and in some cases, biomagnification can be observed (i.e., Se concentrations in predators are higher than in their prey), but it is usually small (biomagnification factors between 1 and 2) [63], unlike e.g., for mercury.

Also unlike for mercury, the biomagnifying Se species is not known to date, and it is quite possible that there is not one specific Se species that is responsible for biomagnification processes because Se in tissues exists in a wide variety of organic species. Even the identity of the Se species taken up into the lowest trophic level of food chains is not unambiguously known.

Selenium in waters is mostly present in inorganic forms, and some microorganisms prefer uptake of selenite, while others prefer selenate, and it remains unclear if the small fraction of "organic" Se in natural waters plays a significant role in Se bioaccumulation. By comparison, "organic" Se is generally much more prevalent in soils and sediments, but again it is not clear if this fraction plays an important role in Se bioaccumulation by soil- or sediment-dwelling organisms, or to what extent inorganic Se species represent the bioavailable Se pool in soils and sediments.

There are extensive recent reviews that summarize the state of knowledge regarding Se bioaccumulation and biomagnification in food chains [64], Se ecotoxicology [65], and Se speciation in plants [66,67] and animals [68]. It is beyond the scope of this review to address the first two aspects, and there is no need to re-review the last two points at the same level that they've been dealt with previously. However, we wish to make the general comment that previous reviews of (organic) Se speciation in tissues (plant or animal) have overall been very uncritical and include references to the occurrence of many organo-Se species which is not backed up by solid analytical evidence. Often, complex metabolic schemes have initially been proposed as conceptual reaction mechanisms, and have over time been "adopted by repetition" as generally acknowledged "facts", when in fact the analytical proof for many intermediate Se species is still outstanding (and may never be produced, due to the instability of certain Se metabolites).

It would be a worthwhile undertaking to review all previous reports on the occurrence of organo-Se species in different kinds of organisms critically with respect to the quality and certainty of the presented analytical evidence, applying the criteria outlined above (under Section 2.1.1), as has been done for Se species in human urine [3]. We wager that the number of discrete organo-Se species (as far as small MW "free" organo-Se species are concerned) actually known (beyond reasonable doubt) to occur in organisms is much smaller than currently assumed, as was demonstrated in the latter example. That notwithstanding, we also want to acknowledge that, since Se is evidently unspecifically-incorporated into proteins [69], there could in fact be an unlimited number of high MW discrete organo-Se species in biota. In the following, we will limit ourselves to the discussion of several key organo-Se species occurring in tissues, and to identifying general differences between certain types of organisms.

2.3.1. *Microorganisms*

Microorganisms play a key role in the biogeochemistry of trace elements because they change the macroscopic chemistry of environmental compartments (e.g., redox potential) and often transform trace element species in the process (intentionally or inadvertently). They are also part of the primary trophic level in many food chains, although the impact of most microbes (except algae, which will be discussed separately in Section 2.3.2) as food sources for higher organisms on Se bioaccumulation and biomagnification are not well characterized. Depending on the environmental compartment, different microorganisms like bacteria, fungi, molds, yeasts, etc. can have significant influence on Se biogeochemistry and speciation.

One of the critical roles played by microorganisms influencing the environmental chemistry of selenium is their capability to convert inorganic Se species to organic (typically: methylated) Se species, including some important volatile methylated Se species. This was first demonstrated by Challenger [70] for molds, which produced volatile methylated Se species from inorganic Se species as substrates. The proposed reaction mechanism consisted of a series of reductions and oxidative methylation reactions [70], based on his experience with arsenic, where As(V) is reduced to As(III), which is subsequently methylated by a methyl-donor (mainly *S*-adenosylmethionine). He assumed that the redox pair Se(VI)/Se(IV) would behave similarly, but most of the proposed intermediates have not been identified to date. Hence, the Challenger model was later revised by taking into account which Se species were actually observed in soils emitting volatile Se species. Doran [71] proposed that selenite is reduced by bacteria in the soil to elemental selenium (Se^0), which would then be methylated to MeSe(II) and DMSe, but this mechanism has also not been verified conclusively yet.

Conclusive studies of microbial interactions with trace element species are very hard to conduct in the actual environment, so most published studies have isolated microorganisms from the environment and carried out metabolic experiments under controlled conditions, mostly as pure cultures in the laboratory. This procedure has two fundamental problems: it is not certain if all relevant microbes are cultured (and in the correct relative abundance), and the supplied substrates (here: Se species) may not match their "natural" substrates well (e.g., for "organic" Se in soils or sediments). For these reasons, the results of controlled laboratory studies should only be transferred to the environment with care. For example, there is a wealth of information about the generation of selenium-containing proteins or selenoproteins in yeast, when grown in highly-concentrated solutions of inorganic Se species, but this medium is obviously not comparable to natural substrates (so these studies will not be discussed further here).

Bacteria are well known for their ability to produce (volatile) methylated Se species, and are the most extensively studied microorganisms in this regard. For example, a selenium-resistant bacterium isolated from Kesterson reservoir produced not only DMSe and DMDSe, but also DMSeS [72]. Other selenium-resistant bacteria isolated from drainage ponds produced small amounts of methylselenol (MeSeH). *Alcaligenes faecalis* isolated from

Table 2. Selenium species produced by fungi and bacteria.

Selenium Species	Microorganism Species
SeCT	*Aspergillus fumigatus*
	Aspergillus terreus
	Penicillium chrysogenum
Se(IV)Cys	*Fusarium sp.*
γGluSeMeSeCys	*Aspergillus terreus*
	Penicillium chrysogenum
SeMet	*Aspergillus fumigatus*
	Aspergillus terreus
Selenobiotin	*Phycomyces blakesleeanus*
SeCys	*Fusarium spp.*
DMSe	*Penicillium chrysogenum*
SeMeSeMet	*Aspergillus fumigatus*
4-Selenouridine	*Escherichia coli*

seawater generated DMSeP, a potential precursor of DMSe [73]. Soil microorganisms were also isolated and investigated for their potential to produce organoselenium compounds. For example, Doran and Alexander [74] found that the soil bacterium *Corynebacterium* produced DMSe from selenate and selenite, elemental selenium, and from several seleno-amino acids. Fungi are also known to contribute to the production of volatile methylated Se species in soils [75], but are generally understudied [76]. A list of identified organoselenium compounds produced by microorganisms is given in Table 2.

2.3.2. *Aquatic Plants*

Plants play a key role in many food chains because they often constitute the first trophic level, so they are "responsible" for the uptake of Se from an abiotic compartment (water, sediment, soil). They limit how much of the total Se load is available for transfer into higher trophic levels, and determine the bioavailability of the accumulated Se to those organisms by their Se metabolome (i.e., in which chemical species Se ends up after it has been metabolized by the plant). In aquatic food chains, plants occur either as algae, which can be free-floating in the water column or be attached to surfaces (sediment, stones), or as macrophytes growing on the sediment surface. Algae generally accumulate Se from the water and show very high bioaccumulation factors; consequently, free-floating microalgae are probably the most extensively studied organisms in the aquatic environment with respect to their Se speciation. They transfer their Se load to small phytoplankton feeders. By contrast, macrophytes tend to accumulate Se from the sediment, and pass their Se load on to larger herbivorous organisms. Aquatic macrophytes may have comparable Se concentrations to phytoplankton, but generally don't show significant Se bioconcentration from the sediment (i.e., their Se concentrations rarely exceed those in the sediment).

A green freshwater microalgae (*Chlorella sp.*), isolated from the effluent of a wetland receiving the Se-bearing discharge from a coal-fired power plant, converted selenate to DMSe very effectively (90% of a 20 μmol/L selenate solution over 24 h) in the absence of sulfate, resulting in volatilization fluxes of 550 ± 100 mg Se/(g algae (dw) · d). The uptake of selenate (and, consequently, the volatilization of DMSe) was significantly reduced in the presence of sulfate (≥ 20 μmol/L), or when the algae were exposed to selenite or SeMet instead of selenate. The resulting DMSe volatilization fluxes were 2–3 orders of magnitude lower than those for selenate in the absence of sulfate, and were comparable to those measured for macroalgae [77]. In another study, the same kind of microalgae (*Chlorella sp.*), this time isolated from saline evaporation ponds, produced DMSe, DMDSe, and DMSeS from selenite [31]. The major Se species in the algal tissue could not be identified, but was suggested to be DMSeP or Me-Se-Met, based on its ^{77}Se NMR spectrum. Trace amounts of SeMet were also identified in the algae by GC-MS after silylation. In a subsequent study on a cyanophyte mat [78], the same volatile Se species were found, but no free SeMet was detected; instead, SeMet was found incorporated into (unspecified) proteins with MW >3.5 kDa. In a third study (at the site of the second study), the authors were able to quantify proteinaceous SeMet in (unspecified) microalgae, and reported that this form of selenium constituted 3–37% of the total Se in the algae [79].

Se speciation in aquatic macrophytes has been studied much less than in algae. Yan et al. [80] performed an operationally-defined fractionation of Se in edible seaweed, and found protein-bound Se to be the major Se species (30–32% of TSe) in seaweed exposed to high selenite concentration ($200\,mg\,L^{-1}$), while the same plant grown in sea water with natural Se concentration had 48% of its TSe in the protein-bound fraction. Other organic Se fractions in the seaweed included, in decreasing relative concentration, "lipid Se" (20–22% with Se exposure versus 6% without), "polysaccharide Se" (14–15% versus 10%) and "small organic Se" (2–6% versus 23%). While the exact identity of the separated Se species is unknown and the performance of the used operational fractionation was not documented, it is interesting to note that a large fraction of the Se taken up by the plant was not in inorganic forms, and most of the "organic" species were not water-soluble, but soluble in less polar solvents, providing motivation to study such plants with more sophisticated analytical methods.

Wu and Guo [81] reported the occurrence of free SeMet in two aquatic macrophytes exposed to selenate, along with ten-fold lower concentrations of SeCys and SeMeSeCys; interestingly, no $(SeCys)_2$ was found. The Se amino acids were determined as their heptafluorobutyric acid esters by GC-MS after extraction from the plant tissue with $0.1\,mol\,L^{-1}$ HCl [82]. Interestingly, the study also showed a highly significant increase of operationally-defined "organic" Se in the culture medium at very low absolute concentrations ($0.5–3.6\,ng\,L^{-1}$) with increasing TSe concentration in the plant [81], indicating that the plants may have been releasing some of the formed Se amino acids back into the water. In comparison to microalgae, though, macroalgae were shown to release much smaller amounts of volatile methylated Se species [77].

2.3.3. *Terrestrial Plants*

Plants take up different Se species by different pathways. Whereas selenate competes with sulfate for the sulfate transporter [83], there is evidence that selenite may be taken up competitively via the phosphate transporter [84], and it remains unclear if and how organoselenium compounds are taken up. Once taken up by the plant, inorganic selenium species transform into a suite of different organic Se species. Selenium can accumulate in plants as (unaltered) inorganic Se species, as free selenoamino acids, or as SeCys or SeMet incorporated in proteins. Contrary to fish and mammals, the majority of the selenium that has been taken up by plants is not incorporated into proteins. Plants also excrete volatile Se species. Figure 1 illustrates the major transformation reactions observed in plants.

Generally, in the roots, Se(VI) and Se(IV) are reduced to HSe^- and then subsequently transformed into SeCys, which can either be incorporated unspecifically into selenoproteins, or transformed into SeMet via SeCT and SeHcys. The relative abundance of different Se species depends on the plant species. One of the key selenium species in plants seems to be SeMeSeCys, which is formed either directly by methylation of SeCys or from SeMeSeMet. SeMeSeMet can cleave the Se-C bond and release DMSe directly or transform into DMSeP, which again can release DMSe.

The variety of selenium species with their differences in mobility, bioavailability, and toxicity makes selenium speciation in plants another vibrant field of research. Many controlled exposure studies have been carried out using micro- and mesocosms in which plants have been exposed to different concentrations of the most commonly occurring selenium species. Most knowledge about selenium speciation in plants comes from those experiments, rather than from the analysis of naturally-occurring plants. A list of selenium species isolated from plants can be seen in Table 3.

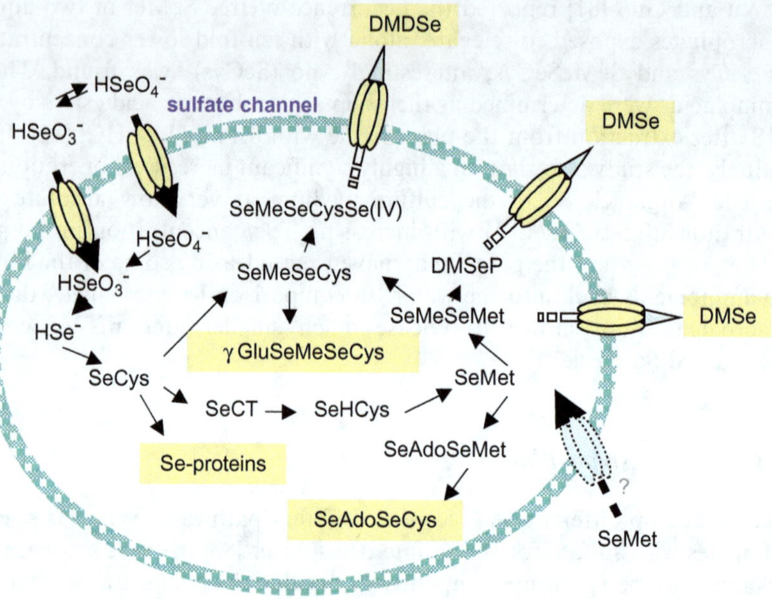

Figure 1. Uptake, transformation, and excretion of Se species in plants. The circle signifies a plant cell. Highlighted Se species accumulate in plants.

Selenium has not been established to be essential for higher plants. Certain plants (*Asteraceae, Brassicasae, Leguminoseae*), however, build up high Se concentrations in their tissues, and can thus be described as selenium hyperaccumulators. For example, *Astragalus bisulcatus* accumulates up to 0.6% selenium in shoots (dw) when growing in its natural habitat [85]. In addition to unmetabolized selenate, SeMeSeCys can also be one of the major selenium species in its leaves [86]. It has been speculated that the enzyme selenocysteine methyltransferase is responsible for the generation of this species from SeCys. More than twenty Se hyperaccumulator plants have been identified to date, and all of them contain not only MeSeCys, but also other methylation products, including SeCT, γGluSeMeSeCys, MeSeOH, γGluSeCT, and SeHcys. Some extraordinary selenium species can be found in members of the *Brassica* family; e.g., *Stanleya pinnata* from a semi-desert (SW USA). In this plant, selenium occurs mainly as the isoselenocyanate species BuNCSe.

Aside from *Brassica* spp., *Allium spp.* are among the most investigated plant species, and SeMeSeCys, SeMet, and SeMeSeMet are the major Se species in those plants [87,88]. Interesting is also that selenium uptake into garlic (*Allium sativum*), a selenium accumulator, was enhanced by growing it together with mycorrhiza, a symbiotic fungus [89]. A maximum concentration of 1 mg g^{-1} TSe was reached in garlic in these experiments, when selenate

Table 3. Selenium species identified in plants.[a]

Selenium Species	Plant Species
SeCT	*Astragalus pectinatus* *Astragalus praleongus* *Brassica oleracea capitata* *Lecythis ollaria* *Morinda reticulate* *Neptunia amplexicaulis* *Stanleya pinnata*
SeMeSeCys	*Allium cepa* *Allium sativum* *Allium tricoccum* *Astragalus bisulcatus* *Astragalus crotalariae* *Astragalus praleongus* *Brassica oleracea botrytis* *Brassica oleracea capitata* *Melilotus indica* *Oonopsis condensate* *Phaseolus lunatus*
SeCys	*Vigna radiata*
γGluSeMeSeCys	*Allium cepa* *Allium sativum* *Astragalus bisulcatus* *Phaseolus lunatus*
SeMet	*Allium tricoccum* *Brassica juncea* *Brassica oleracea capitata* *Melilotus india*
SeMeSeCysSe(IV)	*Brassica oleracea capitata*
γGluSeCT	*Astragalus pectinatus*
γGluSeMet	*Allium sativum*
SerSeCysSG	*Thunbergia alata*
$SePC_2$	*Thunbergia alata*
Selenosugars	*Astragalus racemosus*
BuNCSe	*Stanleya pinnata*
Selenosinigrin	*Stanleya pinnata* *Amoracia laphifolia*

[a]Information taken mainly from ref. [68].

was used as the substrate. The major selenium species was γGluSeMeSeCys, with significant amounts of MeSeCys and SeMet. No SePrSeCys or SeAllylSeCys were found, although the analogue sulfur compounds are synthesized by garlic in high concentration.

Plants not only accumulate selenium in their biomass, but they can also excrete selenium efficiently by volatilization [90]. This process was first described more than 35 years ago for a fungus *Penicillium* [91], but Lewis et al. [92] later also observed that cabbage leaves released selenium in a volatile form. It has been recognized that this process is a detoxification pathway for plants, since the uptake process by plants does not seem to be regulated, although the volatilization rate can be influenced by the uptake of selenium. Furthermore, Zayed and Terry [93] determined that selenate uptake into *Brassica* spp. (and the subsequent production of DMSe) was reduced in the presence of increasing sulfate concentrations. It is however not clear whether selenium excretion is regulated specifically or the excretion happens via the sulfate pathway. The main volatile metabolite for selenium excluders or non-accumulating plants is DMSe, while hyperaccumulating plants tend to produce large amounts of DMDSe as well. Although DMDSe is less volatile than DMSe, it contains two Se atoms per molecule, hence it is a more efficient way of releasing selenium into the air. Some reports even show the volatilization of mixed selenenyl sulfides, such as DMSeS and MeSSeSPr [94,95]. Wetland plants, which are technically both aquatic and terrestrial species, have received particular interest regarding their ability to volatilize Se, since they are used extensively in treatment wetlands. A comparative study measured the Se volatilization efficiency of 20 different wetland plants and found that selenite was volatilized more than twice as effectively as selenate, but that selenate accumulates more in the shoots of the plants [96].

Plants generate phytochelatins, oligopeptides made from γ-glutamic acid cysteinyl units, with different endgroups such as glycine, when they are exposed to elevated amounts of toxic trace elements, such as arsenic and cadmium. It is believed that phytochelatins are responsible for detoxifying these trace elements by binding them to the SH groups of their cysteines. So far, it is unclear if plants react similarly when exposed to elevated levels of selenium, but it seems that plants form selenium complexes with biothiols such as those phytochelatins [97]. The roots extract of *Thunbergia alata* contained at least six different complexes from which only two have been identified ($SePC_2$, SerSeCysSG) after 24 h exposure to 1 mg Se(IV) L^{-1} [97].

2.3.4. Mushrooms

The selenium concentration in edible and wild mushrooms can vary by two orders of magnitude, although most species have a total selenium concentration below 1 mg g^{-1} (dw) in their edible parts [98]. In an earlier study by Piepponen, Pellinen, and Hattula [99], selenium seemed to be bound to low molecular weight (<6 kDa) organic molecules, or occurred in its inorganic forms, in *King Bolete* and *Champignon* mushrooms. Only 20% of the selenium was bound to proteins, chitin and polysaccharides, while only 10% were in the lipid phase or bound to nucleic acids. The species *A. pescaprae* contained mainly selenite with small amounts of SeCys, while the mushroom *King Bolete* contained up to 7.5% of its total selenium as SeCys and 1% as SeMet [100].

2.3.5. *Detritivorous Organisms*

In terrestrial and benthic food chains, detritivores may (partially) replace plants as the first trophic level. This could have important consequences for the mechanisms and magnitude of Se bioaccumulation, since these organisms are exposed to very different Se species (specifically "organic" and elemental Se) than plants, which take up dissolved Se species from water or pore water. Also, Se concentrations in sediments are several orders of magnitude higher than in waters, which may lead to significant differences in Se bioaccumulation and speciation between benthic and pelagic food webs. For example, it was demonstrated [101] that clams (*Macoma balthica*) can take up elemental Se and particulate organic Se from sediments. In the cytosol of a different clam species (*Corbicula fluminea*), Se was present predominantly in the MW fraction <10 kDa, but significant amounts of Se were also observed in the >600 kDa MW fraction [102]. In the tissue of a third clam species (*Donax spp.*), it was shown that the small MW organo-Se species SeMet, $(SeCys)_2$ and SeEt were not present, but 29% of the TSe was present in the form of an unidentified, presumably organic, Se species [103].

In a study of Se speciation in different types of organisms in saline evaporation ponds [79] demonstrated that macroinvertebrates had higher relative concentrations of proteinaceous Se ($42 \pm 11\%$ of TSe) than microphytes ($25 \pm 16\%$), while proteinaceous SeMet concentrations (18 ± 7 versus $16 \pm 11\%$) and TSe (14 ± 9 versus 12 ± 6 mg/g) were comparable between the two groups of organisms, indicating the macroinvertebrates incorporate Se into proteins differently (both with respect to the resulting Se species and to magnitude) than microphytes. A XANES study of Se speciation in aquatic insects also demonstrated that Se was present predominantly ($>80\%$) in the form of organic selenides, with monoselenides typically more abundant than diselenides. An interesting side observation was made in this study when caddisfly pupa and larva were compared; the pupa was the only insect studied that contained an additional organic Se species (30%), which matched the XANES spectrum of $(CH_3)_3Se^+$ [104]. Crickets fed a diet containing 100% SeMet contained 16% of their TSe concentration as $(SeCys)_2$ (and the rest as SeMet), which proves significant metabolism of SeMet [105].

2.3.6. Herbivorous Organisms

On the second trophic level, organisms that feed predominantly on plant material are exposed to a different Se speciation pattern in their diet than organisms who consume mostly animal tissues. Specifically, plants produce certain Se species that are not encountered in animals (e.g., phytochelatin complexes), produce volatile organo-Se species, and tend to have less protein-bound Se than animals. This should result in certain general Se speciation pattern differences between herbivores and carnivores. However, it is unlikely that the similarities in Se speciation between different herbivorous organisms are very pronounced, given that they range from small aquatic insects and fish feeding on phytoplankton to large ruminants like cows, which were incidentally the first organisms for which Se poisoning was postulated.

Brine shrimp (*Artemia*), who feed mostly on microalgae, were found to contain on average $44 \pm 12\%$ of their TSe as proteinaceous Se, while their diet contained only $25 \pm 16\%$ proteinaceous Se [79]. Interestingly, the fraction of proteinaceous SeMet was comparable between both types of organisms (18 ± 5 versus $16 \pm 11\%$ of TSe), indicating that herbivorous organisms are either able to incorporate certain non-proteinaceous Se species in plants into their own proteins, or that they assimilate proteinaceous Se from plants very effectively and convert some of the assimilated proteinaceous SeMet into other proteinaceous Se species. It is generally assumed that selenoamino acids are passed on from prey organisms to their predators, and that proteins are completely disassembled into their individual amino acids in this step. Plants tend to have SeMet as the predominant selenoamino acid, which can be recycled into new proteins in animals, or converted to SeCys, while animals do not synthesize "new" SeMet, so SeCys tends to be the dominant selenoamino acid in animals [68].

The total selenium content in sheep and cattle depends on the selenium content in the soil [106] because that determines the TSe concentration in their feed plants. A recent review by Dumont et al. [107] covers the occurrence of organoselenium species in tissue of farmed animals. Most selenium in the muscle tissue of these animals can be found in the protein fraction, where selenium is incorporated into proteins in the form of SeCys ("selenoproteins") and SeMet ("selenium-containing proteins"). These groups of Se-bearing proteins are distinguished because SeMet substitutes randomly for the structurally very similar methionine (and is thus "unwanted" by the organism), while SeCys is incorporated specifically and is genetically encoded. Selenoproteins (in which Se is intentionally incorporated) are divided into group I, where SeCys is located at the N terminal (examples are glutathione peroxidases and selenoprotein P), and group II, which has SeCys located in the C terminal (e.g., thioredoxin reductases).

2.3.7. *Carnivorous Organisms*

Carnivorous organisms are generally exposed to larger fractions of proteinaceous Se in their diet than their herbivorous counterparts, but the diet's "signature" is not necessarily retained in the predator. Lizards feeding on Se-enriched crickets (SeMet and $(SeCys)_2 = 84$ and 16% of TSe) had altered selenoamino acid composition in some tissues (liver: 100% SeMet; testis: 80% SeMet and 20% selenite) than their prey, but retained the unaltered composition in follicles, demonstrating the higher organisms reprocess selenoamino acids [105]. This study also showed distinctly different patterns of Se associated with proteins in different tissues: while liver tissues contained four distinct MW fractions containing Se (35–133 kDa), testis only showed three fractions (41–338 kDa), confirming that processing and synthesis of Se-bearing proteins is tissue- and gender-specific. Similarly, the eggs of water-using birds contained very high fractions of proteinaceous SeMet [79].

Selenium in fish tissues is mainly bound to proteins, and the distribution between different forms of proteinaceous Se depends on the fish species, as shown by gel electrophoresis [108] or size exclusion chromatography coupled to ICP-MS [109]. The main selenium-containing amino acid in fish is often SeMet [110], but Fan et al. [79] found an interesting difference in this regard between different types of fish: while bottom-dwelling fish (catfish and carp) had remarkably low concentrations of proteinaceous SeMet ($7 \pm 7\%$ of TSe, compared to $46 \pm 18\%$ proteinaceous Se), mosquito fish had much higher concentrations of proteinaceous SeMet ($24 \pm 6\%$) and somewhat higher concentrations of proteinaceous Se ($58 \pm 12\%$), which is likely related to the habitat of their main food sources (sediment versus water column). Interestingly, TMSe has also been identified in the enzymatic extract of trouts, although its origin in the protein fraction is unclear.

In marine mammals and seabirds, selenium concentrates in the liver, but in contrast to metals that show the same behavior (e.g., cadmium), selenium does not bind to low molecular weight proteins, such as metallothioneins (MTs), there. For example, most hepatic selenium in porpoises is actually insoluble and not in the cytosolic fraction [111]. The livers of Dall's porpoises, caught off the coast of Japan, were investigated for mercury and selenium speciation, and it was suggested that selenium forms insoluble HgSe (which would explain the low Se solubility in hepatic tissues), but no direct analytical evidence was given [112]. When the total mercury concentration in the liver was above a certain threshold level, the [Se]/[Hg] ratio was close to unity. The authors suggested that this observation might be indicative of an antagonistic interaction between selenium and mercury [112].

2.3.8. Humans

Selenium is essential for humans and has been shown to decrease the incidence of certain types of cancer. The recommended daily intake is approximately 30–60 μg, but the soils in many countries do not contain enough Se to produce the required Se concentrations in the human diet. Therefore, efforts are underway to enrich our diet in Se, either via Se supplements or via adding Se to deficient soils. Likewise, there is considerable research effort dedicated to the elucidation of human selenium metabolism, in order to find a good biomarker to measure the selenium status of humans and mammals. Most information on human Se metabolism is derived from exposure studies of humans and rats to selenium-enriched yeast, a popular nutritional supplement.

Although most selenium is excreted in urine, significant amounts of DMSe (so far the only volatile selenium species detected in human breath) are exhaled in response to different selenium intake levels [113]. Consequently, indoor air contains measurable concentrations of DMSe [114]. For a long time, Se methylation was believed to be the sole metabolic pathway leading to Se elimination from the human body, either via DMSe exhalation or through urinary excretion of trimethylselenonium (TMSe) [115]. However, TMSe is usually only a minor selenium metabolite in urine [3], while three selenosugars – two galactosamines, MeSeGalNAc (selenosugar 1) and $MeSeGalNH_2$ (selenosugar 3), and one glucosamine, MeSeGluNAc (selenosugar 2) (Table 1) – seem to be the major metabolites [116]. There are, however, enormous individual differences: in the urine of volunteers with elevated selenite intake (200 μg), TMSe was only a trace metabolite in five cases (with selenosugar 1 being the main metabolite), but it was the major metabolite in one volunteer. This demonstrates that much is still unknown about how humans metabolize Se.

3. ORGANOTELLURIUM COMPOUNDS

3.1. Organotellurium Compounds in the Environment

The diversity of organotellurium compounds in abiotic environmental compartments and biota is small compared to the rich carbon-selenium

Table 4. Structures of tellurium and organotellurium compounds.

Name	Abbreviation	Structure
Tellurium	Te^0	Te^0
Telluride	Te^{2-}	Te^{2-}
Tellurate (telluric acid)	Te(VI)	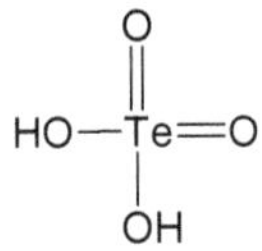
Tellurite (tellurous acid)	Te(IV)	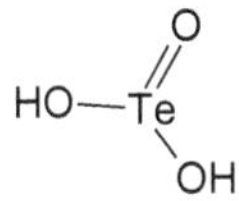
Methyltellurol	MeTeH	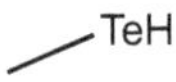
Dimethyltelluride	DMTe	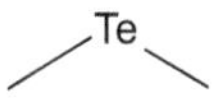
Dimethylditelluride	DMDTe	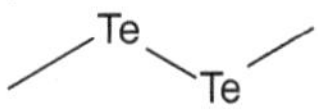
Dimethyltellurenyl sulfide	DMTeS	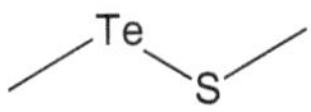
Diethyltelluride	DETe	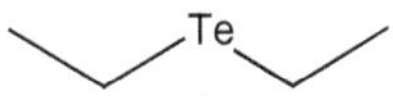
Trimethyltelluronium	$TMTe^+$	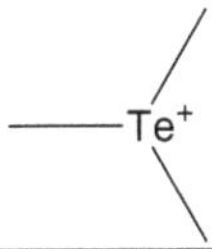

chemistry. So far, the tellurium chemistry in the environment is limited to simple methylated tellurides (Table 4). Dimethyltelluride (DMTe) is the only organo-Te species that has been measured in environmental samples. It has been identified and quantified at concentrations of 10–100 ng L^{-1} in geothermal waters [36]. The water was analyzed by purge-and-trap GC-ICPMS. Surprisingly high concentrations of DMTe were found in the gases from municipal waste deposits and in the headspace of sludge fermentors at municipal sewage treatment plants. Both gases contained methane and carbon dioxide, and DMTe concentrations up to the μg m^{-3} level. [59,117]. Gases from polluted soils also showed the occurrence of DMTe [118].

Kösters et al. [119] identified the presence of DMTe in gas samples created by performing hydride generation on an aqueous slurry of a solid sample consisting of a mixture of organic household waste, contaminated soil and an inorganic Te salt. The authors did not prove, though, whether this Te species was originally present in the solid sample, so it is possible that oxidized dimethylated Te species were the precursor to DMTe, or even that this species was possibly an artefact generated by reaction between inorganic Te species and organic matter in waste or soil during the hydride generation reaction. Likewise, Grüter et al. [120] treated soils from municipal landfills by hydride generation and detected three volatile Te species by GC-ICP-MS. One matched the retention time of DMTe, but no explanation was given for the other two signals obtained.

Tellurium concentrations in ambient waters are at least one order of magnitude lower than those of Se [121]. This is caused by the lower absolute abundance of Te and by its higher affinity to the solid phase, relative to Se [121]. This is especially pronounced in oxic waters, where Te partitioning to soils is three orders of magnitude higher than for Se, but even in reducing soil-water systems, Te still partitions to the solid phase at least ten times more than Se [121]. In these experiments, formation of elemental Se and Te was observed under reducing conditions by XAS, but there was no evidence of association between NOM and Se or Te in the solid phase, which may have been due to the fact that no reference compounds that could serve as a model of Se- or Te-NOM were included in the processing of the XAS spectra.

Since there is already no analytical evidence of the existence of discrete organo-Se species in ambient waters (aside from volatile methylated species), it is not surprising that no such evidence exists for Te either, given its much lower absolute concentrations. LC-ICP-MS methods have been developed for the speciation analysis of only the inorganic species tellurite and tellurate [122], and these methods have not demonstrated the existence of any other (organic) Te species in ambient waters, soils or sediments.

The industrial use of tellurium includes its inorganic compositions in the semiconductor industry, the use of organotellurium compounds as stabilizers for PVC and rubber [123], and as catalysts in chemical synthesis [124]. No studies have identified any of these anthropogenic organotellurium compounds in the environment. Klinkenberg et al. [124] reported that in petrochemical waste waters, most of the total tellurium present (89%) was neither tellurite nor tellurate, but composed of two major and up to eight minor unidentified Te species. These species showed retention in reversed-phase HPLC, so it is likely that they were neutral organic Te species. These apparent organo-Te species were converted to volatile Te compounds (assumed to be DMTe) during biological treatment, and converted to tellurite and/or tellurate under strongly alkaline conditions (pH 12.5; 2 hours reaction time) via other unidentified intermediates.

3.2. Occurrence in Biological Samples

Most information on the interaction between organisms and Te species was generated by laboratory studies with pure cultures of bacteria and fungi,

Table 5. Organotellurium species produced by microorganisms.[a]

Tellurium Species		Microorganisms
MeTeH	Bacteria	*Escherichia coli* JM109 (modified with 3.8 kb chromosomal DNA from *Geobacillus stearothermophilus*)
DMTe	Bacteria	*Pseudomonas fluorescens* K27
		Rhodospirillum rubrum G9
		Rhodospirillum rubrum S1
		Rhodobacter capsulatus
		Rhodocyclus tenuis
		Clostridium collagenovorans
		Desulfovibrio gigas
		Methanobacterium formicicum
	Fungi	*Acremonium falciforme*
		Penicillium chrysogenum
		Penicillium citrinum
		Penicillium sp. (probably *notatum*)
		Penicillum sp.
		Scopulariopsis brevicaulis
DMDTe	Bacteria	*Rhodotorula spp.*
	Fungi	*Acremonium falciforme*
		Penicillium citrinum
DMTeS	Bacteria	*Rhodotorula spp.*

[a]Information taken mainly from ref. [134].

which had been inoculated with different tellurium species as substrates. A number of bacteria and fungi have been shown to produce detectable amounts of organotellurium species, mainly DMTe. Interestingly, with the exception of the gram-positive marine bacterium *Rhodoturola spp.* [125], only fungi have produced dimethylditelluride (DMDTe) so far (see Table 5). Recently, tellurite-resistant strains were isolated from marine sources and tested for the production of volatile tellurium species [125]. The bacteria generated DMTe and DMDTe, but also the less volatile dimethyltellurenyl sulfide (DMTeS). The substrates used in most microbial cultures were mainly tellurite, but *Rhodospirillum rubrum* also generated DMTe from elemental metallic tellurium (Te^0) [126]. Although the generation of DMTe has been discussed to be a detoxification mechanism, it is not clear why those bacteria methylate non-toxic elemental tellurium.

The fungi *Penicillium sp.* generate DMTe directly from tellurate, which suggests that tellurate might be reduced in the cell similarly as selenate [127]. Gharieb et al. [127] exposed two species of soil fungi to tellurite and found very different behavior. *Penicillium citrinum* showed very little Te uptake and no Te volatilization, while *Fusarium spp.* took up around 50% of the Te from a 1 mmol L^{-1} tellurite solution over 2 weeks, and volatilized 0.16%. The identity of the volatile Te species was not confirmed, but it was trapped completely on activated charcoal, from which the authors deduce that it may have been DMTe. Both fungi produced large amounts of elemental Te by reduction. The growth of *Penicillium citrinum* was not affected by 1 mmol L^{-1} tellurite, but the culture pH dropped from 6 to 2.7; by contrast, the growth of *Fusarium spp.* was reduced in the presence of tellurite, but here the culture pH increased to 6.8. The authors suggest that the acidic pH in the *Penicillium citrinum* culture may have been a reason for the lack of Te volatilization because the optimal pH for the microbial formation of volatile methylated Se species has been reported to be in the range 7.7–8.0 [128]. Another possible reason for the observed differences in Se volatilization is that *Penicillium spp.* apparently require the presence of Se to volatilize Te [91].

Duck manure compost released also diethyltelluride (DETe) besides the methylated tellurides [59]. In genetically modified *E. coli* JM109, which express the gene 3.8 kb chromosomal DNA from *Geobacillus stearothermophilus V,* DMTe, DMDTe, DMTeS, and methyltellurol (MeTeH) were identified [129]. Although the incorporation of tellurium into recombinant proteins has been achieved by the inoculation of *E. coli* with the tellurium analogue of SeMet [130], this telluro amino acid has not been identified to occur in the natural environment.

The biochemistry of tellurium in mammals is characterized by the formation of DMTe. DMTe is exhaled as well as excreted in sweat and urine. The pungent smell of this compound makes the exposure of humans to elevated levels of tellurium easily detectable, although a thorough characterization by mass spectrometry has not been done on breath [131]. Recently, rats administered tellurite have generated TMTe as a major metabolite in urine [132,133].

Ogra et al. [133] suggest that dimethylated Te species are incorporated into red blood cells when rats are fed tellurite. However, the analytical evidence presented is questionable for two reasons. First, the species of Te in red blood cells could only be measured after extraction of the cells with H_2O_2. Several products of the oxidation of DMTe with H_2O_2 were measured by ESI-MS, but the assigned chemical structures do not match the observed m/z ratios, no MS-MS confirmation of the proposed structures was performed, and the Te species extracted from the red blood cells were not measured by ESI-MS to confirm their match with the oxidation products of DMTe. Second, all products of DMTe co-eluted in the used chromatographic separation, and were ill-resolved from tellurate in standard solution samples. Although the red blood cell extract showed a co-eluting peak with the oxidation products of DMTe, there is no evidence that the retention times in the cell extract were unchanged over a standard solution. Furthermore, a mixed-mode (size exclusion + reversed phase + cation exchange) HPLC column was employed in these studies, which has the advantage that compounds which interact with the stationary phase in more than one mode are unlikely to co-elute, but the disadvantage that two completely different compounds who each interact with the stationary phase in a different mode (but only in one) can co-elute. Therefore, without further analytical evidence, we feel that the conclusions by the authors are unsubstantiated at this time.

ABBREVIATIONS

For the abbreviations and structures of the selenium and tellurium species see Tables 1 and 4.

AAS	atomic absorption spectroscopy
AEC	anion exchange chromatography
AES	atomic emission spectroscopy
AFS	atomic fluorescence spectroscopy
DOM	dissolved organic matter
dw	dry weight
EXAFS	extended X-ray absorption fine structure spectroscopy
FFF	field flow fractionation
GC	gas chromatography
GC-ICPMS	gas chromatography coupled to ICP-MS
GC-MS	gas chromatography-mass spectrometry
GF	gel filtration
GPC	gel permeation chromatography
HA	humic acid
HS	humic substance
ICP-MS	inductively coupled plasma-mass spectrometry
IPC	ion pairing chromatography
LC	liquid chromatography
MT	metallothionein
MW	molecular weight
NMW	nominal molecular weight
NOM	natural organic matter
OC	organic carbon
PVC	polyvinyl chloride
QC	quality control
SEC	size exclusion chromatography
SEP	sequential extraction procedure
SSHG	selective sequential hydride generation

TISe	total inorganic selenium (selenite + selenate)
TMAH	tetramethylammonium hydroxide
TSe	total selenium
UF	ultrafiltration
XANES	X-ray absorption near-edge spectroscopy
XAS	X-ray absorption spectroscopy

REFERENCES

1. N. Wiberg: Lehrbuch der Anorganischen Chemie, 91st-100th ed., Walter de Gruyter, Berlin, 1985.
2. N. Kamei-Ishikawa, Y. Nakamaru, K. Tagami and S. Uchida, *J. Environ. Radioact.*, 2008, **99**, 993.
3. K. A. Francesconi and F. Pannier, *Clin. Chem.*, 2004, **50**, 2240.
4. G. A. Cutter, *Science*, 1982, **217**, 829.
5. D. Wallschläger and J. London, *J. Anal. At. Spectrom.*, 2004, **19**, 1119.
6. J. L. Fio and R. Fujii, *Soil Sci. Soc. Am. J.*, 1990, **54**, 363.
7. Y. W. Chen, M. D. Zhou, J. Tong and N. Belzile, *Anal. Chim. Acta*, 2005, **545**, 142.
8. G. A. Cutter, *Anal. Chim. Acta*, 1978, **98**, 59.
9. A. Tessier, P. G. C. Campbell and M. Bisson, *Anal. Chem.*, 1979, **51**, 1, 79, 844.
10. L. L. Oram, D. G. Strawn, M. A. Marcus, S. C. Takra and G. Moller, *Environ. Sci. Technol.*, 2008, **42**, 6830.
11. C. A. Ponce de Léon, K. DeNicola, M. Montes Bayón and J. A. Caruso, *J. Environ. Monit.*, 2003, **5**, 435.
12. G. Song, E. H. Novotny, A. J. Simpson, C. E. Clapp and M. H. B. Hayes, *European J. Soil Sci.*, 2008, **59**, 505.
13. Y. W. Chen, L. Li, A. D'Ulivo and N. Belzile, *Anal. Chim. Acta*, 2006, **577**, 126.
14. G. M. Peters, W. A. Maher, J. P. Barford and V. G. Gomes, *Water, Air, Soil Pollut.*, 1997, **99**, 275.
15. D. A. Martens and D. L. Suarez, *Environ. Sci. Technol.*, 1997, **31**, 133.
16. N. Belzile, Y. W. Chen and R. Xu, *Appl. Geochem.*, 2000, **15**, 1439.
17. M. T. Wright, D. R. Parker and C. Amrhein, *Environ. Sci. Technol.*, 2003, **37**, 4709.
18. I. J. Pickering, G. N. George, V. Van Fleet-Stalder, T. G. Chasteen and R. C. Prince, *J. Bioinorg. Inorg. Chem.*, 1999, **4**, 791.
19. D. G. Beak, R. T. Wilkin, R. G. Ford and S. D. Kelly, *Environ. Sci. Technol.*, 2008, **42**, 1643.
20. V. Van Fleet-Stalder, T. G. Chasteen, I. J. Pickering, G. N. George and R. C. Prince, *Appl. Environ. Microbiol.*, 2000, **66**, 4849.
21. R. Andrahennadi, M. Wayland and I. J. Pickering, *Environ. Sci. Technol.*, 2007, **41**, 7683.
22. A. C. Scheinost, R. Kirsch, D. Banerjee, A. Fernandez-Martinez, H. Zaenker, H. Funke and L. Charlet, *J. Contaminant Hydrol.*, 2008, **102**, 228.

23. E. Tessier, D. Amouroux, G. Abril, E. Lemaire and O. F. X. Donard, *Biogeochem.*, 2002, **59**, 183.
24. J. Meija, M. Montes Bayón, D. L. Le Duc, N. Terry and J. A. Caruso, *Anal. Chem.*, 2002, **74**, 5837.
25. R. Atkinson, S. M. Aschmann, D. Hasegawa, E. T. Thompson-Eagle and W. T. Frankenberger, *Environ. Sci. Technol.*, 1990, **24**, 1326.
26. H. J. Wen and J. Carignan, *Atmos. Environ.*, 2007, **41**, 7151.
27. R. M. Rael and W. T. Frankenberger, *Wat. Res.*, 1996, **30**, 422.
28. G. A. Cutter and L. S. Cutter, *Mar. Chem.*, 1995, **49**, 295.
29. J. H. Ansede, P. J. Pellechia and D. Yoch, *Environ. Sci. Technol.*, 1999, **33**, 2064.
30. J. E. Conde and M. S. Alaejos, *Chem. Rev.*, 1997, **97**, 1979.
31. T. W. M. Fan, A. N. Lane and R. M. Higashi, *Environ. Sci. Technol.*, 1997, **31**, 569.
32. D. C. Reamer and W. H. Zoller, *Science*, 1980, **208**, 500.
33. D. Amouroux and O. F. X. Donard, *Mar. Chem.*, 1997, **58**, 173.
34. W. A. Maher, (personal communication).
35. X. Diaz, W. P. Johnson and D. L. Naftz, *Sci. Total Environ.*, 2009, **407**, 2333.
36. A. V. Hirner, J. Feldmann, E. Krupp, R. Grumping, R. Goguel and W. R. Cullen, *Org. Geochem.*, 1998, **29**, 1765.
37. X. Diaz, W. P. Johnson and W. A. Oliver, *Environ. Sci. Technol.*, 2009, **43**, 53–59.
38. D. Hansen, P. J. Duda, A. Zayed and N. Terry, *Environ. Sci.Technol.*, 1998, **32**, 591.
39. H. B. Ross, *Tellus*, 1985, **37B**, 78.
40. B. W. Mosher and R. A. Duce, *J. Geophys. Res.*, 1987, **92**, 13289.
41. N. R. Bottino, C. H. Banks, K. J. Irgolic, P. Micks, A. E. Wheeler and R. A. Zingaro, *Phytochem.*, 1984, **23**, 2445.
42. D. Amouroux, C. Pecheyran and O. F. X. Donard, *Appl. Organomet. Chem.*, 2000, **14**, 236.
43. T. Ferri and P. Sangiorgio, *Anal. Chim. Acta*, 1996, **321**, 185.
44. C. Bruggeman, A. Maes and J. Vancluysen, *Appl. Geochem.*, 2007, **22**, 1371.
45. X. M. Guo, R. E. Sturgeon, Z. Mester and G. K. Gardener, *Environ. Sci. Technol.*, 2003, **37**, 5645–5650.
46. P. H. Masscheleyn, R. D. Delaune and W. H. Patrick, *Environ. Sci. Technol.*, 1990, **24**, 91.
47. D. A. Martens and D. L. Suarez, *J. Environ. Qual.*, 1997, **26**, 424.
48. D. J. Velinsky and G. A. Cutter, *Anal. Chim. Acta*, 1990, **235**, 419.
49. Y. Q. Zhang and J. N. Moore, *Environ. Sci. Technol.*, 1996, **30**, 2613.
50. T. R. Kulp and L. M. Pratt, *Geochim. Cosmochim. Acta*, 2004, **68**, 3687.
51. P. T. Zawislanski, S. M. Benson, R. Terberg and S. E. Borglin, *Environ. Sci. Technol.*, 2003, **37**, 2415.
52. F. E. Huggins and G. P. Huffman, *Int. J. Coal. Geol.*, 1996, **32**, 31.
53. P. Shah, V. Strezov, K. Prince and P. F. Nelson, *Fuel*, 2008, **87**, 1859–1869.
54. K. W. Riley, D. H. French, N. A. Lambropoulos, O. P. Farrell, R. A. Wood and F. E. Huggins, *Int. J. Coal Geol.*, 2007, **72**, 72.
55. Y. Zhang and W. T. Frankenberger, *Environ. Sci. Technol.*, 2000, **34**, 776.

56. W. T. Frankenberger and U. Karlson, *Geomicrobiol. J.*, 1994, **12**, 265.
57. G. S. Banuelos and Z. Q. Lin, *Environ. Poll.*, 2007, **150**, 306.
58. P. Pinel-Raffaitin, C. Pecheyran and D. Amouroux, *Atmos. Environ.*, 2008, **42**, 7786.
59. J. Feldmann, *J. Anal. At. Spectrom.*, 1997, **12**, 1069.
60. J. Feldmann and A. V. Hirner, *Intern. J. Environ. Anal. Chem.*, 1995, **60**, 339.
61. M. Lenz, M. Smit, P. Binder, A. C. van Aelst and P. N. L. Lens, *J. Environ. Qual.*, 2008, **37**, 1691.
62. D. A. Martens and D. L. Suarez, *Sequential Extraction of Selenium Oxidation States*, in *Environmental Chemistry of Selenium*, Ed. W. T. Frankenberger and R. A. Engberg, Marcel Dekker, New York, 1998, p. 61–80.
63. C. E. Schlekat, D. G. Purkerson and S. N. Luoma, *Environ. Toxicol.Chem.*, 2004, **23**, 3003–3010.
64. J. F. Jasonsmith, W. Maher, A. C. Roach and F. Krikowa, *Marine Freshwat. Res.*, 2008, **59**, 1048.
65. D. B. D. Simmons and D. Wallschläger, *Environ. Toxicol. Chem.*, 2005, **24**, 1331.
66. Z. Pedrero and Y. Madrid, *Anal. Chim. Acta*, 2009, **634**, 135–152.
67. T. Rezanka and K. Sigler, *Phytochem.*, 2008, **69**, 585–606.
68. N. V. C. Ralston, C. R. Ralston, J. L. Blackwell and L. J. Raymond, *Neurotoxicol.*, 2008, **29**, 802.
69. L. Ouerdane and Z. Mester, *J. Agricult. Food Chem.*, 2008, **56**, 11792.
70. F. Challenger, *Chem. Rev.*, 1945, **36**, 315.
71. J. W. Doran, *Adv. Microb. Ecol.*, 1982, **6**, 1.
72. T. G. Chasteen, in *Environmental Chemistry of Selenium*, Ed. W. T. Frankenberger and R. A. Engberg, Marcel Dekker, New York, 1998, Ch. 29.
73. J. H. Ansede, P. J. Pellechia and D. C. Yoch, *Appl. Environ. Microbiol.*, 2001, **67**, 3134.
74. J. W. Doran and M. Alexander, *Appl. Environ. Microbiol.*, 1977, **33**, 31.
75. W. T. Frankenberger and U. Karlson, *J. Ind. Microbiol.*, 1995, **14**, 226.
76. J. M. Brady, J. M. Tobin and G. M. Gadd, *Mycol. Res.*, 1999, **103**, 299.
77. P. M. Neumann, M. P. De Souza, I. J. Pickering and N. Terry, *Plant Cell Env.*, 2003, **26**, 897.
78. T. W. M. Fan, R. M. Higashi and A. N. Lane, *Environ. Sci. Technol.*, 1998, **32**, 3185.
79. T. W. M. Fan, S. J. Hen, D. E. Hinton and R. M. Higashi, *Aq. Toxicol.*, 2002, **57**, 65.
80. X. J. Yan, L. Zheng, H. M. Chen, W. Lin and W. W. Zhang, *J. Agricult. Food Chem.*, 2004, **52**, 6460.
81. L. Wu and X. Guo, *Ecotoxicol. Environ. Safety*, 2002, **51**, 22.
82. X. Guo and L. Wu, *Ecotoxicol. Environ. Safety*, 1998, **39**, 207.
83. N. Terry, A. M. Zayed, M. P. Souza and A. S. Tarun, *Annu. Rev. Plant Physiol. Plant Biol.*, 2000, **51**, 401.
84. J. L. Hopper and D. R. Parker, *Plant Soil*, 1999, **210**, 199.
85. H. G. Byers, *Selenium Occurrence in Certain Soils in the United States*, with a discussion of related topics, Second Report, *U.S. Dept. Agric. Technol. Bull.*, 1936, 530.

86. I. J. Pickering, C. Wright, B. Bubner, D. Ellis, M. W. Persans, E. Y. Yu, G. N. George, R. C. Prince and D. E. Salt, *Plant Physiol.*, 2003, **131**, 1460.
87. M. Montes-Bayon, M. J. D. Molet, E. B. Gonzalez and A. Sanz-Medel, *Talanta*, 2006, **68**, 1287.
88. T. D. Grant, M. Montes-Bayon, D. LeDuc, M. W. Fricke and J. A. Caruso, *J. Chromatogr.*, 2004, **1026A**, 159.
89. E. H. Larsen, R. Lobinski, K. Burger-Meyer, M. Hansen, R. Ruzik, L. Mazurowska, P. H. Rasmussen, J. J. Sloth, O. Scholten and C. Kirk, *Anal. Bioanal. Chem.*, 2006, **385**, 1098.
90. J. Meija, M. Montes-Bayon, D. L. LeDuc, N. Terry and J. A. Caruso, *Anal. Chem.*, 2002, **74**, 5837.
91. R. W. Fleming and M. Alexander, *Appl. Microbiol.*, 1972, **24**, 424.
92. B. G. Lewis, C. M. Johnson and T. C. Broyer, *Plant Soil*, 1974, **40**, 107.
93. A. M. Zayed and N. Terry, *J. Plant Physiol.*, 1992, **140**, 646.
94. X. J. Cai, P. Uden, E. Block, X. Zhang, B. D. Quimby and J. J. Sullivan, *J. Agric. Food Chem.*, 1994, **42**, 2081.
95. X. J. Cai, E. Block, P. C. Uden, X. Zhang, B. D. Quimby and J. J. Sullivan, *J. Agric. Food Chem.*, 1995, **43**, 1754.
96. E. A. H. Pilon-Smits, M. P. De Souza, G. Hong, A. Amini, R. C. Bravo, S. T. Payabyb and N. Terry, *J. Environ. Qual.*, 1999, **28**, 1011.
97. K. Bluemlein, *PhD Thesis*, University of Aberdeen, 2008.
98. J. Falandysz, *J. Environ. Sci. Health*, 2008, **26**, 256.
99. S. Piepponen, M. J. Pellinen and T. Hattula, in *Trace Element – Analytical Chemistry in Medicine and Biology*, Ed. P. Bratter and P. Schramel, W. de Gruyter & Co, Berlin, 1984, pp. 159–166.
100. Z. Slekovec, J. T. Van Elteren, U. D. Woroniecka, K. J. Kroon, I. Falonga and A. R. Byrne, *Biol. Trace Elem. Res.*, 2000, **75**, 139.
101. S. N. Luoma, C. Johns, N. S. Fisher, N. A. Steinberg, R. S. Oreland and J. R. Reinfelder, *Environ. Sci. Technol.*, 1992, **26**, 485.
102. C. Adam-Guillermin, E. Fournier, M. Floriani, V. Camilleri, J. C. Massabuau and J. Garnier-Laplace, *Environ. Sci. Technol.*, 2009, **43**, 2112.
103. J. L. Gomez-Ariza, M. A. C. de la Torre, I. Giradles, D. Sanchez-Rodas, A. Velasco and E. Morales, *Appl. Organomet. Chem.*, 2002, **16**, 265.
104. R. Andrahennadi, M. Wayland and I. J. Pickering, *Environ. Sci. Technol.*, 2007, **41**, 7683.
105. J. M. Unrine, B. P. Jackson and W. A. Hopkins, *Environ. Sci. Technol.*, 2007, **26**, 3601.
106. J. W. Finley, *J. Anim. Sci.*, 2000, **77**, 1.
107. E. Dumont, F. Vanhaecke and R. Cornelis, *Anal. Bioanal. Chem.*, 2006, **385**, 1304.
108. G. Oenning, *Food Chem.*, 2000, **68**, 133.
109. P. Moreno, M. A. Quijano, A. M. Gutierrez, M. C. Perez-Conde and C. Camara, *Anal. Chim. Acta*, 2004, **524**, 315.
110. A. I. Cabanero, C. Carvalho, Y. Madrid, C. Batoreu and C. Camara, *Biol. Trace Elem. Res.*, 2005, **103**, 17.
111. T. Ikemoto, T. Kunito, Y. Anan, H. Tanaka, N. Baba, N. Mityazaki and S. Tanabe, *Environ. Toxicol. Chem.*, 2004, **23**, 2008.

112. J. Yang, T. Kunito, S. Tanabe and M. Miyazaki, *Environ. Poll.*, 2007, **148**, 669.
113. D. Kremer, G. Ilgen and J. Feldmann, *Anal. Bioanal. Chem.*, 2005, **383**, 509.
114. C. Pecheyran, B. Lalere and O. F. X. Donard, *Environ. Sci. Technol.*, 2000, **34**, 27.
115. H. E. Ganther, *J. Am. Coll. Toxicol.*, 1986, **5**, 1–5.
116. B. Gammelgaard, C. Gabel-Jensen, S. Sturup and H. R. Hansen, *Anal. Bioanal. Chem.*, 2008, **390**, 1691–1706.
117. J. Feldmann, *ACS Symp. Ser.*, 2003, **835**, 128.
118. A. V. Hirner, E. Krupp, F. Schulz, M. Koziol and W. Hofmeister, *J. Geochem. Explor.*, 1998, **64**, 133.
119. J. Kösters, R. A. Diaz-Bone, B. Planer-Friedrich, B. Rothweiler and A. V. Hirner, *J. Mol. Struct.*, 2003, **661**, 347.
120. U. M. Grüter, J. Kresimon and A. V. Hirner, *Fresenius J. Anal. Chem.*, 2000, **368**, 67.
121. T. Harada and Y. Takahashi, *Geochim. Cosmochim. Acta*, 2008, **72**, 1281.
122. C. Y. Kuo and S. J. Jiang, *J. Chromatogr.*, 2008, **1181A**, 60.
123. L. Engman, D. Stern and B. Stenberg, *J. Poly. Sci.*, 1996, **59**, 1365–1369.
124. H. Klinkenberg, S. van der Wal, C. de Koster and J. Bart, *J. Chromatogr.*, 1998, **794A**, 219.
125. P. R. L. Ollivier, A. S. Bahrou, S. Marcus, T. Cox, T. M. Church and T. E. Hanson, *Appl. Environ. Microbiol.*, 2008, **74**, 7163.
126. V. Van Fleet-Stalder and T. G. Chasteen, *J. Photochem. Photobiol.*, 1998, **43B**, 193.
127. M. M. Gharieb, M. Kierans and G. M. Gadd, *Mycol. Res.*, 1999, **103**, 299.
128. W. T. Frankenberger and U. Karlson, in *Selenium in the Environment*, Ed. W. T. Frankenberger and S. Benson, Marcel Dekker Inc., New York, 1994, pp. 369–388.
129. J. W. Swearingen Jr, M. A. Araya, M. F. Plishker, C. P. Saavedra, C. C. Vasquez and T. G. Chasteen, *Anal. Biochem.*, 2004, **331**, 106.
130. N. Budisa, B. Steipe, P. Demange, C. Eckerskorn, J. Kellermann and R. Huber, *Eur. J. Biochem.*, 1995, **230**, 788.
131. A. Taylor, *Biol. Trace Elem. Res.*, 1996, **55**, 231.
132. Y. Ogra, R. Kobayashi, K. Ishiwata and K. T. Suzuki, *J. Anal. At. Spectrom.*, 2007, **22**, 153.
133. Y. Ogra, R. Kobayashi, K. Ishiwata and K. T. Suzuki, *J. Inorg. Biochem.*, 2008, **102**, 1507.
134. T. G. Chasteen and R. Bentley, *Chem. Rev.*, 2003, **103**, 1.

Met. Ions Life Sci. **2010**, *7*, 365–401

11

Organomercurials. Their Formation and Pathways in the Environment

Holger Hintelmann

Department of Chemistry, Trent University, Peterborough ON K9J 7B8, Canada
<hhintelmann@trentu.ca>

Metal Ions in Life Sciences, Volume 7 Edited by Astrid Sigel, Helmut Sigel, and Roland K. O. Sigel

Published by the Royal Society of Chemistry, www.rsc.org DOI: 10.1039/9781849730822-00365

ABSTRACT: The most important mercury species in the environment is monomethylmercury (MMHg), the topic of this chapter. This organic mercury compound is normally not released into the environment but formed by natural processes. Mercuric mercury (Hg^{2+}) is methylated by bacteria and to a lesser extent through abiotic pathways. Highest rates of formation are found in anoxic aquatic environments. Terrestrial systems are mostly irrelevant for MMHg production and not a concern. Most productive environments are sediments, wetlands, and coastal marshes, but also the anoxic hypolimnion of lakes and anaerobic microhabitats like the rhizosphere of floating macrophytes. Prime suspects for methylation are sulfate-reducing bacteria, although also iron reducers have lately been identified as capable mercury methylators. What makes methylmercury such an insidious contaminant is its enormous biomagnification potential. Methylmercury is accumulated by more than seven orders of magnitude from sub ng/L concentrations in water to over 1,000,000 ng/kg in piscivorous fish, which are the main concern from a human health point of view. Since methylmercury is a very potent neurotoxin, particularly small children, pregnant women, and women in childbearing age are advised to either limit their fish consumption to a few meals per week or to select fish species known to have low levels of methylmercury. Formation of methylmercury is counteracted by other bacteria, which are capable of demethylating methylmercury. This process is regulated by an inducible *mer* operon system and serves as a detoxification mechanism in polluted environments. The other naturally occurring organic mercury species, dimethylmercury (DMHg), is only present at very low levels at great depths in the world oceans. However, it might be an important and very mobile pre-cursor for methylmercury in marine and polar ecosystems.

KEYWORDS: Bioaccumulation · demethylation · dimethylmercury · mercury · methylation · methylmercury

1. INTRODUCTION

Mercury is a persistent pollutant with unique chemical and physical characteristics, making this trace element one the most highly studied of all times. A distinctive feature is its high vapor pressure in elemental form, which is the main reason for the rapid global dispersion from point sources. Combined with its trait to be converted into organometal compounds of high toxicity, namely monomethylmercury, it creates a scenario for global concern.

While all mercury compounds are highly toxic, this element is an exceptional contaminant, because its most harmful species, methylmercury, is not actually discharged into the environment, but naturally generated from mercuric mercury. Apart from point sources such as mining operations or industrial activities, which discharge inorganic mercury and cause at times severe local pollution, the major concern with mercury lies in the formation of organic methylmercury in aquatic environments. Methylmercury shows up as the most common contaminant in fish all over the world and drives most of the mercury research. Many countries have issued advisories to manage the consumption of fish, representing the main entry of methylmercury into the human diet. While the problem is clearly identified, the solution is less obvious. Numerous studies have been conducted to elucidate the factors controlling methylmercury formation and biomagnification. While the latter is fairly well understood, the former is not. Decades of research have unearthed an impressive amount of often, alas, contradictory, circumstantial evidence, based on which scientists are trying to compose a theoretical framework of methylmercury in the environment.

Considering the massive literature dealing with mercury in the environment, this chapter will not venture into analytical [1–6] and toxicological [7,8] aspects of MMHg, which are described in some excellent reviews elsewhere (see also Chapters 2 and 12). The organomercury issue will be approached from a dual source and sink point of view. After a general introduction to mercury speciation, it starts with looking at processes that either generate or decompose organomercury species in the environment. The second section considers the mobility and the fate of mercury species in the natural environment to describe their occurrence in and movement through the ecosystem.

2. SPECIATION OF ORGANOMERCURY COMPOUNDS

In metal speciation, it has now long been accepted that the total metal content in a given sample is not a reliable predictor for its toxicity, mobility or bioavailability and thus, should not be used for risk assessment purposes. Instead, it is much more useful to know the actual concentration of individual metal species. This is of particular importance for mercury, which shows enormous physical-chemical differences among mercury species (see Table 1). For the purpose of this review, only compounds having one or more covalent Hg-carbon bonds qualify as an organomercury species. By this definition, complex ions composed of mercuric Hg and organic compounds (e.g., dissolved organic matter, DOM) are not considered an organomercurial. This leaves a rather limited assortment of compounds, some of

Table 1. Physical and chemical properties of selected mercury compounds.

	Hg^0	$HgCl_2$	HgS	CH_3HgCl	CH_3CH_2HgCl	CH_3HgCH_3
Melting point (°C)	−39	277	584 (subl)	167 (subl)	192 (subl)	na
Boiling point (°C)	357	303	---	---	---	96
Vapor pressure (Pa)	0.18	9.0×10^{-3}	negligible	1.13	0.4	8.3×10^{3}
Water solubility (g/L)	5.5×10^{-5}	66	2×10^{-24} (theory)	5	1.5×10^{-3}	2.95
Henry's Law coefficient	0.32	3.7×10^{-5}	na	1.6×10^{-5}	na	0.31
Octanol/water coefficient	4.2	0.1–3.3	na	1.7–2.5	na	180

subl: sublimation temperature
na: not available or impossible to calculate from the data provided in the original source
Compiled from [212–220].

Table 2. Chemical formulas and 3D-structures of common organomercurials.

Common name	Chemical formula	Most common ligands
methylmercury	$CH_3–Hg^+$	Cl^-, (Br^-), (I^-), OH^-, cys, RS^-, COO^-
ethylmercury	$CH_3–CH_2–Hg^+$	as methylmercury; thiosalicylate (in thiomersal)
thiomersal (thimerosal)	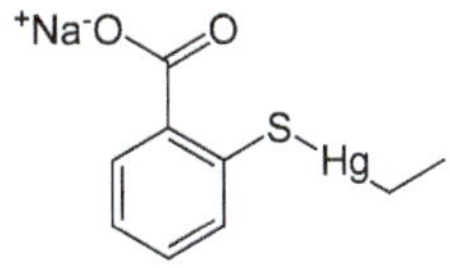	none
dimethylmercury	CH_3HgCH_3	none

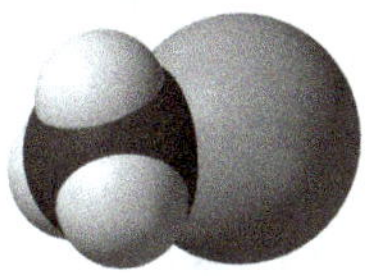

methylmercury

ethylmercury

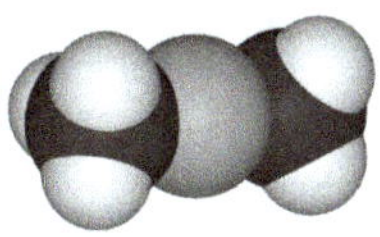

dimethylmercury

thiomersal [8]

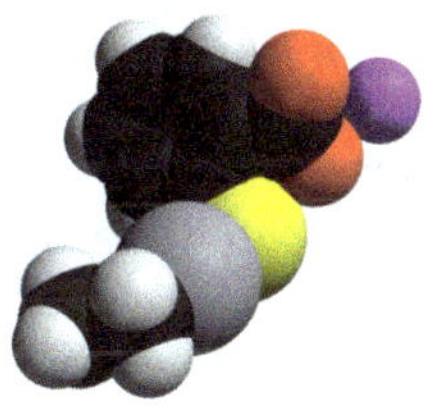

Structures are based on Wikipedia information.

which are shown in Table 2, including monomethylmercury (MMHg), monoethyl- or ethylmercury (EtHg), dimethylmercury (DMHg), and aromatic mercury compounds such as phenylmercury for discussion.

2.1. Monomethylmercury

Most commonly referred to as "methylmercury", the actual species of concern is the methylmercury cation CH_3Hg^+. However, this cation is essentially absent in the environment. CH_3Hg^+ is virtually always coordinated to other ligands and it is the variety of those MMHg complexes, which are responsible for the complex and manifold behavior commonly ascribed to methylmercury. CH_3Hg^+ exhibits rich, but straightforward coordination chemistry in aqueous environments. Being the simplest "soft" Lewis acid, it is almost always coordinated to a single ligand, leading to 1:1 complexes with other soft Lewis bases. A comprehensive tabulation of formation constants for a wide range of complexes was early on established [9], demonstrating the high affinity of MMHg for sulfur containing ligands. Other important ligands from an environmental point of view are halogens, hydroxide and some amine and oxygen containing functional groups. While multinuclear complexes of the nature $[CH_3HgL_2]^-$ are possible, they are hardly relevant for natural systems. Owing to the very high affinity of MMHg towards thiols, there is consensus that virtually all of the MMHg in aquatic systems will be bound to such groups (e.g., sulfur in DOM or cysteine groups of proteins in biota). This has recently been verified experimentally for DOM, soils, and fish [10–12].

Chemically, MMHg is surprisingly stable. Hot, concentrated acids mineralize MMHg very slowly, e.g., MMHg has a half-life of 300 days in 1 M H_2SO_4. Strong oxidizing reagents such as permanganate, halogens or peroxides are necessary for efficient breakdown. The Hg-C bond is also prone to easy photochemical cleavage in the presence of UV and visible light. The other effective pathway of degradation in the environment is microbial demethylation, which is very effective in sediments.

2.2. Dimethylmercury

DMHg is the only peralkylated mercury species of relevance occurring in the environment. The molecule has a very high vapor pressure, which is even higher than that of elemental Hg, and unlike MMHg, is always hydrophobic (the hydrophilic/lipophilic character of MMHg is modulated and controlled by its ligands). Like MMHg, it is easily photodegraded, but relatively stable towards chemicals except strong oxidizing reagents.

2.3. Other Organomercurials

Ethylmercury is the only other monoalkylmercury compound besides MMHg that was ever found in the environment. There is no known pathway of biotic formation, which, combined with its environmental instability, is probably the reason for the rare occurrence of EtHg. Even if discharged under some unique situations, it is not very persistent, readily decomposes and has no history of (bio)accumulation [13,14]. Despite its presumed toxicity, it is still widely used for preservation of vaccines, where ethylmercury is added in form of thiomersal (or thimerosal: sodium ethylmercurithiosalicylate; see Table 2 for chemical structure), a very effective antiseptic. It is mostly used in multi-dose vaccines outside North America and Europe, where it is only applied in some specific single-use vaccines, but not any more in routine childhood vaccination schedules. The administration of EtHg to children in form of thimerosal (typically 25 μg per vaccination) is highly controversial and under suspicion to be a co-factor for the increased occurrences of autism [15]. However, no conclusive prove has been established to date [16].

Aromatic (e.g., phenylmercury) as well as other alkylmercury compounds (e.g., ethoxyethylmercury) have been used in the past as pesticides and/or fungicides. Due to their historical heavy use in some countries such as Scandinavia, it led to environmental problems in these countries. In fact, the observation of increased mercury levels in Swedish birds triggered intense Hg research in this country. Today, the use of these organomercurials is banned as pesticides.

3. FORMATION OF ORGANOMERCURY COMPOUNDS

Much of our knowledge about mercury distribution and cycling in the environment is still incomplete. Natural processes convert inorganic mercury into the potent toxin MMHg. Although we understand many aspects of mercury geochemistry, we still lack thorough knowledge to fully explain and forecast MMHg formation in the environment. There is, however, consensus that total Hg concentrations are not a good predictor for MMHg levels [17], and that site-specific factors control mercury methylation. The first step in mercury bioaccumulation is its methylation, a process that we have come to realize is mostly mediated by bacteria. In most simple terms, methylmercury production is the product of microbial activity and Hg(II) bioavailability:

$$\text{MMHg} = (\text{bioavailable Hg}^{2+}) \times (\text{bacterial activity}) \qquad (1)$$

Unfortunately, both of these factors are incredibly difficult to quantify, as we will see below. To complicate matters, bacteria do not only produce

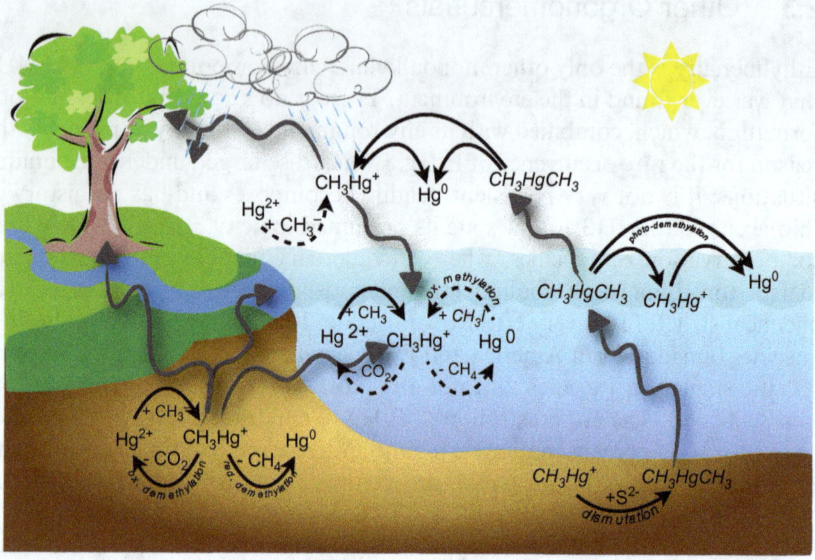

Figure 1. Summary of main methylation and demethylation pathways and locations, where they predominate. Solid arrows indicate major processes, while dashed arrows indicate reactions of minor or uncertain importance. The wiggly arrows shows cross-compartmental fluxes of methylmercury and dimethylmercury. Note that processes shown are unidirectional and not equilibrium reactions, the reverse reaction is always mediated by different groups of bacteria or reagents. Hence, environmental concentrations are usually not equilibrium, but steady state concentrations.

methylmercury, but microbial decomposition of methylmercury is also a crucial process. It is timely to evaluate what we know about sites of mercury methylation and demethylation, the organisms involved in those processes, and the factors controlling these processes. We should also inquire about how newly methylated mercury is released into the aquatic environment and transferred to the next trophic level. Figure 1 provides a schematic overview regarding main methylation and demethylation pathways and locations, where they predominate.

3.1. Biotic Formation of Methylmercury

After the first studies concluded that most mercury methylation is driven by microorganisms [18], research was immediately initiated to find the specific bacteria responsible for this process. As a result of those initial investigations during the 1970s and early 1980s, a large set of potential methylators was identified, including sulfate-reducing bacteria (SRB), which still appear to be most important for mercury methylation in many environments. However, at least in some circumstances there is evidence that SRB are not the only or the main mercury methylators [19,20].

3.1.1. *Biological Control of Mercury Methylation*

Bacteria play a pivotal role in converting Hg(II) to MMHg. Over the years, many microorganisms have been identified of being capable of generating MMHg. However, methylation activity in environments such as sediments is often correlated with the presence and activity of sulfate-reducing bacteria, which are the prime suspects of mercury methylation. SRB are an old, complex, and heterogeneous group of bacteria. Their common trait is the ability to use sulfate as a single final electron acceptor in anaerobic respiration, one of the oldest processes in microbial evolution [21–24]. SRB are not only exceptionally diverse, but also globally distributed [25]. They have been found in most continents and are probably present in every corner of the planet, as long as the right conditions for their growth exist. They can inhabit a wide range of habitats [22], which are not as limited by oxygen as previously thought [26].

The initial idea that SRB are limited by oxygen and sulfate [27] probably biased early investigations of microbial mercury methylation towards marine sediments [28–31]. However, it now seems that active SRB are also present in freshwater sediment, water, and other low oxygen environments. There are recent reports showing significant mercury methylation in floating macrophyte mats in tropical regions [32–34], in the water column of boreal lakes [35,36], and in epilithic biofilms [37]. All these new microenvironments, where mercury methylation is observed, are inhabited by a wide range of new bacteria that could play an important role in mercury methylation. In fact, already some studies suggest that SRB are not the only [19,20] or at all responsible of mercury methylation.

Although there are SRB among at least four phylums of the Eubacteria domain, the best characterized mercury methylating SRB are members of the *Desulfovibrionaceae*, *Desulfobacteriaceae*, and *Desulfobulbaceae* families. While these bacteria are predominantly anaerobic, recent studies have demonstrated that many of them are tolerant to oxygen, which may allow them to facilitate mercury methylation in aerobic environments like the periphyton of macrophytes. Initial investigations regarding the mercury methylation capacity of other bacteria revealed that also *Enterobacter aerogenes* [38], *Clostridium cochlearium* [39], *Neurospora crassa* [40], and *Methanogenic bacterium* [41] are able to produce methylmercury as a resistance pathway to tolerate inorganic mercury. Bacteria such as *Pseudomonas aeruginosa, P. fluorescens, Escherichia coli, Citrobacter, Bacillus subtilis*, and *B. megaterium* apparently do not have the ability to methylate mercury [42], while *Desulfovibrio desulfuricans, Selenomonas ruminantum*, and *Megasphaera elsdenii* are able to demethylate methylmercury [43].

In vitro and *in situ* experiments demonstrated that methanogenic [27,28,31,44] and acetogenic bacteria [28] do not contribute significantly to methylmercury production in sediments. Eventually, consensus emerged that SRB are the most important mercury methylators in marine sediments [27]. The importance of SRB was later extended to other environments [28,30,45,46]. A key evidence identifying SRB was the frequently observed inhibition of mercury methylation in the presence of molybdate, a potent inhibitor of SRB activity. However, recent studies have pointed out that in certain environments molybdate is not completely inhibiting mercury methylation [20,37,47] and the question regarding which bacteria are implicated in mercury methylation has been raised again.

Much of our knowledge stems from culture experiments, where single bacterial strains are tested for their ability to methylate mercury. While these studies are instructive to characterize potential mercury methylating bacteria, they need to be interpreted with caution. Mercury methylation rates often vary according to experimental conditions, among species of the same genus, and generally show significant variability (Table 3).

Variation in the mercury methylation rate by a single strain may be attributed to different factors. In some SRB capable of fermentation it has been observed that methylation activity potentially changes when bacteria switch from fermentative to respiratory growth conditions [48,49]. This may be because hydrogen sulfide produced during respiration interferes with the bioavailability of mercury(II) substrates [49]. The degree of mercury methylation measured for individual bacterial strains should not be the sole defining criterion for the strain's importance as a mercury methylator. Given this, the significance of culture experiments must be carefully considered. It must be emphasized that the measurement of a mercury methylation potential in culture experiments can demonstrate the strain's ability to methylate mercury, however, this does not constitute evidence for this bacteria to also play a role in mercury methylation in the environment. Since new microenvironments are being studied for their role in methylmercury production, new potentially important bacteria for mercury methylation have been suggested, e.g., dissimilatory iron reducing bacteria (IRB) [20,50,51].

3.1.2. Chemical Control of Mercury Methylation

It is often very difficult to separate confounding factors to clearly isolate individual parameters controlling microbial mercury methylation, which is

Table 3. Mercury methylation potentials determined for pure cultures of different sulfur-reducing bacteria. Reported rates depend greatly on experimental conditions such as culture conditions, cell density, and concentration of Hg amendments.

Genus	Degree of methylation	Net methylation		Methylation/sulfate reduction	Source
	%	$pg\,mL^{-1}h^{-1}$	$pg\,cell^{-1}h^{-1}$	Rate ratio	
Desulfovibrio desulfuricans	~0.300	4.48±11.8	$4.2\times10^{-8}\pm12.2\times10^{-8}$	$1.37\times10^{-6}\pm1.8\times10^{-7}$	[30]
Desulfovibrio desulfuricans LS	0.200–37	3.12–770.83	na	na	[48]
Desulfovibrio desulfuricans LS	0.001–0.002	~208–~340	na	na	[44]
Desulfovibrio africanus	7.47	13.89±6.1	$1.2\times10^{-7}\pm5.0\times10^{-8}$	na	[88]
Desulfovibrio vulgaris	<LOD	<LOD	<LOD	na	[88]
Desulfobulbus propionicus ATCC	0.085	1.05±30.4	$4.3\times10^{-9}\pm32.1\times10^{-9}$	$2.80\times10^{-7}\pm9.0\times10^{-8}$	
Desulfobulbus propionicus 1pr3	na	1.9–48.0	na	na	[49]
Desulfobulbus propionicus 1pr3	6.64	11.45±1.6	$1.0\times10^{-6}\pm2.4\times10^{-7}$	na	[88]
Desulfobulbus propionicus MUD	0.303	3.03±0.7	na	na	[88]
Desulfococcus multivorans ATTC	~0.350	4.62±21.4	$2.7\times10^{-7}\pm23.9\times10^{-7}$	$4.60\times10^{-6}\pm1.30\times10^{-6}$	[30]
Desulfococcus multivorans 1be1	6.8	9.60±3.64	$9.06\times10^{-7}\pm3.45\times10^{-7}$	na	[88]
Desulfobacter	~0.120	1.55±21.3	$1.6\times10^{-6}\pm21.5\times10^{-6}$	$4.10\times10^{-6}\pm9.5\times10^{-7}$	[30]
Desulfobacterium	0.472	7.53±9.9	$6.2\times10^{-6}\pm10.4\times10^{-6}$	$2.58\times10^{-5}\pm2.80\times10^{-6}$	[30]

na: Not available or impossible to calculate from the data provided in the original source
LOD: limit of detection

affected by many environmental factors such as mercury concentrations, temperature, organic substrate supply for the methylating bacteria, sulfur speciation, and pH. For example, sulfuric acid deposition leads to acidification and increases sulfate levels. Scandinavia experienced a significant decrease of anthropogenic mercury emissions after the closure of several mercury emitters in central Europe in the mid-nineties, and fish mercury levels seem to be decreasing. However, the decrease in mercury deposition was also paralleled by controls on sulfate deposition. Likewise, eutrophication does not only add nutrients, which boosts microbial activity, but may also alter the pH. Most of our current information on individual factors is gleaned from laboratory studies, which have their own limitation. Only recently, a couple of experiments at the ecosystems scale are beginning to shed some light on those intricately interconnected relationships. Nevertheless, some consensus on common features seems to be emerging.

Often, increased MMHg formation is reported under low pH conditions [52–56]. One possible explanation is that Hg methylating bacteria dominate over other microbes at lower pH [57]. Alternatively, acidification enhances Hg(II) bioavailability making a larger fraction of mercury available to bacteria for methylation.

Sulfate is coming up time and time again as a critical parameter. Considering that SRB are thought to be mainly responsible for methylation, sufficient sulfate must be present to maintain optimum activities. *In vitro* studies have found a direct relationship between sulfate reduction rates and MMHg production [58–61] with optimum sulfate levels for maximum MMHg formation [31,62]. This information is augmented by ecosystems studies [63]. An interesting and maybe counter-intuitive effect is attributed to the product of microbial sulfate reduction, sulfide. High sulfide levels render Hg(II) unavailable by forming solid HgS. This is probably the reason for the upper limit of optimum sulfate concentration, above which too much sulfide is produced. However, in contrast to widely accepted text book chemistry, which would predict quantitative precipitation of Hg(II) by any excess of S^{2-}, dissolved concentrations of Hg are often elevated in anoxic, sulfidic waters relative to aerobic water with no sulfide present. This apparent contradiction is explained by the formation of complex ions and multinuclear, neutral complexes such as HgS_2^{2-}, $Hg(SH)_2$, $Hg(SH)^-$, Hg(SH)(OH), and maybe even HgS^0. Sulfide appears to greatly influence the first factor in equation (1), the bioavailability of Hg(II) for methylation [64,65].

There is now a large body of literature predicting the formation of neutral Hg-sulfur complexes such as HgS(0) and $Hg(SH)_2$ at moderate sulfide concentrations [66,67]. These uncharged Hg-sulfur complexes are thought to be able to penetrate cell membranes and are a potential Hg uptake route into bacteria for subsequent methylation. However, this hypothesis is difficult to test directly. Analytically, the concentrations of the Hg species of interest are well below currently available *in situ* technologies and cannot be measured directly. Therefore, we rely on equilibrium distribution calculations. Unfortunately, some of the required formation constants are only known approximately [65], and others are modeled rather than experimentally determined. The existence and significance of species such as $HgS(0)_{(aq)}$ are still controversial. Nevertheless, the idea of neutral sulfur species is consistent with experimental results showing good correlation between methylation and modeled concentrations of Hg-sulfur complexes. Various studies found good correlations between MMHg production and predicted HgS^0 concentrations based on total sulfide, Hg, and H_3O^+ concentrations [68–71]. Likewise, very high sulfide levels should shift the equilibrium distribution of Hg-thiol species to charged complexes and HgS precipitation to reduce the degree of mercury methylation, which is also observed experimentally.

The second most important factor controlling the bioavailability of Hg^{2+} is the concentration of dissolved organic matter. Like sulfide, also DOM appears to have a complex affect on MMHg formation, affecting it at least on three different levels: (i) the biological activity is enhanced in the presence of fresh DOM, serving as an organic substrate for microbes, which may explain in part enhanced MMHg levels observed in newly created and flooded reservoirs [72,73]; (ii) Hg(II) concentrations in water are commonly well correlated with DOM levels [74,75], enhancing Hg(II) mobility and delivering it to sites of methylation. Likewise, DOM can complex MMHg elevating its total concentration in water and therefore increase bioaccumulation rates [76]; (iii) at the same time, binding of Hg(II) and MMHg by large DOM molecules decreases its bioavailability for methylation reactions and potentially diminished the availability of MMHg for bio-uptake [77,78]. While each of those three effects has been studied and documented in isolation *in vitro*, the overall result in nature is very difficult to predict and field measurements are sometimes contradictory. Like with sulfide, there is probably also a sweet spot for optimum DOM concentrations, at which the factors promoting mercury methylation overcompensate the diminished bioavailability. To complicate matters, previous studies mostly considered bulk concentrations (i.e., quantity) of DOM, but rarely considered the type of DOM (i.e., quality). It is conceivable that DOM with high sulfur content (especially in the form of thiols) binds Hg(II) especially strong and has a relatively larger negative effect on methylation rates [79,80].

Temperature usually enhances bacterial activity. It is therefore not surprising that higher temperatures often promote mercury methylation. Sediments in shallow water typically form more MMHg during warm summers compared to colder winter months [81]. Likewise, tropical environments usually show higher methylation rates. However, higher rates of gross methylation are probably counterbalanced by enhanced rates of bacterial demethylation, and the net methylation rate might not change dramatically, unless temperature shifts change the overall composition of the microbial community or the relative activity of methylating and demethylating bacteria. The potential effect of global warming on MMHg production is therefore uncertain [82]. What appears to be clear though, is that global warming will likely extend the methylating season in arctic and sub-arctic regions, e.g., earlier onset of thawing and later start of freezing during the year. Prolonging the period during which methylmercury can be produced will likely lead to enhanced MMHg levels in local biota and even increased export of MMHg into sub-arctic lakes and arctic oceans.

3.1.3. *Biochemical Pathways of Formation*

Without doubt the easiest and most direct approach to identify, which bacteria are responsible for mercury methylation would be to identify the methylation pathway and the enzymes involved. For example, the relatively easy identification of bacteria able to reduce Hg^{2+} to Hg^0 is possible thanks to the *mer* operon, which is a cluster of genes codifying for the enzymes responsible of such mercury reduction [83]. Unfortunately, unlike bacterial resistance to inorganic mercury, the pathway for mercury methylation is not well understood. In fact, it is not even established beyond doubt if mercury methylation is a detoxification strategy in some bacteria or an accidental process [84].

One proposed mechanism for mercury methylation among SRB, suggests that *Desulfovibrio desulfuricans* LS methylates mercury through a cobalamin (vitamin B_{12}) mediated acetyl-coenzyme A pathway [48,84–86]. This was not surprising because under certain conditions methylcobalamin can spontaneously methylate mercury and may be responsible in large part for the abiotic mercury methylation [87]. So, the presence of methylcobalamin alone could have been responsible for mercury methylation in the *D. desulfuricans* cells, but evidence suggests that methylation is catalyzed by an enzyme [84]. Subsequently, a method of quantifying mercury methylation potential using methyltransferase as indicator was developed [85]. But later, some SRB were found to methylate mercury in an acetyl-coenzyme A independent pathway [88], which means that there could be at least one alternate mechanism for mercury methylation by SRB.

3.2. Abiotic Formation of Methylmercury

Another critical problem in measuring the bacterial potential to methylate mercury is differentiating biotic from abiotic methylation. Several bacteria may appear to methylate mercury because some methylmercury is produced in their presence. However, such methylation could be caused by extra cellular enzymes or other abiotic processes initiated by a bacterial product. Laboratory experiments identified three purely chemical reactions of potential relevance: methylation by methylcobalamin, transmethylation involving other methylated metals, and oxidative methylation.

Methylcobalamin is also able to transfer its methyl group onto Hg(II) in the absence of enzymes. In fact, this reaction is widely used to synthesize methylmercury and isotopically labeled MMHg compounds for analytical purposes [89–91]. There is also some discussion in the literature that it may be produced and released into the environment by microbes, subsequently generating MMHg. While methylcobalamin for this reaction is provided by bacteria, the actual methylation reaction is non-enzymatic and the process should then be considered an abiotic process. Unfortunately, there is no information regarding methylcobalamin levels in natural environments, so the potential importance of this reaction is difficult to assess.

Transmethylation by organometallic compounds such as methyltin, methyllead or methylarsenic species is another possibility [87,92]. This pathway has been proposed to occur in certain contaminated sites and has also been used to synthesize MMHg compounds [93,94]. A recent systematic study shows that both monomethyltin and dimethyltin chlorides are potent Hg(II) methylators. Highest rates were observed at elevated pH and required the presence of chloride. Hence, the authors concluded that this methylation pathway is possibly of importance in oceans [95]. They estimated a potential rate of MMHg formation of 0.5 pg/L/day under typical seawater conditions. Albeit low, this rate could produce as much as 180 pg/L of MMHg per year; a concentration that exceeds measured MMHg levels typically observed in oceans. Hence, this pathway should not be dismissed outright and might require further consideration.

Oxidative methylation of elemental mercury by methyliodide proceeds according to

$$Hg^0 + CH_3I \rightleftharpoons CH_3HgI \qquad (2)$$

Methyliodide is also fairly abundant in seawater. However, it only reacts with Hg(0) and not with Hg^{2+} [96]. Since concentrations of dissolved elemental are much lower than mercuric Hg, this reaction may be less significant and yields under typical environmental conditions are expected to be very low. MMHg production rates in the order of 0.2 pg/L year are estimated [95]. While this reaction is not affected by the water chemistry (i.e., Hg(0) activity is always unity), it appears to be too low to be relevant.

There have been sporadic reports of abiotic methylation facilitated by DOM in freshwater environments [87,97], including pore waters. However, abiotic methylation data were always difficult to delineate from biotic methylation and results are often inconclusive (it is virtually impossible to sterilize sediment or water samples and without changing their chemistry at the same time). Nevertheless, a small contribution to the overall MMHg formation can potentially be attributed to DOM methylation. Recently, it was demonstrated in laboratory experiments that mercury can be methylated by acetic acid, but the authors concluded that this process may contribute at most a few percent of the MMHg concentrations observed in rain water [98].

3.3. Formation of Dimethylmercury

DMHg is clearly a naturally occurring Hg species. Since it is not released or discharged into the environment by any known man-made process, there must be a natural process generating this compound. However, the exact mechanism, by which DMHg is formed, is still shrouded in mystery. Researchers have often speculated that it could be formed by methylation of methylmercury, but no conclusive evidence has emerged so far. The only known formation process is of chemical nature. In the presence of high concentrations of sulfide, MMHg may react with sulfide to form a methylmercury-sulfide complex (which has not been verified, yet) that dismutates into cinnabar and DMHg according to equilibrium (3):

$$2\,CH_3Hg^+ + S^{2-} \rightleftharpoons CH_3Hg\text{-}S\text{-}HgCH_3 \rightleftharpoons HgS_{(s)} + (CH_3)_2Hg \qquad (3)$$

The formation of DMHg has been observed in sulfide amended freshwater and salt marsh sediments [99–101] at sulfide concentrations exceeding 2 mg/kg. However, DMHg was never detected in freshwater systems, so it is unclear if this route is of any significance under natural conditions.

3.4. Formation of Other Organomercurials

There are no known reports of microbial formation of ethylmercury in the natural environment. Under very specific circumstances the formation of some other unusual organomercurials was observed. At the site of a former industrial complex with extremely high levels of Hg contamination a series of organomercury compounds was identified including ethoxyethyl and aromatic Hg species as well as a couple of other unidentified species [102]. However, this should be considered an isolated case. The unusual compounds were only found on-site and not in downstream rivers sediments, suggesting that these compounds are either immobile or quickly degraded in the environment.

4. DEGRADATION OF ORGANOMERCURIALS

Although mercury methylation is the process most frequently studied, methylmercury demethylation is equally important in regulating net production and standing pools of MMHg in the environment. In many environments, both processes balance out to a steady state concentration of MMHg. Known environmental sinks for MMHg include bacterial and photochemically induced demethylation, sedimentation, and bio-uptake.

4.1. Bacterial Demethylation

In contrast to methylation, the demethylation process is well understood at the molecular level [103–107]. The biochemical reaction is characterized in detail, distinguishing between an oxidative pathway producing Hg^{2+} and CO_2 and a reductive mechanism leading to CH_4 and Hg^0. The reductive pathway dominates in polluted sediments [108] and is induced by enzymes related to the *mer* operon. It is considered a detoxification mechanism and is found in "broad-spectrum" resistant bacteria. The two-enzyme system consists of a Hg-C bond cleaving organomercurial-lyase and a mercuric reductase, which produces Hg^0. The two-step reaction detoxifies MMHg by eventually converting it to a volatile mercury species that readily leaves the immediate microbial habitat. The oxidative mechanism seems to dominate at normal, non-elevated MMHg concentrations and is associated with methanogenic and sulfate-reducing bacteria [109–111].

However, the exact molecular mechanism or enzymes involved are not characterized in detail. The oxidative pathway is presumably not a detoxification, since the product is still available and toxic to bacteria. Rather, it is thought that bacteria metabolize the methyl group of MMHg. While reductive demethylation appears to dominate in marine environments, oxidative demethylation is more prominent in freshwater sediments [112,113]. Bacterial demethylation rates determined in sediments are very high, potentially turning over the entire MMHg pool within days (calculated MMHg half-lives are less than 2 days) [114,115]. Bacterial demethylation of MMHg in lake water was undetectable (i.e., $<10\%$ per day) [36].

Since bacterial demethylation is only significant in sediments and photo-degradation only active in surface waters, MMHg is a relatively persistent contaminant in lakes, and particularly in oceans.

4.2. Abiotic Degradation of Methylmercury

While methylmercury is chemically susceptible to attack by strong oxidants and concentrated acids, such reagents are not present in the environment. This leaves photo-induced demethylation as the most important methylmercury decomposing process [82,116,117], especially in clear water lakes and the surface of oceans. MMHg is degraded by ultraviolet (100–400 nm) as well as visible light (400–800 nm) [118,119]. The overall decomposition rate is controlled by two factors: (i) the wavelength irradiating MMHg, with shorter wavelengths being more efficient in cleaving the Hg-C bond; and (ii) the intensity of that wavelength.

Depending on the nature of the water body, UV and visible light are attenuated differently. The latter penetrates deeper into water and is therefore affecting a relatively larger volume of dissolved MMHg. Short and long wavelengths are equally important in clear water lakes with relatively little light attenuation. Dark colored lakes, however, have an equalizing effect and the more energetic UV-light is the dominating source for MMHg decomposition. UV-A and UV-B are accountable for approximately 50% of the overall photodemethylation in clear water, and for more than 75% in colored lakes [120]. While natural light penetrates quite deep into clear marine water, it is not expected that MMHg photodegradation would significantly lower the pool of MMHg in oceans, considering their enormous depth. It may, however, contribute to the concentration gradients frequently observed in oceans.

Like MMHg, DMHg is also very susceptible to photodegradation. Owing to the analytical difficulties measuring DMHg, however, we have no experimental evidence for its actual persistence in natural water.

5. DISTRIBUTION AND PATHWAYS OF ORGANOMERCURIALS IN THE ENVIRONMENT

The degree of methylation and demethylation can differ quite dramatically from compartment to compartment, and both spatially and temporally. Figure 2 illustrates for various matrices and sample types the typical range of environmental MMHg concentrations and the fraction of Hg that is present in form of MMHg. However, when interpreting these data, the reader

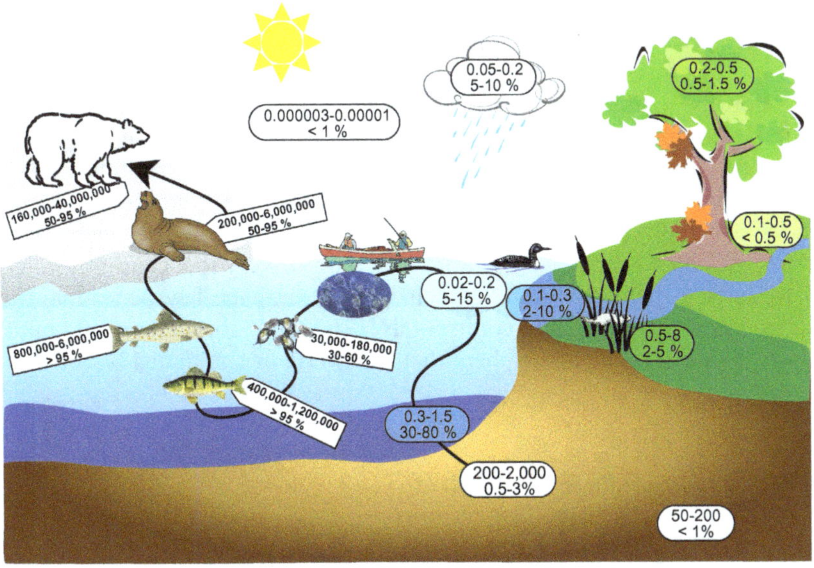

Figure 2. Typical range of methylmercury concentrations and the fraction of Hg that is present as methylmercury in environmental and biological matrices. Concentration units are ng/kg for solids and ng/L for water and air. The arrow illustrates a typical bioaccumulation pathway in the aquatic food chain.

should keep in mind that sites of methylmercury production are not always the location, where MMHg accumulates in the environment. Hence a higher percentage of MMHg in water relative to sediments does not indicate that MMHg was also formed in the water. In many systems, most MMHg is probably generated in sediments, but demethylation rates are also very high in sediments and virtually absent in water. This combination leads to high turnover of MMHg in sediments and a standing MMHg pool, which constitutes only 1% of the total Hg. Demethylation activity in water on the one hand is very low, making the little MMHg escaping form sediment into the overlaying water very persistent in this compartment. An exception for this general rule are probably lakes developing an anoxic hypolimnion.

5.1. Atmosphere

There are very few reports on MMHg measurements in air. Most of our knowledge is indirect and stems form MMHg measurements in precipitation. This scarcity of information is surprising considering the importance of the atmosphere for the global distribution of Hg. While we can safely assume that MMHg species are volatile under the right conditions (MMHgCl has a high vapor pressure, is presumably the dominating species in seawater, and should therefore be emitted into air in form of sea spray), it is also expected that MMHg compounds are not very stable under UV irradiation. This would argue for very low levels of MMHg in air. However, polar regions are in the dark for long periods of the year, potentially allowing the build-up of MMHg in the polar atmosphere, from which it could be distributed and deposited in other regions. Overall, there is an expectation of very low concentrations of MMHg in air (probably less than 10 pg/m^3), and the few occasional measurements reported seem to confirm this [121,122]. If correct, it would put the fraction of Hg in the atmosphere that is MMHg at less than 1%.

5.2. Precipitation

MMHg is regularly found in precipitation ranging from 0.01–0.2 ng/L, which is usually less than 1% of the total Hg in rain water [123]. Concentrations in the summer are often higher than in winter. Although MMHg concentrations in the initial precipitation and during low volume events are typically highest, the total mass of MMHg that gets deposited is usually delivered in high volume events. Regardless, the origin of this MMHg is not well explained. Some studies support the idea that MMHg in precipitation is of local origin, which leads to three potential scenarios of (i) formation of MMHg in lake-effect clouds and fogs, essentially atmospheric mercury methylation in droplets serving as micro reactors [124]; (ii) MMHg emission from surfaces, most likely wetlands or landfills; (iii) upwelling DMHg from deep water photodegrades to MMHg in the atmosphere and is scavenged during precipitation events. However, none of these hypotheses has been thoroughly tested and the origin of MMHg in precipitation is still a mystery.

Nevertheless, circumstantial evidence emerging from recent studies might be useful to narrow down the possibilities. Lately, concentrations of MMHg as high as 0.28 ng/L have been observed in artic snow packs [125]. Since MMHg levels decline with onset of warmer temperatures, it is believed that this MMHg is deposited to the snow rather than produced in the snow [126]. The observation of DMHg in polar oceans strengthens the suggestion that the source of MMHg in polar regions is actually photodegraded DMHg. Attempts to detect Hg methylation directly in snow packs were unsuccessful [127]. On the other hand, high levels of MMHg in polar melt water of up to 0.24 ng/L [125] and seasonal freshwater ponds [128] have been reported.

5.3. Aquatic Systems

Numerous studies have linked MMHg production to anaerobic microbial activity in lake sediments and associated wetlands [129,132,27–28,30,31]. Anaerobic sediments have long been believed to be the main site of mercury methylation. SRB predominate in the top few cm of freshwater sediments, and the zone of highest methylation activity is often found just below the oxic/anoxic transition zone underlying oxygenated water [30,133,134]. Here, MMHg concentrations of over 1 ng/g in sediments are not uncommon (or 0.5 to 2 ng/L of dissolved MMHg in porewater [109,135]). Wetlands are considered a sink of total mercury, but are often a net source of MMHg and suggested to be the principal source of MMHg to lakes, especially when wetland runoff dominates the catchment hydrology [135–143].

Wetland runoff is enhanced in MMHg relative to MMHg in precipitation, runoff from non-wetland regions or the lake water itself. As well, the concentration of MMHg in lake water is often correlated to the wetland areas in the lake catchment [144,145]. Furthermore, studies conducted in wetlands show a high degree of methylation relative to forest soils or even lake sediments. However, to fully assess the importance of wetlands, one needs to construct a thorough mass balance, since high concentrations alone do not guarantee that wetlands are necessarily the principal source of MMHg [146]. In addition, high net formation rates in wetland leading to high concentrations of standing pools of MMHg are only relevant if the wetland is also hydrologically connected to the lake. In other words, it is important that the produced MMHg is also exported from the wetland. Otherwise, it may only be subject to fast internal recycling due to the concurrent and efficient demethylation process. However, even if MMHg is confined, it is still of importance for wildlife and biota living in the wetland. This is of primary concern for ecosystems like coastal marshes, which often serve as food sources for migratory birds exposing them to high levels of MMHg [147,148].

The problem of increased MMHg levels in flooded reservoirs, created for power generation, has long been recognized [149–153]. The flooding of terrestrial soils and vegetation during impoundment releases a pulse of easily accessible inorganic carbon to the aquatic system and bacteria inhabiting the system. Flooded portions of reservoirs typically show a higher degree of Hg methylation compared to non-flooded areas or nearby natural lakes [154]. The pulse of microbial activity combined with presumably temporarily more bioavailable Hg leads to increased MMHg production, which is immediately transferred to the food web, leading to extremely high levels of MMHg in fish for at least 5 years after impoundment. It may take up to 20 years or more until MMHg levels decline, but even very old reservoirs typically show much higher MMHg levels in biota compared to natural lakes in the same area [151].

Oxygen depletion in the hypolimnion of stratified lakes creates a redox transition zone similar to those identified in lake sediments. It is reasonable to assume that this might be a zone of high sulfate reduction which has been shown to effectively methylate mercury [35,36,146,155]. Methylmercury directly produced in the water column quickly accumulates in the anoxic hypolimnion to high concentrations. While it may not be readily available to the foodweb (little life in anoxic water), it eventually mixes into the overlaying oxic water column during lake turnover, providing fresh MMHg for bio-uptake. The fraction of Hg(II) conversion in the water column is comparable to that in lake sediments, but since the substrate concentration in water is lower than in sediments, the methylmercury production expressed as mass per volume is also significantly lower, but more than compensated for by the large volume of water compared to the thin active sediment layer, in which methylation proceeds. In addition, MMHg produced in the water column is directly bioavailable. Even if subsurface sediments produce large amounts of MMHg, once generated, it must migrate somehow into the overlying water to be available for the pelagic foodchain. The benthic foodchain, however, would be more immediately affected by sedimentary MMHg formation. It is therefore suggested that water column methylation is a significant, but often overlooked source of MMHg in lakes, especially in those developing an anoxic hypolimnion. The potential to methylated Hg coincides with an accumulation of MMHg in the hypolimnion [36,156]. As well, SRB were recently isolated from the hypolimnetic water [157]. Other locations of potential significance for aqueous MMHg production are epilithic biofilms [37,158] and periphyton associated to macrophyte roots [33,47,159–162]. Although these environments are often found in oxygenated water, they sustain anoxic microhabitats, which house SRB [159] and have been shown to produce MMHg.

5.4. Terrestrial Environment and Vegetation

Uplands can be important areas to deliver bioavailable Hg to methylation zones in wetlands [140,163]. While forest soils are known to store large pools of Hg(II), they are not very effective in methylating Hg(II). Consequently, concentrations of MMHg in soils are typically much lower than corresponding levels in sediments and wetland peats [164]. However, compared to Hg(II), the mobility of MMHg in forested catchments is greater and especially high volume runoff events are responsible for increased MMHg flux from watersheds [163,165], with MMHg transport facilitated by dissolved organic and particulate matter. Maximum concentrations of MMHg in surface waters are often found during warmer months [81,163], coinciding with increased microbial activity and low flow conditions. Leaching of MMHg from forest soil also increases after soil disturbances (e.g., following the use of heavy machinery and clear-cutting). MMHg export quadrupled in affected forested catchments in Sweden and Finland [166]. Nevertheless, the mass of MMHg exported from terrestrial uplands is often a minor contributor to MMHg in lakes.

Owing to the low MMHg productivity, most terrestrial vegetation shows only low levels of MMHg. Although root uptake of MMHg appears to be slow [167], concentration in cattail foliage showed a diurnal pattern and changed with water concentrations of MMHg [168]. Observed correlations between MMHg concentrations in soil and green plant tissue strengthen the hypothesis that plants can mobilize MMHg via their root system [169]. Of particular concern are rice plants. Recent studies have demonstrated that rice paddies are very effective sites of methylation. This is expected as rice paddies resemble wetlands and marshy environments, which are known to be productive MMHg ecosystems. Especially rice grown in Hg-polluted regions can accumulate very high levels of MMHg, causing abnormally high exposure to humans [170]. Levels of over 100 ng/g have been measured in the edible portion of rice, which is 10–100 fold higher than in other crop plants. Data from non-Hg-polluted areas is rare; hence, the general risk of MMHg exposure via rice consumption is unclear. Considering the enormous importance of rice as a main food source for a large fraction of the world's population, this potential pathway of MMHg exposure could be of critical importance and deserves special attention.

Generally, MMHg in non-crop plants and bioaccumulation of MMHg in terrestrial food chains is normally not considered to be a significant problem and was therefore rarely investigated. MMHg levels in vegetation at pristine sites range from 0.1–1.5 ng/g [171], with levels of up to 100 ng/g at mining impacted locations [172]. A terrestrial food chain study showed some bioaccumulation of MMHg in a forested ecosystem [172]. Since the fraction of Hg that is MMHg is normally less than 1% in soils, but >1–2% in vegetation, it also points to a moderate MMHg bioaccumulation from soil to plant. While Hg(II) hyper-accumulating plants have been reported, no MMHg hyper-accumulating species are known. However, it is suggested that genetically engineered macrophytes (trees, grasses, shrubs) might be used to degrade MMHg at polluted sites [173].

The forest canopy has an amplifying effect of scavenging MMHg from air in foliage. Although it is not clear if leaf and needles actively take up MMHg from air or simply serve as surface for physical adsorption, litterfall has been identified a source of MMHg to forested ecosystems [167]. Likewise, concentrations of MMHg in throughfall (i.e., rain water collected under trees) are significantly higher compared to MMHg in precipitation collected in the open. For example, estimates of MMHg deposition in the boreal forest region of Canada are 0.4, 0.4–09, and 0.7 mg/ha for precipitation, throughfall, and litterfall, respectively [174].

5.5. Bioaccumulation

Mercury is the most common contaminant of fish in many regions of the world. All mercury in fish tissue is essentially MMHg and responsible for consumption advisories in thousands of lakes because of mercury levels, which are deemed unsafe. This is especially of concern for populations, which rely heavily on fish as their main food source [175–177].

Methylmercury has a remarkable bioaccumulation potential. Concentrations in water are often near the detection limit (e.g., 0.05 ng/L), but can be biomagnified to over 1 mg/kg in fish occupying high trophic positions. An additional biomagnification step occurs in piscivorous wildlife such as loon, otter, seals or polar bears. Because of the ubiquitous nature of Hg and mercury methylation, elevated amounts of Hg are reported even in remote, undeveloped areas with no local sources of pollution.

There is only sporadic information on MMHg levels in the lower food chain and measurements of MMHg in phytoplankton are virtually non-existent. Most measurements have been conducted on zooplankton, showing a range of 30–400 ng/g of MMHg (dry weight, the corresponding wet weigh is difficult to estimate due to near impossible determination of water content in zooplankton) [178–180]. Owing to the great importance of fish as a food source, the overwhelming number of measurements are on fish. Small freshwater species have as little as 10–300 ng/g (fish-MMHg concentrations are usually expressed in Hg per wet weight mass; the equivalent dry weight concentrations are approximately 4–5 fold larger). This can easily increase in piscivorous fish to over 1000 ng/g (wet weight), even in non-polluted areas [181,182]. Fish from flooded reservoirs or Hg-contaminated areas are often reported to even exceed this level [183,184]. Mercury in fish increases with age and is often manifested in the good correlation between Hg concentration and size (age). However, age is the more important factor as can be seen in some northern Quebec lakes, where fish grow very slowly. In those lakes, relatively small fish have high Hg concentrations for their size. On the other hand, in fast growing environments (aquaculture, highly productive natural lakes) large fish have relatively low mercury levels, owing to bio-dilution of accumulated MMHg.

The largest MMHg biomagnification step occurs at the first step of the foodchain, when MMHg is transferred from water into plankton [185–187]. Presumably, the uptake of MMHg is facilitated by diffusion of its uncharged chloride complex, CH_3HgCl, which has a high lipid solubility and high membrane permeability. The accumulation of MMHg is therefore maximized by conditions that favor formation of the CH_3HgCl species, such as low pH and high chloride concentration. Subsequently, MMHg is assimilated by planktonic organisms and passed on to its predators, where it is equally well retained. Retention is due to the high lipophilic nature of non-polar CH_3HgCl and/or the high affinity of the CH_3Hg^+ cation to thiols, namely cysteine groups in proteins [12]. This explains the high concentration of MMHg in muscle tissue of fish.

It should be noted that uptake of MMHg from water into organisms is only significant at the planktonic level. Higher organisms almost exclusively get their MMHg from food ingestion and additional uptake from the surrounding water is negligible [188]. There is usually an excellent correlation between the trophic level of an organism [189] (as indicated by its $\partial^{15}N$ status) and MMHg concentrations. Consequently, ecosystems with extra trophic levels lead to higher MMHg concentration in fish.

In the presence of mysids, a small planktivoric freshwater shrimp, fish accumulate significantly higher Hg concentrations compared to fish in nearby mysid-free lakes [190]. The proportion of Hg that is MMHg is consistently amplified during the bioaccumulation process. Originally the fraction of Hg that is MMHg is approximately 10% in water, increases to 30–50% in zooplankton, and finally to more than 95% in fish of almost any kind. MMHg is only very slowly eliminated from fish. Estimates of MMHg half-lives vary from as low as four weeks to more than one year [191]. Often, fast rates of elimination are only obtained under acute exposure scenarios, while the longest half-lives are more typical for natural MMHg levels. Considering this slow rate of elimination, it is clear that a lowering of MMHg in the environment will only gradually reduce MMHg concentrations in older fish having already accumulated significant concentrations.

Fish eating mammals effectively accumulate MMHg. Good correlations exist between MMHg exposure and levels in fur and brain tissue of otter and mink, raising the possibility that some otter populations are already experiencing clinical symptoms judging by their brain-Hg levels of over 5 mg/kg [192,193]. Arctic mammals such as seals, walrus, beluga and polar bears are at the very top of the food chain and accumulate the highest concentrations of MMHg [194–197]. However, polar bears feeding on ringed seals actually have lower MMHg concentrations than their prey, which suggests a potential detoxification mechanism (methylation ?) in polar bears [198].

Loon in northeastern US and Canada are particularly vulnerable. They are feeding almost entirely on fish and live in regions suffering from acidification, which exacerbates the MMHg problem [199,200]. Their exposure to MMHg is high enough to cause reproductive impairment in some populations in New England and the Canadian Maritimes [201].

5.6. Dimethylmercury

As mentioned earlier, DMHg was never detected in freshwater or terrestrial systems. The only place where it seems to exist naturally is in deep oceans, where it was detected every time, when a measurement was attempted. Once formed in deep oceans it may resurface in coastal regions with upwelling waters, e.g., the Pacific US coast [202]. Owing to its high volatility and favorable Henry's Law coefficient, DMHg has the potential to degas from oceans into the atmosphere. Once exposed to light, it easily degrades and might be an important source for atmospheric MMHg. Positive marine DMHg sightings include the Mediterranean [203,204], Atlantic [205], Pacific [206], and most recently also the Arctic ocean.

Maximum DMHg levels are usually found below the oxycline or in deep ocean waters, suggesting formation in the low oxygen zone. While the origin of DMHg is unknown, a microbial source of DMHg is suspected. 60 pg/L of DMHg were measured near the Strait of Gibraltar [204], and an average of 40 pg/L in deep waters of the Eastern and 18 pg/L in the Western Mediterranean, with no DMHg at the surface [203]. Likewise, DMHg was only found at levels of up to 20 pg/L in the deep South and equatorial Atlantic Ocean [205], and again no DMHg (<2pg/L) at the surface. DMHg levels in the Arctic ocean were as high as 110 pg/L at depths below 600 m and 5–10 pg/L at the surface [207]. Flux estimates suggest that as much as 40 ng/m^2/day of DMHg may volatilize from Arctic marine waters during the ice-free season [207], which would be sufficient DMHg to explain a significant fraction of the high levels of MMHg that is observed in snow packs close to the ocean's edge. Although DMHg is very susceptible to degradation by UV light one needs to consider that polar regions are in the dark for long periods. This would allow a significant accumulation and long-range transport of atmospheric DMHg, before it is deposited or photodegraded.

One of the only documented terrestrial sources of organomercury compounds is fugitive emission from landfill sites. 40–50 ng/m^3 and 10 ng/m^3 of DMHg have been measured on average at various US [208] and Chinese [209] sites, respectively, suggesting that landfills could act as a bioreactor forming methylated Hg species. Once formed, DMHg easily volatilizes due to its high vapor pressure and might contribute (after degradation) to MMHg deposition at continental sites with no other known sources of atmospheric MMHg emissions. DMHg was also detected in crude oil [210].

5.7. Other Organomercurials

There are a few isolated occurrences of EtHg. They usually coincide with discharge of EtHg from water from industrial operations. EtHg of up to

2–8 ng/g was detected in river sediments near an industrial site, where organometal compounds, including a variety of organomercurials, were synthesized for over two centuries [14]. However, EtHg was only found at the surface and no other organomercury compounds were detected. EtHg was also found nearby a factory producing ethyllead additives for gasoline, likely a trans-ethylation reaction with Et_4Pb [13,211]. A variety of unusual organomercury species was found at an industrial site of a former acetaldehyde and chlor-alkali plant and identified as ethylmercury, methoxyethylmercury, ethoxyethylmercury and phenylmercury [102].

6. CONCLUDING REMARKS AND FUTURE DIRECTIONS

Over 20,000 papers, of which almost 3000 dealt with MMHg, were published on mercury research in the past decade. This impressive number not only demonstrates the tremendous scientific interest, but also the societal significance of Hg. As well, it implies that a number of questions are still unresolved.

For one, investigators are still seeking the holy grail of Hg research, i.e., a tool that allows the determination of *in situ* methylation rates. Currently, our predictions are based on operationally defined methods, making comparisons between studies and forecasting for specific environments very difficult. Likewise, the measurement of demethylation activity was often neglected in the past, presumably due to a lack of sensitive and robust analytical methods.

A robust predictive model to calculate net methylmercury formation, incorporating the effects of DOM, pH, temperature, general water chemistry, and bacterial activity, is still sorely needed for accurate risk assessment. There is some hope that modern methods of molecular microbiology will revolutionize our approach to study and characterize bacterial communities and eventually succeed in identifying the bacterial methylation process. Once established it may be valuable in quantifying mercury methylating bacteria and their activity. The second knowledge gap lies in the reliable identification and determination of the Hg fraction that is bioavailable for bacterial methylation. While theoretical models now exist, we are still lacking the experimental tools to directly quantify this fraction.

An ecosystem that came more and more into focus over the past decade is the arctic and sub-arctic region, which is considered particularly vulnerable. The open question is, if and how climate change and global warming will affect mercury cycling, methylmercury formation and biomagnification. Related to this concern is MMHg in the world's oceans, which are another emerging ecosystem seeing increased Hg research activities. MMHg does not seem to bioaccumulate to the same degree in marine as in freshwater system and MMHg concentrations are approximately an order of magnitude lower. However, we currently have no good grasp on where and how MMHg is generated in oceans. Considering the importance of marine fish as world-wide food staple, it would be critical to allow long term forecasts of MMHg in marine fish.

ABBREVIATIONS

cys	cysteine
DMHg	dimethylmercury
DOM	dissolved organic matter
EtHg	(mono)ethylmercury
IRB	iron-reducing bacteria
MMHg	monomethylmercury
SRB	sulfate-reducing bacteria

REFERENCES

1. P. Rodríguez-González, J. M. Marchante-Gayón, J. I. García Alonso and A. Sanz-Medel, *Spectrochim. Acta, Part B*, 2005, **60**, 151–207.
2. M. Leermakers, W. Baeyens, P. Quevauviller and M. Horvat, *Trends Anal. Chem.*, 2005, **24**, 383–393.
3. J. E. Sánchez Uría and A. Sanz-Medel, *Talanta*, 1998, **47**, 509–524.
4. H. Hintelmann, in *Sample Preparation for Trace Element Analysis*, Ed. Z. Mester and R. Sturgeon, Elsevier, Amsterdam, 2003, pp. 1063–1080.
5. E. Björn, T. Larsson, L. Lambartsson, U. Skyllberg and W. Frech, *Ambio*, 2007, **36**, 443–451.
6. T. W. Clarkson and L. Magos, *Crit. Rev. Toxicol.*, 2006, **36**, 609–662.
7. D. Mergler, H. A. Anderson, L. H. M. Chan, K. R. Mahaffey, M. Murray, M. Sakamoto and A. H. Stern, *Ambio*, 2007, **36**, 3–11.
8. J. G. Melnick, K. Yurkerwich, D. Buccella, W. Sattler and G. Parkin, *Inorg. Chem.*, 2008, **47**, 6421–6426.
9. G. Schwarzenbach and M. Schellenberg, *Plant and Soil*, 1965, **48**, 28–46.
10. U. Skyllberg, P. R. Bloom, J. Qian, C. -M. Lin and W. F. Bleam, *Environ. Sci. Technol.*, 2006, **40**, 4174–4180.
11. S. -J. Yoon, L. M. Diener, P. R. Bloom, E. A. Nater and W. F. Bleam, *Geochim. Cosmochim. Acta*, 2005, **69**, 1111–1121.
12. H. H. Harris, I. J. Pickering and G. N. George, *Science*, 2003, **301**, 1203.
13. A. Jerneloev and G. Wennergren, *Studies of Concentrations of Methyl Mercury in Sediments from the St. Clair System and Rate of Biological Methylation in*

Incubated Samples of Sediments, Swedish Water and Air Pollution Research Institute, Stockholm, 1980, p. 1–13.
14. H. Hintelmann and R. D. Wilken, *Appl. Organometal. Chem.*, 1993, **7**, 173–180.
15. D. A. Geier, P. G. King, L. K. Sykes and M. R. Geier, *Indian. J. Med. Res.*, 2008, **128**, 383–411.
16. S. K. Parker, B. Schwartz, J. Todd and L. K. Pickering, *Pediatr. Res.*, 2004, **114**, 793–804.
17. C. A. Kelly, J. W. M. Rudd, V. St. Louis and A. Heyes, *Water, Air, Soil Pollut.*, 1995, **80**, 715–724.
18. S. Jensen and A. Jernelov, *Nature*, 1969, **223**, 753–754.
19. L. L. Loseto, S. D. Siciliano and D. R. Lean, *Environ. Toxicol. Chem.*, 2004, **23**, 17–23.
20. E. J. Fleming, E. E. Mack, P. G. Green and D. C. Nelson, *Appl. Environ. Microbiol.*, 2006, **72**, 457–464.
21. M. Wagner, A. J. Roger, J. L. Flax, G. A. Brusseau and D. A. Stahl, *J. Bacteriol.*, 1998, **180**, 2975–2982.
22. D. A. Stahl, S. Fishbain, M. Klein, B. J. Baker and M. Wagner, *Antonie Van Leeuwenhoek*, 2002, **81**, 189–195.
23. Y. Shen, R. Buick and D. E. Canfield, *Nature (London)*, 2001, **410**, 77–81.
24. D. E. Canfield and R. Raiswell, *Am. J. Sci.*, 1999, **299**, 697–723.
25. J. R. Perez-Jimenez and L. J. Kerkhof, *Appl. Environ. Microbiol.*, 2005, **71**, 1004–1011.
26. L. K. Baumgartner, R. P. Reidb, C. Duprazc, A. W. Dechod, D. H. Buckleye, J. R. Spearf, K. M. Przekopa and P. T. Visscher, *Sediment. Geol.*, 2006, **185**, 131–145.
27. G. C. Compeau and R. Bartha, *Appl. Environ. Microbiol.*, 1985, **50**, 489–502.
28. K. -R. Pak and R. Bartha, *Appl. Environ. Microbiol.*, 1998, **64**, 1013–1017.
29. J. L. Macalady, E. E. Mack, D. C. Nelson and K. M. Scow, *Appl. Environ. Microbiol.*, 2000, **66**, 1479–1488.
30. J. K. King, J. E. Kostka, M. E. Frischer and F. M. Saunders, *Appl. Environ. Microbiol.*, 2000, **66**, 2430–2437.
31. C. C. Gilmour, E. A. Henry and R. Mitchell, *Environ. Sci. Technol.*, 1992, **26**, 2281–2287.
32. M. Miranda, J. Guimaraes, M. Roulet, D. Acha, S. Coelho- Souza, J. Mauro and V. Iniguez, *Mat. Geoenviron.*, 2004, **52**, 1218–1220.
33. J. B. N. Mauro, J. R. D. Guimarães, H. Hintelmann, C. J. Watras, E. A. Haack and S. A. Coelho-Souza, *Anal. Bioanal. Chem.*, 2002, **374**, 983–989.
34. D. Acha, V. Iniguez, M. Roulet, J. R. D. Guimarães, R. Luna, L. Alanoca and S. Sanchez, *Appl. Environ. Microbiol.*, 2005, **71**, 7531–7535.
35. C. Eckley, C. J. Watras, H. Hintelmann, A. Kent and O. Regnell, *Can. J. Fish. Aquat. Sci.*, 2005, **62**, 400–411.
36. C. Eckley and H. Hintelmann, *Sci. Total Environ.*, 2006, **368**, 111–125.
37. M. Desrosiers, D. Planas and A. Mucci, *Environ. Sci. Technol.*, 2006, **40**, 1540–1546.
38. M. K. Hamdy and O. R. Noyes, *Appl. Environ. Microbiol.*, 1975, **30**, 424–432.
39. H. S. Pan-Hou and N. Imura, *Arch. Microbiol.*, 1982, **131**, 176–177.

40. L. Landner, *Nature (London)*, 1971, **230**, 452–454.
41. J. M. Wood, F. S. Kennedy and C. G. Rosen, *Nature (London)*, 1968, **220**, 173–174.
42. H. W. Holm and M. F. Cox, *Appl. Microbiol.*, 1975, **29**, 491–494.
43. S. Kosak and C. W. Forsberg, *Appl. Environ. Microbiol.*, 1979, **38**, 626–636.
44. G. C. Compeau and R. Bartha, *Appl. Environ. Microbiol.*, 1987, **53**, 261–265.
45. J. K. King, J. E. Kostka, M. E. Frischer, F. M. Saunders and R. A. Jahnke, *Environ. Sci. Technol.*, 2001, **35**, 2491–2496.
46. K. M. Batten and K. M. Scow, *Microb. Ecol.*, 2003, **46**, 429–441.
47. J. R. D. Guimarães, J. B. N. Mauro, M. Meili, M. Sundbom, A. L. Haglund, S. A. Coelho-Souza and L. D. Hylander, *J. Environ. Managem.*, 2006, **81**, 95–100.
48. S. C. Choi and R. Bartha, *Appl. Environ. Microbiol.*, 1993, **59**, 290–295.
49. J. M. Benoit, C. C. Gilmour and R. P. Mason, *Appl. Environ. Microbiol.*, 2001, **67**, 51–58.
50. K. A. Warner, E. E. Roden and J. C. Bonzongo, *Environ. Sci. Technol.*, 2003, **37**, 2159–2165.
51. E. J. Kerin, C. C. Gilmour, E. Roden, M. T. Suzuki, J. D. Coates and R. P. Mason, *Appl. Environ. Microbiol.*, 2006, **72**, 7919–7921.
52. Y. H. Lee and H. H. Environ, *Toxicol. Chem.*, 1990, **9**, 833–841.
53. R. J. M. Hudson, S. A. Gherini, W. F. Fitzgerald and D. B. Porcella, *Water, Air, Soil Pollut.*, 1995, **80**, 265–272.
54. L. Håkanson, A. Nilsson and T. Andersson, *Environ. Pollut.*, 1988, **49**, 145.
55. D. J. Spry and J. G. Wiener, *Environ. Pollut.*, 1991, **71**, 243.
56. C. J. Watras, K. A. Morrison, O. Regnell and T. K. Kratz, *Limnol. Oceanogr.*, 2006, **51**, 257–270.
57. S. Winch, T. Praharaj, D. Fortin and D. R. S. Lean, *Sci. Total Environ.*, 2008, **392**, 242–251.
58. S. Han, A. Obraztsova, P. Pretto, D. D. Deheyn, J. Gieskes and B. M. Tebo, *Mar. Chem.*, 2008, **111**, 214–220.
59. J. K. King, F. M. Saunders, R. F. Lee and R. A. Jahnke, *Environ. Toxicol. Chem.*, 1999, **18**, 1362–1369.
60. M. Marvin-DiPasquale and J. L. Agee, *Estuaries*, 2003, **26**, 1517–1528.
61. J. K. King, S. M. Harmon, T. T. Fu and J. B. Gladdena, *Chemosphere*, 2002, **46**, 859–870.
62. J. M. Benoit, C. C. Gilmour, R. P. Mason and A. Heyes, *Environ. Sci. Technol.*, 1999, **33**, 951–957.
63. J. D. Jeremiason, D. R. Engstrom, E. B. Swain, E. A. Nater, B. M. Johnson, J. E. Almendinger, B. A. Monson and R. K. Kolka, *Environ. Sci. Technol.*, 2006, **40**, 3800–3806.
64. E. M. Sunderland, F. A. P. C. Gobas, B. A. Branfireun and A. Heyes, *Mar. Chem.*, 2006, **102**, 111–123.
65. J. A. Jay, F. M. Morel and H. F. Hemond, *Environ. Sci. Technol.*, 2000, **34**, 2196–2200.
66. A. S. Mehrotra and D. L. Sedlak, *Environ. Sci. Technol.*, 2005, **39**, 2564–2570.
67. J. A. Jay, K. J. Murray, C. C. Gilmour, R. P. Mason, F. M. Morel, A. L. Roberts and H. F. Hemond, *Appl. Environ. Microbiol.*, 2002, **68**, 5741–5745.

68. C.-C. Lin and J. A. Jay, *Environ. Sci. Technol.*, 2007, **41**, 6691–6697.
69. A. Drott, L. Lambertsson, E. Björn and U. Skyllberg, *Environ. Sci. Technol.*, 2007, **41**, 2270–2276.
70. J. M. Benoit, C. C. Gilmour and R. P. Mason, *Environ. Sci. Technol.*, 2001, **35**, 127–132.
71. K. E. Paquette and G. R. Helz, *Environ. Sci. Technol.*, 1997, **31**, 2148–2153.
72. L. Lambertsson and M. Nilsson, *Environ. Sci. Technol.*, 2006, **40**, 1822–1829.
73. C. R. Hammerschmidt and W. F. Fitzgerald, *Environ. Sci. Technol.*, 2004, **38**, 1487–1495.
74. U. Skyllberg, J. Qian, W. Frech, K. Xia and W. F. Bleam, *Biogeochem.*, 2003, **64**, 53–76.
75. S. P. Shawn, P. Chadwick, C. L. Babiarz, J. P. Hurley and D. E. Armstrong, *Sci. Total Environ.*, 2006, **368**, 177–188.
76. O. Regnell and T. Hammar, *Can. J. Fish. Aquat. Sci.*, 2004, **61**, 2014–2023.
77. R. T. Drexel, M. Haitzer, J. N. Ryan, G. R. Aiken and K. L. Nagy, *Environ. Sci. Technol.*, 2002, **36**, 4058–4064.
78. T. Barkay, M. Gillman and R. R. Turner, *Appl. Environ. Microbiol.*, 1997, **63**, 4267–4271.
79. H. Hintelmann, P. M. Welbourn and R. D. Evans, *Environ. Sci. Technol.*, 1997, **31**, 489–495.
80. C. L. Miller, R. P. Mason, C. C. Gilmour and A. Heyes, *Environ. Toxicol. Chem.*, 2007, **26**, 624–633.
81. J. Canário, V. Branco and C. Vale, *Environ. Pollut.*, 2007, **148**, 380–383.
82. C. R. Hammerschmidt, W. F. Fitzgerald, C. H. Lamborg, P. H. Balcom and C. -M. Tseng, *Environ. Sci. Technol.*, 2006, **40**, 1204–1211.
83. A. M. Osborn, K. D. Bruce, P. Strike and D. A. Ritchie, *FEMS Microbiol. Rev.*, 1997, **19**, 239–262.
84. S. C. Choi, T. Chase Jr and R. Bartha, *Appl. Environ. Microbiol.*, 1994, **60**, 1342–1346.
85. S. D. Siciliano and D. R. Lean, *Environ. Toxicol. Chem.*, 2002, **21**, 1184–1190.
86. S. C. Choi, T. Chase and R. Bartha, *Appl. Environ. Microbiol.*, 1994, **60**, 4072–4077.
87. J. H. Weber, *Chemosphere*, 1993, **26**, 2063–2077.
88. E. B. Ekstrom, F. M. Morel and J. M. Benoit, *Appl. Environ. Microbiol.*, 2003, **69**, 5414–5422.
89. C. Bancon-Montigny, L. Yang, R. E. Sturgeon, V. Colombini and Z. Mester, *Appl. Organometal. Chem.*, 2004, **18**, 57–64.
90. R. C. R. Martin-Doimeadios, T. Stoichev, E. Krupp, D. Amouroux, M. Holeman and O. F. X. Donard, *Appl. Organometal. Chem.*, 2002, **16**, 610–615.
91. H. Hintelmann and R. D. Evans, *Fresenius J. Anal. Chem.*, 1997, **358**, 378–385.
92. R. Ebinghaus, H. Hintelmann and R. D. Wilken, *Fresenius J. Anal. Chem.*, 1994, **350**, 19–35.
93. M. Rahman, H. M. S. Kingston and S. Bhandari, *Appl. Organometal. Chem.*, 2003, **17**, 913–920.
94. H. Hintelmann, *Chemosphere*, 1999, **39**, 1093–1105.

95. V. Celo, D. R. S. Lean and S. L. Scott, *Sci. Total Environ.*, 2006, **368**, 126–127.
96. V. Celo and S. L. Scott, *Inorg. Chem.*, 2005, **44**, 2507–2512.
97. H. Nagase, Y. Ose, T. Sato and T. Ishikawa, *Sci. Total Environ.*, 1984, **32**, 147–156.
98. K. Gårdfeldt, J. Munthe, D. Stromberg and O. Lindqvist, *Sci. Total Environ.*, 2003, **304**, 127–136.
99. P. J. Craig and P. A. Moreton, *Mar. Pollut. Bull.*, 1984, **15**, 406–408.
100. D. Wallschläger, H. Hintelmann, R. D. Evans and R.-D. Wilken, *Water, Air, Soil Pollut.*, 1995, **80**, 1325–1329.
101. J. H. Weber, R. Evans, S. H. Jones and M. E. Hines, *Chemosphere*, 1998, **36**, 1669–1687.
102. H. Hintelmann, M. Hempel and R.-D. Wilken, *Environ. Sci. Technol.*, 1995, **29**, 1845–1850.
103. A. O. Summers, *Annu. Rev. Microbiol.*, 1986, **40**, 607–634.
104. C. T. Walsh, M. D. Distefano, M. J. Moore, L. M. Shewchuk and G. L. Verdie, *FASEB J.*, 1988, **2**, 124–130.
105. J. B. Robinson and O. H. Tuovinen, *Microbiol. Rev.*, 1984, **48**, 95–124.
106. S. Silver and T. K. Misra, *Annu. Rev. Microbiol.*, 1988, **42**, 717–743.
107. J. G. Melnick and G. Parkin, *Science*, 2007, **317**, 225.
108. J. K. Schaefer, J. Yagi, J. R. Reinfelder, T. Cardona, K. M. Ellickson, S. Tel-Or and T. Barkay, *Environ. Sci. Technol.*, 2004, **38**, 4304–4311.
109. M. E. Hines, J. Faganeli, I. Adatto and M. Horvat, *Appl. Geochem.*, 2006, **21**, 1924–1939.
110. M. C. Marvin-DiPasquale and R. S. Oremland, *Environ. Sci. Technol.*, 1998, **32**, 2556–2563.
111. M. C. Marvin-DiPasquale, J. L. Agee, C. McGowan, R. S. Oremland, M. Thomas, D. P. Krabbenhoft and C. C. Gilmour, *Environ. Sci. Technol.*, 2000, **34**, 4908–4916.
112. R. S. Oremland, C. W. Culbertson and M. R. Winfrey, *Appl. Environ. Microbiol.*, 1991, **57**, 130–137.
113. R. S. Oremland, L. G. Miller, P. Dowdle, T. Connell and T. Barkay, *Appl. Environ. Microbiol.*, 1995, **61**, 2745–2753.
114. H. Hintelmann, K. Keppel-Jones and R. D. Evans, *Environ. Toxicol. Chem.*, 2000, **19**, 2204–2211.
115. H. Hintelmann, B. Dimock, J. Zhu, H. L. Fredrickson and E. P. H. Best, *Mat. Geoenviron.*, 2004, **51**, 1066–1067.
116. P. Sellers and C. A. Kelly, *Limnol. Oceanogr.*, 2001, **46**, 623–631.
117. P. Sellers, C. A. Kelly, J. W. M. Rudd and R. A. MacHutchon, *Nature (London)*, 1996, **380**, 694–697.
118. K. Gårdfeldt, J. Sommar, D. Strömberg and X. Feng, *Atmos. Environ., Part A*, 2001, **35**, 3039–3047.
119. J. Chen, S. O. Pehkonen and C.-J. Lin, *Water Res.*, 2003, **37**, 2496–2504.
120. I. Lehnherr and V. L. St.Louis, Environ. Sci. Technol., 2009, submitted.
121. J. H. Huanga and O. Klemm, *Atmos. Environ., Part A*, 2004, **38**, 5013–5023.
122. Y. H. Lee, I. Wängberg and J. Munthe, *Sci. Total Environ.*, 2003, **304**, 107–113.

123. R. P. Mason, N. M. Lawson and K. A. Sullivan, *Atmos. Environ., Part A*, 1997, **31**, 3541–3550.
124. C. R. Hammerschmidt, C. H. Lamborg and W. F. Fitzgerald, *Atmos. Environ., Part A*, 2007, **41**, 1663–1668.
125. V. L. St Louis, M. J. Sharp, A. Steffen, A. May, J. Barker, J. L. Kirk, D. J. A. Kelly, S. E. Arnott, B. Keatley and J. P. Smol, *Environ. Sci. Technol.*, 2005, **39**, 2686–2701.
126. N. Lahoutifard, M. Sparling and D. R. S. Lean, *Atmos. Environ., Part A*, 2005, **39**.
127. D. Foucher, O. Clarisse and H. Hintelmann, unpublished observations, 2007.
128. R. Pongratz and K. G. Heumann, *Chemosphere*, 1998, **36**, 1935–1946.
129. E. T. Korthals and M. R. Winfrey, *Appl. Environ. Microbiol.*, 1987, **53**, 2397–2404.
130. C. C. Gilmour and E. A. Henry, *Environ. Pollut.*, 1991, **71**, 131–169.
131. S. M. Ullrich, T. W. Tanton and S. A. Abdrashitova, *Crit. Rev. Eviron. Tech.*, 2001, **31**, 241–293.
132. B. A. Branfireun, N. T. Roulet, C. A. Kelly and J. W. M. Rudd, *Global Biogeochem. Cycl.*, 1999, **13**, 743–750.
133. R. Duran, M. Ranchou-Peyruse, V. Menuet, M. Monperrus, G. Bareille, M. S. Goñi, J. C. Salvado, D. Amouroux, R. Guyoneaud, O. F. X. Donard and P. Caumette, *Environ. Pollut.*, 2008, **156**, 951–958.
134. H. Hintelmann, R. D. Evans and J. Y. Villeneuve, *J. Anal. At. Spectrom.*, 1995, **9**, 619–624.
135. B. A. Branfireun, D. P. Krabbenhoft, H. Hintelmann, R. J. Hunt, J. P. Hurley and J. W. M. Rudd, *Water Resour. Res.*, 2005, **41**, W06016/06011.
136. D. P. Krabbenhoft, J. M. Benoit, C. L. Babiarz, J. P. Hurley and A. W. Andren, *Water, Air, Soil Pollut.*, 1995, **89**, 425–433.
137. Y. -H. Lee, K. H. Bishop, J. Munthe, Å. Iverfeldt, M. Verta, H. Parkman and H. Hultberg, *Biogeochem.*, 1998, **40**, 125–135.
138. C. T. Driscoll, J. Holsapple, C. L. Schofield and R. Munson, *Biogeochem.*, 1998, **40**, 137–146.
139. J. W. M. Rudd, *Water, Air, Soil Pollut.*, 1995, **80**, 697–713.
140. V. L. St. Louis, J. W. M. Rudd, C. A. Kelly, K. G. Beaty, N. S. Bloom and R. J. Flett, *Can. J. Fish. Aquat. Sci.*, 1994, **51**, 1065–1076.
141. B. A. Branfireun, A. Heyes and N. T. Roulet, *Water Resour. Res.*, 1996, **32**, 1785–1795.
142. V. L. St. Louis, J. W. M. Rudd, C. A. Kelly, K. G. Beaty, R. J. Flett and N. T. Roulet, *Environ. Sci. Technol.*, 1996, **30**, 2719–2729.
143. B. D. Hall, G. R. Aiken, D. P. Krabbenhoft, M. Marvin-DiPasquale and C. M. Swarzenski, *Environ. Pollut.*, 2008, **154**, 124–134.
144. O. Lindqvist, K. Johansson, M. Aastrup, A. Andersson, L. Bringmark, G. Hovsenius, L. Håkanson, Å. Iverfeldt, M. Meili and B. Timm, *Water, Air, Soil Pollut.*, 1991, **55**, R11.
145. P. Porvari and M. Verta, *Environ. Pollut.*, 2003, **123**, 181–191.
146. C. J. Watras, K. A. Morrison, A. Kent, N. Price, O. Regnell, C. Eckley, H. Hintelmann and T. Hubacher, *Environ. Sci. Technol.*, 2005, **39**, 4747–4758.

147. M. C. Marvin-DiPasquale, J. L. Agee, R. M. Bouse and B. E. Jaffe, *California Environ. Geol.*, 2003, **43**, 260–267.
148. G. Liu, Y. Cai, T. Philippi, P. Kalla, D. Scheidt, J. Richards, L. Scinto and C. Appleby, *Environ. Pollut.*, 2008, **153**, 257–265.
149. A. Tremblay, M. Lucotte and R. Schetagne, *Sci. Total Environ.*, 1998, **213**, 307–315.
150. A. Tremblay, L. Cloutier and M. Lucotte, *Sci. Total Environ.*, 1998, **219**, 209–221.
151. S. Montgomery, M. Lucotte and I. Rheault, *Sci. Total Environ.*, 2000, **260**, 147–157.
152. L. D. Hylander, J. Janina Gröhn, M. Troppa, A. Vikström, H. Wolpher, E. Silvac, M. Meilie and L. J. Oliveirac, *J. Environ. Managem.*, 2006, **81**, 155–166.
153. T. W. Therriault and D. C. Schneider, *Environ. Pollut.*, 1998, **101**, 33–42.
154. M. Mailman, L. Stepnuk, N. Cicek and R. A. Bodaly, *Sci. Total Environ.*, 2006, **368**, 224–235.
155. T. Matilainen and M. Verta, *Can. J. Fish. Aquat. Sci.*, 1995, **52**, 1597–1608.
156. R. C. Harris, J. W. M. Rudd, M. Amyot, C. Babiarz, K. Beaty, P. Blanchfield, R. A. Bodaly, B. Branfireun, C. C. Gilmour, J. Graydon, A. Heyes, H. Hintelmann, J. Hurley, C. A. Kelly, D. Krabbenhoft, S. Lindberg, R. Mason, M. Paterson, C. Podemski, A. Robinson, K. Sandilands, G. Southworth, V. St. Louis and M. Tate, *Proc. Natl. Acad. Sci. USA*, 2007, **104**, 16586–16591.
157. D. Acha-Cordero and H. Hintelmann, 2009, submitted.
158. L. B. Cleckner, C. C. Gilmour, J. P. Hurley and D. P. Krabbenhoft, *Limnol. Oceanogr.*, 1999, **44**, 1815–1825.
159. D. Achá, H. Hintelmann and J. Yee, *FEMS Microbiol. Ecol.*, 2009, submitted.
160. J. R. Guimarães, M. Meili, L. D. Hylander, E. de Castro e Silva, M. Roulet, J. B. Mauro and R. de Lemos, *Sci. Total Environ.*, 2000, **261**, 99–107.
161. J. R. D. Guimarães, M. Meili, O. Malm and M. de Souza Brito, *Sci. Total Environ.*, 1998, **21**, 165–175.
162. S. A. Coelho-Souza, J. R. D. Guimarães, J. B. N. Mauro, M. R. Miranda and S. M. F. O. Azevedo, *Sci. Total Environ.*, 2006, **264**, 188–199.
163. C. L. Babiarz, J. P. Hurley, J. M. Benoit, M. M. Shafer, A. W. Andren and D. A. Webb, *Biogeochem.*, 1998, **41**, 237–257.
164. D. Schwesig, G. Ilgen and E. Matzner, *Water, Air, Soil Pollut.*, 1999, **113**, 141–154.
165. H. Hintelmann, R. Harris, A. Heyes, J. P. Hurley, C. A. Kelly, D. P. Krabbenhoft, S. Lindberg, J. W. M. Rudd, K. J. Scott and V. L. St.Louis, *Environ. Sci. Technol.*, 2002, **36**, 5034–5040.
166. P. Porvari, M. Verta, J. Munthe and M. Haapanen, *Environ. Sci. Technol.*, 2003, **37**, 2389–2393.
167. D. Schwesig and O. Krebs, *Plant Soil*, 2003, **253**, 445–455.
168. L. Fay and M. S. Gustin, *Wetlands*, 2007, **27**, 1056–1065.
169. R. Dombaiova, J.-H. Huang and E. Matzner, *J. Plant Nutr. Soil Sci.*, 2005, **168**, 238–240.
170. G. Qiu, X. Feng, P. Li, S. Wang, G. Li, L. Shang and X. Fu, *J. Agric. Food Chem.*, 2008, **56**, 2465–2468.

171. M. Mailman and R. A. Bodaly, *Environ. Pollut.*, 2008, **154**, 124–134.
172. A. Gnamu, R. Byrne and M. Horvat, *Environ. Sci. Technol.*, 2000, **34**, 3337–3345.
173. S. P. Bizily, C. L. Rugh, A. O. Summers and R. B. Meagher, *Proc. Natl. Acad. Sci. USA*, 1999, **96**, 6808–6813.
174. J. A. Graydon, V. L. St. Louis, H. Hintelmann, S. E. Lindberg, K. A. Sandilands, J. W. M. Rudd, C. A. Kelly, B. D. Hall and L. D. Mowat, *Environ. Sci. Technol.*, 2009, in press.
175. C. J. S. Passos, D. S. Da Silva, M. Lemire, M. Fillion, J. R. D. Guimarães, M. Lucotte and D. Mergler, *J. Exp. Sci. Environ. Epidem.*, 2008, **18**, 76–87.
176. J. Van Oostdam, S. G. Donaldson, M. Feeley, D. Arnold, P. Ayotte, G. Bondy, L. Chan, E. Dewaily, C. M. Furgal, H. Kuhnlein, E. Loring, G. Muckle, E. Myles, O. Receveur, B. Tracy, U. Gill and S. Kalhok, *Sci. Total Environ.*, 2005, **351**, 165–246.
177. P. Johansen, D. Muir, G. Asmund and F. Riget, *Sci. Total Environ.*, 2004, **331**, 189–206.
178. E. Garcia, R. Carignan and D. R. S. Lean, *Environ. Monit. Assess.*, 2007, **131**, 1–11.
179. M. J. Paterson, P. Blanchfield, C. Podemski, H. Hintelmann, R. Harris, N. Ogrinc, J. M. W. Rudd and K. A. Sandilands, *Can. J. Fish. Aquat. Sci.*, 2006, **63**, 2213–2224.
180. M. Kainz, K. Telmer and A. Mazumder, *Sci. Total Environ.*, 2006, **368**, 271–282.
181. B. A. Monson, *Environ. Sci. Technol.*, 2009, **43**, 1750–1755.
182. J. M. Lepak, H. A. Shayler, C. E. Kraft and B. A. Knuth, *BioScience*, 2009, **59**, 174–181.
183. M. Fujiki and S. Tajima, *Water Sci. Technol.*, 1992, **25**, 133.
184. J. W. M. Rudd, M. A. Turner, A. Furutani, A. L. Swick and B. E. Townsend, *Can. J. Fish. Aquat. Sci.*, 1983, **40**, 2206–2217.
185. W. F. Fitzgerald, C. H. Lamborg and C. R. Hammerschmidt, *Chem. Rev.*, 2007, **107**, 641–662.
186. C. T. Driscoll, Y. J. Han, C. Y. Chen, D. C. Evers, K. F. Lambert and T. M. Holsen, *BioScience*, 2007, **57**, 17–28.
187. P. C. Pickhardt and N. S. Fisher, *Environ. Sci. Technol.*, 2007, **41**, 125–131.
188. B. D. Hall, R. A. Bodaly, R. J. P. Fudge, J. M. W. Rudd and D. M. Rosenberg, *Water, Air, Soil Pollut.*, 1999, **100**, 13–24.
189. M. S. Bank, E. Chesney, S. J. A. Maage and D. Senn, *Ecol. Appl.*, 2007, **17**, 2100–2110.
190. G. Cabana, A. Tremblay, J. Kalff and J. B. Rasmussen, *Can. J. Fish. Aquat. Sci.*, 1994, **51**, 381–389.
191. J. L. A. Van Walleghem, P. J. Blanchfield and H. Hintelmann, *Environ. Sci. Technol.*, 2007, **41**, 5895–5901.
192. K. Klenavic, L. Champoux, O. B. M. P.-Y. Daoust, R. D. Evans and H. E. Evans, *Environ. Pollut.*, **156**, 359–366.
193. A. M. Scheuhammer, M. W. Meyer, M. B. Sandheinrich and M. W. Murray, *Ambio*, 2007, **36**, 12–18.

194. B. M. Braune, P. M. Outridge, A. T. Fisk, D. C. G. Muir, P. A. Helm, K. Hobbs, P. F. Hoekstra, Z. A. Kuzyk, M. Kwan, R. J. Letcher, W. L. Lockhart, R. J. Norstrom, G. A. Stern and I. Stirling, *Sci. Total Environ.*, 2005, **351**, 4–56.
195. W. L. Lockhart, G. A. Stern, R. Wagemann, R. V. Hunt, D. A. Metner, J. DeLaronde, B. Dunn, R. E. A. Stewart, C. K. Hyatt, L. Harwood and K. Mount, *Sci. Total Environ.*, 2005, **351**, 391–412.
196. L. Poissant, H. H. Zhang, J. Canário and P. Constant, *Sci. Total Environ.*, 2008, **400**, 173–213.
197. R. R. Dietz, F. Riget, E. W. Born, C. Sonne, P. P. Grandjean, M. Kirkegaard, O. M. T. G. Asmund, A. Renzoni, H. Baage and C. Andreasen, *Environ. Sci. Technol.*, 2006, **40**, 1120–11125.
198. N. Basu, A. M. Scheuhammer, C. Sonne, R. J. Letcher, E. W. Born and R. Dietz, *Environ. Toxicol. Chem.*, 2009, **28**, 133–140.
199. M. S. Bank, J. R. Burgess, D. C. Evers and C. S. Loftin, *Environ. Monit. Assess.*, 2007, **126**, 105–115.
200. M. G. Mitro, D. C. Evers, M. W. Meyer and W. H. Piper, *J. Wildlife Manag.*, 2008, **72**, 665–673.
201. N. M. Burgess and M. W. Meyer, *Ecotoxicol.*, 2008, **17**, 83–91.
202. C. H. Conaway, F. J. Black, M. Gault-Ringold, J. T. Pennington, F. P. Chavez and A. R. Flegal, *Environ. Sci. Technol.*, 2009, **43**, 1305–1309.
203. M. Horvat, J. Kotnik, M. Logar, V. Fajon, T. Zvonaric and N. Pirrone, *Atmos. Environ., Part A*, 2003, **37**, S93–S108.
204. D. Cossa, J. -M. Martin, K. Takayanagis and J. Sanjuan, *Deep-Sea Res. II*, 1997, **44**, 721–740.
205. R. P. Mason and K. A. Sullivan, *Deep-Sea Res. II*, 1999, **46**, 937–956.
206. R. P. Mason and W. F. Fitzgerald, *Nature (London)*, 1990, **347**, 457–459.
207. J. L. Kirk, V. L. St.Louis, H. Hintelmann, I. Lehnherr, B. Else and L. Poissant, *Environ. Sci. Technol.*, 2008, **41**, 6433–6441.
208. S. E. Lindberg, G. Southworth, E. M. Prestbo, D. Wallschläger, M. A. Bogle and J. Price, *Atmos. Environ., Part A*, 2005, **39**, 249–258.
209. F. Xinbin, T. Shunlin, L. Zhonggen, W. Shaofeng and L. Lian, *Chin. Sci. Bull.*, 2004, **49**, 2068–2072.
210. S. M. Wilhelm, L. Liang and D. Kirchgessner, *Energy Fuels*, 2006, **20**, 180–186.
211. M. Hempel, J. Kuballa and E. Jantzen, *Fresenius J. Anal. Chem.*, 2000, **366**, 470–475.
212. S. P. Wasik, in *Organometals and Organometalloids: Occurence and Fate in the Environemnt*, ACS Symp. Series, Vol. 82, American Chemical Society, Washington, DC, 1978, pp. 314–326.
213. G. Yarwood and H. Niki, *A Critical Review of Available Information on Transformation Pathways for Mercury Species in the Atmospheric Environment*, prepared for Atmospheric Environmental Service, Environment Canada, Downs*view*, ON M3H 5T4, Canada, 1990.
214. D. R. Lide, *Handbook of Chemistry and Physics*, 86th edn., CRC Press, Boca Raton, FL, 2005.
215. B. C. Faust, *Environ. Toxicol. Chem.*, 1992, **11**, 1373–1376.

216. M. A. Major, D. H. Rosenblatt and K. A. Bostian, *Environ. Toxicol. Chem.*, 1991, **10**, 5–8.
217. S. Okouchi and S. Sasaki, *Bull. Chem. Soc. Jpn.*, 1985, **58**, 3401–3402.
218. R. P. Mason, J. R. Reinfelder and F. M. Morel, *Environ. Sci. Technol.*, 1996, **30**, 1835–1845.
219. K. Tajima, F. Kai and T. Kawahara, *Bull. Chem. Soc. Jpn.*, 1980, **53**, 2114–2119.
220. W. H. Schroeder, G. Yarwood and H. Niki, *Water, Air, Soil Pollut.*, 1991, **56**, 653–666.

Met. Ions Life Sci. **2010**, *7*, 403–434

12

Toxicology of Alkylmercury Compounds

Michael Aschner*, [a] ***Natalia Onishchenko,*** [b] ***and Sandra Ceccatelli*** [b]

[a] Vanderbilt University School of Medicine, Department of Pediatrics, Pharmacology, and the Kennedy Center for Research on Human Development, Nashville, TN 37232, USA
*corresponding author: <michael.aschner@vanderbilt.edu>

[b] Karolinska Institute, Department of Neuroscience, SE-17177 Stockholm, Sweden
<natalia.onishchenko@ki.se>
<sandra.ceccatelli@ki.se>

Metal Ions in Life Sciences, Volume 7 Edited by Astrid Sigel, Helmut Sigel, and Roland K. O. Sigel

Published by the Royal Society of Chemistry, www.rsc.org DOI: 10.1039/9781849730822-00403

ABSTRACT: Methylmercury is a global pollutant and potent neurotoxin whose abundance in the food chain mandates additional studies on the consequences and mechanisms of its toxicity to the central nervous system. Formulation of our new hypotheses was predicated on our appreciation for (a) the remarkable affinity of mercurials for the anionic form of sulfhydryl (-SH) groups, and (b) the essential role of thiols in protein biochemistry. The present chapter addresses pathways to human exposure of various mercury compounds, highlighting their neurotoxicity and potential involvement in neurotoxic injury and neurodegenerative changes, both in the developing and senescent brain. Mechanisms that trigger these effects are discussed in detail.

KEYWORDS: ethylmercury · mechanisms · mercury · methylmercury · neurodegenerative diseases · neurodevelopment · neurotoxicity

1. INTRODUCTION

Mercury is a global pollutant with no environmental boundaries. Even the most stringent control of Hg pollution from manmade sources will *not* eliminate human exposure to potentially toxic quantities, given its ubiquitous presence in the environment. The largest global repository for Hg is found in ocean sediments, estimated to contain a total of about 10^{17} g of Hg, mainly in the form of HgS [1]. Ocean waters contain around 10^{13} g, soils and freshwater sediments 10^{13} g, the biosphere 10^{11} g (mostly in land biota), the atmosphere 10^{8} g, and freshwater on the order of 10^{7} g. This budget excludes "unavailable" Hg in mines and other subterranean repositories. A more recent estimate of the global atmospheric repository by Fitzgerald et al. [2]

suggests a level 50 times the previous estimate of Nriagu [1]. According to EPA reports [3,4], recent estimates of total annual natural and anthropogenic mercury emissions are about 4,400 to 7,500 metric tons, with Asia accounting for 53% of the total emissions, followed by Africa (18%), Europe (11%), North America (9%), Australia (6%), and South America (4%). Roughly 2/3 of the total emissions are anthropogenic, mainly from coal combustion and industrial uses. The United States accounts for 3% of all global anthropogenic emissions, with the power sector accounting for 1% of the total. Coal-fired electric power plants are the largest source of human-caused mercury air emissions in the United States. These power plants account for about 40% of total US manmade mercury emissions.

Mercury exists in nature mainly as three different molecular species: elemental (Hg^0), inorganic (Hg^{2+}) and organic ($MeHg^+$). Mercury is released into the environment from both natural and anthropogenic sources [3,4] and it participates in a dynamic cycle in the biosphere, where Hg^0 is photochemically oxidized and deposited to terrestrial and aquatic systems by rainfall and dry deposition. Initially, most of the Hg deposited to terrestrial systems is sequestered by soil and vegetation. A large fraction is reduced to Hg^0 and evaporates back into the atmosphere.

Nearly all fish contain detectable amounts of $MeHg^+$ [5]. In general, there is little information on the balance between methylation and demethylation processes in aquatic systems, and the ecology and genetics of microbial communities within aquatic redox transition zones in the subsurface environment is poorly understood. In the marine ecosphere [6,7] and the upper sedimentary layers of sea and lake beds, sulfate-reducing bacteria readily methylate a portion of the inorganic mercury by the action of microorganisms [8] forming the highly toxic species, $MeHg^+$. The enrichment of $MeHg^+$ in the aquatic food chain is not uniform and is dependent-upon the Hg content in the water and bottom sediments, pH and redox potential of the water, fish species and age, and size of the fish. In addition, environmental conditions, such as anoxia, favor the growth of microorganisms, increasing the methylation rate of Hg [9] and by inference its accumulation in fish. The methylated form, $MeHg^+$, is rapidly taken up by living organisms in the aquatic environment and biomagnified through the food chain reaching concentrations in fish 10,000–100,000 times greater than in the surrounding water [3,10]. The bioaccumulation of Hg in aquatic life is an issue of global human health and ecological risk because Hg input into aquatic systems from atmospheric deposition and terrestrial sources is converted to highly toxic, bioaccumulative $MeHg^+$ by the action of microorganisms. Once methylated, human exposure to $MeHg^+$ occurs predominantly from fish consumption [11–13].

Much higher aqueous concentrations of Hg occur at numerous Superfund sites in the USA, where inorganic Hg contaminates ground and surface water, resulting in $MeHg^+$ contamination of aquatic organisms and other consumers. Hg-contaminated Superfund sites are typically located where metallic Hg was used in large quantities and spilled or discharged as solid or liquid waste. Sites include (1) abandoned chloralkali facilities (examples in the USA include, LaVaca Bay in Texas; North Fork Holston River in Virginia; Penobscot River in Maine; Onondaga Lake in New York), (2) historic Hg, silver, and gold-mining sites (Carson River, Nevada; Clear Lake, California), (3) battery manufacturing plants (Abbotts Creek, North Carolina), and (4) industrial facilities where Hg was used as a solvent (East Fork Poplar Creek, Tennessee) or catalyst (South River, Virginia) [14]. Remedial actions at some sites have been successful at reducing inputs of inorganic Hg to surface waters but have not been successful in reducing waterborne concentrations to levels typical of aquatic systems unimpacted by point sources of Hg. Mercury bioaccumulation in aquatic organisms residing in lakes and reservoirs has often proved responsive to reductions in waterborne Hg inputs [14], but Hg bioaccumulation in stream ecosystems feeding those reservoirs has remained problematic. Effectively reducing $MeHg^+$ concentrations to safe levels in contaminated aquatic ecosystems may require that source control actions at such sites reduce waterborne total Hg concentrations to levels approaching natural background (<5–10 ng/L). At many sites, Hg inputs into surface water originate from groundwater and contaminated soils and often remain too diffuse to be cost-effectively controlled to the degree needed to achieve such low Hg concentrations in affected aquatic systems. Alternative strategies that block the bioaccumulation of Hg in such systems without requiring controls on inorganic Hg inputs have the potential to save tens of millions of dollars in treatment/remediation expenditures, while achieving significant reductions in human and ecological risks.

The total number of fish advisories for Hg in the USA increased from 2,436 in 2004, to 2,682 in 2005, and 3,080 in 2006 (http://www.epa.gov/waterscience/fish/advisories/2006/tech.html#mercury). Forty-eight states in the USA have issued fish advisories, and 80% of all advisories have been issued, at least in part, due to Hg contamination. To put this in perspective, a total of 14,035,676 lake acres and 882,428 river miles were under advisory for Hg in 2005. In 2006, these numbers increased to 14,177,175 lake acres and 882,963 river miles, representing an 8% and 15% increase, respectively, between 2004 and 2006. Currently, 23 states have issued statewide advisories for mercury in freshwater lakes and/or rivers. Twelve states have statewide advisories for Hg in their coastal waters. Hawaii has a statewide advisory for Hg in marine fish.

2. MERCURY SPECIES OF RELEVANCE TO HUMAN HEALTH

2.1. Elemental Mercury

Elemental or metallic mercury (Hg^0), which occurs in liquid form, volatilizes with heating and becomes more hazardous to humans. Acute Hg^0 vapor exposure induces serious respiratory problems, including dyspnea, associated with increased excitability, whereas chronic exposure affects mostly the central nervous system provoking a variety of alterations and symptoms, such as tremors, polyneuropathy, delusions, hallucinations, loss of memory, insomnia, and neurocognitive disorders [15]. There is some concern about the release of mercury vapor from amalgam used for dental fillings, however, the evidence that dental amalgam can have adverse health effects is limited (http://ec.europa.eu/health/ph_risk/committees/04_scenihr/scenihr_cons_07_en.htm). A few amalgam bearers with excessive chewing habits, such as ex-smokers using nicotine chewing gum may be exposed to levels of Hg^0 at the safe limits [16]. Because of the recognized high susceptibility of developing organisms to Hg, several countries have introduced a precautionary approach whereby amalgam fillings should be avoided in pregnant women and children (http://www.env-health.org/IMG/pdf/HEA_009-07.pdf). Still, it is reassuring that a recent study showed that there are no differences in the neuropsychological performances between children with amalgam and other types of dental fillings [17].

2.2. Inorganic Mercury

Inorganic Hg was largely used in medical products, such as topical antiseptic, vermifuges, skin-lightening creams, and teething powders. Mercury salts are extremely toxic to kidneys, causing severe renal dysfunctions including tubular necrosis and glomerulonephritis. Acrodynia, characterized by painful extremities and also known as pink disease, can also be induced in response to mercury as reported in children exposed to mercurial chloride calomel-containing teething powders [18]. Another immunotoxic response that has been associated to exposure to inorganic mercury is the Kawasaki syndrome. Patients present a variety of signs and symptoms including skin lesions and rashes, peripheral extremity changes, fever, and photophobia [15]. Skin sensitization with contact dermatitis has been described in conjunction to inorganic, but also organic, mercury exposure. Interestingly, subjects prone to skin reactions have a higher prevalence of glutathione *S*-transferase depletion [19].

2.3. Organic

2.3.1. *Methylmercury*

As evidenced originally in Minamata Bay (Japan), $MeHg^+$ that accumulates in the tissue of shellfish and fish is readily consumed by wildlife and humans. However, the risk from dietary exposure to $MeHg^+$ is not limited to islanders with high consumption of fish. The EPA's Mercury Study Report to Congress [3] estimated that 7% of women of childbearing age have blood Hg concentrations greater than those equivalent to the reference dose (RfD). Based on the prevalence in the overall U.S. population of women of reproductive age and the number of U.S. births each year, an estimated 300,000 newborns each year may have increased risk of learning disabilities associated with *in utero* exposure to $MeHg^+$ (http://www.epa.gov/mercury/exposure.htm#meth). Almost all people have at least trace amounts of $MeHg^+$ in their tissues, reflecting its widespread presence in the environment and human exposure through the consumption of fish and shellfish. Exposure scenarios vary in relationship to geographical location, urban or rural environment, lifestyles and dietary habits, and occupational settings. These factors overlie differences in life-stage and genetics that influence background disease occurrence and impose differential sensitivity to Hg exposure. $MeHg^+$ is a proven neurotoxin whose effects differ according to developmental stage.

2.3.2. *Ethylmercury*

Ethylmercury thiosalicylate (chemical structure, $C_9H_9HgNaO_2S$) is also known under the trade names thimerosal, thiomersal, merthiolate, mercurothiolate, merfamin, mertorgan, and merzonin [20]. It is best known for its

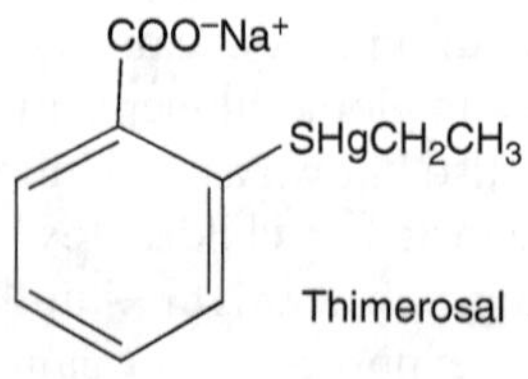

Thimerosal

role as a preservative in vaccines (since the 1930s) after a series of studies in several animal species and humans provided assurance for its safety and effectiveness [21]. Thimerosal in concentrations of 0.001% (1 part in 100,000) to 0.01% (1 part in 10,000) has been shown to be effective in clearing a broad spectrum of pathogens. A vaccine containing 0.01%

thimerosal as a preservative contains 50 micrograms of thimerosal per 0.5 mL dose or approximately 25 micrograms of mercury per 0.5 mL dose. After approximately 70 years of safe practice and a long record of effectiveness in preventing bacterial and fungal contamination of vaccines with only minor local reactions at the site of injection, in 2001 the use of thimerosal was questioned as a potential toxic hazard to infants [22]. Though it is still used in developing countries, where advantages of multiple use vials outweigh thimerosal's putative toxicity [23], as well as in certain vaccines and medications, it was removed from the US market in 2001.

The Word Health Organization (WHO) [24], the US Environmental Protection Agency (EPA) [3], the US Agency for Toxic Substances and Disease Registry [4], and the US Food and Drug Administration [25] have assessed the risk associated with $MeHg^+$ in diet and have published a series of recommendations for safe exposures to this metal. These recommendations encompass a safety margin and range from 0.7 μg $MeHg^+$/kg of body weight per week (EPA) to 3.3 μg $MeHg^+$/kg of body weight per week (WHO). The range of recommendations reflects varying safety margins, differing emphasis placed on various sources of data, the different missions of the agencies and the population that the guideline is intended to protect. All guidelines, however, fall within the same order of magnitude. If applied to a female infant in the lowest 5th percentile of weight between birth and 14 weeks, the period during which most infant vaccines are administered, these guidelines translate to limits of safe total $MeHg^+$ exposure of 34 μg and 159 μg, per the EPA and WHO safe exposure limits, respectively. An infant generally receives 3 doses of diphtheria/tetanus/pertussis (DTaP) vaccine or a total of 75 μg of $EtHg^+$ during the first 14 weeks of life [25]. If the hepatitis B vaccine is added to the immunization schedule during the first 14 weeks of life, the maximum exposure to $EtHg^+$ is 112.5 μg. If *Haemophilus influenzae* type b conjugate (Hib) vaccine is added during the same time, the total $EtHg^+$ dose reaches 187.5 μg. Thus, some infants receiving vaccines according to the recommended schedule will receive doses of mercury exceeding the cutoff levels established by regulatory agencies.

Most human exposures to $EtHg^+$ are in the form of thimerosal, and tissue disposition patterns of mercury in experimental animals after equivalent doses of either $EtHg^+$ chloride or thimerosal are the same [26]. Accordingly, it appears that the thiosalicylic acid anion attached to $EtHg^+$ in the thimerosal plays no role in influencing the fate of $EtHg^+$ in the body. Thus, thimerosal rapidly dissociates to release $EtHg^+$ [27,28], which is the active species of concern.

Preceding its usage as a vaccine preservative, $EtHg^+$ compounds, in the form of diethylmercury were used in the treatment of syphilis as early as the 1880s. Later on, in the twentieth century, the fungicidal properties of the short-chain alkylmercury compounds were fully recognized, leading to commercialization of agricultural applications containing $EtHg^+$. A variety of organic mercury compounds were subsequently used to prevent seed-borne diseases of cereal [29,30]. $EtHg^+$ fungicides were effectively and safely used for decades, nonetheless, several poisoning outbreaks have occurred in developing countries. Two outbreaks occurred in rural Iraq in 1956 and 1960 upon misuse of the fungicide $EtHg^+$ toluene sulfonilamide [30]. Having missed the planting season, the $EtHg^+$ containing grains were used by the farmers' families for baking bread. Hundreds of cases of severe poisoning with fatal outcomes ensued. $EtHg^+$ poisonings have also been reported in China as recently as the 1970s after farmers consumed the rice treated with $EtHg^+$ chloride intended for planting [31].

3. NEUROTOXICITY OF MERCURY SPECIES

3.1. Organic

3.1.1. Methylmercury

The neurotoxic effects of $MeHg^+$ are well documented in both humans and experimental animals. Most of the knowledge comes from the mass health disasters occurred in Minamata in the late 1950s, where people were intoxicated by consumption of fish from waters severely contaminated by mercury discharged from local industries [32]. Another mass poisoning took place in Iraq in the early 1970s. Hundreds of people died and several thousands became ill from eating bread made from grain treated with an organomercury pesticide [33]. In the adult brain, $MeHg^+$ poisoning induces distinct damage in the visual cortex, with loss of neurons from the second through the fourth layer of the calcarine cortex, and in the cerebellar granule layer, with selective loss of granule cells. Axonal damage associated with secondary myelin disruption of the sensory branch of the peripheral nerve with preservation of the motor branch can also occur [34]. It may take several weeks before clinical signs, including visual abnormalities, sensory impairment of the extremities, tremor, cerebellar ataxia, muscle weakness, hearing loss, and mental deterioration become manifest.

The developing nervous system is extremely sensitive to $MeHg^+$ exposure, which may give a diffuse and widespread damage. Exposure to high levels may result in cerebral palsy, deafness, blindness, delayed speech, ataxia, and mental retardation as it was found in infants and children in Minamata [35,36]. Studies conducted in Iraq reported that maternal exposure during pregnancy was associated with increased muscle tone and exaggerated deep tendon reflexes in children (maternal hair Hg levels higher than 180 parts per million (ppm)) or retarded development of motor and speech skills (maternal hair Hg levels less than (180 ppm) [37].

The neuropathological changes induced by exposure to high level of $MeHg^+$ exhibit similarities across different species. Reduced brain size, gliosis, damage to cortical areas and basal ganglia have been reported in human and non-human primates as well as small mammals [38]. Interestingly, the monkey's cerebellum seems to be insensitive to the toxic effects of $MeHg^+$, in contrary to what observed in humans and rodents [38–40]. The degree of brain damage depends from the Hg levels, which in rodents are clearly correlated with the seriousness of the neurodegenerative process.

Also exposure to chronic lower levels of $MeHg^+$ produces adverse effects in the developing nervous system as shown by epidemiological and experimental studies. Studies on the Faroe Islands population have revealed that 2-week-old infants prenatally exposed to $MeHg^+$ through maternal fish consumption, resulting in a cord-blood mercury concentration ranging from 1.9 to 102 μg/L, had decreased neurological optimality score, which was used for evaluation of muscle tone and reflexes [41]. Children and adolescents (7 and 14 years old, respectively) with high levels of Hg in cord blood at birth (22.9 μg/L and 4.27 ppm Hg in the maternal hair) showed alterations in motor, attention and verbal tests, and delays in brainstem auditory-evoked potentials [42–44]. In New Zealand, a study was performed on a group of children whose mothers were identified as frequent fish consumers (that had eaten at least three fish/seafood meals per week during pregnancy and had maternal hair Hg level ranging form 6 to 86 ppm) [45,46]. A dose-response relationship was established between mean maternal hair $MeHg^+$ levels and performance of 4-year-old children on the Denver Developmental Screening Test. Poorer scores on full-scale IQ, language development, visual-spatial and gross motor skills in 6-year-old children were associated with maternal hair Hg concentrations in the range of 13–15 ppm [45,47].

Developmental neurotoxic effects were observed in many experimental studies performed in different species. Prenatal exposure of non-human primates (*Macaca fascicularis*) to 50 μg/kg/day altered parameters of cognitive and social development during infancy [38], but continued observation of the animals did not find long-term deficits in adult learning and memory abilities [48]. Chronic exposure to $MeHg^+$ starting *in utero* and continued for 4 years was shown to cause long-term impairments in visual, auditory, and somatosensory function in monkeys [49–51].

Various protocols implementing short high dose or continual low dose treatments during prenatal and postnatal periods have been used in rodent studies. Prenatal high-dose (2.5–6 mg/kg) $MeHg^+$ exposure impaired development of reflexes, such as righting and negative geotaxis, as well as walking and swimming ability [52–54]. There are contradicting reports on changes in locomotor activity in rats and mice after prenatal exposure to

$MeHg^+$ [55]. Interestingly, decreased exploratory activity in $MeHg^+$-exposed animals exhibiting normal motor function has been found in several studies [56–58]. Reported effects of developmental exposure to $MeHg^+$ also include deterioration of spatial learning and memory retention, as well as impairments of reference and working memory, depending on the exposure protocol and the resulting brain Hg concentrations. The type of behavioral tests performed as well as the age of the animals tested also appear to be critical factors [59,60]. Recent studies have also reported depression-like behavior in adult male mice exposed to $MeHg^+$ during prenatal and early postnatal periods [57,61].

3.1.2. Ethylmercury

Several studies have reported on the neurotoxicity of thimerosal. A patient who ingested 83 mg/kg thimerosal (41mg Hg/kg) in a suicide attempt had 14,000 mg/L blood mercury and developed anurea, coma, polyneuropathy, and respiratory failure. He had a complete recovery with no permanent brain damage [62]. Death has been reported in two boys in a family of four members who ate meat from a butchered hog that had been fed seed treated with ethylmercuric chloride [63]. The clinical, electrophysiological, toxicological, and, in two of the patients the pathological data, showed that when ingested, this organic mercury compound has a very high toxicity, not only for the brain, but also for the spinal motor neurons, peripheral nerves, skeletal muscles, and myocardium. Notably, all four members of this family had blood mercury levels exceeding 1,000 μg/L, and for the two boys that succumbed to the poisoning, peak mercury blood concentrations were estimated at 9,600 μg/L. However, given the delay between mercury consumption and the onset of symptoms, the amount of organic mercury ingested in these cases is difficult to ascertain.

A large-scale poisonings with $EtHg^+$ also occurred in Iraq in 1956 and 1960 [33,64]. Thirty-one pregnant women were victims of poisoning; 14 women died from ingesting wheat flour from seeds treated with $EtHg^+$ *p*-toluene sulfonanilide [64]. Infants were born with blood mercury concentrations of 2500 μg/L and suffered severe brain damage. Additional reports of acute toxicity associated with $EtHg^+$ exposure included the administration of immune globulin (γ globulin) [65] and hepatitis B immune globulin [65], choramphenicol formulated with 1000 times the proper dose of thimerosal as a preservative [66], thimerosal ear irrigation in a child with tympanostomy tubes [67] and thimerosal treatment of omphaloceles in infants. The total doses of thimerosal administered in these reports of acute toxicity ranged from $\sim$3 mg/kg to several hundred mg/kg. While these case studies of accidental and intentional poisonings clearly led to toxicity (ranging from local necrosis, acute hemolysis, disseminated intravascular coagulation, acute renal tubular necrosis, and central nervous system injury including obtundation, coma, and death), they merely corroborate an important toxicological principle that the dose makes the poison, and offer no value in evaluating the risk associated with exposure to thimerosal in vaccinations (a topic well beyond the scope of this book chapter).

Rothstein and Hayes [68] evaluated the metabolism of mercury in the rat following intravenous (IV) or intramuscular (IM) injection, using radioactive mercury (^{203}Hg) as a tracer. The authors report that in the first few hours after IV injection, a large fraction of mercury is taken up by the liver, but this was rapidly (few days) cleared via fecal excretion. The kidney was the major site of deposition. By the IM route, the clearance from the site of injection took about 2 weeks. No particularly large amounts of mercury appeared in the liver. By the end of 2 weeks, as in the IV studies, most of the mercury was localized in the kidney. The pharmacokinetics of $EtHg^+$ has been extensively studies by Magos and his colleagues [69–71]. With respect to its accumulation in the brain, distinct differences in the pharmacokinetics of $EtHg^+$ and $MeHg^+$ exist. Magos et al. [71] examined the disposition of $EtHg^+$ versus $MeHg^+$ in rats administered the respective chloride salts. Rats were treated with 8 mg/kg of methylmercuric or ethylmercuric chloride or 9.6 mg/kg of ethylmercuric chloride. This is consistent with other studies where it has been shown that given identical doses, more total mercury is also deposited in the brain of mice [72] after the administration of $MeHg^+$ compared to $EtHg^+$. Thus the weight of evidence establishes that at equimolar doses, $MeHg^+$ exposure result in higher brain levels of the organic species than treatment with $EtHg^+$.

Observations by Pichichero et al. [73] on levels of mercury in samples of stool and urine indicate that substantial excretion of mercury is taking place via the fecal route upon the administration of $EtHg^+$. Urinary excretion of $EtHg^+$ appeared to be negligible. Thus, $EtHg^+$ appears to behave like $MeHg^+$ with fecal excretion accounting for most of the elimination from the body. The absorption rate and initial distribution volume of total mercury are also reported to be generally similar after $EtHg^+$ injections and oral $MeHg^+$ exposure [74]. In other words, peak total blood mercury levels after a single exposure to either $EtHg^+$ or $MeHg^+$ are very similar, implying that the organic mercury compounds behave similarly in the early hours after exposure.

As pointed out by Burbacher et al. [74], there is a significant difference in blood half-times between $MeHg^+$ and $EtHg^+$ in infant monkeys. This is associated with a remarkable accumulation of blood mercury during repeated exposure to $MeHg^+$. Although the initial blood mercury concentration (at 2 days after the first dose) did not differ between the two groups, the peak blood mercury concentration in the $MeHg^+$-exposed infant monkeys rose to a level nearly three times higher than in the thimerosal monkeys after the fourth dose. Levels of organic mercury were lower in the brains of infant monkeys exposed to $EtHg^+$ compared to those exposed orally to $MeHg^+$ consistent with studies in mice [72] and rats [71] (for further details see below).

The brain half-times also differed; the clearance half-times for organic mercury in the brain were 58 days on average after oral $MeHg^+$ exposure versus 14 days after injection of $EtHg^+$ [74]. In addition, the blood clearance of total mercury was 5.4-fold higher after intramuscular $EtHg^+$ than after oral $MeHg^+$ exposure, implying that mercury was cleared at a much faster rate in infant monkeys dosed with thimerosal versus $MeHg^+$. There are additional significant differences in the pharmacokinetic behavior between $MeHg^+$ and $EtHg^+$. The kinetics of clearance of total mercury in the blood compartment is quite different for the two species [74]. The one-compartment model best described blood concentrations after $MeHg^+$ exposure, while a two-compartment model best described blood concentrations after $EtHg^+$ exposure. Thus, $EtHg^+$ will be cleared from the blood much faster compared to $MeHg^+$. If the data from infant monkeys predict half-times in brain as well as they do for whole blood, then most of the organic mercury would be expected to clear from brain in a 2-month period. This would not be true for the inorganic species (Table 1), as it was noted that a much higher proportion of inorganic mercury is found in the brains of $EtHg^+$-treated infant monkeys than in the brains of $MeHg^+$ exposed monkeys (up to 71% versus 10%), with absolute inorganic mercury concentrations in the brains of the $EtHg^+$-exposed monkeys reaching levels twice as high as in the $MeHg^+$-treated monkeys. These findings are consistent with the dealkylation of $EtHg^+$ to the inorganic mercury species.

The idea that the inorganic species of mercury is the damaging species of alkylmercurials has also been advanced. It has been proposed that latency period associated with $MeHg^+$ exposure might be due to the slow production and accumulation of the divalent inorganic mercury in the brain over

Table 1. The inorganic mercury is different from the organic species, which are characterized by a mercury-carbon bond.

Chemical name:	Elemental mercury[a]	Mercuric chloride	Mercurous chloride	Methylmercuric chloride	Ethylmercuric chloride
Molecular formula:	Hg^0	$HgCl_2$	Hg_2Cl_2	CH_3HgCl	C_2H_5HgCl
Molecular structure:		Cl-Hg-Cl	Cl-Hg-Hg-Cl	CH_3-Hg-Cl	C_2H_5-Hg-Cl

[a]Also known as metallic mercury.

periods of months [75]. However, as reported [76], one would expect the buildup of inorganic mercury to be faster at higher levels of $MeHg^+$ exposure, resulting in a shorter latency period. This is contrary to evidence published in the literature [71,76]. Corroborating the studies in infant monkeys [74], Magos et al. [71] also noted that the concentration of brain inorganic mercury was significantly lower in the brains of rats treated with $MeHg^+$ compared with rats dosed the same amount of $EtHg^+$. Brain damage was inherent to the $MeHg^+$-treated rats, whereas rats dosed with $EtHg^+$ showed no evidence of brain damage. The dose of $EtHg^+$ necessary to elicit brain damage had to be increased to the borderline of a lethal dose [71]. Thus, it would appear that inorganic mercury derived from the decomposition of alkylmercury does not play an important role in the etiology of $MeHg^+$-induced neurotoxicity. This conclusion is also supported by case reports of victims of methyl and $EtHg^+$ poisoning. For example, and as described earlier, a patient who ingested 83 mg/kg thimerosal (41 mgHg/kg) had 14,000 mg/L blood mercury. Nevertheless, there were no signs of anuria, polyneuropathy or respiratory failure, with full recovery absent permanent brain damage within months of exposure [62]. Conversely, exposure in a worker to $MeHg^+$, resulting in blood mercury levels of 1840 mg Hg/L led to severe intoxication, and the patient remained ataxic, dysarthric and with constricted visual fields [19].

4. MECHANISMS OF NEUROTOXICITY

The neurotoxic effects of $MeHg^+$ have been linked to multiple mechanisms based on different molecular targets. Among them are proteins and peptides bearing cysteines that are particularly susceptible to structural and functional modification by $MeHg^+$ because of its high affinity for thiol groups. Below, we briefly review the most relevant MeHg molecular effects.

4.1. Apoptosis and Necrosis

Both apoptotic and necrotic cell death can be induced by $MeHg^+$, depending on the cell type and the exposure conditions (dose and duration) [59]. In contrast to necrosis, apoptosis is an energy-dependent, highly regulated process characterized by the activation of signaling pathways leading to specific cleavage of proteins and DNA, condensation of the nucleus, cell shrinkage, and engulfment by phagocytic cells [77].

Depending on the cell type, different signaling pathways are activated in $MeHg^+$-induced apoptosis. In neural stem cells $MeHg^+$ induces apoptosis via the mitochondrial pathway, as shown by Bax activation, cytochrome *c* translocation, and caspase activation [78]. The calcium-dependent protease calpain is also activated and the full protection achieved by pre-treating the $MeHg^+$ exposed cells with caspases and calpain inhibitors points to a parallel activation of both pathways [78]. In contrast, other types of cells, such as neuroblastoma, glioblastoma, cerebellar granule cells, and hippocampal HT22 cells undergo caspase-independent apoptosis when exposed to $MeHg^+$ [78–83]. In astrocytoma cells $MeHg^+$ induces lysosomal alterations that precede a decrease in mitochondrial membrane potential. This points to lysosomal membranes as target of $MeHg^+$ and lysosomal hydrolytic enzymes as executor/regulator factors in cell death induced by $MeHg^+$ [59,84].

4.2. Oxidative Stress

Excessive formation of reactive oxygen species (ROS), as well as impaired antioxidant defenses contribute significantly to the onset of $MeHg^+$ neurotoxicity [59]. In fact, both in *in vivo* and in *in vitro* models have provided evidence for the occurrence of oxidative stress-related intracellular events, including increased lipid peroxidation, superoxide and hydrogen peroxide amounts, impaired superoxide dismutase (SOD), glutathione (GSH) reductase, and GSH peroxidase activities, as well as decreased GSH levels [85]. In agreement, antioxidants have been successfully used in cases of $MeHg^+$ poisoning in humans [80,85–89], as well as in *in vivo* and *in vitro* experimental models to reduce the ROS production and protect against $MeHg^+$ induced cell death [90]. Alterations in mitochondrial functions [91] seems to play a critical role in the onset of oxidative stress induced by $MeHg^+$, as proved by the protective effects of Mn-SOD, suggesting that superoxide anions formed in the mitochondria might be involved in the mechanism of $MeHg^+$ cytotoxicity [92–95].

4.3. Calcium Homeostasis

Increased intracellular Ca^{2+} levels after exposure to $MeHg^+$ have been observed in many cell types, including neural cells, and the protective action exerted by Ca^{2+} chelators or Ca^{2+} channel blockers point to a critical role of Ca^{2+} in the mechanism of $MeHg^+$ toxicity [90]. The initial mobilization of Ca^{2+} from intracellular stores and the entry of extracellular Ca^{2+} through plasma membrane voltage-gated channels [78,96,97] result in a Ca^{2+} overload and altered intracellular Ca^{2+} compartmentalization that can lead to activation of degradative enzymes, perturbation of mitochondrial function and exacerbate the damage caused by ROS with subsequent cell death [77]. Intracellular Ca^{2+} is involved in cell cycle, cell migration and differentiation, thus $MeHg^+$ may affect these processes by altering Ca^{2+} homeostasis [98].

4.4. Microtubules

Cytoskeletal components, especially microtubules, are $MeHg^+$ targets mainly because of the SH groups present in tubulin. As a consequence, depolymerization of existing microtubules occurs and microtubules assembly is inhibited [79,99]. Impairments in the cytoskeleton affect many crucial cellular processes, including cell survival, proliferation, differentiation and migration, which have all been shown to be altered by $MeHg^+$. The occurrence of cell death in $MeHg^+$-exposed neuronal cells has also been linked to cytoskeletal breakdown [100,101], in particular to destruction of mitotic spindles that results in cell cycle arrest [102]. In addition, neuropathological findings, such as reduced brain size observed in postmortem brains of infants exposed *in utero* to $MeHg^+$ during the Iraqi outbreak, may also be explained by disruption of microtubule function [103,104].

4.5. Neurotransmission

Alterations in different neurotransmitter systems have been reported after $MeHg^+$ exposure and it is conceivable that an imbalance in neurotransmission can be behind the neurotoxic effects of $MeHg^+$. Interferences with synthesis, uptake, release, and degradation of neurotransmitters have been reported in various experimental models. The major systems shown to be affected by $MeHg^+$ are the glutamatergic, cholinergic, and dopaminergic ones.

4.5.1. Glutamatergic

The involvement of a glutamate-mediated excitotoxic mechanism in $MeHg^+$ neurotoxicity is supported by consistent experimental data. $MeHg^+$ accumulates mostly in astrocytes where it causes cell swelling and inhibits excitatory amino acid uptake [105]. Uptake of both L-glutamate and D-aspartate gets significantly reduced in astrocyte cultures exposed to concentrations of $MeHg^+$ as low as 10^{-5} M [106]. Increased levels of glutamate in the extracellular space may lead to excitotoxic neurodegeneration [107]. In agreement, co-exposure to non-toxic concentrations of $MeHg^+$ and glutamate induces neuronal lesions typical of excitotoxic damage [105]. Additional support for the theory that excitotoxicity mediates, at least in part, $MeHg^+$ neurotoxicity is provided by the protective effects exerted by competitive *N*-methyl D-aspartate (NMDA) antagonist, D-2-amino-5-phosphonovaleric acid (a competitive NMDA antagonist), and 7-chlorokynurenic acid (an antagonist at the glycine site associated with the NMDA receptor) on $MeHg^+$-induced neurotoxicity [89].

4.5.2. Cholinergic

Muscarinic receptors represent a target for $MeHg^+$. Chronic ingestion of low doses of $MeHg^+$ (0.5 or 2 mg/kg per day for 16 days) significantly increases muscarinic cholinergic density in rat hippocampus and cerebellum, but not in the cerebral cortex, with no changes in receptor affinity [103]. Interestingly, this is a delayed effect that appears 2 weeks after the end of the exposure, which might be seen as a compensatory mechanism for the $MeHg^+$-induced inhibition of acetylcholine synthesis occurring at an earlier stage of exposure. Also $MeHg^+$ developmental exposure affects the cholinergic system: oral exposure of rat dams to 1 mg from gestational day 7 to postnatal day 7 (PND7) causes a delayed (PND21) enhancement of the number of cortical and cerebellar muscarinic receptors both in dams and offspring. This increase was more relevant in dams than in pups, in agreement with the higher Hg levels present in the adult brains as compared to the developing ones (7–9 μg/g versus 1.5–1.7 μg/g in offspring) [108].

Some *in vitro* studies have suggested the involvement of cholinergic neurotransmission alterations in $MeHg^+$-induced cell death. Activation of muscarinic M3 receptors has been reported to contribute to the elevated intracellular Ca^{2+} levels in cerebellar granule cells [109].

4.5.3. Dopaminergic

$MeHg^+$ causes inhibition of dopamine (DA) uptake [110] that seems to be at least in part, associated with a blocking of the DA uptake system [111]. Systemic or intrastriatal administration of different doses of $MeHg^+$ produced significant increases in the release of DA from rat striatum [112].

Several studies have shown that $MeHg^+$ developmental exposure affects the dopaminergic system. Delayed effects on a number of brain dopaminergic parameters including DA levels, DA turnover and synaptosomal DA uptake, at weaning occur in rat offspring following *in utero* exposure to 1 mg

$MeHg^+$/kg/day [113]. However, other studies did not show any changes in the offspring regional brain levels of DA in weaning [114] or adult [115] rats after long-term maternal exposure. Transient effects on DA receptor number associated with behavioral dysfunctions are reported in rat pups exposed to a single high-dose of $MeHg^+$ at late stage of gestation [56,116,117]. The important role of striatal dopaminergic neurotransmission in locomotor control is well known. Behavioral changes indicative of altered dopaminergic neurotransmission have been reported after chronic perinatal exposure to low doses of $MeHg^+$ (0.5 mg/kg/ day) in pre-pubertal as well as in adult male rats [116]. The behavioral alterations correlate to a significant reduction in D2 receptor binding in the caudate putamen. Dopamine neurons are implicated in a number of neurological pathologies, including Parkinson's disease, schizophrenia, attention deficits, motor control, and perception. The toxic effects of $MeHg^+$ on the developing dopaminergic system might predispose individuals to the onset of pathological conditions later in life.

5. MERCURY AND NEURODEGENERATIVE DISORDERS: A LITERATURE SURVEY

5.1. Parkinson's Disease

Parkinson's disease (PD) is a neurodegenerative disorder associated predominantly with motor skills and speech impairment. The disease belongs to a group of conditions called movement disorders, and it is characterized by muscle rigidity, tremor, a slowing of physical movement (bradykinesia) and, in extreme cases, a loss of physical movement (akinesia). Decreased stimulation of the motor cortex by the basal ganglia is responsible for PD-associated primary symptoms, and at the morphological levels this is associated with the insufficient formation and action of dopamine in the substantia nigra pars compacta. Secondary symptoms may include high level cognitive dysfunction and subtle language problems. A classic symptom of mercury poisoning, as with PD, is fine tremor of the hands. However, $MeHg^+$-induced tremor (as seen in Minamata disease) is physiologically distinct in frequency and amplitude from PD-associated tremor, with tremor frequency being significantly higher for $MeHg^+$ exposure versus PD [118].

The first study to test the hypothesis that a high level of body burden of mercury is associated with an increased risk of Parkinson's disease was reported in 1989 [119]. The study was conducted in Singapore, where 54 cases of idiopathic PD and 95 hospital-based controls, matched for age, sex, and ethnicity were evaluated. After adjusting for potential confounding factors, including dietary fish intake, medications, smoking and alcohol consumption, the authors reported on a dose-response association between PD and blood mercury levels. Similar associations were reported for hair and urinary mercury levels. A second epidemiological and clinical study was performed on dental technicians and reported in 2007 [120]. Corroborating the earlier study, 4 of the 14 tested technicians revealed postural tremor and one had a diagnosis of PD, along with a high prevalence of extrapyramidal signs and symptoms in this group. As acknowledged by the authors, there were several limitations to the study, namely: (a) the absence of a control group; (b) lack of exposure assessment or biological markers of neurotoxins present in the workplace itself; (c) the small number of individuals studied; and (d) the study did not specifically screen the subjects for essential tremor, thus it is biased towards self-reporting. Dantzig [121] examined patients with PD for cutaneous eruptions and blood mercury levels and reported that of the PD patients, 13/14 had Grover's disease and detectable blood mercury. Only 2/14 control patients had detectable blood mercury levels. The study was conducted in a small group of individuals and these findings will have to be confirmed in larger cohorts.

Using a combination of approaches to systematic case finding in the Faroe Islands, Wermuth et al. [122] reported on an age-adjusted prevalence of idiopathic PD as high as 183.3 per 100,000 persons in 1995. A follow-up study by the same authors [123] suggested a high prevalence of idiopathic PD and total parkinsonism of 187.6 and 233.4 per 100,000 inhabitants, respectively. The reported age-adjusted prevalence of PD in this population is approximately twice as high compared to the available data from Norway and Denmark. While no explanation for this high prevalence exists at this time, the authors suggest marine pollutants, such as $MeHg^+$, or other environmental risks and interactions with genetic predisposition may underlie the findings. It is also noteworthy that a recent study [124] reported on no significant association between PD and prenatal $MeHg^+$ exposure, establishing that prenatal $MeHg^+$ exposure does not appear to be an important risk factor that might explain the doubling of the prevalence of PD in this population.

The role of dental amalgam in PD was also evaluated [125]. This case-control study compared 380 German PD patients with 379 neighborhood controls and 376 regional controls. On average, PD patients reported a higher number of amalgam fillings than both neighborhood controls and regional controls. Limitations of this study include the usage of prevalent cases and amalgam exposure data that are solely based on interview and subject to bias. Dental records were not utilized [126].

The New Zealand Defense Force conducted a large scale study on the health effects of dental amalgams between 1977 and 1997 [126]. The final cohort contained 20,000 people, 84% of them males. Associations with medical diagnostic categories, particularly disorders of the nervous system and kidney were examined, however, the authors noted insufficient cases for investigation of associations between dental amalgams and PD. Gorell et al. [127] also found no significant association of PD with any occupational exposure to mercury.

In summary, the published literature is inconclusive. While few, and mostly small scale studies are suggestive of an association between mercury and PD, limitations inherent to them include the choice of prevalent cases, inadequate control recruitment methods, lack of confirmation of case diagnoses, in general (with the exception of [125]) small numbers of subjects, as well as inadequate exposure data. The larger studies [125,127–129] failed to uncover an association between mercury and PD. Nevertheless, the possibility remains that differences in ethnic or racial groups, or different routes of mercury exposure (e.g., ingestion of contaminated foods or medications) may account for the variability in the studies thus far [127]. Data substantiating elevated mercury levels in tissues derived for PD autopsied tissue could not be found.

5.2. Alzheimer's Disease

Alzheimer's disease (AD) is the most common type of dementia for which there is no known cure. In its most common form, it afflicts individuals over 65 years old; a less prevalent early-onset form also exists. AD is characterized by progressive memory loss. As the disease advances, symptoms include confusion, anger, mood swings, language breakdown, long-term memory loss, and the general withdrawal of the patient as his or her senses decline. The etiology of AD is poorly understood. At the morphological levels the disease is associated with loss of neurons and synapses in the cerebral cortex and certain subcortical regions, leading to gross atrophy of the affected regions, including degeneration in the temporal lobe and parietal lobe, and parts of the frontal cortex and cingulate gyrus. Both amyloid plaques and neurofibrillary tangles characterized by mostly insoluble deposits of amyloid-β protein and cellular material outside and around neurons are seen. While earlier disease familial onset is mainly explained by three genes, later age of disease onset representing most cases of AD has yet to be explained by a purely genetic model. Mercury has been evaluated in several studies as a potential etiologic factor in AD.

As with PD, studies exist both in support and against a role for mercury in AD. It was proposed that the genetic risk factor for the development of AD is increased by the presence of the apolipoprotein E4 allele whereas the apolipoprotein E2 allele reduces the risk of developing AD [130,131]. Notably, a statistical shift toward the at-risk apolipoprotein E4 groups was found in AD patients with dental amalgam fillings [130]. Elevated mercury concentrations have been reported in autopsied brain regions of AD patients [132,133]. An ecological study in Canada also found a correlation between the prevalence of AD and edentulism prevalence [134]. The authors interpreted this as evidence against an association between amalgam fillings (in people with teeth) and AD. However, a converse interpretation could be applied – that prevalence of edentulism is a marker of higher caries rates and, therefore, higher amalgam filling prevalence [126,135]. Concentration of mercury (and other metals) in plasma and cerebrospinal fluid (CSF) were also recently determined [136] by inductively coupled plasma mass spectrometry (ICP-MS) in 173 patients with AD and 54 healthy controls. Total plasma mercury concentrations were significantly higher in subjects with AD compared with controls. However this association was absent in the CSF, thus the significance of elevated blood mercury levels is elusive. Finally, a trend towards statistical difference in mercury content was noted by Cornett et al. [137]. Mercury levels in autopsied brain regions of AD subjects were generally higher compared to controls. However, variability in mercury levels in both AD and control subjects precluded the AD versus control difference from reaching statistical significance.

Mutter et al. [138] in a brief review made the following observations: (1) no metal other than mercury is capable to produce every single change in the nervous system of animals and in cell tests that is typical for AD, including the increase of β-amyloid and the formation of neurofibrillar tangles; (2) the presence of aluminum and/or other metals in the brain along with mercury may lead to synergistic toxic effects; (3) elevated mercury levels were found indeed in brains of deceased AD patients; (4) the development of AD may require 30–50 years before its clinical effects are manifest, hence there is a potential that many of mercury's effects would be masked in early studies on its effects; (5) since approximately 95% of all AD cases are triggered by exogenic factors and the disease is pandemic in developing countries, reflecting a rise in the use of dental amalgams; (6) AD risk is augmented with the incidence of dental decay; and, finally, (7) the presence of the apolipoprotein E subtype (Apo-E-4 allele) represents a major risk factor for developing AD.

Negative associations between AD and mercury as a pathogenic factor also exist. An AD case-control study by Saxe and colleagues [139] assessed the association with dental amalgam exposure. The study involved 68 postmortem cases and 33 controls drawn from a volunteer brain donation program. Detailed dental histories were obtained from dental records and X-rays. Specimens from the cerebral cortex of the brain were analyzed for mercury. Three indices of amalgam exposure, based on event (i.e., amalgam placement, repair or removal), location in the mouth, and time in the mouth, were developed. The study concluded that no statistical association could be ascertained between exposure indices and either AD or mercury concentrations in parts of the brain. The study had high-quality dental history data, but was limited by a small number of subjects. The New Zealand Defense Force mentioned within the context of PD (see above) also mined for possible associations between dental amalgams and AD [128]. The authors noted insufficient cases for investigation of associations between dental amalgams and AD. No association between brain Hg levels and dental amalgam and no differences in dental amalgam experience were also noted in an earlier study [140]. Accordingly, while the results are mixed, there does not appear to be strong evidence and support for the hypothesis that mercury derived from dental amalgam or other sources is a major contributor to the pathogenesis of AD.

5.3. Amyotrophic Lateral Sclerosis

Amyotrophic lateral sclerosis (ALS) is characterized by deterioration of anterior horn cells in the spinal cord that leads to loss of muscle strength and respiratory problems, commonly with fatal outcome. Both genetic and environmental etiologies likely contribute to ALS. Pesticides and herbicides, rotenone, cocaine, amphetamine, and electrical injury, as well as cockpit occupation [141] have all been suggested to potentially trigger ALS.

ALS cases related to mercury intoxication and professional exposure have also been reported. Brown, as early as 1954 [142], reported on chronic mercurialism as a potential cause of the clinical syndrome of ALS. This was followed by a report of Kantarjian [143] on a syndrome clinically resembling ALS following chronic mercurialism. Barber [144] described two employees in a mercuric oxide manufacturing plant, which progressed to develop neurologic changes unknown at the time to be associated with exposure to inorganic or elemental mercury vapor. Their symptoms, physical findings and laboratory studies were consistent with those in ALS patients. Notably, all symptoms and laboratory findings were reversed and returned to normal values, respectively, after three months in a mercury-free work environment [144]. ALS-like symptoms were also described in a nurse accidentally injected with mercury [145] and other cohorts of metalloid-exposed individuals [146]. Exposure to elemental mercury was also reported to be associated with a syndrome resembling ALS in a case study of a 54-year-old man exposed to mercury. The syndrome resolved as his urinary mercury levels fell [147].

However, negative associations between ALS and mercury exposure have also been reported. A retrospective case-control study of occupational heavy metal exposure in 66 ALS patients and 66 age- and sex-matched controls failed to document an association between a number of heavy metal exposure (including mercury) and the pathogenesis of ALS in this patient population [148]. A recent study by the ALS CARE study group also failed to establish that metals, including mercury, present a significant risk factor for ALS [141]. Thus, it has yet to be determined that mercury plays a role in the etiology of anterior horn-cell dysfunction, associated with ALS.

5.4. Others

5.4.1. *Multiple Sclerosis*

Multiple sclerosis (MS) is an autoimmune condition in which the immune system attacks the CNS, leading to demyelination. It may cause numerous physical and mental symptoms, and often progresses to physical and cognitive disability. Both the brain and spinal cord white matter are affected in the course of the disease, with destruction of oligodendrocytes, the CNS myelinating cells. MS results in a thinning or complete loss of myelin and effectively the conduct of neuronal electrical signals. The inflammatory process associated with MS is triggered by T cells, which recognize myelin as foreign and attack it as if it were an invading virus.

The interested reader is referred to a comprehensive review article on the relationship between mercury exposure and ALS [149]. The first study on mercury exposure and MS was reported by Craelius [150], where a correlation between the disease and dental caries was noted. Other studies also reported a positive association between MS and dental amalgam fillings. Several studies [128,151,152] reported elevated relative risk for MS and amalgam fillings. However, McGrother et al. [153] found no such correlation. Aminzadeh and Etminan [154] in a meta analysis study also reported a slight and consistent, yet non-significant increase in the odds ratio for the risk of MS among amalgam users. The authors suggest that their investigation was limited by the availability of only four studies, all suffering from great heterogeneity. Though the data thus far are reassuring, future consideration on amalgam restoration size and surface area along with the duration of exposure are needed to better define a potential link between amalgam and MS. An association between exposure to organic mercury exposure and MS has not been reported.

5.4.2. *Skogholt's Disease*

Skogholt's disease is a hereditary neurological disease that was recently reported in a Norwegian family. It is characterized by a demyelination disorder affecting both the central and the peripheral nervous system. The onset of symptom varies from before 30 to after 50 years of age, and the disease is uniformly gradually progressive. The disease is characterized by gradual loss of distal sensation, distal atrophy of extremity muscles or weakness of muscles in all extremities, unsteady gait, dysarthria, cognitive slowness, and memory impairment. The disease was shown to be associated with increased levels of Cu, Fe, and Zn in the CSF of Skogholt patients compared to controls, however, no changes in concentrations of mercury were noted [155].

In summary, current data does not support hypotheses linking mercury exposure with neurodegenerative diseases (Alzheimer's disease, amyotrophic lateral sclerosis, multiple sclerosis, Parkinson's disease, and Skogholt's disease). Another issue of great debate is associated with the role vaccine-derived $EtHg^{+}$ in the etiology of autism and other developmental neurocognitive syndromes. Despite compelling scientific evidence against a causal association, it remains one of the most contentious health controversies in recent years. The issue is deemed to be beyond the scope of this review.

5.4.3. Neurodevelopmental Alterations

As mentioned previously, prenatal exposures to low concentrations of $MeHg^{+}$ occurring in populations with a high intake of seafood and freshwater fish have been correlated to a three-point decrement in intelligence quotient (IQ) [156] and impairments in memory, attention, language, and visuospatial perception in exposed children [44]. Another study provided discordant results [157]. Factors, such as exposure to polychlorinated biphenyls (PCBs) [158] as well as aspects related to samples and data analyses [159] have been taken into account to explain the discrepancy.

Autism has also been linked to Hg exposure via thimerosal in vaccines. However, recent publications have concluded that there is no link between thimerosal and autism or other neurological or psychological outcomes [160,161].

6. GENERAL CONCLUSIONS

This chapter addresses the pathways of mercury compounds to humans, as well as their neurotoxicity, both in young and adult animals. The effects of $MeHg^{+}$ were tragically revealed in large numbers of poisonings in Japan and Iraq. The clinical picture varies both with the severity of exposure and the age of the individual at the time of exposure. In adults, the most dramatic sites of injury are the neurons of the visual cortex and the small internal granular cell neurons of the cerebellar cortex, whose massive degeneration results in blindness and marked ataxia. In children, particularly those exposed to $MeHg^+$ *in utero*, the neuronal loss is widespread, and in settings of greatest exposure, it produces profound mental retardation and paralysis.

The effects of $EtHg^+$ are discussed as well, but it needs to be kept in mind that extrapolation on the pharmacokinetics of $EtHg^+$ from the $MeHg^+$ data is unwarranted. While the scientific literature supports the concept that $MeHg^+$ is a potent and well known developmental neurotoxin, the assertion that $EtHg^+$ leads to developmental abnormalities is hypothetical and unsubstantiated, resting on indirect and incomplete information, primarily from analogies with $MeHg^+$. This approach is not surprising, as until recently there was sparse information on the disposition of $EtHg^+$ as compared to $MeHg^+$. However, results from the few studies that have provided a direct comparison between these compounds have established that extrapolation of $EtHg^+$'s disposition and toxic potential from the $MeHg^+$ literature is flawed, as distinct differences exist with respect to the pharmacokinetic behavior of the two organomercurials. Key observations to substantiate this statement include the following: (1) mercury clears from the body much faster after the administration of $EtHg^+$ than after the administration of $MeHg^+$; (2) the brain-to-blood mercury concentration ratio established for $MeHg^+$ will overestimate mercury in the brain after exposure to $EtHg^+$; and (3) because $EtHg^+$ decomposes much faster than $MeHg^+$, the risk of brain damage is less for $EtHg^+$ than for $MeHg^+$.

As noted in the chapter, though great strides have been made in better understanding the molecular mechanisms of $MeHg^+$ neurotoxicity, it remains unknown why and what precise mechanisms account for its neurodevelopmental effects. Puzzling are also the specific effects of $MeHg^+$ in the adult brain, under conditions of homogenous distribution. Finally, the role of mercury in the etiology of neurodegenerative disorders is also not well substantiated. These and other areas on neurotoxic research should be further assessed so we may better understand the neurotoxicity and risk associated with various mercury compounds.

ACKNOWLEDGMENTS

This review was partially supported by grants from NIEHS 10563, ES007331, DoD W81XWH-05-1-0239, and the Gray E.B. Stahlman Chair of Neuroscience (MA) as well as grants from The Swedish Research Council, The Swedish Research Council for Environment, Agricultural Sciences and Spatial Planning (FORMAS), The European Commission (FOOD-CT-2003-506143) (SC).

ABBREVIATIONS

AD	Alzheimer's disease
Apo-E	apolipoprotein E
ALS	amyotrophic lateral sclerosis
CNS	central nervous system
CSF	cerebrospinal fluid
DA	dopamine
DTaP	diphtheria/tetanus/pertussis
FDA	U.S. Food and Drug Administration
EPA	U.S. Environmental Protection Agency
$EtHg^+$	ethylmercury
GSH	glutathione
Hib	*Haemophilus influenzae* type b conjugate
ICP-MS	inductively coupled plasma mass spectrometry
IM	intramuscular
IQ	intelligence quotient
IV	intravenous
$MeHg^+$	methylmercury
MS	multiple sclerosis
NMDA	*N*-methyl D-aspartate
PD	Parkinson's disease
PND	postnatal day
ppm	parts per million
RfD	reference dose
ROS	reactive oxygen species
SH	thiol group
SOD	superoxide dismutase
WHO	World Health Organization

REFERENCES

1. J. O. Nriagu, *Nature*, 1979, **279**, 409–411.
2. W. F. Fitzgerald, D. R. Engstrom, C. H. Lamborg, C. M. Tseng, P. H. Balcom and C. R. Hammerschmidt, *Environ. Sci. Technol.*, 2005, **39**, 557–568.
3. *Office of Air Quality Planning & Standards and Office of Research and Development. Volume V: Health Effects of Mercury and Mercury Compounds, EPA-452/R-97-007*, US EPA, Mercury Study Report to Congress,1997.
4. *Toxicological Profile for Mercury*, U.S. Department of Health and Human Services, Public Health Service, Center for Disease Control, Atlanta, GA, Agency for Toxic Substance and Disease Registry, 2003.

5. T. W. Clarkson, J. B. Hursh, P. R. Sager and J. P. Syversen, *Mercury*, 1988, 199–246.
6. P. M. Stokes and C. D. Wren, *Bioaccumulation of Mercury by Aquatic Biota in Hydroelectric Reservoirs: A Review and Consideration of Mechanisms, in Lead, Mercury, Cadmium, and Arsenic in the Environment*, Ed. T. C. Hutchinson and V. M. Meera, John Wiley and Sons, Chichester, UK, 1987, pp. 255–278.
7. M. M. Veiga, J. A. Meech and N. Onate, *Nature*, 1994, **368**, 816–817.
8. S. Jensen and A. Jernelov, *Nature*, 1969, **223**, 753–754.
9. A. Boudou, R. Maury-Brachet, M. Coquery, G. Durrieu and D. Cossa, *Environ. Sci. Technol.*, 2005, **39**, 2448–2454.
10. J. G. Wiener, D. P. Krabbenhoft, G. H. Heinz and A. M. Scheuhammer, *Ecotoxicology of Mercury*, in *Handbook of Ecotoxicology*, Ed. D. J. Hoffman, B. A. Rattner and G. A. Burton Jr., CRC Press, Boca Raton, USA, 2003, pp. 409–463.
11. T. W. Clarkson, *Crit. Rev. Clin. Lab. Sci.*, 1997, **34**, 369–403.
12. L. R. Kamps, R. Carr and H. Miller, *Bull. Environ. Contam. Toxicol.*, 1972, **8**, 273–279.
13. D. J. Spry and J. G. Wiener, *Environ. Pollut.*, 1991, **71**, 243–304.
14. G. R. Southworth, R. R. Turner, M. J. Peterson and M. A. Bogle, *Chemosphere*, 1995, **30**, 779–787.
15. G. Guzzi and C. A. La Porta, *Toxicology*, 2008, **244**, 1–12.
16. G. Sallsten, J. Thoren, L. Barregard, A. Schutz and G. Skarping, *J. Dent. Res.*, 1996, **75**, 594–598.
17. D. C. Bellinger, F. Trachtenberg, L. Barregard, M. Tavares, E. Cernichiari, D. Daniel and S. McKinlay, *J. Am. Med. Ass.*, 2006, **295**, 1775–1783.
18. L. R. Goldman and M. W. Shannon, *Pediatrics*, 2001, **108**, 197–205.
19. L. Magos and T. W. Clarkson, *Ann. Clin. Biochem.*, 2006, **43**, 257–268.
20. *Joint statement of the American Academy of Pediatrics and the United States Public Health Service*, American Academy of Pediatrics (AAP) and U.S. PHS, *Pediatrics*, 1999, 104, 568–569..
21. H. M. Powell and W. A. Jamieson, *Am. J. Hyg.*, 1931, **13**, 296–310.
22. L. K. Ball, R. Ball and R. D. Pratt, *Pediatrics*, 2001, **107**, 1147–1154.
23. *Recommendations from the Strategic Advisory Group of Experts,* World Health Organization (WHO), *Weekly Epidemiol. Rec.*, 2002, 77, 305–316.
24. *Environmental Health Criteria 101: Methylmercury. Geneva: International Program on Chemical Safety*, World Health Organization (WHO), 1990.
25. Thimerosal in vaccines, *http://www.fda.gov/CbER/vaccine/thimerosal.htm*, US Food and Drug Administration, 2007.
26. T. Suzuki, T. Takemoto, H. Kashiwazaki and T. Miyama, *Metabolic Fate of Ethylmercury Salts in Man and Animal*, in *Mercury, Mercurials and Mercaptans*, Ed. M. W. Miller and T. W. Clarkson, C. Charles Thomas, Springfield, IL, 1973, pp. 209–232.
27. M. J. Reader and C. B. Lines, *J. Pharm. Sci.*, 1983, **72**, 1406–1409.
28. M. Tan and J. E. Parkin, *Int. J. Pharm.*, 2000, **208**, 23–34.
29. S. Al-Damluji, *M. J. Faculty Med.*, 1962, **4**, 82–103.
30. M. A. Jalili and A. H. Abbasi, *Br. J. Ind. Med.*, 1961, **18**, 303–308.

31. J. Zhang, *Am. J. Ind. Med.*, 1984, **5**, 251–258.
32. T. Takeuchi, T. Kambara, N. Morikawa, H. Matsumoto, Y. Shiraishi and H. Ito, *Acta Pathol. Jpn*, 1959, **9**, 769–783.
33. F. Bakir, S. F. Damluji, L. Amin-Zaki, M. Murtadha, A. Khalidi, N. Y. al-Rawi, S. Tikriti, H. I. Dahahir, T. W. Clarkson, J. C. Smith and R. A. Doherty, *Science*, 1973, **181**, 230–241.
34. T. Takeuchi, *Acta Pathol. Jpn*, 1982, **32**, 73–99.
35. B. H. Choi, *Prog. Neurobiol.*, 1989, **32**, 447–470.
36. M. Harada, *Teratology*, 1978, **18**, 285–288.
37. D. O. Marsh, G. J. Myers, T. W. Clarkson, L. Amin-Zaki, S. Tikriti, M. A. Majeed and A. R. Dabbagh, *Clin. Toxicol.*, 1981, **18**, 1311–1318.
38. T. M. Burbacher, G. P. Sackett and N. K. Mottet, *Neurotoxicol. Teratol.*, 1990, **12**, 65–71.
39. L. W. Chang, K. R. Reuhl and G. W. Lee, *Environ. Res.*, 1977, **14**, 414–423.
40. V. P. Markowski, C. B. Flaugher, R. B. Baggs, R. C. Rawleigh, C. Cox and B. Weiss, *Neurotoxicology*, 1998, **19**, 879–892.
41. U. Steuerwald, P. Weihe, P. J. Jorgensen, K. Bjerve, J. Brock, B. Heinzow, E. Budtz-Jorgensen and P. Grandjean, *J. Pediatr.*, 2000, **136**, 599–605.
42. F. Debes, E. Budtz-Jorgensen, P. Weihe, R. F. White and P. Grandjean, *Neurotoxicol. Teratol.*, 2006, **28**, 536–547.
43. P. Grandjean, P. Weihe, R. F. White and F. Debes, *Environ. Res.*, 1998, **77**, 165–172.
44. P. Grandjean, P. Weihe, R. F. White, F. Debes, S. Araki, K. Yokoyama, K. Murata, N. Sorensen, R. Dahl and P. J. Jorgensen, *Neurotoxicol. Teratol.*, 1997, **19**, 417–428.
45. T. Kjellstrom, P. Kennedy, S. Wallis, A. Stewart, L. Friberg, B. Lind, P. Witherspoon and C. Mantell, in *National Swedish Environmental Protection Board*, Solna, Sweden, 1989.
46. T. Kjellstrom, P. Kennedy, S. Wallis and C. Mantell, in *National Swedish Environmental Protection Board*, Solna, Sweden, 1986.
47. K. S. Crump, T. Kjellstrom, A. M. Shipp, A. Silvers and A. Stewart, *Risk Anal.*, 1998, **18**, 701–713.
48. S. G. Gilbert, T. M. Burbacher and D. C. Rice, *Toxicol. Appl. Pharmaco.l*, 1993, **123**, 130–136.
49. D. C. Rice, *Toxicol. Sci.*, 1998, **44**, 191–196.
50. D. C. Rice and S. G. Gilbert, *Toxicol. Appl. Pharmacol.*, 1995, **134**, 161–169.
51. D. C. Rice and S. G. Gilbert, *Toxicol. Appl. Pharmacol.*, 1990, **102**, 151–163.
52. M. A. Geyer, R. E. Butcher and K. Fite, *Neurobehav. Toxicol. Terato. l*, 1985, **7**, 759–765.
53. J. L. Jacobson, S. W. Jacobson, G. G. Fein, P. M. Schwartz and J. K. Dowler, *Dev. Psychol.*, 1984, **20**, 523–532.
54. C. Watanabe, K. Yoshida, Y. Kasanuma, Y. Kun and H. Satoh, *Environ. Res.*, 1999, **80**, 208–214.
55. C. S. Roegge and S. L. Schantz, *Neurotoxicol. Teratol.*, 2006, **28**, 260–277.
56. A. D. Rossi, E. Ahlbom, S. O. Ogren, P. Nicotera and S. Ceccatelli, *Exp. Brain Re.s*, 1997, **117**, 428–436.

57. N. Onishchenko, C. Tamm, M. Vahter, T. Hokfelt, J. A. Johnson, D. A. Johnson and S. Ceccatelli, *Toxicol. Sci.*, 2007, **97**, 428–437.
58. M. R. Carratu, A. Coluccia, A. M. Modafferi, P. Borracci, S. Scaccianoce, M. Sakamoto and V. Cuomo, *Bull. Environ. Contam. Toxicol.*, 2008, **81**, 539–542.
59. C. Johansson, A. F. Castoldi, N. Onishchenko, L. Manzo, M. Vahter and S. Ceccatelli, *Neurotox. Res.*, 2007, **11**, 241–260.
60. A. F. Castoldi, N. Onishchenko, C. Johansson, T. Coccini, E. Roda, M. Vahter, S. Ceccatelli and L. Manzo, *Regul. Toxicol. Pharmacol.*, 2008, **51**, 215–229.
61. N. Onishchenko, N. Karpova, F. Sabri, E. Castren and S. Ceccatelli, *J. Neurochem.*, 2008, **106**, 1378–1387.
62. R. Pfab, H. Muckter, G. Roider and T. Zilker, *J. Toxicol. Clin. Toxicol.*, 1996, **34**, 453–460.
63. I. Cinca, I. Dumitrescu, P. Onaca, A. Serbanescu and B. Nestorescu, *J. Neurol. Neurosurg. Psychiatry*, 1980, **43**, 143–149.
64. F. Bakir, H. Al-Shahristani, N. Y. Al-Rawi, A. Khadouri and A. W. Al-Mufti, *Bull. World Health Organ.*, 1976, **53**(suppl), 129–132.
65. J. A. Lowell, S. Burgess, S. Shenoy, M. Peters and T. K. Howard, *Lancet*, 1996, **347**, 480.
66. J. H. Axton, *Postgrad. Med. J.*, 1972, **48**, 417–421.
67. J. Rohyans, P. D. Walson, G. A. Wood and W. A. MacDonald, *J. Pediatr.*, 1984, **104**, 311–313.
68. A. Rothstein and A. L. Hayes, *J. Pharmacol. Exp. Ther.*, 1960, **130**, 166–176.
69. T. W. Clarkson and L. Magos, *Crit. Rev. Toxicol.*, 2006, **36**, 609–662.
70. L. Magos, *J. Appl. Toxicol.*, 2003, **23**, 263–269.
71. L. Magos, A. W. Brown, S. Sparrow, E. Bailey, R. T. Snowden and W. R. Skipp, *Arch. Toxicol.*, 1985, **57**, 260–267.
72. G. J. Harry, M. W. Harris and L. T. Burka, *Toxicol. Lett.*, 2004, **154**, 183–189.
73. M. E. Pichichero, E. Cernichiari, J. Lopreiato and J. Treanor, *Lancet*, 2002, **360**, 1737–1741.
74. T. M. Burbacher, K. S. Grant, D. B. Mayfield, S. G. Gilbert and D. C. Rice, *Toxicol. Appl. Pharmacol.*, 2005, **208**, 21–28.
75. M. Vahter, N. K. Mottet, L. Friberg, B. Lind, D. D. Shen and T. Burbacher, *Toxicol. Appl. Pharmacol.*, 1994, **124**, 221–229.
76. B. Weiss, T. W. Clarkson and W. Simon, *Environ. Health Perspect.*, 2002, **110**, 851–854.
77. S. Orrenius, B. Zhivotovsky and P. Nicotera, *Nat. Rev. Mol. Cell Biol.*, 2003, **4**, 552–565.
78. C. Tamm, J. Duckworth, O. Hermanson and S. Ceccatelli, *J. Neurochem.*, 2006, **97**, 69–78.
79. A. F. Castoldi, S. Barni, I. Turin, C. Gandini and L. Manzo, *J. Neurosci. Res.*, 2000, **59**, 775–787.
80. E. Dare, M. E. Gotz, B. Zhivotovsky, L. Manzo and S. Ceccatelli, *J. Neurosci. Res.*, 2000, **62**, 557–565.
81. E. Dare, A. M. Gorman, E. Ahlbom, M. Gotz, T. Momoi and S. Ceccatelli, *Neurotoxicity Res.*, 2001, **3**, 501–514.

82. E. Dare, W. Li, B. Zhivotovsky, X. Yuan and S. Ceccatelli, *Free Radical Biology and Medicine*, 2001, **30**, 1347–1356.
83. E. Fonfria, E. Dare, M. Benelli, C. Sunol and S. Ceccatelli, *Eur. J. Neurosci.*, 2002, **16**, 2013–2016.
84. M. Aschner, T. Syversen, D. O. Souza, J. B. Rocha and M. Farina, *Braz. J. Med. Biol. Res.*, 2007, **40**, 285–291.
85. T. W. Clarkson, L. Magos, C. Cox, M. R. Greenwood, L. Amin-Zaki, M. A. Majeed and S. F. Al-Damluji, *J. Pharmacol. Exper. Ther.*, 1981, **218**, 74–83.
86. S. Belletti, G. Orlandini, M. V. Vettori, A. Mutti, J. Uggeri, R. Scandroglio, R. Alinovi and R. Gatti, *J. Neurosci. Res.*, 2002, **70**, 703–711.
87. C. Johansson, R. Tofighi, C. Tamm, M. Goldoni, A. Mutti and S. Ceccatelli, *Toxicol. Lett.*, 2006, **167**, 183–190.
88. P. Kaur, M. Aschner and T. Syversen, *Neurotoxicology*, 2006, **27**, 492–500.
89. S. T. Park, K. T. Lim, Y. T. Chung and S. U. Kim, *Neurotoxicology*, 1996, **17**, 37–45.
90. T. L. Limke, S. R. Heidemann and W. D. Atchison, *Neurotoxicology*, 2004, **25**, 741–760.
91. A. Naganuma, K. Miura, T. Tanaka-Kagawa, J. Kitahara, Y. Seko, H. Toyoda and N. Imura, *Life Sci.*, 1998, **62**, PL157–161.
92. W. D. Atchison and M. F. Hare, *FASEB J.*, 1994, **8**, 622–629.
93. T. A. Sarafian, *J. Neurochemistry*, 1993, **61**, 648–657.
94. M. Sakamoto, N. Ikegami and A. Nakano, *Pharmacol. Toxicol.*, 1996, **78**, 193–199.
95. M. S. Marty and W. D. Atchison, *Toxicol. Appl. Pharmacol.*, 1997, **147**, 319–330.
96. E. M. Faustman, R. A. Ponce, Y. C. Ou, M. A. Mendoza, T. Lewandowski and T. Kavanagh, *Environ. Health Perspect.*, 2002, **110**, 859–864.
97. H. Komuro and P. Rakic, *J. Neurobiol.*, 1998, **37**, 110–130.
98. D. G. Vogel, R. L. Margolis and N. K. Mottet, *Toxicol. Appl. Pharmacol.*, 1985, **80**, 473–486.
99. A. F. Castoldi, T. Coccini, S. Ceccatelli and L. Manzo, *Brain Res. Bull.*, 2001, **55**, 197–203.
100. R. A. Ponce, T. J. Kavanagh, N. K. Mottet, S. G. Whittaker and E. M. Faustman, *Toxicol. Appl. Pharmacol.*, 1994, **127**, 83–90.
101. K. Miura, N. Koide, S. Himeno, I. Nakagawa and N. Imura, *Toxicol. Appl. Pharmacol.*, 1999, **160**, 279–288.
102. B. H. Choi, L. W. Lapham, L. Amin-Zaki and T. Saleem, *J. Neuropathol. Exper. Neurol.*, 1978, **37**, 719–733.
103. T. Coccini, G. Randine, A. F. Castoldi, P. Grandjean, G. Ostendorp, B. Heinzow and L. Manzo, *Neurotoxicology*, 2006, **27**, 468–477.
104. T. Coccini, G. Randine, S. M. Candura, R. E. Nappi, L. D. Prockop and L. Manzo, *Environ. Health Perspect.*, 2000, **108**, 29–33.
105. M. Aschner, C. P. Yao, J. W. Allen and K. H. Tan, *Neurochemistry Int.*, 2000, **37**, 199–206.
106. M. Aschner, Y. L. Du, M. Gannon and H. K. Kimelberg, *Brain Res.*, 1993, **602**, 181–186.

107. N. Brookes, *Toxicology*, 1992, **76**, 245–256.
108. T. L. Limke, J. J. Bearss and W. D. Atchison, *Toxicol. Sci.*, 2004, **80**, 60–68.
109. L. R. Faro, J. L. do Nascimento, J. M. San Jose, M. Alfonso and R. Duran, *Neurochem. Res.*, 2000, **25**, 225–229.
110. H. Kumolainen and J. Tuomisto, *Acta Pharmacol. Toxicol.*, 1981, **48**, 214–222.
111. J. J. Bonnet, S. Benmansour, N. Amejdki-Chab and J. Costentin, *Eur. J. Pharmacol.*, 1994, **266**, 87–97.
112. J. Bartolome, W. L. Whitmore, F. J. Seidler and T. A. Slotkin, *Life Sci.*, 1984, **35**, 657–670.
113. A. F. Castoldi, F. Blandini, G. Randine, A. Samuele, L. Manzo and T. Coccini, *Brain Res.*, 2006, **1112**, 91–98.
114. H. Lindstrom, J. Luthman, A. Oskarsson, J. Sundberg and L. Olson, *Environ. Res.*, 1991, **56**, 156–169.
115. R. Cagiano, M. A. De Salvia, G. Renna, E. Tortella, D. Braghiroli, C. Parenti, P. Zanoli, M. Baraldi, Z. Annau and V. Cuomo, *Neurotoxicol. Teratol.*, 1990, **12**, 23–28.
116. E. Dare, S. Fetissov, T. Hokfelt, H. Hall, S. O. Ogren and S. Ceccatelli, *Naunyn Schmiedebergs Arch. Pharmacol.*, 2003, **367**, 500–508.
117. L. Gimenez-Llort, E. Ahlbom, E. Dare, M. Vahter, S. Ogren and S. Ceccatelli, *Environ. Toxicol. Pharmacol.*, 2001, **9**, 61–70.
118. H. Biernat, S. A. Ellias, L. Wermuth, D. Cleary, E. C. de Oliveira Santos, P. J. Jorgensen, R. G. Feldman and P. Grandjean, *Neurotoxicology*, 1999, **20**, 945–952.
119. C. H. Ngim and G. Devathasan, *Neuroepidemiology*, 1989, **8**, 128–141.
120. E. Fabrizio, N. Vanacore, M. Valente, A. Rubino and G. Meco, *BMC Neurol.*, 2007, **7**, 24.
121. P. I. Dantzig, *J. Occup. Environ. Med.*, 2006, **48**, 656.
122. L. Wermuth, P. Joensen, N. Bunger and B. Jeune, *Neurology*, 1997, **49**, 426–432.
123. L. Wermuth, S. Bech, M. S. Petersen, P. Joensen, P. Weihe and P. Grandjean, *Acta Neurol. Scand.*, 2008, **118**, 126–131.
124. M. S. Petersen, P. Weihe, A. Choi and P. Grandjean, *Neurotoxicology*, 2008, **29**, 591–595.
125. A. Seidler, W. Hellenbrand, B. P. Robra, P. Vieregge, P. Nischan, J. Joerg, W. H. Oertel, G. Ulm and E. Schneider, *Neurology*, 1996, **46**, 1275–1284.
126. M. N. Bates, *Int. J. Hyg. Environ. Health*, 2006, **209**, 309–316.
127. J. M. Gorell, C. C. Johnson, B. A. Rybicki, E. L. Peterson, G. X. Kortsha, G. G. Brown and R. J. Richardson, *Neurotoxicology*, 1999, **20**, 239–247.
128. M. N. Bates, J. Fawcett, N. Garrett, T. Cutress and T. Kjellstrom, *Int. J. Epidemiol.*, 2004, **33**, 894–902.
129. J. M. Gorell, C. C. Johnson, B. A. Rybicki, E. L. Peterson, G. X. Kortsha, G. G. Brown and R. J. Richardson, *Neurology*, 1997, **48**, 650–658.
130. M. E. Godfrey, D. P. Wojcik and C. A. Krone, *J. Alzheimers Dis.*, 2003, **5**, 189–195.
131. J. Mutter, J. Naumann, C. Sadaghiani, R. Schneider and H. Walach, *Neuro. Endocrinol. Lett.*, 2004, **25**, 331–339.

132. W. D. Ehmann, W. R. Markesbery, M. Alauddin, T. I. Hossain and E. H. Brubaker, *Neurotoxicology*, 1986, **7**, 195–206.
133. C. M. Thompson, W. R. Markesbery, W. D. Ehmann, Y. X. Mao and D. E. Vance, *Neurotoxicology*, 1988, **9**, 1–7.
134. J. P. Lund, P. Mojon, M. Pho and J. S. Feine, *Age Ageing*, 2003, **32**, 228–229.
135. S. R. Saxe, D. A. Snowdon, M. W. Wekstein, R. G. Henry, F. T. Grant, S. J. Donegan and D. R. Wekstein, *J. Am. Dent. Assoc.*, 1995, **126**, 1495–1501.
136. L. Gerhardsson, T. Lundh, L. Minthon and E. Londos, *Dement. Geriatr. Cogn. Disord.*, 2008, **25**, 508–515.
137. C. R. Cornett, W. R. Markesbery and W. D. Ehmann, *Neurotoxicology*, 1998, **19**, 339–345.
138. J. Mutter, J. Naumann and C. Guethlin, *Crit. Rev. Toxicol.*, 2007, **37**, 537–549discussion 551–532.
139. S. R. Saxe, M. W. Wekstein, R. J. Kryscio, R. G. Henry, C. R. Cornett, D. A. Snowdon, F. T. Grant, F. A. Schmitt, S. J. Donegan, D. R. Wekstein, W. D. Ehmann and W. R. Markesbery, *J. Am. Dent. Assoc.*, 1999, **130**, 191–199.
140. D. Wenstrup, W. D. Ehmann and W. R. Markesbery, *Brain Res.*, 1990, **533**, 125–131.
141. B. R. Brooks, *Amyotroph. Lateral Scler. Other Motor Neuron. Disord.*, 2000, **1**, S19–26.
142. I. A. Brown, *AMA Arch. Neurol. Psychiatry*, 1954, **72**, 674–681.
143. A. D. Kantarjian, *Neurology*, 1961, **11**, 639–644.
144. T. E. Barber, *J. Occup. Med.*, 1978, **20**, 667–669.
145. S. Schwarz, I. Husstedt, H. P. Bertram and K. Kuchelmeister, *J. Neurol. Neurosurg. Psychiatry*, 1996, **60**, 698.
146. R. D. Currier and A. F. Haerer, *Arch. Environ. Health*, 1968, **17**, 712–719.
147. C. R. Adams, D. K. Ziegler and J. T. Lin, *J. Am. Med. Assoc.*, 1983, **250**, 642–643.
148. L. S. Gresham, C. A. Molgaard, A. L. Golbeck and R. Smith, *Neuroepidemiology*, 1986, **5**, 29–38.
149. J. Praline, A. M. Guennoc, N. Limousin, H. Hallak, B. de Toffol and P. Corcia, *Clin. Neurol. Neurosurg.*, 2007, **109**, 880–883.
150. W. Craelius, *J. Epidemiol. Community Health*, 1978, **32**, 155–165.
151. D. Bangsi, P. Ghadirian, S. Ducic, R. Morisset, S. Ciccocioppo, E. McMullen and D. Krewski, *Int. J. Epidemiol.*, 1998, **27**, 667–671.
152. I. Casetta, M. Invernizzi and E. Granieri, *Neuroepidemiology*, 2001, **20**, 134–137.
153. C. W. McGrother, C. Dugmore, M. J. Phillips, N. T. Raymond, P. Garrick and W. O. Baird, *Br. Dent. J.*, 1999, **187**, 261–264.
154. K. K. Aminzadeh and M. Etminan, *J. Public Health Dent.*, 2007.
155. K. Gellein, J. H. Skogholt, J. Aaseth, G. B. Thoresen, S. Lierhagen, E. Steinnes, T. Syversen and T. P. Flaten, *J. Neurol. Sci.*, 2008, **266**, 70–78.
156. T. Kjellstrom, P. Kennedy, S. Wallis, A. Stewart, L. Friberg, B. Lind, P. Witherspoon and C. Mantell, *Physical and Mental Development of Children with Prenatal Exposure to Mercury from Fish. Stage 2. Interviews and Psychological Tests at Age 6, Report 364*, National Swedish Environmental Protection Board., Solna, Sweden, 1989.

157. G. J. Myers, P. W. Davidson, C. Cox, C. F. Shamlaye, D. Palumbo, E. Cernichiari, J. Sloane-Reeves, G. E. Wilding, J. Kost, L. S. Huang and T. W. Clarkson, *Lancet*, 2003, **361**, 1686–1692.
158. M. P. Longnecker, M. S. Wolff, B. C. Gladen, J. W. Brock, P. Grandjean, J. L. Jacobson, S. A. Korrick, W. J. Rogan, N. Weisglas-Kuperus, I. Hertz-Picciotto, P. Ayotte, P. Stewart, G. Winneke, M. J. Charles, S. W. Jacobson, E. Dewailly, E. R. Boersma, L. M. Altshul, B. Heinzow, J. J. Pagano and A. A. Jensen, *Environ. Health Perspect.*, 2003, **111**, 65–70.
159. A. H. Stern, J. L. Jacobson, L. Ryan and T. A. Burke, *Environ. Health*, 2004, **3**, 2.
160. S. K. Parker, B. Schwartz, J. Todd and L. K. Pickering, *Pediatrics*, 2004, **114**, 793–804.
161. J. Heron and J. Golding, *Pediatrics*, 2004, **114**, 577–583.

Met. Ions Life Sci. **2010**, *7*, 435–463

13

Environmental Bioindication, Biomonitoring, and Bioremediation of Organometal(loid)s

John S. Thayer

Department of Chemistry, University of Cincinnati, Cincinnati, OH 45221-0172, USA
< thayerj@ucmail.uc.edu >

Metal Ions in Life Sciences, Volume 7 Edited by Astrid Sigel, Helmut Sigel, and Roland K. O. Sigel

Published by the Royal Society of Chemistry, www.rsc.org DOI: 10.1039/9781849730822-00435

ABSTRACT: Environmentally occurring organometal(loid)s have generated some severe health and safety problems. Consequently, scientists have been investigating various organisms to show the presence of such compounds (bioindicators), to follow their movement through the environment (biomonitors), and to remove them (bioremediators). Examples of such organisms and the mechanisms of their action(s) are discussed. Also mentioned are those organisms that form organometal(loid)s as a way of removing toxic inorganic species.

KEYWORDS: Bioindicator · biomarker · biomonitor · bioremediation · organometallic compound · organometalloid compound · microbial remediation · phytoremediation

1. INTRODUCTION

1.1. Terminology

The use of living organisms to trace, monitor and clean up environmental contaminants has expanded greatly in recent years [1,2], generating a specialized vocabulary, that includes the following terms:

Bioindicator: "an organism (or part of an organism or a community of organisms) that contains information on the quality of the environment (or a part of the environment)" [3,4].

Biomarker: "measurable biological parameters at the suborganismal ... level in which structural or functional changes indicate environmental influences" [3]. In practice, bioindicators are measured in organism populations, while biomarkers are measured in single organisms. The term **sentinel species** describes a bioindicator used to warn of the initial appearance of a specific environmental pollutant in a defined ecosystem.

Biomonitor: a bioindicator "that contains information on the quality of the environment" [3]. Biomonitors may be laboratory-bred bioindicators exposed to the natural environment for some period (*active biomonitoring*) or naturally occurring bioindicators present in the ecosystem (*passive biomonitoring* [3]).

Biomonitoring usually involves systematic investigation of bioindicators within some specified area for a particular period of time, using a particular biomarker.

Bioremediation: the use of living organisms to remove pollutants from the environment. Plants are frequently used (**phytoremediation** [5]), as are microbes, algae and a variety of other organisms. **Bioscavengers** are chemical compounds which react with **xenobiotics** (foreign compounds) within a single organism, to decompose xenobiotics (**biodegradation** [6]) or otherwise render them harmless.

1.2. Scope of Article

Much has been written on the occurrence, movement, and environmental effects of organometallic and organometalloidal compounds [7–13]. This article will consider specific compounds that have become, or may become, a threat to human health, such as:

Organomercury compounds. Methylmercuric compounds have caused major poisonings over the last half century [9–11,14–18]. Inorganic mercury compounds undergo methylation through biological action [14,16] to form CH_3HgX, which enter and moves through food chains and webs, eventually reaching toxic concentrations.

Organotin compounds. Tri-*n*-butyltin (TBT) compounds, used as antifouling agents for hulls of watercraft, have become a major problem in marine and estuarine environments [19–21]. Other organotin compounds, especially phenyl-, methyl-, and *n*-octyltin, along with mixed alkyltin compounds have been found [20,22,23].

Organolead compounds. Tetraethyllead has been used since the 1920s (and in some places is still used), as an additive to gasoline. Methyl- and ethyllead (especially triethyllead) compounds, are frequently detected in the environment [24,25].

Organoarsenic compounds. Occurrence of organoarsenicals in the environment arises largely through organismal metabolism of arsenites and arsenates [10,26–28]. Salts of methylarsonic and dimethylarsinic (cacodylic) acids, used in agriculture, provide another entry route. Organoarsenicals have been used in warfare; they and their degradation products provide still another entry route [29].

Organophosphorus compounds. In this article, the term "organophosphorus" refers specifically to compounds having one or more phosphorus-carbon bond(s). Most such compounds are phosphonates of general formula $RP(:O)O_2^{2-}$ (e.g., ciliatine, where $R = NH_2CH_2CH_2$-) [30–33] that occur, or enter into various living organisms [30,31]. Extensive agricultural use of two organophosphorus compounds, glyphosate (N-(phosphonomethyl)glycine) [34,35] and glufosinate (phosphinothricin) [35], had led to their introduction into the environment. In addition, nerve gases containing P-C linkages have also been detected.

Other organometallic or organometalloidal compounds occur in the natural environment. Some are toxic, but have not become widespread; these, too, will be discussed later.

2. BIOMARKERS AND BIOINDICATORS

2.1. Biomarkers

2.1.1. Introduction

Various compounds have been used or proposed as biomarkers [36–38]. They have been divided [36] into three categories: (i) biomarkers of exposure; (ii) biomarkers of effect; (iii) biomarkers of susceptibility. Biomarkers provide an early warning – a biochemical signal that some toxic effect is occurring in one organism before the entire population becomes affected. Biomarkers for ten metals have been listed [39]. The few specific biomarkers proposed for organometallic compounds fall into the first category.

2.1.2. Organotin Compounds

Environmentally occurring examples of organotins have already been mentioned. TBT compounds are the most toxic, and they have been the

Table 1. Biomarkers proposed for tri-*n*-butyltin poisoning.

Organism name			
Common	Scientific	Biomarker	Reference
Cultivated clam	*Tapes philippinarum*	amoebocytic index phagocytic index lysosomal activity index	[45]
Clam	*Coelomactra antiquata*	cytochrome P_{450} level	[46]
Blue mussel	*Mytilus edulis* other *Mytilus* spp	acetylcholinesterase glutathione-*S*-transferase catalase activities thiobarbituric acid-reactive substances	[47,48]
Red snapper	*Lutjanus argentimaculatus*	echinocytes; multinuclei	[49]

primary target of biological/ecological investigation. The two biomarkers generally used for TBT are **imposex** (imposition of male sexual characteristics on female gastropods) and **intersex** (corresponding effects in bivalves) [40–44]. These conditions can be measured quantitatively by one of three indices [40]: the relative penis size index (RPSI) or the vas deferens stage index (VDSI) (for imposex) and the intersex stage index (ISI) (for intersex). Such indices enable quantitative comparisons among different group studies.

Other proposed biomarkers are shown in Table 1 [45–49]. All involve marine organisms, because TBT poisonings have all developed in water, primarily (though not exclusively) in oceans and harbors.

2.1.3. *Other Organometal(loid)s*

Organophosphorus compounds have been studied in relation to their toxicity towards humans; some examples are listed in Table 2 [50–53]. Biomarkers for mercury exposure have been reviewed [54], and human umbilical cords have been proposed as a biomarker for methylmercury [55]. The total arsenic content of human fingernails has been suggested as a biomarker for organoarsenic poisoning [56].

Table 2. Biomarkers proposed for organophosphorus poisoning.

Compound	Organism name: Common	Organism name: Scientific	Biomarker	Reference
Soman	rat	(Sprague-Dawley)	fluoride regeneration, miosis	[50]
Sarin	rat	(Sprague-Dawley)	urinary 3-nitrotyrosine and 8-hydroxy-2′-deoxyguanosine	[51]
Sarin	guinea pig	not listed	phosphorylated tyrosine/albumin	[52]
Soman	guinea pig		phosphorylated tyrosine/albumin	[52]
Cyclosarin	guinea pig		phosphorylated tyrosine/albumin	[52]
Tabun	guinea pig		phosphorylated tyrosine/albumin	[52]
Glyphosate	mosquito fish	*Gambusia yucatana*	cholinesterase activity	[53]

2.2. Bioindicators

2.2.1. Introduction

Although more numerous than biomarkers, bioindicators used for organometallic compounds are still less numerous than those used for pure inorganic or organic compounds. Applications of bioindicators have been reviewed [1–2,57]. As with biomarkers, TBT and other organotin compounds have had the greatest number of bioindicators used or proposed. Methylmercury is second, and other organometals are much less commonly represented.

2.2.2. Organotin Compounds

Organisms used or proposed as bioindicators for organotin compounds appear in Table 3 [58–80]. These are all aquatic organisms, primarily marine invertebrates. Imposex and intersex, depending on the organism, serve as the principal biomarkers (Table 1).

Table 3. Organisms used or proposed as bioindicators for organotin compounds.

Organism name		
Common	Scientific	Reference
	GASTROPODS	
Dog whelk	*Nucella lapillus*	[58–62]
Rock shell	*Thais clavigera*	[62–65]
Marine snail	*Conus betulinus*	[65]
Neogastropod	*Hinia reticulata*	[66]
Snail	*Adelomelon brasiliana*	[67]
Whelk	*Morula granulata*	[68]
Whelk	*Nassarius reticulatus*	[69]
Whelk	*Stramonita haemastoma*	[70]
Mud snail	*Hydrobia ulvae*	[71]
Ramshorn snail	*Marisa cornuarietis*	[72]
	PELECYPODS	
Periwinkle	*Littorina littorea*	[71,73]
Blue mussel	*Mytilus edulis*	[74]
Soft-shelled clam	*Mya arenaria*	[75]
Freshwater mussel	*Elliptio complanata*	[76]
Freshwater mussel	*Anodonta woodiana*	[77]
	OTHER INVERTEBRATES	
Amphipod	*Caprella* spp	[78]
Daphnia	*Daphnia magna*	[79]
	FISHES	
European flounder	*Platichthys flesus*	[80]
Chinese rare minnow	*Gobiocypris rarus*	[80]

2.2.3. *Methylmercuric Compounds*

Organisms used or proposed as bioindicators for methylmercuric compounds are listed in Table 4 [81–95]. Methylmercuric compounds are more widely distributed throughout the environment than organotins, resulting in a larger variety of bioindicator organisms. The mink, *Mustela vison*, has been proposed as a sentinel species [90].

2.2.4. *Other Organometallic Compounds*

At present, few bioindicator organisms for other organometallic compounds are known. The mussel *Mytilus galloprovincialis* has been suggested for use in detecting trimethyllead and other organolead compounds [96]. The

Table 4. Organisms used or proposed as bioindicators for methylmercuric compounds.

Organism name		
Common	Scientific	Reference
Lichen	*Hypogymnia physodes*	[81]
Water hyacinth	*Eichhornia crassipes*	[82]
Sea purslane	*Halimone portulacoide*	[83]
Mussel	*Mytilus galloprovincialis*	[84–86]
Mussel	*Perna perna*	[87]
Earthworm	*Eisenia foetida*	[88]
Mosquito	*Ochlerotatus* spp	[89]
Audouin's gull	*Larus audouinii*	[90]
Cliff swallow	*Petrochelidon pyrrhonota*	[91]
Sharptailed sparrow	*Ammodramus caudacutus*	[92]
Diamondback terrapin	*Malaclemys terrapin*	[93]
Mink	*Mustela vidon*	[94]

dandelion *Taraxacum officinale* was investigated as a potential bioindicator for methylcyclopentadienylmanganese tricarbonyl (now used as a gasoline additive) and its decomposition products [97]. Growing concern over organophosphorus and organoarsenic nerve gases will very probably lead to bioindicators being developed for these compounds and their metabolites.

3. BIOMONITORS

3.1. Introduction

Theory and applications of biological monitoring (biomonitoring) have been presented in detail [2]. Increasing awareness of organometallic compounds in the environment and the resulting health hazards [7–10] has resulted in development of biomonitors specifically for them. To date, this effort has concentrated on organotins and organomercurials. Chemical warfare agents that contain organo derivatives of arsenic and phosphorus are also receiving attention. Other organometal(loid)s, as awareness of their presence and hazardous effects increases, will certainly get greater attention in the future.

Environmental organometal monitoring, whether biological or not, are becoming more and more systematic. Problems in this area have been discussed [1,98,99]. Biomonitoring has been used to investigate metal pollution in natural waters [100].

3.2. Organotin Compounds

Various organotin compounds occur in natural waters and sediments [19–21]. Zebra mussels (*Dreissena polymorpha*) were used to measure the absorption of nine different organotin compounds [101]. Caged dogwhelks (*N. lapillus*) were active biomonitors of TBT pollution at various locations [102]. Imposex occurrence in the sensitive neogastropod *Hinia reticulata* (*Nassarius reticulatus*) served to monitor TBT occurrence in two estuaries in Portugal [103]. In a pilot "Freshwater Mussel Watch Project", the mussel *A. woodiana* was used as biomonitor around the Taihu Lake region of China [77]. Both dogwhelks and periwinkles (*Littorina littorea*) were employed to determine persistence of TBT in Halifax Harbour, Canada [104].

The use of imposex as a biomonitoring tool has been called into question [41]. Although TBT-containing antifouling paints have been restricted or banned in numerous countries, TBT still persists in many locations. Investigation of the Hérault River watershed showed total organotin levels of 0.51 ($\pm$0.02) – 71 ($\pm$2) ng Sn/L, compared to a proposed maximum allowable concentration of 1.5 ng/L [105]. International guidelines and collaborative efforts have been established to deal with organotin pollution in marine waters [106,107].

3.3. Organomercury Compounds

Human biomonitoring has often been employed for environmental methylmercuric compounds [14–16]. One study used human hair for this purpose [108]. Another study proposed the compound N-acetylcysteine as both biomonitoring agent and antidote [109]. The risk versus benefit problem for consumption of fish that may contain methylmercuric species has been discussed [110].

Environmental biomonitoring of methylmercuric compounds has been reviewed [111]. Cysteine complexes of methylmercuric compounds have been proposed as a generic toxicological model for fishes [112]. The need for thorough, systematic and continuing biomonitoring in various areas has been expressed frequently, e.g., watersheds in Brazil [113]; a National Park in America [114].

3.4. Organophosphorus Compounds

Certain organophosphorus compounds have been developed as weapons of warfare, and have received increased attention in recent years because of their use by terrorist groups [115–118]. Despite the variety of biomarkers shown in Table 2, most biomonitoring has been done on humans [117]. A biosensor using *Daphnia magna* provided a method to detect organophosphorus nerve gases in drinking water [119].

Decomposition of sarin, soman, cyclosarin and VX (Figure 1) would yield methylphosphonic acid, $CH_3PO_3H_2$, its esters and other derivatives. These have been detected by various analytical techniques [120,121].

Another organophosphorus compound deliberately introduced into the environment is glyphosate (*N*-phosphonomethylglycine), widely used as a nonselective herbicide [122]. While a bioindicator has been proposed [53] (Table 2), glyphosate is not usually tracked by biomonitoring. Decomposition of glyphosate in aerated water is shown by the following equation:

$$H_2O_3PCH_2NHCH_2CO_2H + H_2O \xrightarrow{[O_2]} NH_2CH_2PO_3H_2 + HOCH_2CO_2H$$

The first product, aminomethylphosphonic acid, is stable and has been reported many times in waters and soils where glyphosate has been used [122].

$CH_3P(:O)(F)OCH(CH_3)_2$

Sarin

$CH_3P(:O)(F)OC_6H_{13}$

Cyclosarin

$CH_3P(:O)(F)OCH(CH_3)C(CH_3)_3$

Soman

$ClCH{=}CHAsCl_2$

Lewisite

$CH_3P(:O)(OCH_2CH_3)SCH_2CH_2N[CH(CH_3)_2]_2$

VX

$NCP(:O)(OCH_2CH_3)N[CH(CH_3)_2]_2$

Tabun

Figure 1. Chemical formulas of nerve gases.

3.5. Organoarsenic Compounds

The primary organoarsenical subject to biomonitoring is Lewisite (Figure 1). Lewisite has been prepared and stored in substantial quantities [29]. Leakage of stored Lewisite has caused toxicity problems [123,124], leading to development of techniques for its disposal [29]. Lewisite is included among various chemical warfare agents reviewed for biomonitoring [125]. Binding of Lewisite to human hemoglobin has been proposed as a biomonitoring technique [125].

Another class of organoarsenicals used in chemical warfare are arylarsenic derivatives, e.g., $(C_6H_5)_2AsX$ (X = -Cl, -CN) [126–128]. These usually are oxidized to phenylarsonic oxide or As(V) oxide, but can exist for considerable periods of time in the natural environment. Phenylarsonic acid entered the drinking water of a Japanese community [127,128]. No biomonitors have been proposed for these species as of this writing.

3.6. Other Organometal(loid)s

Various additional organometal(loid)s have been detected in the natural environment, usually in localized areas. Methylantimony compounds [129] have been reported in natural waters and in landfill gases. Similarly, methyl derivatives of bismuth and cadmium have been detected in environmental samples [9]. These compounds have only been discovered relatively recently; no biomonitors have been developed for them. Thallium is a special case. Tl(I), as Tl^+ salts, is extremely toxic [130–132]. One paper reported that Tl(III) was more toxic to algae of the genus *Chlorella* than was Tl(I) [133]. Thallium occurs in the environment and has undergone biomonitoring [1,134,135]. Recently, workers reported finding $(CH_3)_2Tl^+$ in environmental samples [136–140]. This ion underwent bioaccumulation in plankton [139], diatoms, and chlorophytes [140]. The only toxicity study reported for $(CH_3)_2Tl^+$ indicated that dimethylthallium ion, in contrast to methylmercuric and trimethyllead ions, was *less* toxic than inorganic thallous ion [141]. Given the reported bioaccumulation of this ion [140], and the likelihood of its moving through a food web and/or undergoing demethylation to Tl^+, dimethylthallium should be considered a potential health hazard and deserves more complete investigation.

Two industrially important organometalloids also occur in the environment. Silicones, especially polydimethylsiloxanes $[(CH_3)_2SiO]_x$, enter by various routes [142–145], and may cause environmental damage despite very low water solubility and bioavailability [145]. No bioindicators have yet been suggested, though abiotic monitoring continues. Complexes between triphenyl- or alkyldiphenylboranes and amines (usually pyridine) are active ingredients in antifouling preparations, providing a possible entry route [146–149]. Triphenylborane-pyridine underwent slow abiotic degradation in water [150]. Whether triphenylborane or its derivatives become environmentally significant remains uncertain, but the possibility exists.

4. BIOREMEDIATION

4.1. Introduction

4.1.1. Concepts and Terminology

"Bioremediation" has been defined as "the process of judiciously exploiting biological processes to minimize an unwanted environmental impact; usually it is the removal of a contaminant from the biosphere" [151]. Bioremediation is discussed in detail elsewhere [5,6,151]. This article will only consider specific application to organometal(loid)s.

The presence of organic groups bonded to a metal or metalloid atom usually changes the toxicity. How it changes depends primarily on the specific metal or metalloid involved, and, to a lesser extent, on the nature of the organic group used. The present situation may be summarized as follows:

(i) Metals: Hg, Sn, and Pb show greater toxicity as organo derivatives. This also seems to be true for Bi; aromatic Bi compounds have been studied for their cytotoxicity [142–154]. Tl may be an exception (see Section 3.6), but too little is known to be certain. This is also true for Cd, Ge, and Po. No organoindium compounds have yet been reported in the environment.

(ii) Metalloids: As and Se oxides/oxyanions are more toxic than the methyl derivatives. This may also be true for Te.

(iii) Organic Groups: The toxic effects vary substantially. For metals, the alkyl compounds tend to be more toxic than analogous aryl compounds.

4.1.2. Chemistry of Bioremediation

Probably the most common route of bioremediation involves cleavage of the metal(loid)-carbon bond. Such cleavage occurs one bond at a time, and the intermediate species can usually be detected. Complete cleavage produces the element itself, which may remain as such or be converted to an inorganic compound, such as an oxide. Both the element and the intermediate forms may undergo subsequent reactions! Ultimate products will depend on the element, the organism performing the bioremediation, and the specific conditions involved.

Another route is sequestration, where some chelating agent binds the organometal(loid) moiety and sequesters it. Thiols (e.g., glutathione) are often used by organisms for this purpose, especially for metals; for metalloids, hydroxyl groups can serve the same purpose.

Excretion can be used by organisms to dispose of a toxic moiety. In higher organisms, excretion may occur in urine. Some organometal(loid)s may bind to chelates to aid their excretion. Another route, common to all organisms, is volatilization. The most common examples are the permethyl compounds (e.g., $(CH_3)_3As$, $(CH_3)_2Hg$, etc.) [17,28]. Formation of trimethylarsine by fungi led to the development of the concept of biological methylation (biomethylation) [123]. Depending on circumstances, more than one of these routes may be used for a particular organometal(loid).

4.2. Phytoremediation

4.2.1. *Introduction*

The use of plants to remove toxic substances from soils, waters, and air is a well-developed subject [5,155–158]. The growing occurrence of organometal(loid)s in the natural environment, along with their employment in agriculture, has resulted in their being studied for phytoremediation [159]. Both terrestrial and aquatic plants can be used, depending on the ecosystem involved. Bacterial genes have been added to certain plants to enhance their remediation abilities [160–164]; such plants are termed "transgenic plants".

Plants used for rhizoremediation will be discussed in Section 4.5.

4.2.2. *Arsenic*

Phytoremediation of arsenic has an extensive literature ([9,10,28,165–171] and references therein), but most of these deal with "inorganic arsenic" (arsenite and arsenate salts in varying combinations). Some plants accumulate very high levels of arsenic and are termed "hyperaccumulators" [165,168]. During phytoremediation, these plants often generate methylarsenic compounds, usually methylarsonates and dimethylarsinates, although others have also been reported [169]. Two studies revealed differing bioaccumulation behavior of plants towards methylarsenicals versus inorganic arsenicals: duckweed (*Spirodela polyrhiza*) accumulated arsenate ion via the phosphate uptake route [172], whereas dimethylarsinate accumulation followed a different route. The arsenic hyperaccumulators *Pteris vittata* and *P. cretica*, along with arsenic-tolerant *Boehmeria nivea*, showed greater toxicity and lower bioaccumulation towards dimethylarsinate than towards arsenate [173]. Phytoremediation has been employed, generally in conjunction with other techniques, for treatment of arsenic pollution caused by chemical warfare agents [29,124].

4.2.3. *Mercury*

Phytoremediation of organomercury derivatives usually involves genetic engineering, specifically the addition of *mer A* and *mer B* genes [159,160,174–179]. These genes code for the enzymes mercuric ion reductase and organomercurial lyase, respectively (Section 4.3.2). Certain transgenic plants thus treated are shown in Table 5. Such plants tend to be more resistant to organomercury poisoning than corresponding varieties having neither or only one of the genes [175–178,180–182]. Careful studies on tobacco plants showed that such treatment follows uptake by roots and translocation into stem and leaves [177].

Table 5. Plants used for the bioremediation of organomercury compounds.

Plant name		
Common	Scientific	Reference
–	*Arabidopsis thaliana*	[175]
Tobacco	*Nicotiana tabacum*	[175–177]
Rice	*Oryza sativa*	[178]
Eastern cottonwood	*Populus deltoides*	[180,181]
Salt marsh cordgrass	*Spartina alterniflora*	[182]

4.2.4. *Selenium*

Selenium resembles mercury in that phytoremediation involves formation of organo derivatives. Plants remove selenium from soils by a combination of volatilization and/or sequestration in plants. *Hydrilla verticillata* formed and volatilized R_2Se (R = methyl, ethyl) and $(CH_3)_2Se_2$ [183]. Perennial ryegrass (*Lolium perenne*) removed radioactive ^{75}Se from a contaminated water table [184]. Transgenic Indian mustard (*Brassica juncea*) plants receiving selenocysteine lyase or selenocysteine methyltransferase showed enhanced ability to concentrate selenium relative to their wild counterparts [185]. Both the presence of insects [186] and of sulfate ion [187] affected phytoremediating abilities of plants.

4.2.5. *Other Metals*

"Organophosphorus" pesticides (i.e., phosphate esters) can undergo phytoremediation by transgenic plants [188,189]. The use of plants to remove phosphonates has not been reported, although workers have investigated the mechanism of absorption and decomposition of glyphosate (*N*-phosphonomethylglycine), a widely used herbicide [190,191]. Microbial decomposition is more common for these compounds (see Section 4.3.4).

Willow trees will grow on tributyltin-contaminated sludge and may have potential for phytoremediation [192,193]. Various plants were tested for growth and tin bioaccumulation on tributyltin-containing sediments [194]. Barley (*Hordeum vulgare*) proved to be the most effective of these, removing tin while not accumulating any in the grain, and growing well despite the presence of salt [194].

Inorganic thallium undergoes phytoextraction by kale (*Brassica oleracea acephala)* and related species [195]. Bioaccumulation of dimethylthallium ion by algae (cf. Section 3.6) [140] suggests a possible bioremediation application. Thus far, dimethylthallium has been reported only in aquatic environments.

4.3. Microbial Remediation

4.3.1. *Introduction*

This form of remediation usually involves microorganisms [5,196], and usually proceeds by cleavage of metal(loid)-carbon linkages. Such bond breaking proceeds through enzymatic interactions, though mechanistic details remain sparse. Most reports involve sediments (marine and freshwater), along with terrestrial soils. To date, there has been relatively little deliberate use of microbes for organometal(loid) bioremediation. The special case of rhizoremediation, which involves bacteria on plant roots, will be discussed in Section 4.5.

4.3.2. *Mercury*

Various species of sediment bacteria cleave the Hg-C linkage in CH_3-Hg compounds [197]. Such bacteria have been proposed and tested for the removal of methylmercury from sediments [198–201]. The process involves two steps:

$$H^+ + CH_3Hg^+ \rightarrow CH_4 + Hg^{2+}$$

$$Hg^{2+} \rightarrow Hg^0$$

Both steps involve enzymes, controlled by the *mer* operon found in genes of mercury-resistant bacteria [202–204] (cf. Section 4.2.3). The first enzyme involved is organomercurial lyase (*merB*) [203,205–209]. The following mechanism has been proposed [205,206,209–212]: the methylmercurial binds to thiol groups of two separate cysteine molecules in a protein chain; another amino acid (yet to be identified), donates a proton; the mercury-carbon linkage is cleaved, and methane is released; the bound Hg(II) is transferred to the *merA* center, where it is reduced to Hg(0) [213]. This enzyme is not specific for methylmercurials; it works as well, possibly better, on other organomercurials [205,208]. Microorganisms bearing these genes apparently evolve in ecosystems afflicted by high levels of mercury pollution, and the genes appear in numerous species [205,207,214].

4.3.3. *Tin*

Organotin compounds found in the environment include $R_{3-n}SnX_n$ (n = 0–3; R = methyl, *n*-butyl, *n*-octyl, phenyl). Microbes degrade these compounds by Sn-C bond cleavage, leading to a wide range of organotin species reported [21,57,215].

Tributyltin decomposition has been the most investigated [216–221]. Triphenyltin degradation was enhanced by pyoverdins excreted by *Pseudomonas chlororaphis* [222–225], and triphenyltin chloride was decomposed by pyochelin secreted by *Pseudomonas aeruginosa* [226], ferripyochelin (an iron-pyochelin chelate), enhanced the rate of triphenyltin decomposition [227]. Comparative biodegradation studies in an activated sludge batch reactor showed that dibutyltin degraded faster than tributyltin ($t_{1/2}$ = 5.1 and 10.2 days, respectively), whereas triphenyltin and monobutyltin degraded at a much slower rate [228].

4.3.4. *Phosphorus*

Most reported studies concerned microbial degradation of phosphonates by cleavage of the phosphorus-carbon linkage [228–233]. The simplest phosphonate is methylphosphonic acid, $CH_3PO_3H_2$, formed by decomposition of nerve gases (cf. Section 3.4) and from other sources. Methyl-phosphorus cleavage has been proposed as a source of methane in the oceans [234,235]. Addition of incubated paddy soil to phosphonoacetic acid, $H_2O_3PCH_2$-CO_2H, generated methane and phosphine [236]. *Campylobacter* species caused phosphonate catabolism in various substrates [237]. ^{31}P NMR was used to monitor the degradation of glyphosate by *Spirulina* [238]. Acetyltransferase from *Bacillus licheniformis* was used to study glyphosate resistance [239]. *Acidithiobacillus ferrooxidans* (a chemolithoautotroph) generated a carbon-phosphorus lyase that enabled it to use phosphonates as a phosphorus source [240]. The crystal structure of a carbon-phosphorus lyase in *Escherichia coli* has been reported [241]. Nerve gases, such as sarin and soman, undergo enzymatic degradation [242].

4.3.5. *Arsenic*

Microbial degradation of organoarsenic compounds has been less studied than phosphorus. They have an important role in the biogeochemical cycling of arsenic [243], usually via biomethylation. Bacteria hyper-resistant to arsenic reduce a portion of arsenate to arsenite, but also use other pathways, including biomethylation [244]. As-CH_3 bond cleavage has also been reported: *Mycobacterium neoaurum* demethylated monomethyl derivatives of both As(III) and As(V) [245]; strains of *Pseudomonas putida* from soil contaminated by arsenical chemical warfare agents demethylated methylarsonic acid [246]; microorganisms in anaerobic methanogenic sludge demethylated both mono- and dimethylarsenic(V) compounds [247]. Marine samples of both arsenobetaine [248] and arsenoribofuranosides ("arseno sugars") [249] underwent microbial demethylation under marine conditions. Bacteria that degraded dimethylarsinic acid in Lake Kahokugata (Japan) showed seasonal variations in community composition and activity [250]. Investigations into microbial demethylation of both arsenobetaine and dimethylarsinic acid in organic soil showed that the former underwent demethylation more rapidly [251]; the authors proposed the mechanistic pathway:

$$\begin{aligned}\text{arsenobetaine} &\rightarrow \text{unknown(dimethylarsenoylacetate?)}\\ &\rightarrow \text{dimethylarsinic acid} \rightarrow \text{methylarsonic acid} \rightarrow \text{arsenate}\end{aligned}$$

Phenylarsenic compounds enter the environment through two principal sources: decomposition of abandoned chemical warfare agents [29] and the use of roxarsone (3-nitro-4-hydroxyphenylarsonic acid) as a growth promoter and pesticidal agent in the poultry industry [252]. In an example of the first route, phenylarsenic compounds entered a well providing drinking water to a city in Japan (cf. Section 3.5). Investigation of bacterial attack on triphenylarsine and the corresponding oxide showed that both were first degraded to phenylarsonic acid and subsequently to inorganic arsenic [253].

4.3.6. *Other Metals and Metalloids*

Dimethyldiselenide was converted by soil microbes to dimethylselenide, Se(0), and other methylselenium species [254]. Polydimethylsiloxanes ("silicones") undergo microbial degradation to monomeric $(CH_3)_2Si(OH)_2$, subsequently converted to CO_2, SiO_2, and H_2O; however, abiotic processes accounted for most environmental degradation of these compounds [255]. Soil contaminated by tetraethyllead contained microorganisms that degraded it initially to triethyllead cation, then subsequently to diethyllead and inorganic lead compounds [256].

4.4. Fungal Remediation

The use of fungi in bioremediation has been reviewed [257,258]. While methylmetal compounds are often formed by fungi and used to remove toxic metalloids from soil, their use for remediation of organometals has hitherto been limited. Fungi [259,260], algae [261], and lichens [262] form methylarsenic compounds in the presence of inorganic arsenic, and seem to be able to add additional methyl groups to partially methylated arsenic species. Fungal species degraded organophosphorus compounds by cleavage of the phosphorus-carbon linkage [263,264]. Among the compounds serving as substrate was the herbicide glyphosate [265]. Cells of *Aspergillus terreus* were able to convert various phenylselenium compounds to methylphenylselenide [266].

4.5. Rhizoremediation

Rhizoremediation is a special subclass of microbial remediation, involving microbes on the roots of plants. This combination, and the soil area immediately adjacent to it, is termed the "rhizosphere". Rhizoremediation has been used to treat metal-contaminated sites [267,268]. The root mass of *Spartina alterniflora* converted tetrabromobisphenol to bisphenol [269]. Rhizoremediation apparently involves formation of organometals by the plant, followed by sequestration or volatilization. Pickleweed (*Salicornia bigelovii*) absorbed selenate ion from soil, converted it to organoselenium species which were then emitted as gases [270]. Phosphonates have been used to enhance and protect root-dwelling bacteria [271,272]. A strain of *Pseudomonas fluorescens*, treated with an arsenic-resistant operon, enabled plants to grow in the presence of arsenic compounds [272].

5. CONCLUSIONS

Organometal(loid) compounds occur in the natural environment, whether introduced by humans or formed through biogenic or abiotic processes.

Research efforts into the use of organisms to locate, monitor and/or neutralize such compounds have concentrated on certain ones that have proven toxic to humans: methylmercuric compounds, tri-*n*-butyltin compounds, nerve gases (lewisite, sarin, soman, etc.); others have received little or no attention. Numerous organisms, from microbiota up to and including humans, have been examined for such application. Although voluminous, research on this topic has been rather scattered and is less focused than it might be.

ACKNOWLEDGMENTS

The author wishes to express his appreciation to Mr. John Tebo and the staff of the R. E. Oesper Chemistry-Biology Library of the University of Cincinnati for their assistance in searching out references for this chapter.

REFERENCES

1. *Bioindicators and Biomonitors: Principles, Concepts and Applications*, Ed. B. A. Markert, A. M. Breure and H. G. Zechmeister, Elsevier, Amsterdam, 2003.
2. *Biological Monitoring: Theory and Application*, Ed. M. E. Conti, WIT Press, Southampton (UK), 2008.
3. B. A. Markert, A. M. Breure and H. G. Zechmeister, in Ref. 1, pp.15–17.
4. B. A. Markert, *J. Trace Elem. Biol. Med.*, 2007, **21**, 77–82.
5. *Applied Bioremediation and Phytoremediation*, Ed. A. Singh and O. P. Ward, Springer, Berlin, 2004.
6. M. Alexander, *Biodegradation and Bioremediation*, 2nd edn., Academic Press, San Diego, 1999.
7. J. S. Thayer, *Environmental Chemistry of the Heavy Elements*, VCH, New York, 1995.
8. T. R. Compton, *Occurrence and Analysis of Organometallic Compounds in the Environment*, J. Wiley & Sons, Chichester, UK, 1998.
9. *Organometallic Compounds in the Environment*, 2nd edn., Ed. P. J. Craig, J. Wiley & Sons, Chichester, UK, 2003.
10. *Organic Metal and Metalloid Species in the Environment*, Ed. A. V. Hirner and H. Emons, Springer, Berlin, 2004.
11. E. Dopp, L. M. Hartmann, A.-M. Florea, A. W. Rettenmeier and A. V. Hirner, *Crit. Rev. Toxicol.*, 2004, **34**, 301–333.
12. J. S. Thayer, Chapter 1 of this book.
13. C. F. Harrington, D. S. Vidler and R. O. Jenkins, Chapter 2 of this book.
14. *Mercury Pollution: Integration and Synthesis*, Ed. C. J. Watras and J. W. Huckabee, Lewis/CRC, Boca Raton (FL, USA), 1994.
15. *Dynamics of Mercury Pollution on Regional and Global Scales: Atmospheric Processes and Human Exposures Around the World*, Ed. N. Pirrone and K. R. Mahaffey, Springer, New York, 2005.

16. R. Eisler, *Mercury Hazards to Living Organisms*, Taylor and Francis, Boca Raton, FL, USA, 2006.
17. H. Hintelmann, Chapter 11 of this book.
18. M. Aschner, N. Onishchenko and S. Ceccatelli, Chapter 12 of this book.
19. *Tributyltin: Case Study of an Environmental Contaminant*, Ed. S. J. de Mora, Cambridge University Press, Cambridge, 1996.
20. *Organotin: Environmental Fate and Effects*, Ed. M. A. Champ and P. F. Seligman, Chapman and Hall, London, 1996.
21. T. Gajda and A. Jancsó, Chapter 4 of this book.
22. I. Mersiowsky, R. Brandsch and J. Ejlertsson, *J. Environ. Qual.*, 2001, **30**, 1604–1611.
23. S. K. Mitra, K. Jiang, K. Haas and J. Feldmann, *J. Environ. Monit.*, 2005, **7**, 1066–1068.
24. J. Yoshinaga, in Ref. 9, pp. 151–194.
25. H. G. Abadin and H. R. Pohl, Chapter 5 of this book.
26. *Environmental Chemistry of Arsenic*, Ed. W. T. Frankenberger, Jr., Marcel Dekker, New York, 2002.
27. (a) J. S. Edmonds and K. A. Francesconi, in Ref. 9, pp. 195–222. (b) D. Kuehnelt and W. Goessler, in Ref. 9, pp. 223–276.
28. K. J. Reimer, I. Koch and W. Cullen, Chapter 6 of this book.
29. *Environmental Aspects of Converting CW Facilities to Peaceful Purposes*, Ed. R. R. McGuire and J. C. Compton, Kluwer, Dordrecht, 2002.
30. *The Role of Phosphonates in Living Systems*, Ed. R. L. Hilderbrand, CRC Press, Boca Raton (FL, USA), 1983.
31. J. S. Thayer, *App. Organometal. Chem.*, 1989, **3**, 203–209.
32. B. Nowack, *Water Res.*, 2003, **37**, 2533–2546.
33. B. Nowack, in *Phosphorus in Environmental Technology*, Ed. E. Valsami-Jones, IWA Publishing, London, 2004, pp. 51–61.
34. S. O. Duke and S. B. Powles, *Pest Management Science*, 2008, **64**, 319–325.
35. R. G. Hall, *Phosphorus, Sulfur, Silicon*, 2008, **183**, 258–265.
36. P.-D. Hansen, in Ref. 1, pp. 203–219.
37. M. E. Conti, in Ref. 2, pp. 25–46.
38. P. K. S. Lam and J. S. Gray, *Mar. Pollut. Bull.*, 2003, **46**, 182–186.
39. A. Alimonti and D. Mattei, in Ref. 2, pp. 167–182.
40. I. M. Davies and A. Minchin, *J. Environ. Monit.*, 2002, **4**, 788–790.
41. V. Axiak, D. Micallef, J. Muscat, A. Vella and B. Mintoff, *Environ. Intern.*, 2003, **28**, 743–749.
42. S. M. Evans and G. J. Nicholson, *Sci. Total Environ.*, 2000, **258**, 73–80.
43. A. C. Birchenough, N. Barnes, S. M. Evans, H. Hinz, I. Krönke and C. Moss, *Mar. Pollut. Bullet.*, 2002, **44**, 534–543.
44. M. P. Cajaraville, M. J. Bebianno, J. Blasco, C. Porte, C. Sarasquete and A. Viarengo, *Sci. Total Environ.*, 2000, **247**, 295–311.
45. F. Cima, M. G. Marin, V. Matozzo, L. DaRos and L. Ballarin, *Mar. Pollut. Bull.*, 1999, **39**, 112–115.
46. J.-K. Jeon, M.-H. Lee, D.-J. Kim, W.-J. Shim, J.-R. Oh and S.-H. Lee, *Han'guk Susan Hakhoechi*, 2002, **35**, 185–190; *Chem. Abstracts*, 2002, **136**:397184f.

47. M.-H. Devier, S. Augagneur, H. Budzinski, P. Mora, J.-F. Narbonne and P. Garrigues, *Environ. Toxicol. Chem.*, 2003, **22**, 2679–2687.
48. M.-H. Devier, S. Augagneur, H. Budzinski, K. LeMenach, P. Mora, J.-F. Narbonne and P. Garrigues, *J Environ. Monitoring*, 2005, **7**, 224–240.
49. B.-Q. Huang, W.-H. Li, J.-P. Ueng and S.-J. Joung, *Taiwan Shuichan Xuehuikan*, 2006, **33**, 127–137; *Chem. Abstracts*, 2007, **147**, 196375f.
50. P. A. Dabisch, E. A. Davis, J. A. Renner, E. M. Jakubowski, R. J. Mioduszewski and S. A. Thomson, *Inhalation Toxicol.*, 2008, **20**, 149–156.
51. A. W. Abu-Qare and M. B. Abou-Donia, *Toxicol. Lett.*, 2001, **123**, 51–58.
52. N. H. Williams, J. M. Harrison, R. W. Read and R. M. Black, *Arch. Toxicol.*, 2007, **81**, 627–639.
53. J. Rendón-von Osten, A. Ortíz-Arana, L. Guilhermino and A. M. V. M. Soares, *Chemosphere*, 2005, **58**, 627–636.
54. M. Berglund, B. Lind, K. A. Björnberg, B. Palm, Ö. Einarsson and M. Vahter, *Environ. Health: A Global Access Science Source*, 2005, **4**, 20–30.
55. K. Murata, M. Dakeishi, M. Shimada and H. Satoh, *Nippon Eiseigaku Zasshi*, 2007, **62**, 949–959; *Chem. Abstracts*, 2007, **147**, 535729t.
56. B. K. Mandal, Y. Ogra, K. Auzai and K. T. Suzuki, *Toxicol. Appl. Pharmacol.*, 2001, **123**, 51–58.
57. H. Emons, Z. Sebesvari, K. Falk and M. Krachler, in Ref. 10, pp. 71–95.
58. S. M. Evans, E. Kerrigan and N. Palmer, *Mar. Pollut. Bull.*, 2000, **40**, 212–219.
59. M. Quintela, R. Barreiro and J. M. Ruiz, *Sci. Total Environ.*, 2000, **247**, 227–237.
60. P. E. Gibbs, *ICES Tech. Mar. Environ. Sci.* 1999, **24**, 1–29; *Chem. Abstracts*, 2000, **133**, 173084g.
61. R. J. H. Herbert, S. Bray and S. J. Hawkins, *Proc. Mar. Sci.*, 2000, **1**, 307–310.
62. M. M. Santos, C. C. Ten Hallers-Tjabbes, A. M. Santos and N. Vieira, *J. Sea Res.*, 2002, **48**, 217–223.
63. W. J. Shim, S. H. Kahng, S. H. Hong, N. S. Kim, S. K. Kim and J. H. Shim, *Mar. Environ. Res.*, 2000, **49**, 435–451.
64. H-h. Shi, C-j. Huang and W-y. Xie, *Haiyang Huanjing Kexue*, 2002, **21**, 37–41; *Chem. Abstracts,* 2003, **138**, 165000f.
65. S.-X. Zhu, Q-X. Dong, J.-H. Shen, X.-J. Yu, H. Du and C.-J. Huang, *Shengtai Xuebao*, 2005, **25**, 289–297; *Chem. Abstracts*, 2005, **143**, 64904b.
66. M. M. Santos, N. Vieira, M. A. Reis-Henriques, A. M. Santos, J. L. Gomez-Ariza and C. C. Ten Hallers-Tjabbes, *Environ. Int.*, 2004, **30**, 793–798.
67. R. N. Goldberg, A. Averbuj, M. Cledon, D. Luzzatto and N. S. Nudelman, *Appl. Organometal. Chem.*, 2004, **18**, 117–123.
68. T. J. Reitsema and J. T. Spickett, *Mar. Pollut. Bull.*, 1999, **39**, 280–284.
69. M. Rato, A. Sousa, R. Quintã, W. Langston and C. Barroso, *Environ. Toxicol. Chem.*, 2006, **25**, 3213–3220.
70. A. M. Limaverde, A. Rebello Wagener, M. A. Fernandez, A. deL. Scofield and R. Coutinho, *Mar. Environ. Res.*, 2007, **64**, 384–398.
71. J. Oehlmann, I. Ide, B. Bauer, B. Waterman, U. Schulte-Oehlmann, S. Liebe and P. Fioroni, *Texte-Umweltbundesamt*, 1998, **46**, 1–194; *Chem. Abstracts*, 1999, **131**, 218842w.

72. U. Schulte-Oehlmann, C. Bettin, P. Fioroni, J. Oehlmann and E. Stroben, *Ecotoxicol.*, 1995, **4**, 372–384.
73. J. Oehlmann, *ICES Techniques in Mar. Environ. Sci.*, 2004, **37**, 1–22.
74. A. Albalat, J. Potrykus, J. Pempkowiak and C. Porte, *Chemosphere*, 2002, **47**, 165–171.
75. R.-q. Yang, Q.-f. Zhou and G.-b. Jiang, *Chemosphere*, 2006, **63**, 1–8.
76. F. Yang, R. J. Maguire and Y. K. Chau, *Water Quality Res. J. Canada*, 2001, **36**, 805–814.
77. J. Yang, H. Harino, H. Liu and N. Miyazaki, *Bull. Environ. Contam. Toxicol.*, 2008, **81**, 164–168.
78. I. Takeuchi, S. Takahashi, S. Tanabe and N. Miyazaki, *Mar. Environ. Res.*, 2001, **52**, 97–113.
79. J. C. Martins, M. L. Saker, L. F. Olivia Teles and V. M. Vasconcelos, *Environ. Toxicol. Chem.*, 2006, **26**, 1987–1991.
80. Q.-F. Zhou, G.-B. Jiang and J.-Y. Liu, *Arch. Environ. Contamin. Toxicol.*, 2002, **42**, 332–337.
81. V. Lupšina, M. Horvat, Z. Jeran and P. Stegnar, *Analyst*, 1992, **117**, 673–675.
82. B. B. Panda, D. L. Das, M. Lenka and K. K. Panda, *Mutation Res.*, 1988, **206**, 275–279.
83. M. Válega, A. I. Lillebø, M. E. Pereira, I. Caçador, A. C. Duarte and M. A. Pardal, *Chemosphere*, 2008, **73**, 1224–1229.
84. N. Udzak, T. Zvonaric, Z. Kljakovic Gaspic, M. Horvat and A. Baric, *Sci. Total Environ.*, 2000, **261**, 61–68.
85. K. Haraguchi, T. Ando, M. Sato, C. Kawaguchi, T. Tomiyasu, M. Horvat and H. Akogi, *Sci. Total Environ.*, 2000, **261**, 75–89.
86. Z. Kljaković-Gaçspić, N. Odžak, I. Ujević, T. Zvonarić, M. Horvat and A. Barić, *Sci. Total Environ.*, 2006, **368**, 199–209.
87. M. Costa, E. Paiva and I. Moreira, *Sci. Total Environ.*, 2000, **261**, 69–73.
88. J. J. Hinton and M. M. Veiga, *Geochemistry: Exploration, Environment, Analysis*, 2002, **2**, 269–274.
89. C. R. Hammerschmidt and W. F. Fitzgerald, *Environ. Chem.*, 2008, **5**, 127–130.
90. C. Sanpera, R. Moreno, X. Ruiz and L. Jover, *Mar. Pollut. Bull.*, 2007, **54**, 691–96.
91. R. L. Hothem, B. S. Trejo, M. L. Bauer and J. J. Crayon, *Arch. Environ. Contam. Toxicol.*, 2008, **55**, 111–121.
92. W. G. Shriver, D.C. Evers, T. P. Hodgman, B. J. Macculloch and R. J. Taylor, *Environ. Bioindicators*, 2006, **1**, 129–136.
93. G. Blanvillain, J. A. Schwenter, R. D. Day, D. Point, S. J. Christopher, W. A. Roumillat and D. W. Owens, *Environ. Toxicol. Chem.*, 2007, **26**, 1441–1450.
94. N. Basu, A. M. Scheuhammer, S. J. Bursian, J. Elliot, K. Rouvinen-Watt and H. M. Chan, *Environ. Res.*, 2007, **103**, 130–144.
95. M. F. Wolfe, T. Atkeson, W. Bowerman, J. Burger, D. C. Evers, M. W. Murray and E. Zillioux, in *Ecosystem Responses to Mercury Contamination*, Ed. R. Harris, D. P. Krabbenhoft, R. Mason, M. W. Murray, R. Reash and T. Saltman, CRC Press, Boca Raton FL, USA, 2006, 123–190.
96. N. Mikac, M. Branica, Y. Wang and R. M. Harrison, *Environ. Sci. Technol.*, 1996, **30**, 499–508.

97. L. Normandin, G. Kennedy and J. Zayed, *Sci. Total Environ.*, 1999, **239**, 165–171.
98. M. A. Champ and P. F. Seligman, in Ref. 19, pp. 601–614.
99. M. E. Conti, in Ref. 20, pp. 47–79.
100. Q. Zhou, J. Zhang, J. Fu, J. Shi and G. Jiang, *Anal. Chim. Acta*, 2008, **606**, 135–150.
101. J. A. Stäb, M. Frenay, I. L. Freriks, U. A. T. Brinkman and W. P. Coffino, *Environ. Toxicol. Chem.*, 1995, **14**, 2023–2032.
102. M. Quintela, R. Barreiro and J. M. Ruiz, *Sci. Total Environ.*, 2000, **247**, 227–237.
103. M. F. Pessoa, A. Fernando and J. S. Oliveira, *Environ. Toxicol.*, 2001, **16**, 234–241.
104. C. Coray and S. M. Bard, *Water Qual. Res. J. Canada*, 2007, **42**, 111–122.
105. C. Bancon-Montigny, J.-L. Seidel, F. Brissaud and F. Elbaz-Poulichet, *J. Environ. Monitoring*, 2008, **10**, 638–647.
106. M. J. C. Harding and I. M. Davies, *J. Environ. Monit.*, 2000, **2**, 404–409.
107. P. Quevauviller, M. Astruc, R. Morabito, F. Sriese and L. Ebdon, *Trends Anal. Chem.*, 2000, **19**, 180–188.
108. M. Čejchanova, V. Spěváckova, K. Kratzner, K. Wranová, V. Spěvácek and B. Beneš, *Biol. Trace Elem. Res.*, 2008, **121**, 97–105.
109. D. A. Aremu, M. S. Madejczyk and N. Ballatori, *Environ. Health Perspect.*, 2008, **116**, 26–31.
110. S. M. Arnold, T. V. Lynn, L. A. Verbrugge and J. P. Middaugh, *Amer. J. Pub. Health*, 2005, **98**, 393–397.
111. *Ecosystem Responses to Mercury Contamination: Indicators of Change*, Ed. R. Harris, D. P. Krabbenhoft, R. Mason, M. W. Murray, R. Reash and T. Saltman, CRC Press, Boca Raton (FL, USA), 2006.
112. J. S. Kuwabara, Y. Arai, B. R. Topping, I. J. Pickering and G. N. George, *Environ. Sci. Technol.*, 2007, **41**, 2745–2749.
113. A. C. Tomazelli, L. A. Martinelli, F. J. Krug, D. Santos, I. Ruffini, P. B. de Camargo and M. Horvat, *J. Environ. Qual.*, 2007, **36**, 478–486.
114. M. S. Bank, J. R. Burgess, D. C. Evers and C. S. Loftin, *Environ. Monit. Assess.*, 2007, **126**, 105–115.
115. J. Burden, in *Molecules of Death*, Ed. R. H. Waring, G. B. Steventon and S. C. Mitchell, 2nd edn., Imperial College Press, London, 2007, pp. 209–231.
116. S. L. Bartelt-Hunt, D. R. U. Knappe and M. A. Barlaz, *Crit. Rev. Environ. Sci. Technol.*, 2008, **38**, 112–136.
117. D. Noort, H. P. Benschop and R. M. Black, *Toxicol. Appl. Pharmacol.*, 2002, **184**, 116–126.
118. H. N. Nigg and J. B. Knaak, *Rev. Environ. Contam. Toxicol.*, 2000, **163**, 29–111.
119. U. Green, J. H. Kremer, M. Zillmer and C. Moldaenke, *Environ. Toxicol.*, 2003, **18**, 368–374.
120. D. D. Richardson, B. B. M. Sadi and J. A. Caruso, *J. Anal. At. Spectrom.*, 2006, **21**, 396–403.
121. D. D. Richardson and J. A. Caruso, *Anal. Bioanal. Chem.*, 2007, **388**, 809–823.

122. J. E. Franz, M. K. Mao and J. A. Sikarski, Glyphosate: *A Unique Global Herbicide, ACS Monograph 189*, Amer. Chem. Soc., Washington, 1997, pp. 634–637.
123. J. O. Nriagu, in Ref. 25, pp. 1–26.
124. E. Amato, L. Alcaro, I. Corsi, C. Della Torre, C. Farchi, S. Focardi, G. Marino and A. Tursi, *Marine Biology*, 2006, **149**, 17–23.
125. A. Fidder, D. Noort, A. G. Hulst, L. P. A. de Jong and H. P. Benschop, *Arch. Toxicol.*, 2000, **74**, 207–214.
126. T. Bausinger and J. Preuss, *Bull. Environ. Contam. Toxicol.*, 2005, **74**, 1045–1052.
127. S. Hanaoka, E. Nagasawa, K. Nomura, M. Yamazawa and M. Ishizaki, *Appl. Organomet. Chem.*, 2005, **19**, 265–275.
128. M. Ishizaki, T. Yanaoka, M. Nakamura, T. Hakuta, S. Ueno, M. Komuro, M. Shibata, T. Kitamura, A. Honda, M. Doy, K. Ishii, A. Tamaoka and N. Shimojo, *J. Health Sci.*, 2005, 51, 130–137; *Chem. Abstracts*, 2006, **142**, 468729z.
129. M. Filella, Chapter 8 of this book.
130. D. B. Ramsden, in *Molecules of Death*, Ed. R. H. Waring, G. B. Steventon and S. C. Mitchell, 2nd Edn., Imperial College Press, London, 2007, pp. 415–422.
131. A. W. Hayes, *Principles and Methods of Toxicology*, 5th Edn., CRC Press, Boca Raton, (FL, USA), 2008, pp. 415–422.
132. A. L. J. Peter and T. Viraraghavan, *Environ. Int.*, 2005, **31**, 493–501.
133. L. Ralph and M. R. Twiss, *Bull. Environ. Contam. Toxicol.*, 2002, **68**, 261–268.
134. *Thallium in the Environment*, Vol. 29 of *Advances in Environmental Science & Technology*, Ed. J. O. Nriagu, Wiley, New York, 1998.
135. M. Wierzbicka, G. Szarek-Łukaszewska and K. Grodzińka, *Ecotoxicol. Environ. Safety*, 2004, **59**, 84–88.
136. O. F. Schedlbauer and K. G. Heumann, *Anal. Chem.*, 1999, **71**, 5459–5464.
137. O. F. Schedlbauer and K. G. Heumann, *Appl. Organomet. Chem.*, 2000, **14**, 330–340.
138. M. R. Twiss, B. S. Twining and N. S. Fisher, *Can. J. Fish. Aquat. Sci.*, 2003, **60**, 1369–1375.
139. B. S. Twining, M. R. Twiss and N. S. Fisher, *Environ. Sci. Technol.*, 2003, **37**, 2720–2726.
140. M. R. Twiss, B. S. Twining and N. S. Fisher, *Environ. Toxicol. Chem.*, 2004, **23**, 968–973.
141. J. M. Morgan, *Diss. Abs. Int. B.*, 1981, **41**, 2578.
142. C. Stevens, D. E. Powell, P. Makela and C. Karman, *Mar. Pollut. Bull.*, 2001, **42**, 536–543.
143. D. Graiver, K. W. Farminer and R. Narayan, *J. Polymers Environ.*, 2003, **11**, 129–136.
144. P. Boudjouk and J. Thomas, *Chem. Abstracts*, 2007, **146**, 483333s.
145. M. Nendza, *Mar. Pollut. Bull.*, 2007, **54**, 1190–1196.
146. M. Yoshimura and T. Morishita, *Chem. Abstracts*, 2006, **144**, 207378s.
147. C. P. Park and J. H. Park, *Chem. Abstracts*, 2007, **146**, 403744p.
148. B. P. Framroze, *Chem. Abstracts*, 2007, **146**, 516398.
149. S. Masaoka, *Chem. Abstracts*, 2007, **147**, 399902q.

150. X. Zhou, H. Okamura and S. Nagata, *Chemosphere*, 2007, **67**, 1904–1910.
151. *Van Nostrand's Scientific Encyclopedia*, 10th edn., Ed. G. D. Considine and P. H. Kulik, New York, 2008, pp. 675–679.
152. T. Arata, Y. Oyama, K. Tabaru, M. Satoh, H. Hayashi, S. Ishida and Y. Okano, *Environ. Toxicol.*, 2002, **17**, 472–477.
153. G.-C. Wang, J. Xiao, Y.-N. Lu, L. Yu, H.-B. Song, J.-S. Li, J.-R. Cui, R.-Q. Wang, F.-X. Ran and H.-G. Wang, *Appl. Organomet. Chem.*, 2005, **19**, 113–117.
154. U. von Recklinghausen, L. M. Hartmann, S. Rabieh, J. Hippler, A. V. Hirner, A. W. Rettenmeier and E. Dopp, *Chem. Res. Toxicol.*, 2008, **21**, 1219–1228.
155. *Phytoremediation: Transformation and Control of Contaminants*, Ed. S. C. McCutcheon and J. L. Schnoor, Wiley-Interscience, Hoboken (NJ, USA), 2003.
156. *Phytoremediation of Toxic Metals*, Ed. I. Raskin and B. D. Ensteyed, Wiley-Interscience, New York, 2000.
157. *Phytoremediation: Methods and Reviews*, Ed. N. Willey, Humana Press, Totowa (NJ, USA), 2007.
158. *Phytoremediation of Contaminated Soil and Water*, Ed. N. Terry and G. Banuelos, Lewis, Boca Raton (FL, USA), 2000.
159. U. Krämer, *Curr. Opinions Biotechnol.*, 2005, **16**, 133–141.
160. S. F. Doty, *New Phytologist*, 2008, **179**, 318–333.
161. K. Shah and J. M. Nongkynrih, *Biologia Plantarum*, 2007, **51**, 618–634.
162. B. Van Aken, *Trends Biotechnol.*, 2008, **26**, 225–227.
163. S. Eapen, S. Singh and S. F. D'Souza, *Biotechnol. Adv.*, 2007, **25**, 442–451.
164. S. L. Day, *New Phytologist*, 2008, **179**, 318–333.
165. *Managing Arsenic in the Environment*, Ed. R. Naidu, E. Smith., G. Owens, P. Bhattacharya and P. Nadebaum, CSIRO Publ., Enfield (NH, USA), 2006.
166. K. A. Francesconi and D. Kuehnelt, in Ref. 28, pp. 51–94.
167. R. D. Tripathi, S. Srivastava, S. Mishra, N. Singh, R. Tuli, D. K. Gupta and F. J. M. Maathuis, *Trends in Biotechnol.*, 2007, **24**, 158–165.
168. A. A. Meharg and J. Hartley-Whitaker, *New Phytologist*, 2002, **154**, 29–43.
169. M. J. Ruiz-Chancho, J. F. López-Sánchez, E. Schmeisser, W. Goessler, K. A. Francesconi and R. Rubio, *Chemosphere*, 2008, **71**, 1522–1530.
170. X. Y. Xu, S. P. McGrath, A. A. Meharg and F. J. Zhao, *Environ. Sci. Technol.*, 2008, **42**, 5574–5579.
171. O. P. Dhankher, A. C. P. Heaton, Y. Li and R. B. Meagher, in *Subsurface Contamination Remediation*, Ed. E. Berkey and T. Zachry, *ACS Symposium Series 904*, Amer. Chem. Soc., Washington, D.C., 2005, pp. 96–116.
172. M. A. Rahman, H. Hasegawa, K. Ueda, T. Maki, C. Okumara and M. M. Rahman, *Chemosphere*, 2007, **69**, 493–499.
173. H. Ze-Chun, H.-B. Chen, M. Lei, Y.-R. Liu and T.-D. Hu, *Environ. Sci. Technol.*, 2008, **42**, 5106–5111.
174. O. N. Ruiz, H. S. Hussein, N. Terry and H. Daniell, *Plant Physiol.*, 2003, **132**, 1344–1352.
175. A. C. P. Heaton, C. L. Rugh, N.-J. Wang and R. B. Meagher, *J. Soil Contam.*, 1998, **7**, 497–509.
176. S. P. Bizily, T. Kim, M. K. Kandasamy and R. B. Meagher, *Plant Physiol.*, 2003, **131**, 463–471.

177. H. S. Hussein, O. N. Ruiz, N. Terry and H. Daniell, *Environ. Sci. Technol.*, 2007, **41**, 8439–8446.
178. A. C. P. Heaton, C. L. Rugh, T. Kim, N. J. Wang and R. B. Meagher, *Environ. Toxicol. Chem.*, 2003, **22**, 2940–2947.
179. O. P. Dhanker, A. C. P. Heaton, Y. Li and R. B. Meagher, in *Subsurface Contamination Remediation*, Ed. E. Berkey and T. Zachry, ACS Symposium Series 904, Amer. Chem. Soc., Washington, DC, 2005, pp. 96–116.
180. D. Che, R. B. Meagher, C. L. Rugh, T. Kim, A. C. P. Heaton and S. A. Merkle, *In vitro Cell Dev. Biol.-Plant*, 2006, **42**, 228–234.
181. S. Lyyra, R. B. Meagher, T. Kim, A. C. P. Heaton, P. Montello, R. S. Balish and S. A. Merkle, *Plant Biotechnol. J.*, 2007, **5**, 254–262.
182. M. Czakó, X. Feng, Y. He, D. Liang and L. Márton, *Environ. Geochem. Health*, 2006, **28**, 103–110.
183. K. M. Carvalho, M. J. McGettigan and D. F. Martin, *J. Environ. Sci. Health, Part A*, 2001, **36**, 1403–1409.
184. D. J. Ashworth and G. Shaw, *Sci. Total Environ.*, 2006, **370**, 506–514.
185. G. Bañuelos, D. L. Leduc, E. A. H. Pilon-Smits and N. Terry, *Environ. Sci. Technol.*, 2007, **41**, 599–605.
186. D. B. Vickerman, J. T. Trumble, G. N. George, I. J. Pickering and H. Nichole, *Environ. Sci. Technol.*, 2004, **38**, 3581–3586.
187. G. Cruz-Jímenez, J. R. Peralta-Videa, G. de la Rosa, G. Meitzner, J. G. Parsons and J. L. Gardea-Torresdey, *Environ. Chem.*, 2005, **7**, 100–107.
188. J. Gao, A. W. Garrison, C. Hoehamer, C. S. Mazur and N. L. Wolfe, *J. Agric. Food Chem.*, 2000, **48**, 6114–6120.
189. X. Wang, N. Wu, J. Guo, X. Chu, J. Tian, B. Yao and Y. Fan, *Biochem. Biophys. Res. Comm.*, 2008, **365**, 453–458.
190. R. Hilderbrand, in Ref. 30, pp. 139–169.
191. S. O. Duke and S. B. Powles, *Pest Management Sci.*, 2008, **64**, 319–325.
192. G. Ciucani, H. Mosbæk and S. Trapp, *Environ. Sci. Pollut. Res.*, 2004, **4**, 267–272.
193. G. Ciucani, H. Mosbæk and S. Trapp, *Environ. Sci. Pollut. Res.*, 2005, **5**, 327–330.
194. J. Novak and S. Trapp, *Environ. Sci. Pollut. Res.*, 2005, **12**, 332–341.
195. H. Al-Najar, A. Kaschl, R. Schulz and V. Römheld, *Internat. J. Phytoremediation*, 2005, **7**, 55–67.
196. G. M. Gadd, *FEMS Microbiol. Rev.*, 1993, **11**, 297–316.
197. R. S. Oremland, C. W. Culbertson and M. R. Winfrey, *Appl. Environ. Microbiol.*, 1991, **57**, 130–137.
198. F. Baldi, F. Semplici and M. Filippelli, *Water, Soil, Air Pollut.*, 1991, **56**, 465–475.
199. M. Kiyono and H. Pan-Hou, *J. Health Sci.*, 2006, **52**, 199–204.
200. M. Narita, C. C. Huang, T. Koizumi, T. Yamagata and G. Endo, *Water Sci. Technol.*, 2000, **42**, 109–114.
201. E. Saouter, R. Turner and T. Barkay, in *Emerging Technology for Bioremediation of Metals*, Ed. J. L. Means and R. E. Hinchee, Lewis, Boca Raton, FL. USA, 1994, pp. 99–104.

202. A. O. Summers, *Ann. Rev. Microbiol.*, 1986, **40**, 607–634.
203. T. P. Begley, A. E. Walts and C. T. Walsh, *Biochem.*, 1986, **25**, 7192–7200.
204. J. G. Omichinski, *Science*, 2007, **317**, 205–206.
205. M. J. Moore, M. D. DiStefano, L. D. Zydowsky, R. T. Cummings and C. T. Walsh, *Acc. Chem. Res.*, 1990, **23**, 301–308.
206. T. P. Begley and S. E. Ealick, *Curr. Opinion Chem. Biol.*, 2004, **8**, 508–515.
207. J. K. Schaefer, J. Yagi, J. R. Reinfelder, T. Cardona, K. M. Ellickson, S. Tel-Or and T. Barkay, *Environ. Sci. Technol.*, 2004, **38**, 4304–4311.
208. M. Narita, T. Yamagata, H. Ishii, C.-C. Huang and G. Endo, *Appl. Microbiol. Biotechnol.*, 2002, **59**, 86–90.
209. H. Strasdeit, *Angew. Chem. Int. Ed.*, 2008, **47**, 828–830.
210. K. E. Pitts and A. O. Summers, *Biochem.*, 2002, **41**, 10287–10296.
211. J. G. Melnick and G. Parkin, *Science*, 2007, **317**, 225–227.
212. S. M. Miller, *Nature Chem. Biol.*, 2007, **3**, 357–358.
213. G. C. Benison, P. Di Lello, J. E. Shokes, N. J. Cosper, R. A. Scott, P. Legault and J. G. Omichinski, *Biochem.*, 2004, **43**, 8333–8345.
214. L. D. Rasmussen, C. Zawadsky, S. J. Binnerup, G. Øregaard, S. J. Sørensen and N. Kroer, *Appl. Environ. Microbiol.*, 2008, **74**, 3795–3803.
215. F. Cima, P. J. Craig and C. Harrington, in Ref. 9, pp. 101–149.
216. G. M. Gadd, *Sci. Total Environ.*, 2000, **258**, 119–127.
217. S. K. Dubey and U. Roy, *Appl. Organomet. Chem.*, 2003, **17**, 3–8.
218. J. E. Landmeyer, T. L. Tanner and B. Watt, *Environ. Sci. Technol.*, 2004, **38**, 4106–4112.
219. F. Suehiro, T. Kobayashi, L. Nonaka, B. C. Tuyen and S. Suzuki, *Microbial Ecol.*, 2006, **52**, 19–25.
220. G. I. Paton, W. Cheewasedtham, I. L. Marr and J. J. C. Dawson, *Environ. Pollut.*, 2006, **144**, 746–751.
221. J. E. Landmeyer, T. L. Tanner and B. E. Watt, *Environ. Sci. Technol.*, 2004, **38**, 4106–4112.
222. H. Inoue, O. Takimura, H. Fuse, K. Murakami, K. Kamimura and Y. Yamaoka, *Appl. Environ. Microbiol.*, 2000, **66**, 3492–3498.
223. Y. Yamaoka, H. Inoue and O. Takimura, *Appl. Organometal. Chem.*, 2002, **16**, 665–668.
224. Y. Yamaoka, H. Inoue, O. Takimura and S. Oota, *Appl. Organometal. Chem.*, 2001, **15**, 757–761.
225. H. Inoue, O. Takimura, K. Kawaguchi, T. Nitoda, H. Fuse, K. Murakami and Y. Yamaoka, *Appl. Environ. Microbiol.*, 2003, **69**, 878–883.
226. G.-X. Sun, W.-Q. Zhou and J.-J. Zhong, *Appl. Environ. Microbiol.*, 2006, **72**, 6411–6413.
227. G.-X. Sun and J.-J. Zhong, *Appl. Environ. Microbiol.*, 2006, **72**, 7264–7269.
228. A. S. Stasinakis, N. S. Thomaidis, A. Nikolaou and A. Kantifes, *Environ. Pollut.*, 2005, **134**, 431–438.
229. B. Nowack, *Water Res.*, 2003, **37**, 2533–2546.
230. B. J. Singh and A. Walker, *FEMS Microbiol. Rev.*, 2006, **30**, 428–471.
231. J. P. Quinn, A. N. Kulakova, N. A. Cooley and J. W. McGrath, *Environ. Microbiol.*, 2007, **9**, 2392–2400.

232. J. W. McGrath, N. G. Ternan and J. P. Quinn, *Lett. Appl. Microbiol.*, 1997, **24**, 69–73.
233. B. Schink, In *Metal Ions in Biological Systems. Biogeochemical Cycles of Elements*, Ed. A. Sigel, H. Sigel and R. K. O. Sigel, Taylor & Francis, Boca Raton FL USA, 2005, **43**, pp. 131–151.
234. E. D. Ingall, *Nature Geoscience*, 2008, **1**, 419–420.
235. D. M. Karl, L. Beversdorf, K. M. Björkman, M. J. Church, A. Martinez and E. F. DeLong, *Nature Geoscience*, 2008, **1**, 473–478.
236. S. H. Han, Y. H. Zhuang, H. X. Zhang, Z. J. Wang and J. Z. Yang, *Chemosphere*, 2002, **49**, 651–657.
237. G. L. Mendz, F. Mégraud and V. Korolik, *Arch. Microbiol.*, 2005, **183**, 113–120.
238. J. Lipok, T. Owsiak, P. Młynarz, G. Forlani and P. Kafarski, *Enzyme Microbial Technol.*, 2007, **41**, 286–291.
239. D. L. Siehl, L. A. Castle, R. Gorton and R. J. Keenan, *J. Biol.Chem.*, 2007, **282**, 11446–11455.
240. M. Vera, F. Pagliai, N. Guiliani and C. A. Jerez, *Appl. Environ. Microbiol.*, 2008, **74**, 1829–1835.
241. M. A. Adams, Y. Luo, B. Hove-Jensen, S.-M. He, L. M. van Staalduinen, D. L. Zechel and Z. Hia, *J. Bacteriol.*, 2008, **190**, 1072–1083.
242. R. M. Dawson, S. Pantelidis, H. R. Rose and S. E. Kotsonis, *J. Hazardous Materials*, 2008, **157**, 308–314.
243. R. Turpeinen, M. Pantsar-Kallio, M. Häggblom and T. Kairesalo, *Sci. Total Environ.*, 1999, **236**, 173–180.
244. E. Botes, E. Van Heerden and D. Litthauer, *S. African J. Sci.*, 2007, **103**, 279–282.
245. C. R. Lehr, E. Polishchuk, U. Radoja and W. R. Cullen, *Appl. Organometal. Chem.*, 2003, **17**, 831–834.
246. T. Maki, N. Takeda, H. Hasegawa and K. Ueda, *Appl. Organometal. Chem.*, 2006, **20**, 538–544.
247. R. Sierra-Alvarez, U. Yenal, J. A. Field, M. Kopplin, A. J. Gandolfi and J. R. Garbarino, *J. Agric. Food Chem.*, 2006, **54**, 3959–3966.
248. S. Khokiattiwong, W. Goessler, D. N. Pedersen, R. Cox and K. A. Francesconi, *Appl. Organometal. Chem.*, 2001, **15**, 481–489.
249. P. Pengprecha, M. Wilson, A. Raab and J. Feldmann, *Appl. Organometal. Chem.*, 2005, **19**, 819–826.
250. T. Maki, H. Hasegawa and K. Ueda, *Appl. Organometal. Chem.*, 2005, **19**, 231–238.
251. J.-H. Huang, F. Scherr and E. Matzner, *Water, Soil, Air Pollut.*, 2007, **182**, 31–41.
252. D. H. Chapman and D. B. Johnson, *Poult. Sci.*, 2002, **81**, 356–364.
253. M. Köhler, K. Hoffmann, F. Völsgen, K. Thurow and A. Koch, *Chemosphere*, 2001, **42**, 425–429.
254. Y. Zhang and W. T. Frankenberger Jr, *J. Environ. Qual.*, 2002, **31**, 1124–1128.
255. D. Graiver, K. W. Farminer and R. Narayan, *J. Polymers Environm.*, 2003, **11**, 129–136.
256. C. Gallert and J. Winter, *Water Res.*, 2002, **36**, 3130–3140.
257. *Fungi and Environmental Change*, Ed. J. C. Frankland, N. Magan and G. M. Gadd, Cambridge University Press, Cambridge (UK), 1996.

258. *Fungi in Bioremediation*, Ed. G. M. Gadd, Cambridge University Press, Cambridge (UK), 2001.
259. V. M. Dembitsky and T. Rezanka, *Plant Sci.*, 2003, **165**, 1177–1192.
260. Cs. Soeroes, N. Kienzl, I. Ipolyi, M. Dernovics, P. Fodor and D. Kuehnelt, *Food Control*, 2005, **16**, 459–464.
261. H. Hasegawa, Y. Sohrin, K. Seki, M. Sato, K. Norisuye, K. Naito and M. Matsui, *Chemosphere*, 2001, **43**, 265–272.
262. D. Kuehnelt, J. Lintschinger and W. Goessler, *Appl. Organometal. Chem.*, 2000, **14**, 411–420.
263. J. W. McGrath, F. Hammerschmidt and J. P. Quinn, *Appl. Environ. Microbiol.*, 1998, **64**, 356–358.
264. T. Krzyśko-Lupicka, W. Strof, K. Kubś, M. Skorupa, P. Wieczorek, B. Lejczak and P. Kafarski, *Appl. Microbiol. Biotechnol.*, 1997, **48**, 549–552.
265. C. E. DaCosta, J. V. Comasseto, I. H.-S. Crusius, L. H. Andrade and L. M. Porto, *J. Molec. Cat.B: Enzymatic*, 2007, **45**, 135–139.
266. C. H. Wu, T. K. Wood, A. Mulchandani and W. Chen, *Appl. Environ. Microbiol.*, 2006, **72**, 1129–1134.
267. S. P. B. Kamaludeen and K. Ramasamy, *Indian J. Microbiol.*, 2008, **48**, 80–88.
268. B. Ravit, J. G. Ehrenfeld and M. M. Häggblom, *Soil Biol. Biochem.*, 2005, **37**, 1049–1057.
269. A. Lee, Z.-Q. Lin, I. J. Pickering and N. Terry, *Planta*, 2001, **213**, 977–980.
270. P. Wongwathanarat and K. Sivasithamparam, *Biol. Fertil. Soils*, 1991, **11**, 13–17.
271. M. A. Siddiqi and M. Alexander, *Biol. Fertil. Soils*, 1991, **11**, 141–146.
272. R. P. Ryan, D. Ryan and D. N. Dowling, *Lett. Appl. Microbiol.*, 2007, **45**, 668–674.

Met. Ions Life Sci. **2010**, *7*, 465–521

14

Methylated Metal(loid) Species in Humans

Alfred V. Hirner[a] *and Albert W. Rettenmeier*[b]

[a]Institute of Analytical Chemistry, University of Duisburg-Essen, D-45117 Essen, Germany
<alfred.hirner@uni-due.de>

[b]Institute of Hygiene and Occupational Medicine, University of Duisburg-Essen, D-45122 Essen, Germany
<albert.rettenmeier@uni-due.de>

Metal Ions in Life Sciences, Volume 7 Edited by Astrid Sigel, Helmut Sigel, and Roland K. O. Sigel

Published by the Royal Society of Chemistry, www.rsc.org DOI: 10.1039/9781849730822-00465

ABSTRACT: While the metal(loid)s arsenic, bismuth, and selenium (probably also tellurium) have been shown to be enzymatically methylated in the human body, this has not yet been demonstrated for antimony, cadmium, germanium, indium, lead, mercury, thallium, and tin, although the latter elements can be biomethylated in the environment. Methylated metal(loid)s exhibit increased mobility, thus leading to a more efficient metal(loid) transport within the body and, in particular, opening chances for passing membrane barriers (blood-brain barrier, placental barrier). As a consequence human health may be affected. In this review, relevant data from the literature are compiled, and are discussed with respect to the evaluation of assumed and proven health effects caused by alkylated metal(loid) species.

KEYWORDS: alkylated species · biomethylation · humans · metabolism · metal(loid) species · methylated species · toxicology

1. INTRODUCTION

From a biogeochemical point of view a relatively good correlation between the elemental distributions in human serum and seawater [1], particularly for the siderophile and lithophile elements, may not be too surprising and could support the hypothesis that life originates in the ocean [2]. A closer look at this correlation reveals, however, that chalcophile elements (with affinity to sulfur) such as the biologically essential elements zinc, copper, and selenium are enriched relatively to seawater by a factor of up to 5000 [1]. In mammals, these elements and others like iron, molybdenum, or nickel (with affinity to sulfur) are constituents of metalloenzymes and -proteins fulfilling a great variety of important biological functions. In many cases they are interlinked with their high molecular weight organic rest (mass in the kDa range) via coordination to sulfur.

To study biochemical systems with respect to metal(loid)s present, the chemical form of these elements (i.e., elemental speciation) must be known. For such a "metal-assisted functional biochemistry" the term "metallomics" complementary to the already existing fields of genomics, proteomics, and metabolomics has been introduced [1]. Extremely specific and sensitive speciation methods must be available to cope with this important task.

Within the last two decades many sophisticated instrumental techniques for qualitative as well as quantitative analytical metal(loid) speciation in biological matrices have been developed (e.g., [1,3–9]). These instrumental analytical speciation methods are most often based on chromatographic separation followed by on-line detection of the structural composition (usually by electrospray mass spectrometry (ESI-MS) for the identification of the analyte's structure) and of the elemental composition (usually by inductively coupled plasma mass spectrometry (ICP-MS) for the quantification of the analyte element). Common procedures in chromatography are gas and liquid chromatography (GC and HPLC) and capillary and gel electrophoresis (CE and GE). However, analytical aspects will not be discussed in this chapter, the reader is instead referred to the cited literature.

This review will focus on a dozen of metal(loid)s which can be enzymatically methylated in ecosystems including human beings. Methylated metal(loid) species are volatile, amphiphilic, and able to complex with various sulfur-containing peptides and proteins. Thus, they are usually not only more mobile and toxic than their inorganic counterparts [10], they may also play a role as epigenetic factors by interfering with other known important methylation processes in the body such as DNA and histone methylation [11].

For the first time, we will provide comprehensive information with respect to methylated metal(loid)s in the human body. Data on the stability of these species, their disposition and transport within the body following exposure as well as the toxicological consequences thereof will be summarized. Metal(loid) species with longer alkyl chains exhibiting similar properties and toxic effects are only mentioned if appropriate. These industrially produced compounds are only important as exposure factors because metabolic conversion of metal(oid)s to longer-chain alkyl derivatives within the body has not been proven as yet.

2. EXPOSURE OF HUMANS TO ALKYLATED METAL(LOID)S

Numerous alkylated organometal(loid) species are known to occur in the environment [4,10,12–14]. While organic derivatives of arsenic, lead, mercury, selenium, tellurium, and tin with longer-chain alkyl or with aryl residues are usually of anthropogenic origin (e.g., ethyllead, butyltin or phenyl-mercury), methylated species of these elements and additionally of antimony, bismuth, cadmium, germanium, indium, and thallium may also be generated in biological systems (Figure 1). The preferential way of formation of the latter is assumed to be biomethylation (i.e., enzymatic methylation in bacteria and fungi). Generally, alkylation of metal(loid)s increases mobility and toxicity when compared to the respective properties of the corresponding inorganic species [10]. While fully alkylated metal(loid) species are volatile, due to their amphiphilic character partly alkylated species are water- as well as lipid-soluble and, therefore, can accumulate in organisms. An example is the accumulation of monomethylmercury in fish.

Exposed humans receive fully and partly alkylated metal(loid)s via inhalation and ingestion. As detailed in the following sections, methylated species may also be generated by enzymatic methylation in liver, kidneys, and

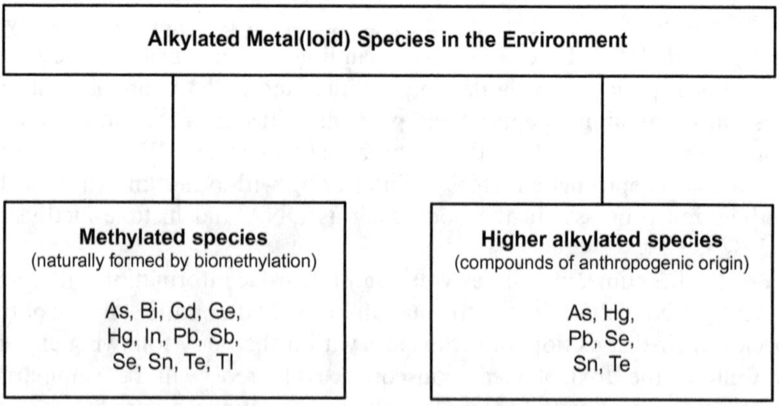

Figure 1. Metal(loid)s found in the environment as alkylated compounds. Compilation based on refs [4,12,14].

Table 1. Concentration (μg/L) of metal(loid)s with proven methylation potential in the environment in the blood of humans and seals.

Metal(loid)	Humans (Bremen FRG)	Humans (France)	Seals (Wadden Sea)	Biomethylation in humans
As	0.1–4	3–18	42–592	+ +
Bi	<0.01–0.02	0.001–0.007		+
Cd	0.1–4	0.1–2	<0.1–1.1	?
Ge		11–20		?
Hg	0.02–16	0.9–8		?
In	<0.01–0.02			?
Pb	5–83	11–63	<0.02–4.5	?
Sb	<0.01–0.1	0.05–0.13		(+)
Se	85–182	89–154	518–2261	+ +
Sn	0.02–0.8	0.1–1.8	<0.06–0.5	?
Te	<0.14	0.11–0.45		(+)
Tl	<0.01–0.05	0.01–0.04		?

Compiled from refs [15–17].

colon (as shown for arsenic, selenium, and bismuth). The potential metal(loid) candidates for undergoing this type of alkylation are those transported in blood and the digestional tract. In Table 1, mammalian blood concentrations of those metal(loid)s which can be methylated in the environment are listed. For comparative purposes, blood concentration ranges of these metal(loid)s obtained from individuals in Germany and in France and, exemplarily, from harbour seals are also presented [15–17].

With the exception of tellurium, the metal(loid) concentrations measured in the German and the French study are within a similar range and are also overlapping with the concentrations determined in blood samples of South African school children (average values for arsenic, lead, and selenium are 1.5, 56, and 176 μg/L, respectively) [18]. Average lead concentrations in blood of school children vary between 13 μg/L (Sweden) and 166 μg/L (Jamaica, urban environment). The metal(loid) concentrations presented in Table 1 exceed in part national reference values. This may be exemplarily illustrated by the German reference values derived for lead in blood of females (70 μg/L), and, regardless of gender, for cadmium (1 μg/L) and mercury (2 μg/L), respectively [91].

Compared to humans harbor seals from the Wadden Sea exhibit lower lead and similar cadmium and tin levels in blood, whereas arsenic and selenium blood concentrations are higher by more than one order of magnitude. Therefore, it was proposed to use seal blood to monitor environmental contamination with metal(loid)s [17].

When the concentrations presented in Table 1 are compared with those determined in Germany and Italy in 1990, the environmental contamination with the above mentioned metal(loid)s is currently lower than at that time, in particular, the contamination with arsenic, lead, and thallium [16]. As expected, cadmium concentrations in blood of smokers are significantly higher than those of non-smokers (geometric means 0.67 and 0. 29 μg/L, respectively). Also, a positive correlation exists between the mercury concentration in blood and the number of amalgam fillings in the teeth (geometric means of mercury blood levels in individuals with and without amalgam fillings are 1.6 and 1.0 μg/L, respectively). With regard to arsenic blood levels there are differences between seafood and non-seafood eaters (geometric means 1.2 and 0.5 μg/L, respectively).

The data in Table 1 indicate that the metal(loid) concentrations in human blood decrease in the order Se > Pb > Ge > As > Hg > Cd > Sn > Te > Sb > Tl > Bi > In. In viewing the potential of the endogenous enzymes to methylate these metal(loid)s, a few aspects have to be considered: For example, it might be extremely difficult to differentiate between an endogenously methylated lead component and a methyllead background arising from the much more abundant anthropogenic sources [10]. Also, reasonable doubts exist about the analytical quality of the germanium data cited in Table 1. (In other extended compilations germanium concentrations are not even mentioned (see e.g., [1]).

If such aspects are taken into account, of all metal(loid)s with proven biomethylation potential in the environment, the only two metal(loid)s being able to perform enzymatic methylation in the human body are among the most abundant metal(loid)s in human blood (arsenic and selenium). Bismuth and likely antimony and tellurium, the other candidates in this respect, are of very low abundance, and the rate and mechanism of their methylation are not yet completely (bismuth) or not at all (antimony and tellurium) known (see below).

There are still no reports on the biomethylation in humans of all the other metal(loid)s listed in Table 1. This holds true even for mercury which is one of the best studied elements in this series and of which the demethylation process has been investigated in detail (see below).

3. DISPOSITION AND TRANSPORT OF METHYLATED METAL(LOID)S IN THE HUMAN BODY

As detailed above, methylated metal(loid) species present in the human body may originate both from external sources and from enzymatic methylation within the body. Nevertheless, appreciable data on the biodisposition (absorption, distribution, metabolism, elimination) of methylated (alkylated) metal(loid) species are only available for arsenic, bismuth, lead, mercury, selenium, and tin. None or only scattered data have been published on the biodisposition of methylated species of antimony, cadmium, germanium, indium, thallium, and tellurium.

3.1. Antimony

External exposure of humans, particularly of landfill and sewage plant workers, to methylated antimony compounds may occur due to the well documented ability of bacteria and fungi to transform inorganic antimony compounds into methylated species [10]. However, studies on the uptake of methylated or other alkylated antimony species by humans have not been performed to date, most likely due to the presumed low toxicity of these species [20]. Respective studies have not even been initiated after Richardson had proposed the "toxic gas hypothesis" as a possible cause of the sudden infant death syndrome (SIDS) [21,22]. As one of the numerous attempts to explain this syndrome, the "toxic gas hypothesis" conveys that microorganisms growing on infants' cot bedding material containing particularly antimony (as a fire retardant) among other elements convert these compounds into volatile toxic species. Evidence of this hypothesis has not been provided to date [23].

Internal exposure to methylated antimony compounds may not only arise from the intake of these species from external sources but also from enzymatic methylation of inorganic antimony within the body. An indication of the latter is the detection of methylated antimony species in urine samples of workers exposed to antimony during the production of batteries and in urine samples of a group of individuals randomly selected from the general population. In the urine samples of the workers trimethylantimony dichloride (Me_3SbCl_2) was detected in a concentration of 0.4–0.57 μg/L, whereas the respective concentrations in the urine samples of two non-exposed individuals were 0.036–0.09 μg/L. The urinary concentrations of tri- and pentavalent antimony in the workers were $<$0.025–0.15 μg/L and 2.0–5.9 μg/L, those of the control persons were $<$0.025 μg/L and $<$0.06 μg/L, respectively [24]. Background concentrations of monomethylantimony and dimethylantimony are in the range of 1.1–4.4 ng/L and 0.9–2.8 ng/L, as measured by Stang et al. [25] in urine samples of 32 not specifically exposed individuals. In contrast to these observations, it was concluded from studies in rats and from a case study of a woman who attempted to commit suicide by the ingestion of an unknown amout of antimony trisulfide that in mammals, unlike arsenic, biomethylation of antimony does not occur [26,27].

If biomethylation of antimony does occur in the human body, it is likely that it proceeds via a mechanism similar to that proposed by Challenger for arsenic. This assumption is based on the following observations: (i) the redox potentials for antimony and arsenic are similar; (ii) the trivalent antimony compounds are much more readily biomethylated than the pentavalent ones; antimony(V) is not reduced in cultures of *Scopulariopsis brevicaulis*; (iii) both di- and trimethylantimony species are found in the medium of cultures of *S. brevicaulis*; and (iv) the methyl group of the methylantimony species produced after the addition of $^{13}CD_3$-L-methionine to cultures of *S. brevicaulis* and potassium antimony tartrate was labelled to approx. 50% [10].

Whether biomethylation of antimony in humans also occurs by the action of bacteria in the human gastrointestinal tract is not known as yet (see discussion in [28]). Similarly to bismuth antimony compounds are poorly absorbed from the gastrointestinal tract, which fosters the exposure of these compounds to the intestinal microflora.

Nothing is known about the transport and half-life of methylated antimony species in the blood and the organ distribution of these compounds.

3.2. Arsenic

The methyl derivatives of arsenic are the most thoroughly investigated compounds among the methylated metal(loid) species when it comes to biodisposition and toxicity. The high interest in arsenic methylation has basically two causes: One is that larger populations in certain areas of the world (e.g., Bangladesh, Taiwan, and Chile) are highly exposed to arsenic due to the geogenic contamination of water and food [29]; the other is the finding that in contrast to previous assumptions some of the methylated arsenic derivatives may seriously contribute to the toxic, in particular to the carcinogenic effects of this metalloid [30]. Since about ten years, a large body of data on the exposure to arsenic and on the toxic properties of arsenic species has been published. Hence, a separate chapter in this book is devoted solely to arsenic to cope with this wealth of information (see E. Dopp et al., Chapter 7).

The following paragraph on the biodisposition of methylated arsenic compounds and the paragraph further below on the toxic properties give just brief summaries of the most important aspects of arsenic methylation and toxicity.

In contrast to water consumption from which arsenic is almost exclusively received in form of its inorganic salts, both inorganic and organic arsenic species are ingested with food. The chemical nature of the arsenic species in food depends on the source: Seafood contains the highest arsenic amount and this mainly in form of arsenobetaine and arsenocholine (marine animals) or in form of arsenosugars (seaweed). Organic arsenic also predominates in fruit and vegetables, whereas meat, poultry, dairy products, and cereals mainly contain arsenic in its inorganic forms [31]. From a toxicological point of view the source of arsenic is important: While most ingested organic arsenic compounds ($MMAs^V$, $DMAs^V$, arsenobetaine, but not arsenoriboses) are less extensively metabolized and more rapidly excreted in urine than the inorganic arsenic species [32,33], the latter, albeit toxic themselves, undergo biotransformation to potentially even more toxic methylated derivatives ($MMAs^{III}$, $DMAs^{III}$) [30].

Following intake, $MMAs^V$ and $DMAs^V$ levels in blood were generally below the limit of detection as long as seafood is not a major constituent of the diet [34,35]. If the latter is the case, $DMAs^V$ and even trimethylated arsenic can be detected in serum [36,37]. $DMAs^V$ has also been found in serum samples of patients with terminal kidney insufficiency [34,35].

The cellular uptake of organic arsenic compounds has been extensively studied by Dopp et al. [38,39]. It appears from these studies that a high methylating capacity of cells favors the degree of uptake and that the trivalent methylarsenic species are more membrane-permeable than the respective pentavalent ones [38,39]. The formation of glutathione complexes seem to play an important role in membrane permeation, in particular alleviating the efflux into the extracellular space [40,41].

Inorganic arsenic and probably also arsenoriboses are extensively metabolized to three- and pentavalent methylarsenic species in liver and kidney. As pointed out by Dopp et al. (see Chapter 7) both the metabolic routes and the role of biotransformation in arsenic toxicity are currently under intensive discussion. Biotransformation products of arsenic are $MMAs^{III}$, $MMAs^V$, $DMAs^{III}$, and $DMAs^V$, whereby $DMAs^V$ and $MMAs^V$ are the major metabolites excreted in urine. Trimethylarsine oxide (TMAsO) has also been found in trace amounts in urine samples after arsenosugars have been consumed [42,43]. Thiolated methylarsenicals, another group of metabolites shown recently to be formed by red blood cells and the liver [44,45], may result from the substitution of oxygen by sulfur subsequently to methylation.

The transfer of the methyl group from the donor *S*-adenosylmethionine (SAM) is accomplished by the catalytic action of arsenite methyltransferase (AS3MT). Varying gene sequences of human As3MT has been considered responsible for the different sensitivity following arsenic exposure [46]. In contrast, dose, age, gender, and smoking seem to contribute only to a negligible extent to the large interindividual variation in arsenic methylation observed in humans [29].

Two mechanisms of arsenic methylation are currently discussed: (i) the mechanism proposed by Challenger in 1945 which involves a series of reductions of pentavalent to trivalent arsenic species and the subsequent oxidative methylation with the sulfur atom of SAM as redox partner [47]; and (ii) methylation of the glutathione-bound trivalent arsenic species without oxidation, a mechanism based on the observation that arsenic-glutathione complexes are preferred substrates for methylation [48]. Independent of the question which of the two mechanisms better reflects reality, glutathione seems to play a role in the methylation of arsenic, probably by reduction of cysteine residues in AS3MT as suggested by Thomas et al. [49].

Methylated arsenic compounds cannot only be formed in liver and kidney, these species may also be produced by microorganisms in the human intestine. Evidence for the potential of bacteria in the gut to methylate inorganic arsenic compounds has been obtained from animal studies [50–52]; and only recently, trimethylarsine, arsine, and hitherto undescribed volatile sulfur-containing arsenic compounds have been discovered in a human colon model [364]. The ability of intestinal microorganisms to metabolize arsenobetaine has also been demonstrated recently [53].

The excretion of elevated amounts of arsenate, $MMAs^{V}$, and $DMAs^{V}$ following consumption of prawn containing arsenic almost exclusively in a trimethylated form indicates that demethylation can also occur [36].

As mentioned above, the major arsenic metabolites in urine are $DMAs^{V}$ and $MMAs^{V}$ (to a lesser degree), which are eliminated in addition to inorganic arsenic. Also dimethyldithioarsinic acid ($DMDTAs^{V}$) and monomethylmonothioarsonic acid ($MMMTAs^{V}$) are regularly found in urine samples of arsenic-exposed humans and animals [53,54]. In several publications the detection of trivalent methylated arsenic metabolites has also been reported. In one paper it was even suggested that $MMAs^{III}$ could serve as an indicator in urine to identify individuals with increased susceptibility to toxic and cancer-promoting effects of arseniasis [55]. Therefore, toxicologists focussed their attention on studies performed during the last five years in which the presence of $MMAs^{III}$ and $DMAs^{III}$ in urine samples of humans exposed to high concentrations of inorganic arsenic (mostly via drinking water) [55–64] and of rats [65,66] were reported. Because of the immense importance of such analytical results, a critical evaluation of the techniques and argumentations used in these studies was needed. The outcome of several critical reviews was that many of the published results seem to be questionable [67,68]. For example, it is unrealistic to report the detection of $DMAs^{III}$ in over two months old urine samples from West Bengal [61,63], while the stability of this arsenic species has been reported not to exceed one day [69]. The same compound has been identified in urine samples from central Mexico [55,62]. Although in this case the samples had been analyzed within six hours after collection, this study was also critisized because it was not strictly differentiated between free and glutathione-complexed $DMAs^{III}$ [67]. This raises the question in general if we know enough about the stability of the latter species in respect to various biochemical binding partners in native blood. In contrast, the existence of the more stable species $MMAs^{III}$ in urine samples of children from Brazil has been demonstrated unequivocally by a multi-step analytical approach [4,64].

3.3. Bismuth

In the environment, the volatile bismuth compound trimethylbismuth has been found to be formed at a relatively high rate by the microflora of anaerobic sewage sludge, even from low concentrations of inorganic bismuth [70] (see also Chapter 9). Here, methylcobalamin is thought to serve as methyl donor in this enzyme-catalyzed methylation. The formation of trimethylbismuth and bismuth trihydride by *Methanobacterium formicicum*, one of the bacteria present in sewage sludge, has also been experimentally demonstrated [71,72]. Some of the microorganisms in sewage sludge are known components of the intestinal microflora in humans.

External exposure to methylated bismuth compounds might affect workers employed in sewage plants or people living nearby such installations. The general population is exposed to bismuth basically in form of inorganic and organic bismuth salts which – due to the presumed low toxicity of these salts – are used as cosmetics and as pharmaceutical products [73]. While the treatment of syphilis and malaria are examples of historical bismuth applications, gastrointestinal disorders such as peptic ulcers are now the major domain of the therapeutic use of bismuth salts.

Dietary bismuth intake by the general population is estimated to be 5 to 20 μg per day. Bismuth absorption from the gastrointestinal tract or when applied to the skin is usually poor, e.g., less than 1% of an oral dose of the three compounds used clinically: colloidal tripotassium dicitrato bismuthate, bismuth subsalicylate, and bismuth citrate [74]. In blood, bismuth associates with plasma proteins and erythrocytes. Bismuth compounds are readily hydrolyzed in aqueous solutions and show a high affinity to sulfur, but also to oxygen and nitrogen. Thus, complexes with both mono- and dianionic thiolate-carboxylate ligands can be formed [75]. Complexes with cysteine, glutathione (GSH), albumin (HSA), lacto- and transferrin, and metallothioneins (MTs) have been detected. It is assumed that ionic bismuth binds specifically to transferrin in preference to albumin [76]. The organ with the highest content has always been found to be the kidney, a likely result of its capacity to induce the expression of metallothionein. In contrast, after intake of trimethylbismuth the concentration of bismuth in the liver was higher than in the kidney, probably due to hydrophobic interactions of the organic ligand [77]. Renal as well as biliary excretion have been reported [78–80]. Absorbed bismuth is excreted rapidly in urine, most of it within one day [81].

The cellular uptake of monomethylbismuth, bismuth citrate (Bi-Cit), and bismuth glutathione (Bi-GS)] was investigated in human hepatocytes, lymphocytes, and erythrocytes [82]. The methylated bismuth compound was better taken up by the cells than Bi-Cit and Bi-GS. All intracellularly detected bismuth species were located in the cytosol of the cells (36% in erythrocytes, 17% in lymphocytes, and 0.04% in hepatocytes). The apparent lower intracellular concentration of bismuth in hepatocytes may be explained by an inhibition of uptake or by the presence of an enhanced efflux mechanism in these cells as described also for arsenic compounds in bacteria, yeast, and mammalian cells [83,84].

The biotransformation and elimination of bismuth have been studied *in vivo* in a pilot study [85] and in a larger volunteer study following ingestion of colloidal bismuth subcitrate (CBS; 215 mg bismuth) as a single oral dose [86].

The bismuth concentration in blood typically increased to a maximum within the first hour following ingestion and subsequently decreased with half-lives of approx. 1.6 ± 0.7 hrs. The rapid appearance of bismuth in blood after oral intake suggests that bismuth can be absorbed from the stomach [87].

Significant variations in the maximum blood bismuth concentrations were observed between the individuals with bismuth concentrations ranging from 1 to 159 μg/L. $68 \pm 16\%$ of the absorbed bismuth were excreted in the first twelve hours after ingestion, mostly with the first urine after ingestion. The maxima of the fecal bismuth concentrations ranged from 0.06 to 2.36 g/kg (wet weight) amounting to a total excretion of typically more than 99% of the ingested bismuth. However, in an accompanying study it was found that only 91–93% of the ingested bismuth are eliminated via feces within five days after ingestion [88]. Thus, some bismuth might be deposited in the body.

Trace levels of the metabolite trimethylbismuth have been detected in blood and in exhaled air samples. Respective concentrations were in the range of up to 2.5 ng/L (blood) and 0.8–458 ng/m^3 (exhaled air; calculated for an average respiratory volume of 0.5 m^3/h).

The high variability observed in bismuth methylation may be either due to a gene polymorphism similar to that found for arsenic methylation in humans [89] or to a varying composition of the intestinal microflora which has been shown to methylate bismuth *ex situ* [90,91].

Although trimethylbismuth in breath was detectable for the first time about two to four hours after ingestion, maximum concentrations were reached after eight hours in most of the study participants. The concentration-versus-time profiles of trimethylbismuth in blood were similar to the corresponding profiles of trimethylbismuth in exhaled air. Also, other volatile methyl and hydride species such as $(CH_3)_2BiH$, CH_3BiH_2, and BiH_3

were detected in exhaled air [85]. These organic hydrides are original to the sample (i.e., no derivatization artefacts) presumably as a product of biohydridization [92]. In addition, trimethylbismuth and (even more) CH_3BiX_2 (counterion X unidentified) were found in blood samples [85].

The data allow the estimation of the elimination routes of bismuth in exhaled air (up to 0.03‰), urine (0.03–1.2%), and feces (>98%). The site of trimethylbismuth production could not be identified in the present study, but the intestinal microflora seems to be involved in this biotransformation if accompanying *ex vivo* studies are taken into consideration: Anaerobic incubation of feces samples obtained from volunteers following ingestion of bismuth demonstrated that intestinal microorganisms are able to methylate bismuth *ex vivo* [85,90,91]. Finally, a strong indication that microbial methylation takes place *in vivo* was the detection of significant amounts of trimethylbismuth in freshly collected feces [88].

However, biomethylation in the colon may not be the sole relevant process, as trimethylbismuth occurs in exhaled air as early as two hours after bismuth ingestion. This points to a relatively rapid methylation process such as enzymatic methylation in the liver. Since the transport into the intestine normally requires more time, it is unlikely that intestinal microorganisms account for trimethylbismuth production during this early period. Moreover, similar time profiles as observed in the present study for trimethylbismuth have been found for the methylated arsenic derivatives which are formed in the liver [93,94].

Though even an abiotic methylation of bismuth by methylcobalamin cannot be ignored [72], two scenarios of bismuth methylation in the human body appear to be the most plausible ones:

(i) A microbial pathway with participation of microorganisms present in the intestine. The evidence obtained from animal studies [50–52] and from a human colon model [364] that bacteria in the gut have the potential to methylate inorganic metal(loid) species, in combination with the fact that bismuth is mainly excreted via feces, strongly supports the hypothesis that methylation of bismuth takes place in the human intestine. After microbial volatilization of trimethylbismuth in the colon, this species diffuses into the blood and is then transferred to the lungs, from where it is exhaled.

(ii) An endogenous enzymatic pathway, in particular in the human liver, as described for arsenic and other elements [89,95], cannot be ruled out. To shed more light upon this potential mechanism, human hepatocytes were exposed to different bismuth species. In the course of these experiments it was found that bismuthcysteine was able to penetrate the cell membrane and was methylated within the cell (Figure 2; [365]).

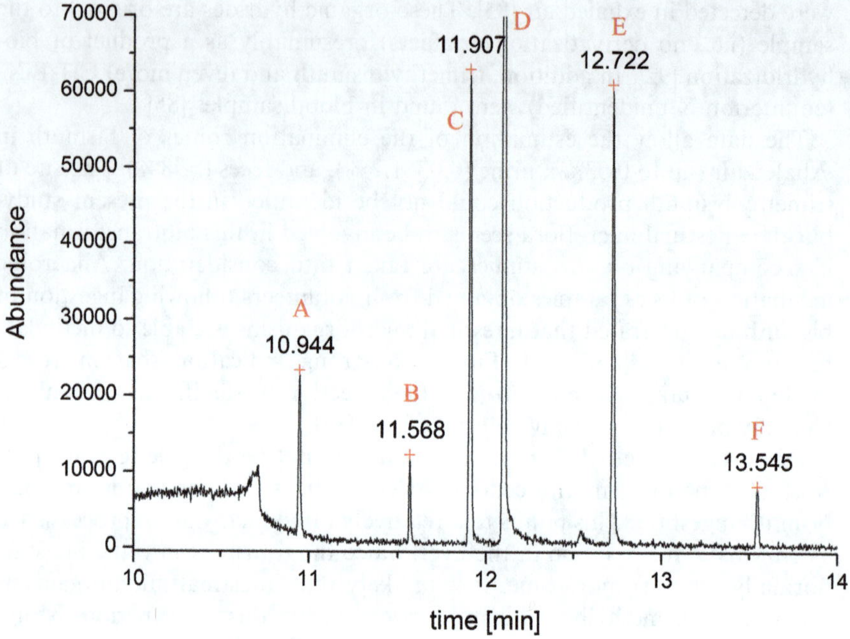

Figure 2. Mass spectrum of ethylated Bi species in HepG2 cells incubated with Bi cysteine. Peak B represents diethylmonomethylbismuth, and peak D triethylbismuth, while peaks A,C, E, and F are siloxanes from an antifoam additive.

In summary, the study of Boertz et al. [86] represents the first *in vivo* study on bismuth biodisposition in humans which includes the analysis of a volatile bismuth species. In addition, the study provides data on total bismuth uptake and elimination which basically confirmed the results of previous studies on bismuth biodisposition [81,87,96].

3.4. Cadmium

Dimethylcadmium occurs only at low concentrations in the environment likely due to its instability in water. Some external exposure of humans may occur because the use of this compound in the semiconductor industry [10].

Very little is known about the biotransformation of inorganic cadmium into organocadmium compounds. It was demonstrated years ago, however, that methylcadmium can be produced if methylcobalamin reacts in an aqueous solution with inorganic cadmium [97].

3.5. Germanium

Though organogermanium compounds are extensively used in the semiconductor industry, no information is available on human exposure to these compounds and their fate in the human body. The highly volatile tetramethylgermanium (boiling point 44.3°C) is commercially available.

The biomethylation of Ge^{IV} to CH_3Ge^{3+}, $(CH_3)_2Ge^{2+}$, and $(CH_3)_3Ge^+$ has been observed under anoxic conditions in the presence of anaerobic bacteria [98].

3.6. Indium

No data on the exposure of humans to methylated indium species and on the biodisposition of these compounds are available.

3.7. Lead

After the prohibition of gasoline containing lead or lead additives as antiknock agents in the 1970s, the external exposure to alkylated lead compounds sharply declined. Until then, the tetraalkylated lead compounds were known to be one of the largest volumes of organic compounds being produced [99] (see also Chapter 5).

The tetraalkyllead compounds, basically tetramethyl- and tetraethyllead, are highly volatile and well lipid-soluble and, thus, are readily absorbed by inhalation and dermal penetration. In an inhalation study with volunteers 51% of the $(CH_3)_4{}^{203}Pb$ inhaled by drawing 10–40 breaths of air containing the compound in a concentration of $1\,mg/m^3$ were absorbed [100]. The absorption of tetramethyllead by the dermal route has been estimated to be approx. 6% [101]. An accident involving transdermal absorption of tetramethyllead has been reported [102]. Due to its lower lipophilicity, the dermal penetration of tetramethyllead is slower than that of tetraethyllead [103].

The absorbed methylated lead species is distributed via the blood over the entire body, but the parent compound and the intermediate dealkylated products are distributed differently according to their lipophilicity. The half-life of methylated lead in blood was found to be 13 seconds [100]. After exposure to tetraethyllead, the highest concentrations of the parent compound and its metabolites, including inorganic lead, have been found in liver and kidneys followed by brain and heart [104].

As with all tetraalkyllead compounds and independent on the route of absorption metabolic degradation of tetramethyllead occurs by cytochrome P450-catalyzed oxidative dealkylation in the liver leading to the formation of the trimethyl, dimethyl, and inorganic lead species [104–108]. The latter is eventually stored in the bones [109]. In rats, the production of the toxic trialkyllead metabolite appears to be fairly rapid (in the order of hours), while the production of the subsequent metabolites is much slower (in the order of weeks). Following inhalation exposure, exhalation of tetraalkyllead compounds is a major pathway of elimination in humans. According to Heard et al. 40% of an inhaled dose of tetramethyllead initially deposited in the lung were exhaled within 48 hrs. The daily elimination via urine and feces was 0.2% [100].

3.8. Mercury

In addition to the incorporation of elemental mercury from amalgam fillings in teeth today's most widespread exposure to mercury is associated with organic species of this element: methylmercury in edible tissues of fish, and ethylmercury as a preservative in vaccines [110]. Health effects of mercury exposure are mainly determined by its chemical form, the dose, the exposure route, and host factors (age, genetic disposition, environmental, and in particular, nutritional aspects). The latter are responsible for different responses to similar doses [111]. While chelators can remove methylmercury and ethylmercury from the body, they cannot reverse the damage to the central nervous system; they may prevent further detoriation, however [112].

A compact overview of the current use, exposure, and clinical manifestations of everyday and accidental use of organic (alkylated) mercury in our societies is given by Clarkson et al. [112] (see also Chapter 12). A synopsis on chelators like DMPS, DMSA, ALA, DHLA, NAC, and GSH and a critical discussion (including the chelation challenge test) can be found elsewhere [113].

3.8.1. Alkylated Mercury Species

Methylmercury cysteine is considerably less toxic than the closely related compound methylmercury chloride, since the Hg-Cl bond is largely covalent and remains intact even in dilute aqueous solutions. Whether the acidic and high chlorine conditions in the human stomach may convert methylmercury cysteine or other methylmercury species to methylmercury chloride, is still a matter of discussion [114]. This points to the question, if methylmercury chloride is a suitable candidate for methylmercury toxicity testing.

Dimethylmercury is rapidly absorbed through the skin even if latex gloves are worn [115]. Tests with disposable latex and vinyl gloves in a new developed permeation cell have shown that already a diluted dimethylmercury solution penetrates these gloves within 15 seconds or less. Nitrile gloves protect from penetration only up to 100 seconds depending on the thickness of the material [366].

When compared to methylmercury, relevant alkylated mercury species such as ethylated and phenylated mercury exhibit lower stability in the human body. Particularly because of the relatively weak C-Hg bond, phenylmercury rapidly decomposes and releases inorganic mercury. Due to its accelerated metabolism ethylmercury appears to be less toxic than methylmercury [116]. Used as a preservative, ethylmercury in form of the water-soluble sodium ethylmercury thiosalicylate (thiomersal) is contained in relatively high concentrations (approx. 10 mg/L) in many commercial products of human plasma, immunoglobulines, and vaccines [117]. Thiomersal rapidly decomposes in the body and releases ethylmercury. Its toxicity is generally regarded as being low, although allergic reactions occur, and symptomatic and even fatal poisonings have been reported. Last but not least in regard to human contact with organic mercury, merbromin (mercurochrome), formerly used as a topical antiseptic for minor skin injuries, has to be mentioned. It is rapidly cleared into the urine, and its accidental ingestion is usually associated with minimal toxicity.

3.8.2. *Thioorganic Ligands*

Based on empirical data it has been proposed [118] that wherever a methylmercury compound has been identified in biological media, it was complexed to –SH-containing ligands. Yet methylmercury rapidly redistributes when novel sulfhydryl groups become available.

These observations can be deeper explained in scientific terms: In general, mercury and its species are known to have a high affinity to reduced sulfur. Methylmercury tends to form 1:1 complexes with thiol-containing small molecules such as GSH and cysteine as well as with the sulfhydryl groups of proteins (in a similar way, mercuric ions form 1:2 complexes). In the living organism, however, these complexes may be labile under certain circumstances as a result of thiol or nucleophilic exchange reactions. The reason for this high importance of sulfur is that affinity constants for thiolate anions are about ten orders of magnitude higher than for O- or N-containing ligands like carboxyl or amino groups [119,120].

In particular, most ionic mercury species are bound to sulfhydryl-containing proteins such as albumin, the most abundant plasma protein, which has a free sulfhydryl group in a terminal cysteinyl residue. Mercury species are transferred from plasma proteins to small molecular weight thiols (glutathione and cysteine) by complex ligand exchange mechanisms. Quantitatively, mercury is bound to albumin in an order of 99% [121]. Thus, the transportable species methylmercury cysteine accounts for less than 1% of plasma mercury and for an even lower proportion of mercury in whole blood.

The amphiphilic methylmercury is apparently able to cross membrane barriers like the placenta or the blood-brain barrier, eventually producing neurological effects. A special pathway through the membrane via the large amino transporter (LAT) system has been proposed for methylmercury cysteine complexes functioning as structural analogs of the essential amino acid methionine ("molecular mimicry") [119,120]. However, a closer look at the L-cysteine/cystine-Hg(II) complexes with the aid of computational chemistry and XANES falsified a detailed mimicry model. Instead, mechanisms involving a less specific mimicry based on structural similarities in amino acid stereochemistry were proposed [122]. While complexes of methylmercury with L-cysteine and D,L-homocysteine but not with D-cysteine, *N*-acetyl-L-cysteine, penicillamine, or GSH have been shown to be substrates for the human L-type large amino acid transporters LAT1 and LAT2 [123], animal experiments have demonstrated the potential of organic anion transporter systems (OAT1 and other OATs) in the renal epithelial transport of *N*-acetylcysteine-*S*-conjugates of methylmercury [124].

3.8.3. *Transport*

Repeated or chronic administration of subtoxic doses of methylmercury increases the intracellular renal and brain content of GSH and the expression of mRNA for γ-glutamylcysteine synthetase, the rate limiting enzyme in GSH synthesis [119,120,125–128]. Methylmercury increased the expression of γ-glutamylcysteine synthetase mRNA specifically in cerebellar and hippocampal regions which are known to be resistant to methylmercury-induced injury [128]. Thus resistance in these brain regions may reflect the ability of specific neuronal cell types to upregulate GSH synthesis.

Like with inorganic mercury, biliary secretion of methylmercury also occurs as the GSH complex. Depletion of hepatic GSH content also decreases the rate of methylmercury efflux into bile [129,130]; most of the biliary methylmercury is in the form of a CysSH-Gly conjugate [131]. Thus, in general, thiol complexes of methylmercury are likely to be processed in the same manner as those of inorganic mercury [132].

During the passage down the biliary tree the methylmercury-glutathione complex is extracellularly hydrolyzed by γ-glutamyl transpeptidase and dipeptidase enzymes releasing methylmercury as a complex with cysteine and homocysteine for reabsorption into the blood [133]. Thus, two main cellular transport mechanisms seem to exist for methylmercury: one for the entry into the cell as a complex with cysteine and homocysteine on the large neutral amino acid carriers, and the other for the exit from the cell as a complex with glutathione on the endogenous glutathione carriers [110,134].

3.8.4. *Metabolism*

Adult men receiving methylmercury excrete 30% of the dose in the feces within 70 days, whereas only about 4% are excreted in the urine [135]. Elimination by feces is also the major excretion path of total mercury. Methylmercury has a long retention time in blood (about 40 to 70 days in adults and as short as one week in infants [111,112]). Its concentration in erythrocytes is about twenty times higher than that in plasma. Therefore, in cases in which mercury concentrations in blood are significantly elevated (e.g., in the mg/L range) while urinary mercury levels are relatively normal, methylmercury may be the cause. Reference values for the general population are in the range of <10 to 20 µg/L, both for mercury in blood and urine [111]. Methylmercury has been shown to react with an AsSe-glutathione complex, and it has been speculated that this species may be formed inside the erythrocytes [136].

Another way to eliminate methylmercury from blood is via uptake in growing hair. Methylmercury concentrations in hair are proportional to the respective concentrations in blood, but are 250 times higher [137]. Keratin is synthesized in hair follicular cells and possesses many cysteine residues that provide ample binding sites and a stable storage for the transported methylmercury [138]. To understand how methylmercury gains entry into the hair follicle is important, as head hair is the most widely used biological indicator for methylmercury exposition. If the same entry mechanism operates for hair follicular cells as has been shown for the endothelian cells of the blood-brain barrier, brain and hair concentrations will be correlated [137]. Consistent with these processes, mercury levels in maternal hair in a population of fish consumers correlate to a high degree with levels in the brain of newborn infants [139].

About 95% of the methylmercury in food are absorbed from the gastrointestinal tract (GI) and are transported via blood to the liver. Methylmercury absorption and disposition should be completed within thirty hours to three days with 5% and 10% ending up eventually in blood and brain [137,140]. Methylmercury undergoes enterohepatic cycling with excretion in bile, reabsorption from the GI tract, and by portal circulation return to the liver [111]. During reabsorption from the GI tract, methylmercury comes into contact with the intestinal microflora which is able to break the C-Hg bond and converts methylmercury to inorganic mercury [141]. This is a rather slow process, probably advancing at a rate of about 1% of the body burden a day [137]. Some demethylation also occurs in phagocytic cells. The underlying biochemical mechanism is still not fully understood, but demethylation in the gut might well constitute an important site for the interaction between diet (e.g., fiber content) and methylmercury accumulation in the body [142].

Methylmercury is also converted to inorganic mercury in the brain [137]. It is possible that the inorganic ion is the ultimate toxic agent responsible for the brain damage. However, experiments on rats comparing methyl- and ethylmercury suggest that the intact methylmercury radical might be the toxic agent [143]. This is in accordance with the observation that in the adult brain methylmercury accumulates in astrocytes and interferes with the glutamate uptake, resulting in high extracellular glutamate concentrations which neurons may not tolerate [118].

Nevertheless, inorganic species account for most if not all of the remaining mercury in the brain of autopsy samples [144]. Therefore, it has been suggested that inorganic mercury released in brain tissue from methylmercury may be the ultimate toxic agent. The long-term stability of this species has however not been discussed [145]. For a more detailed discussion of this issue see a recent review [113].

3.8.5. Nutritional Cofactors

Because of the various biological ligands existing for methylmercury, it is of prime importance to know the methylmercury speciation in fish. XANES spectra of mercury in fish closely resemble only the spectrum of methylmercury cysteine or structurally related species [114]. Thus, cysteine is by far the most likely candidate as the predominant biological thiol, though it is probably part of a peptide (e.g., glutathione) or protein. The advantage of methylmercury cysteine of being of low toxicity is however counterbalanced by its ability to penetrate into brain.

Zinc and selenium have been shown to exert protective effects against mercury toxicity, most likely by the induction of metallothionein and selenoprotein P [113]. Methylmercury does not directly induce MT, but does so upon metabolism to inorganic mercury. Expression of both selenoprotein P and glutathione peroxidase was greatly increased in mercury-exposed persons [146]. These increases were accompanied by elevated selenium concentrations in serum. Selenoproteins play two important roles in protecting against mercury toxicity: First, they may bind more mercury through their highly reactive selenol group, and second, their antioxidative properties help to eliminate the reactive oxygen species induced by mercury *in vivo*.

Selenium and mercury co-accumulation in humans and other mammals is well known [147] and is probably caused by the formation of biologically inert Hg-Se compounds. Selenium and mercury could form Hg-Se complexes in a reducing environment and this 1:1 complex is then bound to plasma selenoprotein P [148].

Diets based on tuna of high mercury content can be fed for long periods without toxic effects in cats and other animals [149]. There is sufficient selenium in tuna to confer protective effects when high enough levels of methylmercury are added to diets to induce toxicity. Vitamin E has a close nutritional relationship with selenium and can decrease methylmercury toxicity when ingested at supranutritional levels. Methylmercury metabolism to a non-toxic Hg-Se complex that accumulates in liver appears to be facilitated in cats, fed tuna compared to those fed pike, out of proportion to the difference in selenium content of the diets. In mice exposed to methylmercury, a 30% bran diet increased the rate of mercury elimination from the body and reduced the amount of mercury in brain [142]. It was proposed that fibers in the diet interrupt the enterohepatic circulation by binding mercury, thus leading to an increased rate of mercury elimination [150]. Using *in vitro* digestion it could be demonstrated that co-consumption of food containing phytochemicals and mercury-containing fish may potentially reduce mercury absorption compared to eating fish alone [151]. Also, other studies seem to point to dietary fibers as potentially enhancing the elimination of methylmercury from the body [152].

3.9. Selenium

Selenium is ingested by humans mainly in form of water-soluble inorganic compounds or as organic derivatives such as selenomethionine (in vegetable products) and selenocysteine (in animal products) [152–158], but exposure may also happen via the dermal route or by inhalation (see also Chapter 10). The absorption is dependent on the selenium status: the higher the selenium content of the daily diet the lower the selenium absorption [159].

It is assumed that the absorbed selenium compounds are reduced to the intermediate selenide which serves as a common source for the synthesis of selenoproteins and selenosugars [160]. In the human genome, 25 genes for selenoproteins have been identified: Examples are glutathione peroxidases, thioredoxin reductases, iodothyronine deiodinases, and selenoprotein P. The functions of the selenoproteins are only partly known [161]. In contrast to selenoproteins, selenosugars are excretion products of selenium. Three different selenosugar species have been identified in human urine samples as yet [162–167].

The intermediate selenide is not only metabolized to selenoproteins and selenosugars but also to methylated derivatives such as monomethylselenide, dimethylselenide, and the trimethylselenonium ion. Donor of the methyl group is *S*-adenosylmethionine [168] which is inducible by organic and inorganic selenium compounds [169]. The formation of dimethylselenide is catalyzed by a microsomal thiol-*S*-methyltransferase [170] and that of the trimethylselenonium ion by a cytosolic thioether-*S*-methyltransferase [171].

Monomethylselenide, the suspected biologically active selenometabolite responsible for the antioxidant activity of selenium, is considered an important intermediate: On the one hand, it can be further methylated to dimethylselenide and the trimethylselenonium ion, on the other hand it is a degradation product of methylselenocysteine and methylseleninic acid which can be subsequently demethylated to selenide [160]. The transformation of methylselenocysteine, a naturally occurring edible product, and of methylseleninic acid, an oxidation product of selenosugar, into monomethylselenide proceeds readily via β-lyase and reduction reactions. Studies in rats indicate that methylselenocysteine is more stable and more efficiently distributed than methylseleninic acid and, therefore, it might be the best monomethylselenide source in most organs [172]. *In vitro* experiments with simultaneous incubation of ^{77}Se-methylseleninic acid and ^{82}Se-selenite in a red blood cell suspension suggest that selenosugars and the trimethylselenonium ion are produced depending on the capacity to convert monomethylselenide to selenide [173].

Based on animal experiments it has been proposed in earlier publications that monomethylselenol is the main metabolite at low dosage (0.1 mg/kg body weight), whereas the trimethylselenonium ion is formed with increasing dose in a dose-dependent manner [174,175] and dimethylselenide only at toxic doses [171,176]. This view is no longer justified given the results of more recent studies. Following the ingestion of a single oral dose of 300 μg ^{77}Se in form of selenite by a volunteer, 11.2% of the compound were found as dimethylselenide in the expired air, and 18.5% of the dose were excreted in urine in form of selenium-containing compounds within ten days after dosage. Most of the dimethylselenide was exhaled within the first two days after application [177]. Using improved HPLC/ICP-MS techniques monomethylselenide has not been found anymore in urine and its detection in the former studies has been ascribed to the use of insufficient analytical procedures [164,178]. To the contrary, the presence of the trimethylselenonium ion has been confirmed, though this metabolite is usually excreted only in trace amounts. There is a marked individual variability in the levels of this metabolite in human urine, and in some individuals it can even be the major urinary elimination product [179]. Apart from the analytical issues there is now general agreement that selenosugars are normally the most important metabolic products of selenium eliminated in urine [162,163,165,167,179,180].

3.10. Tellurium

Although dimethyltelluride is known for a long time as garlic-like odor of mine workers (mistaken as "bismuth breath"), and biomethylation of tellurium by bacteria has been demonstrated in experimental studies, the respective mechanism in humans is not known and analytical species validation is still lacking [10] (see also Chapter 10). According to earlier investigations methylation of tellurium proceeds slowly, and dimethyltelluride is eliminated by exhalation and perspiration and via feces [181–183]. It appears from animal studies that only residual tellurium is metabolized to dimethyltelluride [184,185] which effluxes into the bloodstream and accumulates in red blood cells [186]. Excretion of tellurium in rat urine is in form of trimethyltelluronium [186].

3.11. Thallium

There are no data on human exposure to methylated thallium compounds. One reason for his lack of occurence might be the instability of the trimethylated thallium species.

3.12. Tin

Mono- and dimethyltin compounds are widely distributed in the environment due to anthropogenic entries [187,188] and as a result of microbial transformation (see also Chapter 4). Approximately 5% of the total tin in some rivers in the US and in Germany are present in form of methylated species [189]. One explanation for the high occurence of methylated tin compounds in ports is the degradation of tributyltin and the subsequent biomethylation of the resulting inorganic tin species [190]. The environmental contamination by methylated tin compounds seems to be declining in recent years, however [191].

External exposure of humans to methylated tin compounds may arise from industrial use of mono- and dimethylated tin species, e.g., as stabilizers for PVC. Trimethylated tin species are of minor importance, probably due to the high toxicity, yet these compounds may be present in mono- and dimethyltin preparations (e.g., in mercaptotin acetates) as contaminants [192,193]. In most cases mono- and dimethyltin compounds are produced as mixtures, particularly as intermediates for the synthesis of other methyltin compounds such as methyltin tris(2-ethylhexylmercaptoacetate) and methyltin 2-mercaptoethyltallate [194]. Dimethyltin chloride is also used to improve the quality of glass surfaces. Mono- and dimethyltin compounds are usually produced in closed facilities to prevent release into the environment. Exposure may occur during manual operations such as addition of materials, transport, and collection of samples. For example, if temperatures reach 180 °C-200 °C during the processing of polyvinylchloride, the polymer can decompose and the tin stabilizer can react with released hydrogen chloride causing the formation of small quantities of mono- and dimethyltinthioester chlorides [192,193,195]. In American and Canadian PVC-processing plants, organotin concentrations in the air near extruders ranged from below 0.0001 to up to 0.034 mg/m^3 and during manual operations (e.g., blending) with the tin stabilizer from below 0.0001 to up to 0.102 mg/m^3 (results are reported as total tin) [194].

In view of the large number of methyltin compounds and their mixtures and the lack of data on the individual species, the results obtained for the mono- and dimethyltin species from biodisposition and toxicological studies are presented together. This generalization seems to be justified, since the biological activity of organic tin compounds is mainly determined by the alkyl groups and only to a lesser extent by the ligands. Furthermore, many of the tin-sulfur-bonds present in alkylated tin compounds are hydrolyzed under physiological conditions. This is particularly true if the compounds are incorporated orally. Marked differences in toxicity depending on the ligands may, however, occur following inhalation of or dermal exposure to these compounds.

Methylated tin compounds can be taken up by inhalation, orally, or by dermal penetration. As with other tin compounds, absorption is dependent on the solubility in the physiological media. The better soluble methyltin compounds are better absorbed than the less well soluble higher molecular alkyl- and aryltin compounds [196,197]. Absorption decreases with increasing degree of alkylation.

There are no quantitative data on the exposure to methyltin compounds by inhalation. Evidence of this exposure route comes from reports on strong neurotoxic effects in individuals accidentally exposed to vapors containing trimethyltin species [198–203]. Likewise, no quantitative data are available on the absorption of methyltin compounds from the gastrointestinal tract which appears to be dependent on the ligands. Indications of gastrointestinal absorption are again severe neurological symptoms and even fatalities following intake of methylated tin compounds either accidentally [204] or by unknowingly using organotin-contaminated lard as cooking oil [205]. Evidence of gastrointestinal absorption has likewise been obtained from two-generation studies in animals: Dimethyltin dichloride is much more rapidly absorbed from drinking water than inorganic tin resulting in higher tin concentrations in blood and brain of fetuses. This also shows that the organic tin compound readily crosses the placental barrier, in contrast to inorganic tin which is transferred to the progeny only to a minor extent [206,207].

Dermal exposure to methyltin compounds cause mainly local reactions. If a mixture of dimethyltin dichloride and monomethyltin trichloride (89%:11%; 100 $\mu g/cm^2$) was applied to human epidermis *in vitro*, the maximum absorption rates were 0.015 $\mu g/cm^2/h$ (occlusive) and 0.006 $\mu g/cm^2/h$

(non-occlusive) and the portions absorbed from the applied doses were 1.4% (occlusive) and 0.25% (non occlusive), respectively. The corresponding numbers for the application of a mixture of dimethyltin 2-ethylhexylmercapturic acid and monomethyltin 2-ethylhexlmercapturic acid (100 μL/cm^2) were 0.018 μg/cm^2/h (occlusive) and 0.007 μg/cm^2/h (non-occlusive), respectively, and <0.001% (non-occlusive) and 0.001% (occlusive), respectively [193]. Penetration of dimethyltin compounds through human skin obviously proceeds very slowly.

Following ingestion and depending on their physical and chemical properties methyltin compounds are distributed rapidly in the organs where these compounds reach concentration maxima after different periods of time. In animal studies, the highest tissue concentrations were normally measured in the liver.

Cellular uptake of methyltin compounds was investigated in CHO-9 cells (concentration in medium 0.5 μmol). After an incubation period of one hour dimethyltin dichloride was taken up best, followed by trimethyltin chloride. Monomethyltin trichloride was poorly membrane-permeable, tetramethyltin was not taken up at all. The uptake rate increased with increasing concentration but was relatively enhanced at lower extracellular concentrations. An association of the methyltin compounds to membranes was not observed [208]. According to Arakawa and Wada mono- and dimethyltin compounds are not selectively distributed in the Golgi apparatus and the endoplasmatic reticulum, contrary to dibutyltin compounds. They rationalized this difference by a different affinity to intracellular lipids and lipophilic proteins [209].

There are no data on the biological half-life of methyltin compounds in humans. Following the application of a single dose of 3 mg trimethyltin/kg (1.8 mg tin/kg) to rats the half-life in blood was approx. three days and in brain approx. two days or less [210].

Methylated tin compounds like all alkyltin species are metabolized in the liver by successive oxidative dealkylation catalyzed by microsomal monooxygenases [211]. This metabolic degradation slows down with increasing length of the alkyl chain.

No quantitative data are available on the excretion of methyltin compounds. In general, organic tin compounds are eliminated via bile and feces and to a lesser extent in urine.

4. TOXICOLOGY OF METHYLATED METAL(LOID)S

4.1. Genotoxicity/Carcinogenicity

Half of the 12 metal(loid)s (see Section 3) of which the methylated derivatives are characterized in this chapter, are classified as being carcinogenic or possibly carcinogenic: the International Agency for Research on Cancer (IARC) specifies arsenic [212] and cadmium [213] in carcinogen group 1 (carcinogenic to humans), lead compounds [214] in group 2A (probably carcinogenic to humans), antimony trioxide [215] and mercury and its compounds [216] in group 2 B (possibly carcinogenic to humans), and antimony trisulfide [215] and selenium and its compounds [217] in group 3 (not classifiable as to their carcinogenicity in humans). A similar categorization is made by the German Commission for the Investigation of Health Hazards of Chemical Compounds in the Work Area [218]. Among these metal(loid)s selenium plays a specific role in that it exhibits possibly carcinogenic properties (at high doses) on the one hand and, on the other hand, has been proposed (at lower doses) as dietary supplement with anticancer effects. However, very recently the anticarcinogenic properties of selenium have been seriously challenged by intermediary results of two major epidemiological studies which indicated that selenium supplementation does not decrease cancer risk (see below).

The role of the alkylated and in particular of the methylated derivatives in the ascertained or potential carcinogenic activity of the metal(loid)s in question is largely unknown. There are only a few epidemiological studies in which the carcinogenic risk of humans has been assessed in relation to the intake or the endogenous formation of methylated metal(loid) compounds. And even animal studies on the carcinogenicity of alkyl derivatives of metal(loid)s are scarce. In contrast to the *in vivo* situation quite a few studies have been performed *in vitro* to better understand the role of metal(loid) alkylation and in particular of methylation in the processes leading to cancer.

In general, metal(loid) genotoxicity and carcinogenicity are caused by indirect mechanisms, whereby three mechanisms seem to predominate: (i) induction of oxidative stress, which may cause oxidative DNA damage or trigger signalling cascades leading to the stimulation of cell growth; (ii) inhibition of major DNA repair systems resulting in genomic instability and accumulation of critical mutations; (iii) deregulation of cell proliferation by induction of signalling pathways or inactivation of growth controls such as tumor suppressor genes [219].

In this context, alterations of gene activity, based on phenotypic and not on genotypic differences, named epigenetics [11] deserve a closer look. Epigenetic events participate in the normal process of cell differentiation and phenotype development, but they also contribute to the growth of tumors, e.g., of gastrointestinal neoplasmas [220]. A primary molecular mechanism in epigenetics is the alteration of the chromatin structure by covalent DNA modification, in particular DNA methylation, and histone acetylation: Genes are inactivated when the chromatin is condensed, and expressed when it is opened. Gene-specific hypermethylation is generally involved in the deactivation of tumor suppressor genes, whereas hypomethylation leads to the activation of genes important for cancer development [11].

Unlike the genome, epigenetic processes can be influenced by the environment, the diet or by pharmaceuticals [221]. For example, in New Zealand it is intended to lower the colon cancer incidence by raising the levels of nutrients and phytochemicals by dietary supplementation to positively affect the DNA methylation status [222]. As the change in DNA methylation is affected by the exposure to certain metal(loid)s, elements such as nickel and chromium but also arsenic can be considered epigenetic factors [223].

4.1.1. Arsenic

In contrast to previous assumptions that methylation of arsenic is a detoxification pathway recent *in vitro* studies have indicated that the trivalent methylated metabolites $MMAs^{III}$ and $DMAs^{III}$ are equally or even more genotoxic than the inorganic arsenic species [224–227] and, thus, may contribute to the carcinogenic activity of arsenic. As with the previous section on the biodisposition of arsenic a detailed presentation and discussion of the potential role of methylated metabolites in arsenic-induced genotoxicity and carcinogenicity are given in Chapter 7 of this book. Here, only some findings are summarized.

Methylated arsenic metabolites have been shown to act as mitotic poisons [228,229] and to induce DNA single-strand breaks [230] and sister chromatid exchanges (SCE) [231]. The different types of chromosome damage observed in exposed cells [232,233] suggest that the genetic alterations are likely caused by different mechanisms [225,234–236]. In most genotoxicity assays $MMAs^{III}$ and $DMAs^{III}$ are more potent than inorganic arsenic (both As_i^{III} and As_i^{V}) and the pentavalent methylarsenic species [52,94,224,225,232,237–239]. A strong clastogenic effect including the induction of cell cycle arrest and aneuploidy has also been found in cultured cells exposed to thiodimethylarsinate and dithiodimethylarsenate, arsenic metabolites recently discovered in urine of humans [43,53,54,240]. Volatile arsenic species, potentially generated by bacteria in the human gut, could also contribute to the genotoxic effects of arsenic as indicated by *in vitro* studies and studies in experimental animals [241–243].

The induction of oxidative stress, diminished DNA repair, altered DNA methylation patterns, enhanced cell proliferation, and suppression of p53 have been suggested as mechanisms underlying the genetic damage induced by methylated arsenic species [244]. Particularly the generation of reactive oxygen species seems to play an important role [225,227,239,245,246]. Supportive of this assumption are the $DMAs^{V}$-induced depletion of cellular glutathione and the inhibition of detoxifying enzymes such as glutathione reductase by $MMAs^{III}$ and $DMAs^{III}$ [235], but also the weak SCE induction by these compounds in combination with their potent clastogenicity and cytotoxicity. The oxygen radicals can induce single-strand breaks which may be converted to double-strand breaks, if there is only scant time for DNA repair (by proliferative regeneration) or the repair is inhibited by arsenic [219]. Impairment of DNA repair caused by release of zinc [247], decreased poly(ADP-ribosyl)ation [247], or inhibition of relevant proteins [248,249] have been demonstrated in cells exposed to trivalent methylated arsenicals. Thus, the ability of methylated arsenicals to induce DNA damage and, at the same time, to inhibit DNA repair can lead to the fixation of mutations necessary for cancer induction [219].

There is accumulating evidence from cell culture studies, studies in experimental animals, and also from arsenic-exposed humans that arsenic also alters the DNA methylation pattern and thereby affects the expression of oncogenes and tumor suppressor genes. Interestingly, both hypo- and hypermethylation have been observed. For example, increased cytosine methylation in the p53 promotor was detected in A549 cells, and hypermethylation with the consequences of diminished gene expression of tumor suppressor genes such as p16Nk4a and RASSF1A were found in arsenic-exposed A/J mice [250]. With respect to humans, a dose-dependent hypermethylation of the p53 gene was observed in blood samples of arsenic-exposed skin cancer patients in West Bengal [251]. The underlying mechanisms are still unclear. While hypomethylation may be due to inhibition of DNA-(cytosine-5) methyltransferase as in the instance of cadmium [252] or to the depletion of S-adenosylmethionine, a common cofactor in DNA methylation and arsenic methylation, hypermethylation is not easily understood. Further studies are required to resolve this question. An interesting aspect is in this context that the most important methyl donor for methylation of arsenic, of DNA, and of histone is SAM, regenerated from *S*-adenosylhomocysteine via the methionine cycle. As for the latter compounds selenium-containing analogs exist (SeAM, SeAH, Se-methionine), selenium is also interlinked in these biomethylation processes [11].

4.1.2. Cadmium

As mentioned in the section on biodisposition (3.4), it is uncertain whether methylated cadmium compounds are generated in humans. There are also no data from animal and *in vitro* studies on the genotoxic or carcinogenetic potential of methylated cadmium species. Thus, it is pointless to speculate whether methylated cadmium compounds contribute to the cadmium-induced lung and kidney cancers identified in epidemiological studies [213]. It has been shown, however, that cadmium inhibits DNA-(cytosine-5) methyltransferase and diminishes DNA methylation during cadmium-induced cellular transformation [252]. As described above, decreased DNA methylation is considered to have a tumor-promoting effect, since it is associated with augmented expression of cellular proto-oncogenes [219].

4.1.3. Lead

Inorganic lead compounds have been classified by the IARC as group 2A carcinogens (probably carcinogenic to humans), since long-term animal studies have shown increased tumor incidences in multiple organs including kidneys and brain. In contrast, organic lead compounds have been characterized as not classifiable as to their carcinogenicity to humans (Group 3). The IARC working group emphasizes, however, that organic lead compounds are oxidatively dealkylated in the body, at least in part, to ionic lead both in humans and animals, and that this ionic lead, generated from organic lead, will exert the same toxicities as those associated with inorganic lead exposure. In bacterial test systems, tetramethyl- and tetraethyllead did not induce mutations [214].

4.1.4. Antimony

Antimony is considered a likely lung carcinogen based on epidemiological and animal studies, however, the epidemiology is less conclusive compared to that of arsenic. As supposed by Gebel antimony is methylated to a minor extent if at all [27], thus, it is not clear whether methylation products contribute substantially to the antimony-associated carcinogenicity.

According to Dopp et al. trimethylantimony dichloride in a concentration of up to 1 mM in the incubation medium did not induce micronucleus formation, chromosome aberrations, or sister chromatid exchanges in CHO-9 cells *in vitro* under normal conditions and did not exhibit significant cytotoxicity [253]. Trimethylantimony dichloride was also negative in a plasmid DNA-nicking assay, in contrast to trimethylstibine which as well as stibine showed significant nicking to pBR 322 plasmid DNA. Reaction of trimethylantimony dichloride with either glutathione or L-cysteine to produce DNA-damaging trimethylstibine was observed with a trimethylantimony dichloride concentration as low as 50 μM and L-cysteine or glutathione concentrations as low as 500 and 200 μM, respectively, for a 24 h incubation [254].

4.1.5. Mercury

Methylmercury chloride induced renal adenocarcinomas in male mice in several long-term studies, but not in female mice and not in rats [255–259].

The carcinomas did not develop in castrated male mice but in ovariectomized female mice substituted with testosterone [260] indicating a hormone-dependent mechanism. Based on these studies the IARC classified methylmercury compounds as possibly carcinogenic to humans (Group 2B) [216].

There is some indication that high mercury concentrations in blood resulting from high fish or seal consumption might be correlated with cytogenetic abnormalities [261–264]. However, these studies are not taken as proof of a genotoxic effect of methylmercury in humans.

Positive results were obtained in a variety of short-term tests, for example, in all studies of induction of c-mitosis, sister chromatid exchange, structural chromosomal aberrations and aneuploidy in cultured human lymphocytes, whereas the majority of the bacterial tests were negative. The clastogenicity of methylmercury is most likely due to an impairment of the spindle apparatus, but an involvement of reactive oxygen species as shown for inorganic mercury compounds must also be considered [265]. A review on the immunotoxic effects of mercury compounds including methylmercury which may contribute to the potential carcinogenicity has been published by Moszczynski [266].

So far there is no conclusive evidence of methylmercury-induced carcinogenicity in humans. In a mortality study performed in the Minamata Bay region in Japan which included areas with a high prevalence of methylmercury poisoning excess mortality from cancer of the liver (SMR 2.07; 95% CI: 1.16–43.42) and cancer of the esophagus was found together with an increased risk for chronic liver disease and cirrhosis when the mortality rates were compared with the national cancer registry. A gender-specific evaluation of the results yielded an increased SMR for liver cancer only in men, concomitant with an increased risk for liver cirrhosis. Since alcohol consumption of the people of the region was significantly higher than in the general population in Japan, exposure to methylmercury was not regarded as the cause of the increased cancer-induced mortality [213]. A cohort study of 1657 persons in Sweden with a licence for seed disinfection with organic mercury compounds (among other chemicals) yielded no increased incidence of brain tumors during the observation period of approx. 15 years [213].

4.1.6. Selenium

Selenium is an essential trace element as it is in the form of selenocysteine a structural component of a number of functional proteins such as glutathione peroxidases, thioredoxin reductases, iodothyronine deiodinases, and selenoprotein P. Effects of selenium deficiency are fatigue, inefficiency, hair loss, then hepatic dysfunction, muscular weakness, arthritis, white coloring of the fingernails, and infertility. The Keshan disease, an endemic dilatative cardiomyopathie, and the Kashin-Beck disease, a dystrophic osteoarthrosis and spondylarthrosis, have been associated with selenium deficiency. A daily intake of 30–70 μg of selenium are considered necessary. In contrast to some areas in Eastern Asia there is no evident selenium deficiency in the Western countries. The average selenium intake in Middle Europe is approx. 30–50 μg per day, it is considerably higher in the US population (60–160 μg per day). The therapeutic index of selenium is small (approx. one order of magnitude).

Symptoms of acute selenium poisoning are irritation of the mucous membranes (particularly with selenium hydride), gastrointestinal disorders, and respiratory tract inflammation and, after weeks, potentially hair loss and finger- and footnail injury. Additional target organs are liver, kidney, lung, spleen, thyreoid, and joints. 300 μg/day (Scientific Committee on Food) to 400 μg/day (WHO/FAO/IAEA; Food and Nutrition Board of the US National Academy of Sciences) have been recommended as safe upper intake limit [267].

Initially suspected as a carcinogen, the results of epidemiological and clinical investigations as well as of animal studies revealed that selenium has the potential to prevent cancer when received at levels exceeding nutritional requirements [268]. Especially tumors of the prostate, lung, and colon were thought to be preventable by a regular selenium supplementation, as suggested by the results of the multicenter cancer prevention trial performed by the Nutritional Prevention of Cancer Study Group [269]. Also, an inverse association between serum levels of selenium and the incidence of squamous esophageal and adenomatous gastric cardia cancers were found in a nutritional intervention trial conducted in a Chinese region with epidemic rates of these tumors [270]. These studies not only heightened the interest in additional prevention trials to confirm the results and to include larger populations but also intensified the search for mechanisms by which selenium compounds suppress tumorigenesis [271].

Meanwhile, a variety of mechanisms presumably underlying the protective action of selenium have been proposed [272–276]. Among them are: (i) interference with the cellular redox status by modification of protein thiol groups and methionine mimicry; (ii) effects on cell cycle regulation and apoptosis; (iii) influence on DNA repair and tumor suppressor gene regulation; (iv) effects on signalling pathways; and (v) effects on angiogenesis.

A large number of *in vivo* and *in vitro* studies have been performed to elucidate the role the individual selenium species play in these processes. Basically, these studies revealed that methylation of selenium leads to species which lack some of the toxic effects of selenium compounds like selenite (particularly DNA strand breaks and base damage) [268,277], but retain the chemopreventive properties of the metalloid. Based on these results, a "monomethylselenide pool" (containing monomethylselenide and methylseleninic acid) has been proposed to be responsible for these antitumorigenic properties as counterpart to the "hydrogen selenide" pool which is supplied by selenite and which is made responsible for the DNA damage mediated by reactive oxygen species [278–282]. According to Ganther the "monomethylselenide pool" is supplied by stable methylated selenium species such as methylselenocysteine which serve as a reservoir providing a steady stream of monomethylated selenium to maintain a critical level [271].

The idea of the chemopreventive potency of the "monomethylselenide pool" has been supported by a number of mechanistic studies:

(1) The monomethylselenide precursors induced apoptosis and cell cycle arrest in transformed cells [268,278,283–286]. The monomethylselenide precursor-induced arrest occured in the G1 phase of the cell cycle, wheras exposure of cells to selenite led to an arrest in the S phase [279–280,286–289]. The apoptosis induced by the monomethylselenide precursors is caspase-mediated as demonstrated in DU145 prostate cancer cells [290] and in HL-60 leukemia cells [281].
(2) Methylseleninic acid and methylselenocyanate potently inhibited the cell cycle progression of vascular endothelial cells to the S phase, the endothelial expression of matrix metalloproteinase-2, and the cancer epithelial expression of vascular endothelial growth factor. Half-maximal inhibition of these effects was obtained with concentrations that are within the plasma range of selenium in US adults. In contrast, selenium forms that enter the "hydrogen selenide pool" lacked any inhibitory effect [291,292].

Taken together, these findings support the presence of at least two selenium metabolite pools that induce distinct types of cell cycle, apoptosis, and antiangiogenesis responses. The molecular targets and the pathways underlying these differential responses have not yet been defined, however. In future studies, speciation (profiling) methods have to be applied for the analysis of the selenium metabolites and selenium species in foods and supplements as a prerequisite for the development of mechanism-based selenium status markers for cancer prevention [282].

It has to be noted, however, that the promising prospects of an efficacious cancer prevention by selenium supplementation, nourished by the previous epidemiological studies, have seriously darkened in view of the results of two new studies. In the SELECT study (The Selenium and Vitamin E Cancer Prevention Trial), a double-blind placebo-controlled phase 3 study in which 35 533 men with no prostatic disorder participated, the daily application of 200 μg (in form of L-selenomethionine) had no effect on the development of prostatic cancer. The results were also independent on whether or not vitamin E was simultaneously supplemented. The study was therefore discontinued ahead of schedule [293]. In the other study with 1312 participants no effect of selenium supplementation (200 μg daily) on skin cancer risk was observed.

Apart from this outcome of the study, one third of the participants with the highest initial selenium values (>121.6 ng/mL) had a significant higher risk to develop diabetes type 2 [294]. In view of these results a daily supplementation of 200 μg selenium or more can no longer be recommended for cancer prevention. Further studies are needed to find the right balance between oversupplementation and selenium deficiency in maintaining the protection systems towards DNA damage.

4.1.7. *Bismuth*

To date, bismuth metal or bismuth compounds have not been classified as genotoxic or carcinogenic by the IARC or by any other regulatory agency (e.g., ACGIH, NIOSH, NTP, OSHA, DFG). Recent *in vitro* studies have however indicated that monomethylbismuth exhibits cyto- and genotoxic effects in several human cell systems. Following an exposure period of 24 hrs cytotoxic effects of monomethylbismuth(III) were detectable in erythrocytes at concentrations higher than 4 μM, in hepatocytes at concentrations higher than 130 μM, and in lymphocytes at concentrations higher than 430 μM. In contrast, cytotoxic effects of bismuth citrate (Bi-Cit) or of bismuth glutathione (Bi-GS) were much lower or not detectable even at the maximally applied concentration of 500 μM. Exposure of lymphocytes to monomethylbismuth(III) (250 μM for 1 h and 25 μM/50 μM for 24 hrs) resulted in significantly increased frequencies of chromosomal aberrations (CA) and sister chromatid exchanges, whereas Bi-Cit and Bi-GS induced neither CA nor SCE.

Monomethylbismuth(III) also increased the intracellular production of free radicals in hepatocytes [82]. It appears from these findings that methylation of bismuth observed in human studies [85,86] increases the cyto- and genotoxic potential of ingested bismuth.

Cytotoxic effects have also been observed, when rat thymocytes were exposed to triphenylbismuth [295], after treatment of macrophages with Bi-Cit at 6.25 μM for 24 hrs [296], and again in a macrophages cell line in a time- and dose-dependent manner between 12 and 24 hrs of incubation with Bi-Cit (50 μM) [297]. All these results emphasize the importance of cell type and species identity for bismuth toxicity. Another example are the significant genotoxic effects in bone marrow cells of mice detected after treatment of the animals with bismuth trioxides [298].

The mechanisms underlying the genotoxic activity of organobismuth compounds have not been eludicated as yet but several hypotheses have been proposed. One is the formation of reactive oxygen species by monomethylbismuth(III) which has been demonstrated in the study of von Recklinghausen et al. [82]. However, this formation was only evident in hepatocytes but not in lymphocytes, though chromosome damage was also observed in lymphocytes at non-cytotoxic concentrations [82]. Another hypothesis is based on the fact that bismuth is a powerful metallothionein inducer [297].

MT is a cysteine-rich metal-binding protein which decreases cytotoxicity and induces "hypoxia-like" stress under non-hypoxic conditions. Its functions are transport, metabolism, and detoxification of metals as well as inactivation of radicals. It has been suggested that Bi^{3+} binds strongly to MT, thereby readily displacing Zn^{2+} and Cd^{2+} [299]. Several authors have demonstrated that metals are able to interact with the so-called zinc finger proteins [300,301]. A direct interaction of methylbismuth with DNA, similar to interactions of platinum with nucleic acids, appears to be possible, too [302]. Thus, it may be speculated that monomethylbismuth(III) inhibits DNA repair mechanisms by displacing Zn^{2+} from the zinc finger proteins of DNA repair enzymes leading to increased DNA damage. Undeniably, this discussion raises doubts about the published statement " . . . the element's most exceptional property may well reside in the fact, that . . . it invariably exerts a beneficial influence on human health . . . " [303].

4.1.8. *Tin*

Despite weakly positive results in a few tests the methyltin compounds are probably not genotoxic, as most genotoxicity studies in bacterial and mammalian test systems turned out to be negative. Studies on the carcinogenicity of dimethyltin compounds have not been performed yet. An insufficiently designed 2-year study in rats, in which monomethyltin 2-ethylhexylmercaptoacetate was applied, yielded no significant increase in tumor formation [304]. Taken together, the methyltin compounds are considered not to be carcinogenic.

4.2. Nephrotoxicity

4.2.1. *Mercury*

Inorganic mercury is far more acutely nephrotoxic than is methylmercury. With the latter multiple exposures to large amounts are required to induce renal injury because only the part of methylmercury is effective which has been degraded to inorganic mercury [119,120]. This is in line with the observation that a significant mercury fraction in the kidneys of animals exposed to methylmercury is present in the inorganic form [119,120]. Organic mercury is oxidized prior to or after it has entered the renal tubular epithelial cells or an intracellular conversion of methylated to inorganic mercury can occur. The renal uptake of mercury *in vivo* is very rapid (within a few hours of exposure).

In animals hepatic GSH also plays an important role in the renal accumulation of methylmercury: After administration of methylmercury-GSH to mice renal methylmercury accumulated significantly more than after administration of methylmercury chloride [132]. Therefore, depending on renal cellular thiol status the various thiol conjugates of mercury are either excreted into urine or produce nephrotoxicity [305]. Thus, in renal systems a threshold effect (when exceeding buffer capacities established by metallothioneins and glutathione) is observed: Above that threshold cellular necrosis occurs [119,120] (for nephrocarcinogenicity of methylmercury in mice see above).

4.3. Neurotoxicity

4.3.1. Mercury

The neurotoxic properties of alkylated mercury species (see above) are very different: While dialkylmercury derivatives are considered extremely toxic and methylmercury as being significantly more toxic than inorganic mercury, species such as mercuric selenide or methylmercury cysteine possess a low degree of toxicity. Compared to inorganic species, the distribution of organic mercury compounds in mammals is more diffuse, and the neural (and also the hematopoietic tissue) is affected as primary target organ and not the kidneys [119,120]. Methylmercury has also been linked to an increased risk of myocardial infarction [306].

Following exposure to high doses of methylmercury neurological symptoms such as paresthesia, ataxia, dysarthria, and hearing loss occur after a latency period of several months [115]. While the clinical features of acute methylmercury poisoning have been well described, chronic low-dose exposure to methylmercury is poorly characterized, and its potential role in various chronic disease states remains controversial [113]. However, because of the high potential of methylmercury to damage the brain, there is general agreement to regard this mercury species as a major environmental toxicant [118,307,308].

Because of the passage of methylmercury through the placenta the fetus is at increased risk for methylmercury-induced brain damage. Methylmercury levels in fetal brain have been found to be about five to seven times higher than those in maternal blood [139]. In a study with Swedish mothers and their infants methylmercury concentrations in infant blood were highly associated with those in maternal blood, although being more than twice as high. After delivery, methylmercury concentrations decreased markedly until 13 weeks of life [309]. It was concluded from these findings that exposure to mercury (both inorganic and methylmercury) is higher before birth than during the breast-feeding period, and that methylmercury seems to contribute more than inorganic Hg to the postnatal exposure of the infants via breast milk.

The recommended "safe" intake level of the US EPA is 0.1 μg methylmercury/kg body weight per day, roughly corresponding to one 198 g can (=7 oz) of tuna fish per week. 10 μg methylmercury/g hair has also been proposed as a reference [137].

4.3.2. Tin

Short-chain alkyltin compounds are supposed to exhibit strong neurotoxic effects as shown in animal studies *in vivo* and in *in vitro* studies. Nevertheless, potential health effects following chronic low-dose exposure to these compounds have not been investigated as yet [310], but some information on systemic effects in humans has been obtained from accidental exposure which resulted in the appearance of dramatic behavioral changes, including weakness, aggressive behavior, depression, aggressivity, disorientation, attention deficits, severe memory loss, seizures, and in some instances death [198,200,201,204,205,311]. Recovery from the neurological symptoms was usually slow in the cases who survived [198,201]. Plasmapheresis and application of D-penicillamine neither had an influence on the clinical situation nor on the elimination of tin [200].

The main pathologic findings in a 48-year old woman who died from a multiorgane failure six days after the intake of an unknown amount of trimethyltin were a generalized chromatolysis of the neurons in the brain, spinal cord, and spinal ganglia. Electron microscopy revealed marked accumulation of lysosomal dense bodies and disorganisation of the granular endoplasmic reticulum in the neurons. The findings were similar to those described in experimental intoxications with trimethyltin [204].

A distinguishing feature of organotin toxicity is the high level of specificity that these compounds exhibit toward their biological targets, which make them ideal candidates for studying organotin effects. Being both neurotoxic, trimethyl- and triethyltin induce selective injury to distinct regions of the central nervous system. While trimethyltin damages areas of the limbic system (hippocampus), the neocortex, and the brainstem, triethyltin predominately affects mainly regions of the spinal cord causing massive myelinic edema and demyelination [311–316].

The toxicity of the organotin compounds is directly linked to the number and nature of the organic moiety. Within the methyltin species the neurotoxic effects increase with the degree of methylation where the effects of tetramethyltin are similar to those of trimethyltin. The toxic effects are mainly conveyed by the $R_{(1-3)}Sn^{+}$-cation and are relatively independent on the counter ions [317].

In vitro studies on the molecular mechanisms underlying the trimethyltin-induced neuropathological changes and behavioral deficits indicated that the organotin compound impairs neurite outgrowth and cell viability. The decrease in cell viability was paralleled by a decrease in cell body size, an increase in DNA fragmentation, activation of caspase-9, and cleavage of the caspase substrate poly-ADP-ribose polymerase (PARP). Pharmacological inhibition of caspase activity, p38 stress-responsive protein kinase activity, or oxidative stress prevented trimethyltin-induced cell death. These observations were taken as evidence for a trimethyltin-initiated apoptotic pathway requiring oxidative stress, caspase activation, and p38 protein kinase activity ions [318].

Organotin compounds impair the synthesis and function of proteins in that they bind to amino acids leading to conformational changes [319]. One mechanism postulated for protein-organotin interactions is the formation of covalent bonds between the tin(IV) atom and thiols present in proteins. This mechanism has been corroborated by recent *in vitro* studies showing that vicinal dithiols rather than monothiols are responsible for mediating the biochemical effects of organotin compounds. In particular, it has been shown that both tri- and dialkyltin compounds target dithiols present in mitochondrial proteins, inducing cellular apoptosis. It has been shown that stannin, a mitochondrial membrane protein largely expressed in the hippocampus region sensitizes neuronal cells to trimethyltin intoxication [320]. Stannin has two conserved vicinal cysteines (Cys-32 and Cys-34) that may constitute a trimethyltin binding site. There is a direct correlation between trimethyltin toxicity and the expression of stannin [321]. It was hypothesized that trimethyltin enters the cell, binds to stannin and is dealkylated to dimethyltin which induces a structural change in the protein eliciting the toxic response [322].

4.3.3. *Lead*

Alkyllead compounds exhibit distinct neurotoxic properties as indicated by the neurological and behavioral deficits observed both in animal studies [103,323] and in humans. Following accidental exposure to alkyllead [324], abuse of leaded gasoline [325], or occupational exposure to organolead [326–328] a variety of neurological symptoms and/or behavioral abnormalities have been observed. It appears from a comparison of case reports that tetraethyllead is more neurotoxic than tetramethyllead [102]. According to Walsh and Tilson the neurobehavioral effects (alterations in sensory responsiveness or behavioral reactivity and task-dependent changes in avoidance learning) resemble the sequelae of limbic system damage [329].

In an investigation on the relationship between bone lead concentration (estimated by XRF spectrometry of the tibia) after exposure to organic lead compounds and neurobehavioral test scores in 529 former organolead-exposed workers (on average 16 years since last exposure) the highly exposed workers had significantly lower scores on visuoconstruction tasks, verbal memory, and learning. Peak tibial lead concentrations were associated with a decline in verbal and visual memory, executive function, and manual dexterity. These effects of lead were more pronounced in individuals who had at least one ε 4 allele of the apolipoprotein E4 gene [330]. Apolipoprotein E4 has been implicated in impaired cognitive function and reduced neurite outgrowth and is a risk factor for Alzheimer's disease [331].

The trialkyllead species are the most toxic alkyllead metabolites. Trialkyllead has been shown to inhibit the *in vitro* assembly of microtubules from mammalian brain [332], to induce hypomyelination and to hamper the process of myelin membrane assembly [333], and to decrease the energy level of the cell, presumably by uncoupling oxidative phosphorylation [334]. Inhibition of the ATP synthesis and subsequently cell death has been suggested to be a consequence of the trialkyllead-induced opening of the MTP pore observed in rat liver mitochondria [335].

4.3.4. Arsenic

In addition to the effects on lung, skin, and the hematopoietic systems [336], exposure to arsenic may result in both a central and peripheral neuropathy. Reported effects following occupational or environmental exposure or accidental intoxication include subclinical nerve injuries [337], delirium and encephalopathy [338], peripheral neuropathies [339,340], and symptoms including loss of hearing and taste, blurred vision, tingling and numbness of the limbs, and decrease in muscle strength [341,342]. Furthermore, several investigations revealed that arsenic has an influence on learning, short-term memory, and concentration [343]. In children, chronic exposure to inorganic arsenic via drinking water resulted in a dose-dependent reduction of intellectual functions [344,345]. Alterations in memory and attention have been observed in adolescents after chronic exposure to high levels of arsenic [346].

A pathophysiological finding in patients with arsenic-induced peripheral neuropathy is a reduced nerve conduction velocity [347]. It is assumed that arsenic interacts with cytoskeletal proteins resulting in destruction of axonal cylinders and changes of the cytoskeletal composition which may lead to axonal degeneration. It appears from *in vitro* studies that the arsenic-induced destabilization and disruption of the cytoskeletal framework is in part due to the activation of calpain (calcium-activated cytoplasmatic protease) through influx of Ca^{2+}, which in turn is responsible for the degradation of NF-L (neurofilament light subunit) in a calcium-induced proteolytic process. Arsenic may also affect the phosphorylation of the tau protein (MAP-tau), another important cytoskeletal protein, leading to a deregulation of the tau function which is associated with neurodegeneration. A review of the neurotoxicity of arsenic was published by Vahidnia et al. [348].

The potential role of arsenic metabolites in these neurodegenerative processes was addressed in an *in vitro* study in cell lines derived from the peripheral (ST-8814) and the central (SK-N-SH) nervous system. In this study the effect of inorganic and methylated arsenic species on the expression of several cytoskeletal genes were compared. While As_i^{III} and As_i^{V} did not exhibit any significant effect on either cell line, $MMAs^{V}$ and $DMAs^{V}$ caused significant changes in the expression levels of some of the investigated cytoskeletal genes [349]. Another *in vitro* study performed with hippocampal slices of young (14–21 day-old) and adult (2–4 month-old) rats aimed to find out, whether the dimethylated arsenic metabolites influence the synaptic acitivity. $DMAs^{III}$ blocked the excitatory transmission at the hippocampal Schaffer collateral CA1 synapse in a concentration-dependent manner. The blocking effects were considerably greater in slices taken from young rats compared to those from adult rats. In contrast, $DMAs^{V}$ exerted no effects, neither in young nor in adult rats. The results suggested that the $DMAs^{III}$-induced functional impairment of synaptic activity contributes to the neurotoxicity of arsenic and that the trivalent arsenic species possesses a considerably higher neurotoxic potential than the pentavalent one [350].

4.3.5. Tellurium

The tellurium-induced neuropathies observed in animal studies seem to result from an impaired cholesterol biosynthesis with subsequent destabilization and reduced myelin formation. A likely mechanism of this impairment is the binding of tellurium to vicinal sulfhydryl groups of squalene monoxygenase leading to an inhibition of this microsomal enzyme [351–354]. Studies with purified human squalene monoxygenase have shown that the binding capacity of dimethyltellurium dichloride and dimethyltelluride is higher than that of tellurite. Thus, methylation of tellurium, normally considered a detoxication reaction, may actually yield a more toxic metabolite for this enzyme [355].

4.3.6. *Thallium*

Since it is unknown whether methylated thallium metabolites are formed in humans, it could only be speculated whether such potential derivatives would be involved in the development of the extremely painful sensory neuropathy and the alopecia, the major manifestations of thallium toxicity [356].

4.3.7. *Bismuth*

In addition to special applications in nuclear medicine (e.g., polyaminocarboxylate complexes of α-emitting Bi isotopes of mass 212 and 213 to kill tumor cells, e.g., in leukemia therapy [357,358], bismuth compounds (mainly Bi (sub)salicylate and nitrate complexes, CBS) have been used for a long time in the treatment of microbial infections (syphilis, gastrointestinal disorders) because of their antimicrobial acitivity and presumed low toxicity. A more recent example is the bismuth-based triple therapy (bismuth together with antibiotics) to prevent the growth of *Helicobacter pylori* [359].

The assumption that "bismuth is one of those rare elements considered to be safe because it is non-toxic and non-carcinogenic despite its heavy metal status" [303] must be challenged, however, if bismuth methylation observed in the human volunteer studies [85,86], the results of the recent genotoxicity studies [82], and the available data on acute toxicity of bismuth compounds are considered. Methylation of inorganic bismuth seems to markedly increase the acute toxicity as indicated by the LD_{50} data of BiOCl (22 g/kg, rat, oral) and trimethylbismuth (484 mg/kg, rabbit, oral [10]), respectively. Yet methylation also enhances the lipophilic potency of bismuth which facilitates the crossing of membranes such as the blood-brain barrier. If this change in the physicochemical property of bismuth is taken into account together with observed bismuth-induced neurotoxic effects in animals [360], it may be speculated that the encephalopathies diagnosed in the 1970s in French and Australian patients [81,87,361] were associated with the formation of the volatile trimethylbismuth species. Nearly 1000 of such encephalopathy cases had been reported in France by 1979, of which 72 were fatal [361]. The bismuth levels in the blood of these patients who had ingested up to 20 g bismuth per day over a period of 20 days per month [81,87,361] usually exceeded 100 μg/L and ranged up to 2850 μg/L [362,363]. Menge et al. have suggested that the conversion of bismuth into "soluble neurotoxic compounds" by the intestinal flora may be involved [73]. This theory is supported by the observation that the patients afflicted in the French and Australian epidemics were likely to have had bacterial overgrowth in the intestine [73].

5. GENERAL CONCLUSIONS

Alkylation of metal(loid)s is generating species which are more volatile and amphiphilic, and are able to move more freely and quickly through the human body. While peralkylated compounds because of their vapor pressure may tend to leave the body (e.g., dimethylselenide and -telluride as well as trimethylbismuth are exhaled in breath), partly alkylated species dynamically combine with predominantly sulfur-containing biomolecules like peptides and proteins. In unfavorable cases, the latter can transport metal(loid) species through membrane channels as was demonstrated for methylmercury (mimicring methionine), and, thus, can reach the adult and fetal brain. Another example for the transport of the latter species is its close association with erythrocytes, leading to the long lifespan of methylated mercury in the blood cycle; degradation is only possible by microbial demethylation during colon passage within the enterohepatic cycle.

While the human body is exposed to higher alkylated metal(loid) compounds only externally by industrial products (e.g., butylated tin, ethylated lead, or phenylated mercury), methylated species can be generated additionally inside the body as has been demonstrated for arsenic, bismuth, selenium, and tellurium. Relevant production sites are not only enzymes in the liver (e.g., for arsenic methylation), but also biomethylation by the intestinal flora (e.g., for bismuth). Thus, methylated species will significantly change the metabolism and toxicity of the metal(loid): While ingested arsenic is easily excretable in urine as dimethylarsinic acid, methylated bismuth (in case of bismuth overdose) and mercury (extreme fish eaters) may lead to neurotoxic symptoms. In general, methylation increases the toxicity of metal(loid)s, except in the case of selenium in which the assumed "monomethylselenide pool" is considered a relevant chemopreventive reservoir.

Further research will show if the discussed scenarios will stay as individual cases or are part of larger networks. Eventually, the fact should be reminded that it is still not much known concerning metal(loid) methylation in man (see Table 1).

ABBREVIATIONS

ACGIH	American Conference of Governmental and Industrial Hygienists
ADP	adenosine diphosphate
ALA	alpha lipoic acid
AS3MT	arsenite methyltransferase
ATP	adenosine 5′-triphosphate
CA	chromosomal aberration
CBS	colloidal bismuth subcitrate
CE	capillary electrophoresis
CHO	Chinese hamster ovary (cells)
CI	confidence interval
Cit	citrate
Cys	cysteine
DFG	Deutsche Forschungsgemeinschaft (German Research Foundation)
DHLA	dihydrolipoic acid
$DMAs^{III}$	dimethylarsinous acid
$DMAs^{V}$	dimethylarsinic acid
$DMDTA^{V}$	dimethyldithioarsinic acid
DMPS	2,3-dimercapto-1-propane sulfonic acid
DMSA	dimercaptosuccinic acid
EPA	Environmental Protection Agency
ESI-MS	electrospray mass spectrometry
FAO	Food and Agriculture Organization
GC	gas chromatography
GE	gel electrophoresis
GI	gastrointestinal (tract)
Gly	glycine
GSH	glutathione (reduced form)
HSA	human serum albumin
HPLC	high performance liquid chromatography
IAEA	International Agricultural Exchange Association
IARC	Internation Agency for Research on Cancer
ICP-MS	inductively coupled plasma mass spectrometry
kDa	kilodalton
LAT	large amino acid transporter
LD_{50}	lethal dose for 50% (of animals)
MAP-tau	microtubule-associated-protein tau
$MMAs^{III}$	monomethylarsonous acid
$MMAs^{V}$	monomethylarsonic acid
$MMMTAs^{V}$	monomethylmonothioarsonic acid

mRNA	messenger ribonucleic acid
MT	metallothionein
MTP	mitochondrial transition pore
NAC	*N*-acetylcysteine
NF-L	neurofilament light (subunit)
NIOSH	National Institute for Occupational Safety and Health
NTP	National Toxicology Program
OAT	organic anion transporter
OSHA	Occupational Safety and Health Administration
PARP	poly-ADP-ribose polymerase
PVC	polyvinylchloride
SAH	*S*-adenosylhomocysteine
SAM	*S*-adenosylmethionine
SCE	sister chromatid exchange
SeAH	Se-adenosylhomocysteine
SeAM	Se-adenosylmethionine
SELECT	The Selenium and Vitamin E Cancer Prevention Trial
SIDS	sudden infant death syndrome
SMR	standardized mortality rate
TMAsO	trimethylarsine oxide
WHO	World Health Organisation
XANES	X-ray absorption near-edge structure
XRF	X-ray fluorescence

REFERENCES

1. H. Haraguchi, *J. Anal. At. Spectrom.*, 2004, **19**, 5–14.
2. M. Calvin, *Chemical Evolution: Molecular Evolution towards the Origin of Living Systems on the Earth and Elsewhere*, Oxford University Press, Oxford, UK, 1969.
3. J. Lopez-Barea and J. L. Gomez-Ariza, *Proteomics*, 2006, **6**, S51–S62.
4. A. V. Hirner, *Anal. Bioanal. Chem.*, 2006, **385**, 555–567.
5. J. Huang, X. Hu, J. Zhang, K. Li, Y. Yan and X. Xu, *J. Pharmaceut. Biomed. Anal.*, 2006, **40**, 227–234.
6. D. G. Sar, M. Montes-Bayon, E. B. Gonzales and A. Sanz-Medel, *J. Anal. At. Spectrom.*, 2006, **21**, 861–868.
7. C. G. Hartinger and B. K. Keppler, *Electrophoresis*, 2007, **28**, 3436–3446.
8. Y. Li and W. Maret, *J. Anal. At. Spectrom.*, 2008, **23**, 1055–1062.
9. S. K. Yathavakilla, M. Fricke, P. A. Creed, D. T. Heitkemper, N. V. Shockey, C. Schwegel, J. A. Caruso and J. T. Creed, *Anal. Chem.*, 2008, **80**, 775–782.
10. P. J. Craig, *Organometallic Compounds in the Environment*, 2nd edn., J. Wiley, New York, 2003.

11. K. Wrobel, K. Wrobel and J. A. Caruso, *Anal. Bioanal. Chem.*, 2009, **393**, 481–486.
12. E. Dopp, L. M. Hartmann, A.-M. Florea, A. W. Rettenmeier and A. V. Hirner, *Crit. Rev. Toxicol.*, 2004, **34**, 301–333.
13. A. V. Hirner and H. Emons, *Organic Metal and Metalloid Species in the Environment*, Springer, Berlin, Heidelberg, New York, 2004.
14. P. Pinel-Raffaitin, C. Pecheyran and D. Amouroux, *Atmos. Environ.*, 2008, **42**, 7786–7794.
15. J.-P. Goulle, L. Mahieu, J. Castermant, N. Neveu, L. Bonneau, G. Laine, D. Bouige and C. Lacroix, *Forensic Sci. Int.*, 2005, **153**, 39–44.
16. P. Heitland and H. D. Köster, *J. Trace El. Med. Biol.*, 2006, **20**, 253–262.
17. S. Griesel, A. Kakuschke, U. Siebert and A. Prange, *Sci. Tot. Environ.*, 2008, **392**, 313–323.
18. A. Bazzi, J. O. Nriagu and A. M. Linder, *J. Environ. Monitor.*, 2008, **10**, 1226–1232.
19. M. Wilhelm, U. Ebers and C. Schulz, *Int. J. Hyg. Environ. Health*, 2004, **207**, 69–73.
20. J. Seifter, *J. Pharmacol. Exp. Ther.*, 1939, **66**, 366–377.
21. B. A. Richardson, *Lancet*, 1990, **337**, 670.
22. B. A. Richardson, *J. Forensic Sci.*, 1994, **34**, 199–204.
23. P. J. Fleming, M. Cooke, S. M. Chantler and J. Golding, *Br. Med. J.*, 1994, **309**, 1594–1596.
24. M. Krachler and H. Emons, *J. Anal. At. Spectrom.*, 2001, **16**, 20–25.
25. K. Stang, J. Kresimon, A. Stang, H. Merzenich, A. V. Hirner, K.-H. Jöckel and A. W. Rettenmeier, *Documentations of the 43rd Annual Meeting of the German Society of Occupational Medicine and Environmental Medicine e.V.*, Dresden, 2003, pp. 567–569.
26. R. Bailly, R. Lauwerys, J. P. Buchet, P. Mahieu and J. Konings, *Br. J. Ind. Med.*, 1991, **48**, 93–97.
27. T. Gebel, *Chem. Biol. Interact.*, 1997, **107**, 131–144.
28. J. Meyer, K. Michalke, T. Kouril and R. Hensel, *Syst. Appl. Microbiol.*, 2008, **31**, 81–87.
29. A. Gomez-Caminero, P. Howe, M. Hughes, E. Kenyon, D. R. Lewis, M. Moore, J. Ng, A. Aitio and G. Becking, *Environmental Health Criteria 224, Arsenic and Arsenic Compounds*, WHO, 2001.
30. H. V. Aposhian, *Annu. Rev. Pharmacol. Toxicol.*, 1997, **37**, 397–419.
31. L. J. Yost, R. A. Schoof and R. Aucoin, *Hum. Ecol. Risk Assess.*, 1998, **4**, 137–152.
32. X. C. Le, W. R. Cullen and K. J. Reimer, *Clin. Chem.*, 1994, **40**, 617–624.
33. X. C. Le and M. Ma, *J. Chromatogr. A*, 1997, **764**, 55–64.
34. X. Zhang, R. Cornelis, J. De Kimpe and L. Mees, *Anal. Chim. Acta*, 1996, **319**, 177–185.
35. X. Zhang, R. Cornelis, J. De Kimpe and L. Mees, *J. Anal. At. Spectrom.*, 1996, **11**, 1075–1079.
36. H. Yamauchi and Y. Yamamura, *Bull. Environ. Contam. Toxicol.*, 1984, **32**, 682–687.

37. Y. Shibata, J. Yoshinaga and M. Morita, *Appl. Organomet. Chem.*, 1994, **8**, 249–251.
38. E. Dopp, L. M. Hartmann, U. von Recklinghausen, A. M. Florea, S. Rabieh, U. Zimmermann, B. Shokouhi, S. Yadav, A. V. Hirner and A. W. Rettenmeier, *Toxicol. Sci.*, 2005, **87**, 46–56.
39. E. Dopp, U. von Recklinghausen, L. M. Hartmann, I. Stueckradt, I. Pollok, S. Rabieh, L. Hao, A. Nussler, C. Katier, A. V. Hirner and A. W. Rettenmeier, *Drug Metab. Dispos.*, 2008, **36**, 971–979.
40. S. V. Kala, M. W. Neely, G. Kala, C. I. Prater, D. W. Atwood, J. S. Rice and M. W. Lieberman, *J. Biol. Chem.*, 2000, **275**, 33404–33408.
41. E. M. Leslie, A. Haimeur and M. P. Waalkes, *J. Biol. Chem.*, 2004, **279**, 32700–32708.
42. K. A. Francesconi, R. Tanggaard, C. J. McKenzie and W. Goessler, *Clin. Chem.*, 2002, **48**, 92–101.
43. R. Raml, W. Goessler, P. Traar, T. Ochi and K. A. Francesconi, *Chem. Res. Toxicol.*, 2005, **18**, 1444–1450.
44. K. T. Suzuki, K. Iwata, H. Naranmandura and N. Suzuki, *Toxicol. Appl. Pharmacol.*, 2007, **218**, 166–173.
45. H. Naranmandura and K. T. Suzuki, *Toxicol. Appl. Pharmacol.*, 2008, **227**, 390–399.
46. A. Hernandez and R. Marcos, *Pharmacogenomics*, 2008, **9**, 1113–1132.
47. F. Challenger, *Chem. Rev.*, 1945, **36**, 315–361.
48. T. Hayakawa, Y. Kobayashi, X. Cui and S. Hirano, *Arch Toxicol*, 2005, **79**, 183–191.
49. D. J. Thomas, J. X. Li, S. B. Waters, W. B. Xing, B. M. Adair, Z. Drobna, V. Devesa and M. Styblo, *Exp. Biol. Med.*, 2007, **232**, 3–13.
50. I. R. Rowland and M. J. Davies, *J. Appl. Toxicol.*, 1981, **1**, 273–283.
51. L. L. Hall, S. E. George, M. J. Kohan, M. Styblo and D. J. Thomas, *Toxicol. Appl. Pharmacol.*, 1997, **147**, 101–109.
52. M. J. Mass, A. Tennant, B. C. Roop, W. R. Cullen, M. Styblo, D. J. Thomas and A. D. Kligerman, *Chem. Res. Toxicol.*, 2001, **14**, 355–361.
53. H. Naranmandura, K. Iwata and K. T. Suzuki, *Chem. Res. Toxicol.*, 2007, **20**, 1120–1125.
54. R. Raml, A. Rumpler, W. Goessler, M. Vahter, L. Li, T. Ochi and K. A. Francesconi, *Toxicol. Appl. Pharmacol.*, 2007, **222**, 374–380.
55. O. L. Valenzuela, V. H. Borja-Aburto, G. G. Garcia-Vargas, M. B. Cruz-Gonzalez, E. A. Garcia-Montalvo, E. S. Calderon-Aranda and L. M. del Razo, *Environ. Health Perspect.*, 2005, **113**, 250–254.
56. X. C. Le, M. Ma, X. Lu, W. R. Cullen, H. V. Aposhian and B. Zheng, *Environ. Health Perspect.*, 2000, **108**, 1015–1018.
57. X. C. Le, X. Lu, M. Ma, W. R. Cullen, H. V. Aposhian and B. Zheng, *Anal. Chem.*, 2000, **72**, 5172–5177.
58. H. V. Aposhian, B. Zheng, M. M. Aposhian, X. C. Le, M. E. Cebrian, W. R. Cullen, R. A. Zakharyan, M. Ma, R. C. Dart, Z. Cheng, P. Andrewes, L. Yip, G. F. O'Malley, R. M. Maiorino, W. van Voorhies, S. M. Healy and A. Titcomb, *Toxicol. Appl. Pharmacol.*, 2000, **165**, 74–83.

59. H. V. Aposhian, E. S. Gurzau, X. C. Le, A. Gurzau, S. M. Healy, X. Lu, M. Ma, L. Yip, R. A. Zakharyan, R. M. Maiorino, R. C. Dart, M. G. Tircus, D. Gonzalez-Ramirez, D. L. Morgan, D. Avram and M. M. Aposhian, *Chem. Res. Toxicol.*, 2000, **13**, 693–697.
60. B. K. Mandal, Y. Ogra and K. T. Suzuki, *Chem. Res. Toxicol.*, 2001, **14**, 371–378.
61. L. M. del Razo, M. Styblo, W. R. Cullen and D. J. Thomas, *Toxicol. Appl. Pharmacol.*, 2001, **174**, 282–293.
62. B. K. Mandal, Y. Ogra, K. Anzai and K. T. Suzuki, *Toxicol. Appl. Pharmacol.*, 2004, **198**, 307–318.
63. Z. Wang, J. Zhou, X. Lu, Z. Gong and X. C. Le, *Chem. Res. Toxicol.*, 2004, **17**, 95–103.
64. S. Rabieh, A. V. Hirner and J. Matschullat, *J. Anal. At. Spectrom.*, 2008, **23**, 544–549.
65. Z. Gregus, A. Gyurasics and I. Csanaky, *Toxicol. Sci.*, 2000, **56**, 18–25.
66. X. Lu, L. L. Arnold, S. M. Cohen, W. R. Cullen and X. C. Le, *Anal. Chem.*, 2003, **75**, 6463–6468.
67. K. A. Francesconi and D. Kuehnelt, *Analyst*, 2004, **129**, 373–395.
68. H. R. Hansen, A. Raab, M. Jaspars, B. F. Milne and J. Feldmann, *Chem. Res. Toxicol.*, 2004, **17**, 1086–1091.
69. Z. Gong, X. Lu, W. R. Cullen and X. C. Le, *J. Anal. At. Spectrom.*, 2001, **16**, 1409–1413.
70. J. Feldmann and A. V. Hirner, *Int. J. Environ. Anal. Chem.*, 1995, **60**, 339–359.
71. K. Michalke, E. B. Wickenheiser, M. Mehring, A. V. Hirner and R. Hensel, *Appl. Environ. Microbiol.*, 2000, **66**, 2791–2796.
72. K. Michalke, J. Meyer, A. V. Hirner and R. Hensel, *Appl. Organomet. Chem.*, 2002, **16**, 221–227.
73. H. Menge, M. Gregor, B. Brosius, R. Hobert and A. Lang, *Eur. J. Gastroenterol. Hepatol.*, 1992, **4**, S41–S47.
74. L. A. Tillman, F. M. Drake, J. S. Dixon and J. R. Wood, *Aliment. Pharmacol. Ther.*, 1996, **10**, 459–467.
75. G. G. Briand, N. Burford, M. D. Eelman, N. Aumeerally, L. Chen, T. S. Cameron and K. N. Robertson, *Inorg. Chem.*, 2004, **43**, 6495–6500.
76. H. Sun, H. Li, A. B. Mason, R. C. Woodworth and P. J. Sadler, *J. Biol. Chem.*, 2001, **276**, 8829–8835.
77. T. Sallmann and J. Seifer, *J. Pharm. Exp. Therapeut.*, 1939, **67**, 17–49.
78. M. R. Krigman, T. W. Bouldin and P. Mushak, *Monogr. Pathol.*, 1985, **26**, 58–100.
79. A. J. McLean, P. R. A. Froomes and A. T. Wan, *Clin. Pharmacol. Ther.*, 1989, **45**, 150.
80. L. Z. Benet, *Scand. J. Gastroenterol.*, 1991, **26**, 29–35.
81. H. Sun, L. Zhang and K. Y. Szeto, *Met. Ions Biol. Syst.*, 2004, **41**, 333–378.
82. U. von Recklinghausen, L. M. Hartmann, S. Rabieh, J. Hippler, A. V. Hirner, A. W. Rettenmeier and E. Dopp, *Chem. Res. Toxicol.*, 2008, **21**, 1219–1228.
83. A. D. Kimura, Y. Ishida, T. Wada, H. Yokoyama, N. Mukaida and T. Kondo, *Toxicol. Appl. Pharmacol.*, 2005, **203**, 53–61.

84. Y. F. Lin, A. R. Walmsley and B. P. Rosen, *Proc. Natl. Acad. Sci. USA*, 2006, **103**, 15617–15622.
85. A. V. Hirner, L. M. Hartmann, J. Hippler, J. Kresimon, J. Koesters, K. Michalke, M. Sulkowski and A. W. Rettenmeier, in *Organic Metal and Metalloid Species in the Environment*, Ed. A.V. Hirner and H. Emons, Springer, Berlin, 2004, pp. 181–203.
86. J. Boertz, L. M. Hartmann, M. Sulkowski, J. Hippler, F. Mosel, R. A. Diaz-Bone, K. Michalke, A. W. Rettenmeier and A. V. Hirner, *Drug Metab. Dispos.*, 2009, **37**, 1–7.
87. A. Slikkerveer and F. A. de Wolff, *Med. Toxicol. Adverse Drug Exp.*, 1989, **4**, 303–323.
88. J. Boertz, *Investigation of Bismuth Metabolism via Mass Spectrometric Techniques*, PhD thesis, University of Duisburg-Essen, Germany, 2008.
89. H. V. Aposhian and M. M. Aposhian, *Chem. Res. Toxicol.*, 2006, **19**, 1–15.
90. K. Michalke, J. Meyer and R. Hensel, in *Archea: Evolution, Physiology, and Molecular Biology*, Ed. R. A. Garrett and H. P. Klenk, Blackwell Publishing, 2007, pp. 285–293.
91. K. Michalke, A. Schmidt, B. Huber, J. Meyer, M. Sulkowski, A. V. Hirner, J. Boertz, F. Mosel, P. Dammann, G. Hilken, H. J. Hedrich, M. Dorsch, A. W. Rettenmeier and R. Hensel, *Appl. Environ. Microbiol.*, 2008, **74**, 3069–3075.
92. J. S. Thayer, *Environmental Chemistry of the Heavy Elements*, VCH, New York, 1995.
93. D. Yu, *Regul. Toxicol. Pharmacol.*, 1999, **29**, 128–141.
94. M. Styblo, L. M. del Razo, E. L. LeCluyse, G. A. Hamilton, C. Wang, W. R. Cullen and D. J. Thomas, *Chem. Res. Toxicol.*, 1999, **12**, 560–565.
95. J. S. Thayer, *Appl. Organomet. Chem.*, 2002, **16**, 677–691.
96. U. Klotz, *Clin. Pharmacokinet.*, 2000, **38**, 243–270.
97. J. W. Robinson and E. L. Kiesel, *J. Environ. Sci. Health*, 1981, **A16**, 341–352.
98. B. L. Lewis and H. P. Mayer, in *Metal Ions in Biological Systems*, Ed. H. Sigel and A. Sigel, **Vol. 29**, Marcel Dekker, New York, 1993, pp. 79–99.
99. J. O. Nriagu, *Sci. Total Environ.*, 1990, **92**, 13–28.
100. M. J. Heard, A. C. Wells, D. Newton and A. C. Chamberlain, in *Proceedings of the International Conference on Heavy Metals in the Environment*, London, 1979, pp. 103–108.
101. E. P. Laug and F. M. Kunze, *J. Ind. Hyg. Toxicol.*, 1948, **30**, 256–259.
102. J. Gething, *Br. J. Ind. Med.*, 1975, **32**, 329–333.
103. R. K. Davis, A. W. Horton, E. E. Larson and K. L. Stemmer, *Arch. Environ. Health*, 1963, **6**, 473–479.
104. W. Bolanowska, J. Piotrowski and H. Garczynski, *Arch. Toxicol.*, 1967, **22**, 278–282.
105. J. E. Cremer, *Br. J. Ind. Med.*, 1959, **16**, 191–199.
106. T. Nielsen, K. A. Jensen and P. Grandjean, *Nature*, 1978, **274**, 602–603.
107. D. Ferreira da Silva, U. Schröder and H. Diehl, *Xenobiotica*, 1983, **13**, 583–590.
108. D. Ferreira da Silva and H. Diehl, *Xenobiotica*, 1985, **15**, 789–797.
109. B. S. Schwartz, W. F. Stewart, A. C. Todd and J. M. Links, *Occup. Environ. Med.*, 1999, **56**, 22–29.

110. T. W. Clarkson, J. B. Vyas and N. Ballatori, *Am. J. Ind. Med.*, 2007, **50**, 757–764.
111. K. L. Nuttal, *Ann. Clin. Lab. Sci.*, 2004, **34**, 235–250.
112. T. W. Clarkson, L. Magos and G. J. Myers, *N. Eng. J. Med.*, 2003, **349**, 1731–1737.
113. J. P. K. Rooney, *Toxicol.*, 2007, **234**, 145–156.
114. H. H. Harris, I. J. Pickering and G. N. George, *Science*, 2003, **301**, 1203.
115. D. W. Nierenberg, R. E. Nordgren, M. B. Chang, R. W. Siegler, M. B. Blayney, F. Hochberg, T. Y. Toribara, E. Cernichiari and T. Clarkson, *N. Eng. J. Med.*, 1998, **338**, 1672–1676.
116. L. Magos, *J. Appl. Toxicol.*, 2003, **23**, 263–269.
117. R. Pfab, H. Mucker, G. Roider and T. Zilker, *Clin. Toxicology*, 1996, **34**, 453–460.
118. M. Aschner and T. Syversen, *Ther. Drug Monit.*, 2005, **27**, 278–283.
119. R. K. Zalups, in *Molecular Biology and Toxicology of Metals*, Ed. R. K. Zalups and J. Koropatnick, Taylor & Francis, London, 2000, pp. 234–275..
120. R. K. Zalups, *Pharmacol. Rev.*, 2000, **52**, 113–143.
121. J. Sundberg, B. Ersson, B. Lonnerdal and A. Oskarsson, *Toxicol.*, 1999, **137**, 169–184.
122. R. E. Hoffmeyer, S. P. Singh, C. J. Doonan, A. R. S. Ross, R. J. Hughes, I. J. Pickering and G. N. George, *Chem. Res. Toxicol.*, 2006, **19**, 753–759.
123. T. A. Simmons-Willis, A. S. Koh, T. W. Clarkson and N. Ballatori, *Biochem. J.*, 2002, **367**, 239–246.
124. R. K. Zalups and S. Ahmad, *J. Pharmocol. Exp. Ther.*, 2005, **314**, 1158–1168.
125. J. S. Woods, H. A. Davis and R. P. Baer, *Arch. Biochem. Biophys.*, 1992, **296**, 350–353.
126. J. S. Woods and M. E. Ellis, *Biochem. Pharmacol.*, 1995, **50**, 1719–1724.
127. S. Li, S. A. Thompson, T. J. Kavanagh and J. S. Woods, *Toxicol. Appl. Pharmacol.*, 1996, **141**, 59–67.
128. S. Li, S. A. Thompson and J. S. Woods, *Toxicol. Appl. Pharmacol.*, 1996, **140**, 180–187.
129. N. Ballatori and T. W. Clarkson, *Am. J. Physiol.*, 1983, **244**, G435–G441.
130. Z. Gregus and F. Varga, *Acta Pharmacol. Toxicol.*, 1985, **56**, 398–403.
131. T. Urano, A. Naganuma and N. Imura, *Res. Comm. Chem. Pathol. Pharmacol.*, 1988, **60**, 197–210.
132. A. Naganuma, N. Oda-Urano, T. Tanaka and N. Imura, *Biochem. Pharmacol.*, 1988, **37**, 291–296.
133. W. J. Dutczak and N. Ballatori, *Pharmacol. Exp. Ther.*, 1992, **262**, 619–623.
134. L. E. Kerper, E. M. Mokrzan, T. W. Clarkson and N. Ballatori, *Toxicol. Appl. Pharmacol.*, 1996, **141**, 526–531.
135. J. C. Smith, P. V. Allen, M. D. Turner, B. Most, H. L. Fisher and L. L. Hall, *Toxicol. Appl. Pharmacol.*, 1994, **128**, 251–256.
136. M. Korbas, A. J. Percy, J. Gailer and G. N. George, *J. Biol. Inorg. Chem.*, 2008, **13**, 461–470.
137. T. W. Clarkson, *Environ. Health Perspect.*, 2002, **110**, 11–23.

138. J. Yu, D. W. Yu, D. M. Checkla, I. M. Freedberg and A. P. Bertolino, *J. Invest. Dermatol.*, 1993, **101**, 56S–59S.
139. E. Cernichiari, R. Brewer, G. J. Myers, D. O. Marsh, L. W. Lapham, C. Cox, C. F. Shamlaye, M. Berlin, P. W. Davidson and T. W. Clarkson, *Neurotoxicol.*, 1995, **16**, 705–710.
140. T. G. Kershaw, T. W. Clarkson and P. H. Dhahir, *Arch. Environ. Health*, 1980, **35**, 28–36.
141. I. R. Rowland, *Toxicol. Pathol.*, 1988, **16**, 147–153.
142. I. R. Rowland, A. K. Mallett, J. Flynn and R. J. Hargreaves, *Arch. Toxicol.*, 1986, **59**, 94–98.
143. L. Magos, A. W. Brown, S. Sparrow, E. Bailey, R. E. Snowdon and W. R. Skipp, *Arch. Toxicol.*, 1985, **57**, 260–267.
144. L. E. Davis, M. Kornfeld, H. S. Mooney, K. J. Fiedler, K. Y. Haaland, W. W. Orrison, E. Cernichiari and W. Clarkson, *Ann. Neurol.*, 1994, **35**, 680–688.
145. J. S. Charleston, R. L. Body, R. P. Bolender, N. K. Mottet, M. E. Vahter and T. M. Burbacher, *Neurotoxicol.*, 1996, **17**, 127–138.
146. C. Chen, H. Yu, J. Zhao, B. Li, L. Qu, S. Liu, P. Zhang and Z. Chai, *Environ. Health Perspect.*, 2006, **114**, 297–301.
147. I. Falnoga, M. Tusek-Znidaric, M. Horvat and P. Stegnar, *Environ. Res.*, 2000, **A84**, 211–218.
148. K. T. Suzuki, C. Sasakura and S. Yoneda, *Biochem. Biophys. Acta*, 1998, **1429**, 102–112.
149. H. E. Ganther and M. L. Sunde, *Biol. Trace Elem. Res.*, 2007, **119**, 221–233.
150. M. B. Sobolev, S. B. Khatskel and A. Muradov, *Vopr. Pitan.*, 1999, **68**, 28–30.
151. S. M. Shim, M. G. Ferruzzi, Y. C. Kim, E. M. Janle and C. R. Santerre, *Food Chem.*, 2009, **112**, 46–50.
152. J. W. Ransone, N. M. Scott Jr and E. C Knoblock, *N. Engl. J. Med.*, 1961, **264**, 384–385.
153. J. Högberg and J. Alexander, in *Handbook on the Toxicology of Metals*, Ed. J. Friberg, G. F. Nordberg and V. Vouk, Elsevier, Amsterdam, 1986, pp. 482–520.
154. D. Thomson and M. F. Robinson, *Am. J. Clin. Nutr.*, 1986, **44**, 659–663.
155. WHO (World Health Organization), *Selenium, Environmental Health Criteria Ser.* 58, 1987.
156. B. H. Patterson, O. A. Levander, K. Helzlsouer, P. A. McAdam, S. A. Lewis, P. R. Taylor, C. Veillon and L. A. Zech, *Am. J. Physiol.*, 1989, **257**, R556–R567.
157. J. C. Turner, P. J. Osborn and S. M. McVeagh, *Comp. Biochem. Physiol. A*, 1990, **95**, 297–301.
158. P. Garberg and H. Högberg, in *Arbete och hälsa, Arbetsmiljöinstitutet och författarna*, Ed. B. Beije and P. Lundberg, Solna, Schweden, 1993, pp. 239–265.
159. R. F. Martin, M. Janghorbani and V. R. Young, *Am. J. Clin. Nutr.*, 1989, **49**, 854–861.
160. Y. Ohta and K. T. Suzuki, *Toxicol. Appl. Pharmacol.*, 2008, **226**, 169–177.
161. J. Lu and A. Holmgren, *J. Biol. Chem.*, 2009, **284**, 723–727.
162. Y. Kobayashi, Y. Ogra, K. Ishiwata, H. Takayama, N. Aimi and K. T. Suzuki, *Proc. Natl. Acad. Sci.*, 2002, **99**, 15932–15936.

163. B. Gammelgaard, K. G. Madsen, J. Bjerrum, L. Bendahl, O. Jons, J. Olsen and U. Sidenius, *J. Anal. Atom. Spectrom.*, 2003, **18**, 65–70.
164. B. Gammelgaard and L. Bendahl, *J. Anal. At. Spectrom.*, 2004, **19**, 135–142.
165. L. Bendahl and B. Gammelgaard, *J. Anal. At. Spectrom.*, 2004, **19**, 950–957.
166. B. Gammelgaard, L. Bendahl, N. Wessel Jacobsen and S. Stürup, *J. Anal. At. Spectom.*, 2005, **20**, 889–893.
167. D. Kuehnelt, N. Kienzl, P. Traar, N. H. Le, K. A. Francesconi and T. Ochi, *Anal. Bioanal. Chem.*, 2005, **383**, 235–246.
168. S. K. Tandon, L. Magos and M. Webb, *Biochem. Pharmacol.*, 1986, **35**, 2763–2766.
169. J. G. Dausch and F. R. Fullerton, *Nutr. Cancer*, 1993, **20**, 31–39.
170. H. S. Hsieh and H. Ganther, *Biochem. Biophys. Acta*, 1977, **497**, 205–217.
171. N. M. Mozier, K. P. McConnel and J. L. J. Hoffman, *Biol. Chem.*, 1988, **263**, 4527–4531.
172. K. T. Suzuki, Y. Tsuji, Y. Ohta and N. Suzuki, *Toxicol. Appl. Pharmacol.*, 2008, **227**, 76–83.
173. K. T. Suzuki, K. Kurasaki, S. Ogawa and N. Suzuki, *Toxicol. Appl. Pharmacol.*, 2006, **215**, 189–197.
174. K. T. Suzuki, M. Itoh and M. Ohmichi, *Toxicology*, 1995, **103**, 157–165.
175. M. Itoh and K. T. Suzuki, *Arch. Toxicol.* 1997, **71**, 461–466.
176. S. Vadhanavikit, C. Ip and H. E. Ganther, *Xenbobiotica*, 1993, **23**, 731–745.
177. D. Kremer, G. Ilgen and J. Feldmann, *Anal. Bioanal. Chem.*, 2005, **383**, 509–515.
178. K. A. Francesconi and F. Pannier, *Clin. Chem.*, 2004, **50**, 2240–2253.
179. D. Kuehnelt, D. Juresa, N. Kienzl and K. A. Francesconi, *Anal. Bioanal. Chem.*, 2006, **386**, 2207–2212.
180. D. Kuehnelt, D. Juresa, K. A. Francesconi, M. Fakih and M. E. Reid, *Toxicol. Appl. Pharmacol.*, 2007, **220**, 211–215.
181. H. H. Steinberg, S. C. Massari, A. C. Miner and R. Rink, *J. Ind. Hyg. Toxicol.*, 1942, **24**, 183–192.
182. R. H. de Meio, *J. Ind. Hyg. Toxicol.*, 1947, **29**, 393–395.
183. M. L. Amdur, *Arch. Ind. Health*, 1958, **17**, 665–667.
184. H. A. Schroeder, J. Buckmann and J. Bahlassa, *J. Chron. Dis.*, 1967, **20**, 147–161.
185. A. Taylor, *Biol. Trace Elem. Res.*, 1996, **55**, 231–239.
186. Y. Ogra, R. Kobayashi, K. Ishiwata and K. T. Suzuki, *J. Inorg. Biochem.*, 2008, **102**, 1507–1513.
187. F. E. Brinckman, G. J. Olson and J. S. Thayer, in *Marine and Estuarine Geochemistry*, Ed. A. C. Sigleo, A. Hattori, Lewis Publishers, Inc., Chelsea, 1985, pp. 227–238.
188. R. J. Maguire, *Water Poll. Res. Canada*, 1991, **26**, 243–360.
189. J. T. Byrd and M. O. Andreae, *Science*, 1982, **218**, 565–569.
190. R. S. Braman and M. A. Tompkins, *Anal. Chem.*, 1979, **51**, 12–19.
191. Y. K. Chau, R. J. Maguire, M. Brown, F. Yang and S. P. Batchelor, *Water Quality Res. J. Can.*, 1997, **32**, 453–521.

192. OECD (Organisation for Economic Co-operation and Development), "Monomethyltins", *SIDS Initial Assessment Report for SIAM* 23, 2006.
193. OECD (Organisation for Economic Co-operation and Development) "*Dimethyltins", SIDS Initial Assessment Report for SIAM* 23, 2006.
194. C. Boraiko and J. Batt, *J. Occup. Environ. Hyg.*, 2005, **2**, 73–76.
195. H. A. Muller, *Makromol. Chem. Macromol. Symp.*, 1989, **29**, 197–207.
196. R. D. Kimbrough, *Environ. Health Perspect.*, 1976, **14**, 51–56.
197. I. J. Boyer, *Toxicology*, 1989, **55**, 253–298.
198. E. Fortemps, G. Amand, A. Bomboir, R. Lauwerys and E. C. Laterre, *Int. Arch. Occup. Environ. Health*, 1978, **41**, 1–6.
199. W. D. Ross, E. A. Emmett, J. Steiner and R. Tureen, *Am. J. Psychiatry*, 1981, **138**, 1092–1095.
200. C. Rey, H. J. Reinicke and R. Besser, *Veterin. Hum. Toxicol.*, 1984, **26**, 121–122.
201. N. N. Yanofsky, D. Nierenberg and J. H. Turco, *J. Emerg. Med.*, 1991, **9**, 137–139.
202. R. G. Feldman, R. F. White and I. I. Eriator, *Arch. Neurol.*, 1993, **50**, 1320–1324.
203. M. J. Saary and R. A. House, *Occup. Med. (Lond.)*, 2002, **52**, 227–230.
204. S. Kreyberg, A. Torvik, A. Bjørneboe, W. Wiik-Larsen and D. Jacobsen, *Clin. Neuropathol.*, 1992, **11**, 256–259.
205. J. Gui-Bin, Z. Qun-fang and H. Bin, *Bull. Environ. Contam. Toxicol.*, 2000, **65**, 277–284.
206. E. A. Noland, P. T. McCauley and R. J. Bull, *J Toxicol. Environ. Health*, 1983, **12**, 89–98.
207. E. A. Noland, P. T. McCauley and R. J. Bull, "Dimethyltin Dichloride: Investigations into Its Gastrointestinal Absorption and Transplacental Transfer", Govt. Reports Announcements & Index (GRA&I); p. 12/1983/Issue 20, 1984.
208. E. Dopp, L. M. Hartmann, U. von Recklinghausen, A. M. Florea, B. Shokouhi, S. Rabieh, A. V. Hirner and A. W. Rettenmeier, *Toxicology*, 2007, **232**, 226–234.
209. Y. Arakawa and O. Wada, in *Biological Properties of Alkyltin Compounds,* Vol 9 of Metal *Ions in Biological Systems*, Ed. H. Sigel and A. Sigel, Marcel Dekker, New York, 1993, pp. 101–136.
210. A. W. Brown, R. D. Verschoyle, B. W. Street, W. N. Aldridge and H. Grindley, *J. Appl. Toxicol.*, 1984, **4**, 12–21.
211. ATSDR (Agency for Toxic Substances and Disease Registry), "Toxicological Profile for Tin and Compounds", U.S. Department of Health and Human Services, Atlanta, 2005.
212. IARC (International Agency for Research on Cancer), *in Some Drinking-Water Disinfectants and Contaminants, Including Arsenic*, Vol. 84 of *IARC Monogr. Eval. Carcinog. Risks Hum.*, Lyon, 2004, pp. 1–477.
213. IARC (International Agency for Research on Cancer), in *Cadmium and Cadmium Compounds*, Vol. 58 of *IARC Monogr. Eval. Carcinog. Risks Hum.*, Lyon, 1993, pp. 119–237.
214. IARC (International Agency for Research on Cancer), in *Inorganic and Organic Lead Compounds*, Vol. 87 of *IARC Monogr. Eval. Carcinog. Risks Hum.*, Lyon, 2006, pp. 1–471.

215. IARC (International Agency for Research on Cancer), in *Antimony Trioxide and Antimony Trisulfide*, Vol. 47 of *IARC Monogr. Eval. Carcinog. Risks Hum.*, Lyon, 1989, pp. 291–305.
216. IARC (International Agency for Research on Cancer), in *Mercury and Mercury Compounds*, Vol. 58 of *IARC Monogr. Eval. Carcinog. Risks Hum.*, Lyon, 1993, pp. 239–345.
217. IARC (International Agency for Research on Cancer), in *Selenium and Selenium Compounds*, Vol. 9 of *IARC Monogr. Eval. Carcinog. Risks Hum.*, Lyon, 1975, pp. 245–260.
218. DFG (Deutsche Forschungsgemeinschaft), *List of MAK and BAT-Values 2008, Report No. 44*, Wiley-VCH, Weinheim, 2008.
219. D. Beyersmann and A. Hartwig, *Arch. Toxicol.*, 2008, **82**, 493–512.
220. I. T. Johnson and N. J. Belshaw, *Food Chem. Toxicol.*, 2008, **46**, 1346–1359.
221. Z. Herceg, *Mutagenesis*, 2007, **22**, 91–103.
222. L. R. Ferguson, N. Karunasinghe and M. Philpott, *Environ. Mol. Mutagen.*, 2004, **44**, 36–43.
223. K. Salnikow and A. Zhitkovich, *Chem. Res. Toxicol.*, 2008, **21**, 28–44.
224. M. Styblo, L. M. del Razo, L. Vega, D. R. Germolec, E. L. LeCluyse, G. A. Hamilton, W. Reed, C. Wang, W. R. Cullen and D. J. Thomas, *Arch. Toxicol.*, 2000, **74**, 289–299.
225. T. Schwerdtle, I. Walter, I. Mackiw and A. Hartwig, *Carcinogenesis*, 2003, **24**, 967–974.
226. A. D. Kligerman, C. L. Doerr, A. H. Tennant, K. Harrington-Brock, J. W. Allen, E. Winkfield, P. Poorman-Allen, B. Kundu, K. Funasaka, B. C. Roop, M. J. Mass and D. M. Demarini, *Environ. Mol. Mutagen.*, 2003, **42**, 192–205.
227. E. Dopp, L. M. Hartmann, A.-M. Florea, U. v. Recklinghausen, R. Pieper, B. Shokouhi, A. W. Rettenmeier, A. V. Hirner and G. Obe, *Toxicol. Appl. Pharmacol.*, 2004, **201**, 156–165.
228. G. Endo, K. Kuroda, A. Okamoto and S. Horiguchi, *Bull. Environ. Contam. Toxicol.*, 1992, **48**, 131–137.
229. N. Eguchi, K. Kuroda and G. Endo, *Arch. Environ. Contam. Toxicol.*, 1997, **32**, 141–145.
230. K. Yamanaka, A. Hasegawa, R. Sawamura and S. Okada, *Biochem. Biophys. Res. Commun.*, 1989, **165**, 43–50.
231. G. Endo, K. Kuroda, I. Kiyota and S. Horiguchi, *Mutat. Res.*, 1988, **203**, 370–371.
232. Y. Oya-Ohta, T. Kaise and T. Ochi, *Mutat. Res.*, 1996, **357**, 123–129.
233. L. E. Moore, A. H. Smith, C. Hopenhayn-Rich, M. L. Biggs, D. A. Kalman and M. T. Smith, *Cancer Epidemiol. Biomarkers Prev.*, 1997, **6**, 31–36.
234. A. Basu, J. Mahata, S. Gupta and A. K. Giri, *Mut. Res.*, 2001, **488**, 171–194.
235. D. J. Thomas, M. Styblo and S. Lin, *Toxicol. Appl. Pharmacol.*, 2001, **176**, 127–144.
236. M. F. Hughes, *Toxicol. Lett.*, 2002, **133**, 1–16.
237. J. S. Petrick, F. Ayala-Fierro, W. R. Cullen, D. E. Carter and H. V. Aposhian, *Toxicol. Appl. Pharmacol.*, 2000, **163**, 203–207.

238. M. Sordo, L. A. Herrera, P. Ostrosky-Wegman and E. Rojas, *Teratog. Carcinog. Mutagen.*, 2001, **21**, 249–260.
239. K. T. Kitchin and S. Ahmad, *Toxicol. Lett.*, 2003, **137**, 3–13.
240. R. Raml, W. Goessler and K. A. Francesconi, *J. Chromatogr. A.*, 2006, **1128**, 164–170.
241. K. Yamanaka, M. Hoshino, M. Okamoto, R. Sawamura, A. Hasegawa and S. Okada, *Biochem. Biophys. Res. Commun.*, 1990, **168**, 58–64.
242. K. Kato, K. Yamanaka, A. Hasegawa and S. Okada, *Mutat. Res.*, 2003, **539**, 55–63.
243. P. Andrewes, K. T. Kitchin and K. Wallace, *Chem. Res. Toxicol.*, 2003, **16**, 994–1003.
244. A. Schoen, B. Beck, R. Sharma and E. Duke, *Toxicol. Appl. Pharmacol.*, 2004, **198**, 253–267.
245. S. M. Cohen, L. L. Arnold, M. Eldan, A. S. Lewis and B. D. Beck, *Crit. Rev. Toxicol.*, 2006, **36**, 99–133.
246. S. S. H. Tao and P. M. Bolger, *Food Addit. Contam.*, 1999, **16**, 465–472.
247. I. Walter, T. Schwerdtle, C. Thuy, J. L. Parsons, G. L. Dianov and A. Hartwig, *DNA Repair (Amst.)*, 2007, **6**, 61–70.
248. S. Shen, J. Lee, M. Weinfeld and X. C. Le, *Mol. Carcinog.*, 2008, **47**, 508–518.
249. T. Schwerdtle, I. Walter and A. Hartwig, *DNA Repair (Amst.)*, 2003b, **2**, 1449–1463.
250. X. Cui, T. Wakai, Y. Shirai, K. Hatakeyama K and S. Hirano, *Toxicol. Sci.*, 2006, **91**, 372–381.
251. S. Chanda, U. B. Dasgupta, D. Guhamazumder, M. Gupta, U. Chaudhuri, S. Lahiri, S. Das, N. Ghosh and D. Chatterjee, *Toxicol. Sci.*, 2006, **89**, 431–437.
252. M. Takiguchi, W. E. Achanzar, W. Qu, G. Li and M. P. Waalkes, *Exp. Cell Res.*, 2003, **286**, 355–365.
253. E. Dopp, L. M. Hartmann, A. M. Florea, U. von Recklinghausen, S. Rabieh, B. Shokouhi, A. V. Hirner and A. W. Rettenmeier, *Toxicol. in Vitro*, 2006, **20**, 1060–1065.
254. P. Andrewes, K. T. Kitchin and K. Wallace, *Toxicol. Appl. Pharmacol.*, 2004, **194**, 41–48.
255. H. G. Verschuuren, R. Kroes, E. M. Tonkelaar, J. M. Berkvens, P. W. Helleman, A. G. Rauws, P. L. Schuller and G. J. van Esch, *Toxicology*, 1976, **6**, 107–123.
256. K. Mitsumori, K. Maita, T. Saito, S. Tsuda and Y. Shirasu, *Canc. Lett.*, 1981, **12**, 305–310.
257. K. Mitsumori, K. Maita and Y. Shirasu, *Jpn. J. Vet. Sci.*, 1984, **46**, 549–557.
258. M. Hirano, K. Mitsumori, K. Maita and Y. Shirasu, *Jpn. J. Vet. Sci.*, 1986, **48**, 127–135.
259. K. Mitsumori, M. Hirano, H. Ueda, K. Maita and Y. Shirasu, *Fundam. Appl. Toxicol.*, 1990, **14**, 179–190.
260. M. Hirano, H. Ueda, K. Mitsumori, K. Maita and Y. Shirasu, *Jpn. J. Vet. Sci.*, 1988, **50**, 886–893.
261. S. Skerfving, K. Hansson, C. Mangs, J. Lindsten and N. Ryman, *Environ. Res.*, 1974, **7**, 83–98.

262. H. C. Wulf, N. Kromann, N. Kousgaard, J. C. Hansen, E. Niebuhr and K. Alboge, *Sci. Total Environ.*, 1986, **48**, 81–94.
263. M. V. Monsalve and C. Chiappe, *Environ. Mol. Mutagen.*, 1987, **10**, 367–376.
264. E. Franchi, G. Loprieno, M. Ballardin, L. Petrozzi and L. Migliore, *Mutat. Res.*, 1994, **320**, 23–29.
265. S. de Flora, C. Bennicelli and M. Bagnasco, *Mut. Res.*, 1994, **317**, 57–79.
266. P. Moszczynski, *Int. J. Occup. Med. Environ. Health*, 1997, **10**, 247–258.
267. DFG (Deutsche Forschungsgemeinschaft), in *Selen und seine anorganische Verbindungen, Toxikologische und arbeitsmedizinische Begründungen von MAK-Werten*, Ed. H. Greim, 29. Lieferung, Wiley-VCH, Weinheim, 1999.
268. C. Ip, *J. Nutr.*, 1998, **128**, 1845–1854.
269. L. C. Clark, G. F. Combs Jr, B. W. Turnbull, E. H. Slate, D. K. Chalker, J. Chow, L. S. Davis, R. A. Glover, G. F. Graham, E. G. Gross, A. Krongrad, J. L. Lesher Jr, H. K. Park, B. B. Sanders Jr, C. L. Smith and J. R. Taylor, *J. Am. Med. Assoc.*, 1996, **276**, 1957–1963.
270. S. D. Mark, Y. L. Qiao, S. M. Dawsey, Y. P. Wu, H. Katki, E. W. Gunter, J. F. Fraumeni Jr, W. J. Blot, Z. W. Dong and P. R. Taylor, *J. Natl. Cancer Inst.*, 2000, **92**, 1753–1756.
271. H. E. Ganther, *Carcinogenesis*, 1999, **20**, 1657–1666.
272. B. W. Youn, E. S. Fiala and O. S. Sohn, *Nutr. Cancer*, 2001, **28–33**.
273. C. Jacob, G. I. Giles, N. M. Giles and H. Sies, *Angew. Chem.*, 2003, **42**, 4742–4758.
274. H. Rikiishi, *J. Bioenerg. Biomembr.*, 2007, **39**, 91–98.
275. R. Naithani, *Mini. Rev. Med. Chem.*, 2008, **8**, 657–668.
276. M. I. Jackson and G. F. Combs Jr, *Curr. Opin. Nutr. Metab. Care*, 2008, **11**, 718–726.
277. H. E. Ganther and J. R. Lawrence, *Tetrahedron*, 1997, **53**, 12229–12310.
278. J. Lü, C. Jiang, M. Kaeck, H. E. Ganther, S. Vadhanavikit, C. Ip and H. Thompson, *Biochem. Pharmacol.*, 1995, **50**, 213–219.
279. J. Lü, H. Pei, C. Ip, D. Lisk, H. E. Ganther and H. J. Thompson, *Carcinogenesis (Lond.)*, 1996, **17**, 1903–1907.
280. M. Kaeck, J. Lu, R. Strange, C. Ip, H. E. Ganther and H. J. Thompson, *Biochem. Pharmacol.*, 1997, **53**, 921–926.
281. T. Kim, U. Jung, D. Y. Cho and A. S. Chung, *Carcinogenesis (Lond.)*, 2001, **22**, 559–565.
282. J. Lü and C. Jiang, *Antioxid. Redox Signal.*, 2005, **7**, 1715–1727.
283. C. Ip and H. E. Ganther, *Cancer Res.*, 1990, **50**, 1206–1211.
284. C. Ip, C. Hayes, R. M. Budnick and H. E. Ganther, *Cancer Res.*, 1991, **51**, 595–600.
285. A. C. Wilson, H. J. Thompson, P. J. Schedin, N. W. Gibson and H. E. Ganther, *Biochem. Pharmacol.*, 1992, **43**, 1137–1141.
286. R. Sinha, T. K. Said and D. Medina, *Cancer Lett.*, 1996, **107**, 277–284.
287. R. Sinha and D. Medina, *Carcinogenesis*, 1997, **18**, 1541–1547.
288. Z. Wang, C. Jiang, H. Ganther and J. Lü, *Cancer Res.*, 2001, **61**, 7171–7178.
289. Z. Wang, C. Jiang and J. Lü, *Mol. Carcinogenesis*, 2002, **34**, 113–120.

290. C. Jiang, Z. Wang, H. Ganther and J. X. Lü, *Cancer Res.*, 2001, **61**, 3062–3070.
291. C. Jiang, H. Ganther and J. Lü, *Mol. Carcinogen.*, 2000, **29**, 236–250.
292. J. Lü and C. Jiang, *Nutr. Cancer*, 2001, **40**, 64–73.
293. S. M. Lippman, E. A. Klein, P. J. Goodman, M. S. Lucia, I. M. Thompson, L. G. Ford, H. L. Parnes, L. M. Minasian, J. M. Gaziano, J. A. Hartline, J. K. Parsons, J. D. Bearden rd, E. D. Crawford, G. E. Goodman, J. Claudio, E. Winquist, E. D. Cook, D. D. Karp, P. Walther, M. M. Lieber, A. R. Kristal, A. K. Darke, K. B. Arnold, P. A. Ganz, R. M. Santella, D. Albanes, P. R. Taylor, J. L. Probstfield, T. J. Jagpal, J. J. Crowley, F. L. Meyskens Jr, L. H. Baker and C. A. Coltman Jr, *J. Am. Med. Assoc.*, 2009, **301**, 39–51.
294. S. Stranges, J. R. Marshall, R. Natarajan, R. P. Donahue, M. Trevisan, G. F. Combs, F. P. Cappuccio, A. Ceriello and M. E. Reid, *Ann. Intern. Med.*, 2007, **147**, 217–223.
295. T. Arata, Y. Oyama, K. Tabaru, M. Satoh, H. Hayashi, S. Ishida and Y. Okano, *Environ. Toxicol.*, 2002, **17**, 472–477.
296. J. C. Hutson, *J. Appl. Toxicol.*, 2005, **25**, 234–238.
297. N. E. Magnusson, A. Larsen, J. Rungby, M. Kruhoffer, T. F. Orntoft and M. Stoltenberg, *Cell. Tissue Res.*, 2005, **321**, 195–210.
298. N. Gurnani, A. Sharma and G. Talukder, *Biol. Trace Elem. Res.*, 1993, **37**, 281–292.
299. H. Sun, H. Li, I. Harvey and P. J. Sadler, *J. Biol. Chem.*, 1999, **274**, 29094–29101.
300. A. Hartwig, *Antioxid. Redox Signaling*, 2001, **3**, 625–634.
301. A. Hartwig, M. Asmuss, I. Ehleben, U. Herzer, D. Kostelac, A. Pelzer, T. Schwerdtle and A. Bürkle, *Environ. Health Perspect.*, 2002, **110**, 797–799.
302. B. de Castro, T. J. Kistenmacher and L. G. Marzilli, *Agents Actions*, 1981, **8**(suppl), 435–464.
303. Y. Palmieri, *The Bulletin of the Bismuth Institute*, Grimbergen, Belgium, 1993, ISSN 0379–0401.
304. M. Mosinger, Centre D'Exploitations et de Recherches Medicales, Marseille, Final Report Advastab TM 181 FS (Cincinnati-Milacron), 1975.
305. L. H. Lash, in *Molecular Biology and Toxicology of Metals*, Ed. R. K. Zalups and J. Koropatnick, Taylor & Francis, London, 2000, pp. 382–413..
306. E. Guallar, M. I. Sanz-Gallardo, P. van't Veer, P. Bode, A. Aro, J. Gomez-Aracena, J. D. Kark, R. A. Riemersma, J. M. Martin-Moreno and F. J. Kok, *N. Engl. J. Med.*, 2002, **347**, 1747–1754.
307. T. W. Clarkson and L. Magos, *Crit. Rev. Toxicol.*, 2006, **36**, 609–662.
308. L. Magos and T. W. Clarkson, *Ann. Clin. Biochem.*, 2006, **43**, 257–268.
309. K. A. Björnberg, M. Vahter, B. Berglund, B. Niklasson, M. Blennow and G. Sandborgh-Englund, *Environ. Health Perspect.*, 2005, **113**, 1381–1385.
310. WHO (World Health Organization), *Environmental Health Criteria 15, Tin and Organotin Compounds*, 1980.
311. K. R. Reuhl and J. M. Cranmer, *Neurotoxicology*, 1984, **5**, 187–204.
312. L. W. Chang and R. S. Dyer, *Neurobeh. Toxicol. Teratol.*, 1983, **5**, 673–696.
313. D. E. McMillan and G. R. Wenger, *Pharmacol. Rev.*, 1985, **37**, 365–379.
314. M. Aschner and J. L. Aschner, *Neurosci. Biobehav. Rev..*, 1992, **16**, 427–435.

315. V. C. Karpiak and C. L. Eyer, *Cell Biol. Toxicol.*, 1999, **15**, 261–268.
316. P. S. Mehta, A. Bruccoleri, H. W. Brown and G. J. Harry, *J. Neuroimmunol.*, 1998, **88**, 154–164.
317. S. J. Jenkins, K. Ehman and S. Barone Jr, *Dev. Brain Res.*, 2004, **151**, 1–12.
318. S. J. Jenkins and S. Barone Jr, *Toxicol. Lett.*, 2004, **147**, 63–72.
319. L. G. Costa and R. Sulaiman, *Toxicol. Appl. Pharmacol.*, 1986, **86**, 189–196.
320. S. M. Toggas, J. K. Krady and M. L. Billingsley, *Mol. Pharmacol.*, 1992, **42**, 44–56.
321. T. A. Thompson, J. M. Lewis, N. S. Dejneka, W. B. Severs, R. Polavarapu and M. L. Billingsley, *J. Pharmacol. Exp. Ther.*, 1996, **276**, 1201–1216.
322. B. A. Buck-Koehntop, F. Porcelli, J. L. Lewin, C. J. Cramer and G. Veglia, *J. Organomet. Chem.*, 2006, **691**, 1748–1755.
323. F. Springman, E. Bingham and K. L. Stemmer, *Arch. Environ. Health*, 1963, **6**, 469–472.
324. L. W. Sanders, *Arch. Environ. Health*, 1964, **8**, 270–277.
325. M. Tenenbein, *Hum. Exp. Toxicol.*, 1997, **16**, 217–222.
326. A. Seeber, E. Kiesswetter, B. Neidhart and M. Blaszkewicz, *Neurotox. Teratol.*, 1990, **12**, 653–655.
327. C. S. Mitchell, M. S. Shear, K. I. Bolla and B. S. Schwartz, *J. Occup. Environ. Med.*, 1996, **38**, 372–378.
328. J. M. Balbus, W. Stewart, K. I. Bolla and B. S. Schwartz, *Am. J. Ind. Med.*, 1997, **32**, 544–549.
329. T. J. Walsh and H. A. Tilson, *Neurotoxicology*, 1984, **5**, 67–86.
330. W. F. Stewart, B. S. Schwartz, D. Simon, K. Kelsey and A. C. Todd, *Environ. Health Perspect.*, **110**, 501–505.
331. H. Belinson and D. M. Michaelson, *J. Alzheimers Dis.*, 2009, in press.
332. G. Röderer and K. H. Doenges, *Neurotoxicology*, 1983, **4**, 171–180.
333. G. Konat, *Neurotoxicology*, 1984, **5**, 87–96.
334. M. Bragadin, D. Marton, M. Murgia, V. Rizzoli, G. Scutari and R. Deana, *J. Inorg. Biochem.*, 1998, **69**, 259–262.
335. M. Bragadin, D. Marton and S. Manente, *J. Inorg. Biochem.*, 2007, **101**, 876–878.
336. A. H. Hall, *Toxicol. Lett.*, 2002, **128**, 69–72.
337. B. J. Lagerkvist and B. Zetterlund, *Am. J. Ind. Med.*, 1994, **25**, 477–488.
338. W. E. Morton and G. A. Caron, *Am. J. Ind. Med.*, 1989, **15**, 1–5.
339. D. N. Mazumder, J. das Gupta, A. K. Chakraborty, A. Chatterjee, D. Das and D. Chakraborti, *Bull. World Health Organ.*, 1992, **70**, 481–485.
340. F. Gerr, R. Letz, P. B. Ryan and R. C. Green, *Neurotoxicology*, 2000, **21**, 475–488.
341. K. V. Q. Luong and L. T. H. Nguyen, *Am J. Med. Sci.*, 1999, **317**, 269–271.
342. J. Liu, B. Zheng, H. V. Aposhian, Y. Zhou, M.-L. Chen, A. Zhang and M. P. Waalkes, *Environm. Health Perspect.*, 2002, **110**, 119–122.
343. K. Bolla-Wilson and M. L. Bleecker, *J. Occup. Med.*, 1987, **29**, 500–503.
344. G. A. Wassermann, X. Liu, F. Parvez, H. Ahsan, P. Factor-Litvak, A. van Geen, V. Slavkovich, N. J. Lolacono, Z. Cheng, I. Hussain, H. Momotaj and J. H. Graziano, *Environ. Health*, 2004, **112**, 1329–1333.

345. R. O. Wright, C. Amarasiriwardena, A. D. Woolf, R. Jim and D. C. Bellinger, *Neurotoxicology*, 2006, **27**, 210–216.
346. S.-Y. Tsai, H.-Y. Chou, H.-W. The, C.-M. Chen and C.-J. Chen, *Neurotoxicology*, 2003, **24**, 747–753.
347. P. M. le Quesne and J. P. McLeod, *J. Neurol. Sci.*, 1977, **32**, 437–451.
348. A. Vahidnia, G. B. van der Voet and F. A. de Wolff, *Hum. Exp. Toxicol.*, 2007, **26**, 823–832.
349. A. Vahidnia, R. J. van der Straaten, F. Romijn, J. van Pelt, G. B. van der Voet and F. A. de Wolff, *Toxicol. in Vitro*, 2007, **21**, 1104–1112.
350. K. Krüger, H. Repges, J. Hippler, L. M. Hartmann, A. V. Hirner, H. Straub, N. Binding and U. Mußhoff, *Toxicol. Appl. Pharmacol.*, **225**, 40–46.
351. J. F. Goodrum, T. S. Earnhardt, N. D. Goines and T. W. Bouldin, *J. Neurochem.*, 1990, **55**, 1928–1932.
352. M. Wagner-Recio, A. D. Toews and P. Morell, *J. Neurochem.*, 1991, **57**, 1891–1901.
353. P. Morell, A. D. Toews, H. E. Jurevics and M. Wagner, *J. Neurochem.*, 1995, **65**(suppl), S112.
354. M. Wagner, A. D. Toews and P. Morell, *J. Neurochem.*, 1995, **64**, 2169–2176.
355. B. P. Laden and T. D. Porter, *J. Lipid Res.*, 2001, **42**, 235–240.
356. R. S. Hoffman, *Toxicol. Rev.*, 2003, **22**, 29–40.
357. S. Hassfjell and W. M. Brechbiel, *Chem. Rev.*, 2001, **101**, 2019–2036.
358. T. K. Nikula, M. R. McDevitt, R. D. Finn, C. Wu, R. W. Kozak, K. Garmestani, M. W. Brechbiel, M. J. Curcio, C. G. Pippin, L. Tiffany-Jones, M. W. Geerlings Jr, O. A. Gansow and D. A. Scheinberg, *J. Nucl. Med.*, 1999, **40**, 166–176.
359. N. Yang and H. Sun, *Coord. Chem. Rev.*, 2007, **251**, 2354–2366.
360. J. F. Ross, R. C. Switzer, M. R. Poston and G. T. Lawhorn, *Brain Res.*, 1996, **725**, 137–154.
361. G. Martin-Bouyer, C. Barin, A. Beugnet, J. Cordier and H. Guerbois, *Gastroenterol. Clin. Biol.*, 1978, **2**, 349–356.
362. G. Martin-Bouyer, B. Foulon, H. Guerbois and C. Barin, *Therapie*, 1980, **35**, 307–313.
363. P. Hillemand, M. Palliere, B. Laquais and P. Bouvet, *Semin. Hop. Paris*, 1977, **53**, 1663–1669.
364. R. A. Diaz-Bone, M. Hollmann, O. Wuerfel and D. Pieper, *J. Anal. At. Spectrom.*, 2009, **24**, 808–814.
365. M. Hollmann, J. Boertz, E. Dopp, J. Hippler and A. V. Hirner, *Metallomics*, submitted.
366. J. Hippler and A. V. Hirner, in preparation.

Met. Ions Life Sci. **2010**, *7*, 523–575

Subject Index

A

Metal Ions in Life Sciences, Volume 7 Edited by Astrid Sigel, Helmut Sigel, and Roland K. O. Sigel

Published by the Royal Society of Chemistry, www.rsc.org DOI: 10.1039/9781849730822-00523

B

C

D

E

F

H

I

M

N

O

P

Q

R

T

U

X

Y

Z

www.ingramcontent.com/pod-product-compliance
Lightning Source LLC
LaVergne TN
LVHW010547300926
849162LV00002B/253
* 9 7 8 3 1 1 0 4 4 2 8 0 9 *